国家科学技术学术著作出版基金资助出版
中国科学院中国孢子植物志编辑委员会 编辑

中国淡水藻志

第二十二卷

硅 藻 门

管壳缝目

王全喜 主编

中国科学院知识创新工程重大项目
国家自然科学基金重大项目
（国家自然科学基金委员会 中国科学院 科学技术部 资助）

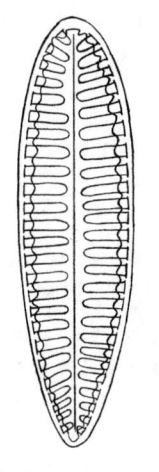

科学出版社
北 京

内 容 简 介

本卷册记述了我国管壳缝目硅藻3科13属189个分类单位，包括149种37变种3变型，每个种类都有详细的形态描述、生境和国内外分布，有些种类附有讨论和说明，图版以高质量的光镜和扫描电镜照片为主，均以我国标本为材料。此外，对管壳缝目硅藻的细胞结构及形态特征、生殖、生态习性及在自然界中的分布做了详细论述，书后附有各级分类群英文检索表、汉英术语对照表、参考文献和名称索引，是我国目前最为全面和系统记述管壳缝目硅藻的专著。

本书可供生物学、植物学、藻类学、水产科学、环境科学、地质学及地理学等领域的科研人员及高等院校有关专业师生参考。

图书在版编目（CIP）数据

中国淡水藻志. 第22卷, 硅藻门. 管壳缝目 / 王全喜主编. —北京：科学出版社，2018.6

（中国孢子植物志）

ISBN 978-7-03-057144-1

I. ①中… II. ①王… III. ①藻类–植物志–中国 ②硅藻门–植物志–中国 IV. ①Q949.2

中国版本图书馆 CIP 数据核字（2018）第 077036 号

责任编辑：韩学哲 孙 青 / 责任校对：郑金红
责任印制：肖 兴 / 封面设计：刘新新

科学出版社 出版
北京东黄城根北街 16 号
邮政编码：100717
http://www.sciencep.com

北京通州皇家印刷厂 印刷
科学出版社发行 各地新华书店经销

*

2018 年 6 月第 一 版　　开本：787×1092　1/16
2018 年 6 月第一次印刷　印张：12　插页：68
字数：324 000
定价：198.00 元
（如有印装质量问题，我社负责调换）

Supported by the National Fund for Academic Publication in Science and Technology

CONSILIO FLORARUM CRYPTOGAMARUM SINICARUM
ACADEMIAE SINICAE EDITA

FLORA ALGARUM SINICARUM AQUAE DULCIS

TIMUS XXII

BACILLARIOPHYTA

Aulonoraphidinales

REDACTOR PRINCIPALIS

WANG QUANXI

**A Major Project of the Knowledge Innovation Program
of the Chinese Academy of Sciences
A Major Project of the National Natural Science Foundation of China**
(Supported by the National Natural Science Foundation of China,
the Chinese Academy of Sciences, and the Ministry of Science and Technology of China)

Science Press
Beijing

《中国淡水藻志》第二十二卷
硅 藻 门
管壳缝目

主 编

王全喜

副主编

尤庆敏

参加者

王全喜　尤庆敏　范亚文
刘　妍　施之新　李海玲

中国孢子植物志第五届编委名单
（2017年5月）

主　　编　魏江春

副 主 编　庄文颖　夏邦美　吴鹏程　胡征宇

　　　　　阿不都拉·阿巴斯

委　　员　（以姓氏笔画为序）

　　　　　丁兰平　王全喜　王幼芳　王旭雷　吕国忠
　　　　　庄剑云　刘小勇　刘国祥　李仁辉　李增智
　　　　　杨祝良　张天宇　陈健斌　胡鸿钧　姚一建
　　　　　贾　渝　高亚辉　郭　林　谢树莲　蔡　磊
　　　　　戴玉成　魏印心

序

中国孢子植物志是非维管束孢子植物志，分《中国海藻志》、《中国淡水藻志》、《中国真菌志》、《中国地衣志》及《中国苔藓志》五部分。中国孢子植物志是在系统生物学原理与方法的指导下对中国孢子植物进行考察、收集和分类的研究成果；是生物物种多样性研究的主要内容；是物种保护的重要依据，对人类活动与环境甚至全球变化都有不可分割的联系。

中国孢子植物志是我国孢子植物物种数量、形态特征、生理生化性状、地理分布及其与人类关系等方面的综合信息库；是我国生物资源开发利用，科学研究与教学的重要参考文献。

我国气候条件复杂，山河纵横，湖泊星布，海域辽阔，陆生和水生孢子植物资源极其丰富。中国孢子植物分类工作的发展和中国孢子植物志的陆续出版，必将为我国开发利用孢子植物资源和促进学科发展发挥积极作用。

随着科学技术的进步，我国孢子植物分类工作在广度和深度方面将有更大的发展，对于这部著作也将不断补充、修订和提高。

中国科学院中国孢子植物志编辑委员会

1984 年 10 月·北京

中国孢子植物志总序

中国孢子植物志是由《中国海藻志》、《中国淡水藻志》、《中国真菌志》、《中国地衣志》及《中国苔藓志》所组成。至于维管束孢子植物蕨类未被包括在中国孢子植物志之内，是因为它早先已被纳入《中国植物志》计划之内。为了将上述未被纳入《中国植物志》计划之内的藻类、真菌、地衣及苔藓植物纳入中国生物志计划之内，出席1972年中国科学院计划工作会议的孢子植物学工作者提出筹建"中国孢子植物志编辑委员会"的倡议。该倡议经中国科学院领导批准后，"中国孢子植物志编辑委员会"的筹建工作随之启动，并于1973年在广州召开的《中国植物志》、《中国动物志》和中国孢子植物志工作会议上正式成立。自那时起，中国孢子植物志一直在"中国孢子植物志编辑委员会"统一主持下编辑出版。

孢子植物在系统演化上虽然并非单一的自然类群，但是，这并不妨碍在全国统一组织和协调下进行孢子植物志的编写和出版。

随着科学技术的飞速发展，人们关于真菌的知识日益深入的今天，黏菌与卵菌已被从真菌界中分出，分别归隶于原生动物界和管毛生物界。但是，长期以来，由于它们一直被当作真菌由国内外真菌学家进行研究，而且，在"中国孢子植物志编辑委员会"成立时已将黏菌与卵菌纳入中国孢子植物志之一的《中国真菌志》计划之内并陆续出版，因此，沿用包括黏菌与卵菌在内的《中国真菌志》广义名称是必要的。

自"中国孢子植物志编辑委员会"于1973年成立以后，作为"三志"的组成部分，中国孢子植物志的编研工作由中国科学院资助；自1982年起，国家自然科学基金委员会参与部分资助；自1993年以来，作为国家自然科学基金委员会重大项目，在国家基金委资助下，中国科学院及科技部参与部分资助，中国孢子植物志的编辑出版工作不断取得重要进展。

中国孢子植物志是记述我国孢子植物物种的形态、解剖、生态、地理分布及其与人类关系等方面的大型系列著作，是我国孢子植物物种多样性的重要研究成果，是我国孢子植物资源的综合信息库，是我国生物资源开发利用、科学研究与教学的重要参考文献。

我国气候条件复杂，山河纵横，湖泊星布，海域辽阔，陆生与水生孢子植物物种多样性极其丰富。中国孢子植物志的陆续出版，必将为我国孢子植物资源的开发利用，为我国孢子植物科学的发展发挥积极作用。

中国科学院中国孢子植物志编辑委员会
主编　曾呈奎
2000年3月北京

Preface to the Cryptogamic Flora of China

Cryptogamic Flora of China is composed of *Flora Algarum Marinarum Sinicarum*, *Flora Algarum Sinicarum Aquae Dulcis*, *Flora Fungorum Sinicorum*, *Flora Lichenum Sinicorum*, and *Flora Bryophytorum Sinicorum*, edited and published under the direction of the Editorial Committee of the Cryptogamic Flora of China, Chinese Academy of Sciences （CAS）. It also serves as a comprehensive information bank of Chinese cryptogamic resources.

Cryptogams are not a single natural group from a phylogenetic point of view which, however, does not present an obstacle to the editing and publication of the Cryptogamic Flora of China by a coordinated, nationwide organization. The Cryptogamic Flora of China is restricted to non-vascular cryptogams including the bryophytes, algae, fungi, and lichens. The ferns, a group of vascular cryptogams, were earlier included in the plan of *Flora of China*, and are not taken into consideration here. In order to bring the above groups into the plan of Fauna and Flora of China, some leading scientists on cryptogams, who were attending a working meeting of CAS in Beijing in July 1972, proposed to establish the Editorial Committee of the Cryptogamic Flora of China. The proposal was approved later by the CAS. The committee was formally established in the working conference of Fauna and Flora of China, including cryptogams, held by CAS in Guangzhou in March 1973.

Although myxomycetes and oomycetes do not belong to the Kingdom of Fungi in modern treatments, they have long been studied by mycologists. *Flora Fungorum Sinicorum* volumes including myxomycetes and oomycetes have been published, retaining for *Flora Fungorum Sinicorum* the traditional meaning of the term fungi.

Since the establishment of the editorial committee in 1973, compilation of Cryptogamic Flora of China and related studies have been supported financially by the CAS. The National Natural Science Foundation of China has taken an important part of the financial support since 1982. Under the direction of the committee, progress has been made in compilation and study of Cryptogamic Flora of China by organizing and coordinating the main research institutions and universities all over the country. Since 1993, study and compilation of the Chinese fauna, flora, and cryptogamic flora have become one of the key state projects of the National Natural Science Foundation with the combined support of the CAS and the National Science and Technology Ministry.

Cryptogamic Flora of China derives its results from the investigations, collections, and classification of Chinese cryptogams by using theories and methods of systematic and evolutionary biology as its guide. It is the summary of study on species diversity of cryptogams and provides important data for species protection. It is closely connected with human activities, environmental changes and even global changes. Cryptogamic Flora of China is a comprehensive information bank concerning morphology, anatomy, physiology, biochemistry, ecology, and phytogeographical distribution. It includes a series of special monographs for using the biological resources in China, for scientific research, and for teaching.

China has complicated weather conditions, with a crisscross network of mountains and rivers, lakes of all sizes, and an extensive sea area. China is rich in terrestrial and aquatic cryptogamic resources. The development of taxonomic studies of cryptogams and the publication of Cryptogamic Flora of China in concert will play an active role in exploration and utilization of the cryptogamic resources of China and in promoting the development of cryptogamic studies in China.

C. K. Tseng
Editor-in-Chief
The Editorial Committee of the Cryptogamic Flora of China
Chinese Academy of Sciences
March, 2000 in Beijing

《中国淡水藻志》序

中国是一个陆地国土面积960万平方公里的大国，地跨寒带、温带、亚热带和热带，不仅有陆地和海洋，还有5000多个岛屿，大陆地形十分复杂，海拔自西向东由高而低。中国西部海拔在5000 m以上的土地面积占全国总面积的25.9%（其中世界最高峰珠穆朗玛峰海拔为8848 m），往东依次为：海拔2000—3000 m的占7%，海拔1000—2000 m的占25%，海拔500—1000 m的占16.9%，东部和东北部及沿海地带海拔都在500 m以下，约占25.2%。这其间山地、高原、盆地、平原和丘陵等等连绵起伏。中国又是一个河流丰富的国家，仅流域面积超过100平方公里的就有50 000条以上；几条大的河流自西向东或向南流入大海。我国的湖泊也很多，已知的天然湖泊，面积在1平方公里以上的即有2800个，人工湖86 000个，还有难以计数的塘堰、水池、溪流、沟渠、沼泽、泉水等。这些地理特征使得我国各地在日照、气温和降水等方面有极大的差异，产生了种类丰富的植物。我国已知的高等植物，包括苔藓、蕨类和种子植物超过30 000种。无数的大小水坑，包括临时积水、稻田、水井，还有地下水、温泉、湿地、草场，以及表面多少覆盖有土壤的或潮湿的岩石、道路和建筑物等，形成无法计算、情况各异的小生境，生长着各种藻类。

中国的淡水藻类，早期是由外国专家采集和研究的。其中，最先于1884年由俄国专家J.Istvanffy发表的一种绿球藻的报告，是由N.M.Przewalski在蒙古采得标本而由圣彼得堡植物园主任K.Maximovicz研究的。其后德国的Schauinsland和Lemmermann采集和研究了长江中下游的藻类（1903年，1907年）。瑞典学者和探险家Sven-Hedin曾在1893—1901年和1927—1933年，几次到我国新疆、青海、甘肃、西藏和北京，其所得材料分别由Wille（1900，1922）、Borge（1934）和Hustedt（1922，1927）研究发表。1913—1914年，奥地利的植物学家Handel-Mazzatti曾深入我国云南、贵州、四川、湖南、江西、福建6省，所得藻类由H.Skuja于1937年正式发表。前东吴大学任教的美籍教授Gee于1919年发表了他研究苏州和宁波藻类的文章。俄国的Skvortzow自1925年起即定居我国，直到20世纪60年代，他采集和研究过我国东北数省的藻类，还为各地的许多专家研究过不少的中国标本。

中国科学家所发表的第一篇淡水藻类学论文，是1916—1921年毕祖高的题为"武昌长湖之藻类"一文，分4次在当时的《博物学杂志》上刊登。其后有王志稼（1893—1981年）、李良庆（1900—1952年）、饶钦止（1900—1998年）、朱浩然（1904—1999年）

和黎尚豪（1917—1993年）。到1949年，除西藏、宁夏、西康（今四川）外，所采标本大体上已遍及全国各个省、自治区和直辖市。研究的类群主要是蓝藻、绿藻、红藻、硅藻，兼及轮藻、黄藻和金藻。饶钦止还建立了腔盘藻科（Coelodiscaceae 1941），即今之饶氏藻科（Jaoaceae 1947）；又发现了两种采自四川的褐藻（1941年）：层状石皮藻（*Lithoderma zonata*）和河生黑顶藻（*Sphacelaria fluviailis*）。

1949年后，中国的藻类学发展很快，研究人员增加，所采标本遍及全国，研究的类群不断增加。1979年饶钦止出版的《中国鞘藻目专志》中记述了在中国采集的2属301种，81变种和33变型，其中的96种，38变种和32变型的模式标本产于中国[①]。

1964年我国决定编写《中国藻类志》。1973年，编写工作正式开始。其后《中国藻类志》决定采用曾呈奎院士建立的分类系统，将藻类分成如下12门（Division）：①蓝藻门（Cyanophyta），②红藻门（Rhodophyta），③隐藻门（Cryptophyta），④甲藻门（Dinophyta），⑤黄藻门（Xanthophyta），⑥金藻门（Chrysophyta），⑦硅藻门（Bacillariophyta），⑧褐藻门（Phaeophyta），⑨原绿藻门（Prochlorophyta），⑩裸藻门（Euglenophyta），⑪绿藻门（Chlorophyta）和⑫轮藻门（Charophyta）。1984年，为了工作方便，又决定将《中国藻类志》分为《中国海藻志》和《中国淡水藻志》两大部分，各自分开出版。由于各类群在我国原有的工作基础不一致，"志"的编写工作又由不同的主编负责进行，工作进度和交稿时间难以统一安排，因此《中国淡水藻志》的卷册编序，决定不以门、纲、目等分类学类群的次序为序，而以出版先后为序，即最先出版者为第一卷，以下类推。种类较多，必须分成若干册出版者，即在同一卷册号之下再分成若干册，依次编成册号。

1988年，由饶钦止主编的《中国淡水藻志》第一卷"双星藻科"（Zygnemataceae）出版，此卷记录本科藻类9属347种，其中有219种的模式标本产于中国。到1999年，已先后出版6卷。这6卷中，所有的描述和附图，除极少数例外，几乎全是根据中国的标本作出的，所采标本覆盖了全国省、自治区、直辖市的80%—100%。轮藻门、蓝藻门和褐藻门的分类系统经过了主编修订。包括鞘藻目在内，上述已出版的各类群中，中国记录的种的数目，绝大多数均占全国已知种数的40%以上，如色球藻纲的蓝藻已超过80%。特有种（endemic species）在许多类群中也很显著，如鞘藻目和双星藻科的中国特有种几乎占国内已记录的一半！

中国的淡水藻类，种类十分丰富，并有自己的区系特点。但是目前在编写和出版《中国淡水藻志》时，还存在一些问题。

第一，已出版的6个卷册，由于原来各类群的研究基础不同，所达到的水平和质量

[①] 刘国祥与毕列爵于1993年正式报道了采自武汉的勃氏枝鞘藻（*Oedocladium prescotti* Islam），至此鞘藻目（科）所含的3个属，在中国已全有报道。

也不一样。例如，对有些省（自治区），所记种类太少，有一个省甚至只有一种；有许多报道较早的种类，特别是早期由外国专家发表的，已难以看到模式标本；还有许多种类，只在较早时期报告过一次，但描述非常简单，甚至没有附图，并且还未能第二次采到。对这些情况，我们尽量在适当的地方加以说明，更希望再版时有所改进。

第二，在 12 门藻类植物中，除原绿藻外，每一门都有淡水种类。但到目前为止，还有许多类群，尤其是门以下的某些纲、目和科，我国还没有开始进行调查研究，有的几乎是空白。金藻门、隐藻门、甲藻门还有许多种类是由动物学家进行研究的。

第三，藻类分类学是一门既古老又年轻的科学。百多年来，已积累了非常丰富的、极有价值的科学知识，但也存在很多问题。由于不断有许多新属种被发现；新的研究手段，特别是电镜研究、培养和分子生物学的研究，在增加了很多新知识的同时，也使藻类的系统学和分类学出现许多新问题。只有把传统的形态分类学与近代新兴的科学研究手段结合起来，才能使藻类分类学得到长足进步，才能编写出更高质量的《中国淡水藻志》。

总之，我们已取得不少成绩，但肯定还有缺点和不足，希望国内外读者不吝赐教。

毕列爵（湖北大学，武汉 430062）
胡征宇（中国科学院水生生物研究所，武汉 430072）
1997 年 8 月 18 日

FLORA ALGARUM SINICARUM AQUAE DULCIS
FOREWORD

China is a big country with an area of 9,600,000 km^2, covering not only land and ocean, but also 5 thousand islands, with a territory across the cold, temperate, subtropical and tropical belts of the northern Hemisphere. The topography of China is very complicated. In the main, the land runs from high to low gradually along the direction from the west to the east. Of the whole area of the country, 25.9% in the western part are at an altitude of 5,000 m, and then successively from the west to the east, 7% at 2,000 to 3,000 m, 25% at 1,000 to 2,000 m, 16.9% at 500 to 1,000 m, and 25.2% in the eastern, north-eastern and coastal regions below 500 m. There are countless rises and falls of the land to make the various topographical reliefs into mountains, plateaus, basins, plains and mounts. China is a country full of rivers and rivulets too. There are over 50,000 rivers with their basins of 100 km^2. The principal rivers overflow from the west to the eastern or southern seas of the country. The lakes and ponds are also numerous. The number of ever-known natural lakes of an area more than 1 km^2 is no less than 2,800, and the artificial reservoirs are believed to be 86,000. And the ponds, pools, streams, ditches, swamps and springs are uncountable. All the above fundamental characteristics comprehensively lead to a very complicated variation of the sunshine, temperature and precipitation in different localities in China, and thus produce a very rich flora of higher plants, including the bryophytes, ferns and seed plants of more than 30,000 species. In addition, there are innumerable pits of different size marshes, grasslands and rocks, roads and buildings with more or less moisture or soil, all of which forms quite a big number of niches for the freshwater algae inhabitants.

Chinese freshwater algae was collected and studied by foreign experts in the earlier years. The first paper published was written by Russian scientist (J.Istvanffy) in 1884 and the specimens were collected by Russian Military Officer N.M. Przewalski from Mongolia and studied by K. Maximovicz. Later two Germany phycologists, H.Schauinsland and E.Lemmermann, collected and studied the algae of the middle and lower reaches of Yangtze River (1903,1907). Sven-Hedin, a Swedish scholar and explorer, traveled through Xinjiang, Qinghai, Gansu, Xizang (Tibet), and Beijing for several times in 1893—1901 and 1927—1933. The specimens he obtained were studied and published separately by N. Wille (1900,1922), O. Borge (1934), and F. Hustedt (1922, 1927). In 1913—1914, the famous Austrian botanist H.Handel-Mazzatti collected Chinese plants thoroughly in his journey in Yunnan, Guizhou, Sichuan, Hunan, Jiangxi and Fujian Provinces. Among those, the algal material were published formally by the phycologist, H. Skuja (1937). About the same period, N. Gee, an American teacher of the Soochou University, Suzhou, Jiangsu province published his paper about the freshwater algae from Suzhou and Ningbo, Zhejiang province. And B. V. Skvortzow, a Russian naturalist, settled from Russia to China in 1925 till the 1960s of the 20th century. He collected and studied tremendous algal materials both collected from the NE-provinces from China and those presented by a number of experts from various localities of China.

The first paper of Chinese freshwater algae titled as "Algae from Changhu Lake, Wuchang, Hubei" by Bi Zugao, was published in *Journal of Natural History* separately in 4 volumes in 1916—1921. From then on, Wang Chichia (1893—1981), Li Liangching (1900—1952), Jao Chinchih (1900—1998), Zhu Haoran (1904—1999) and Li Shanghao (1917—1993) were the successors. Up to 1949, specimens were collected almost over all the provinces, municipalities and autonomous regions of China with few exceptions as Xizang (Tibet) and Ningxia. The groups were examined carefully concerning the cyanophytes, chlorophytes, rhodophytes, diatoms; and at the same time some attention has been given to charophytes, xanthophytes and chrysophytes too. By C. C. Jao, a new family, the Coelodiscaceae (1941), now the Jaoaceae (1947) was established, and two very rare freshwater brown algae, *Lithodera zonata* and *Sphacelaria fluviatilis* were discovered (1941).

The development of phycology in China was more rapid than ever from 1949 on. The faculties were enlarged, specimens were obtained over all the country and the group's studies were increased. In 1979, Jao published his monograph *Monographia Oedogoniales Sinicae*. In his big volume Jao described 301 species, 81 varieties and 33 forms belonging to 2 of the 3 of the world genera from China. Among them, the types of 96 species, 38 varieties and 32 forms are inhabited in this country[1].

In 1964 a resolution of editing the *Flora of Chinese Algae* was made by the Chinese phycologists. The work was actually put into being since 1973. It was decided in 1978 that the system published by Academician Tseng Chenkui would be adopted in the FLORA. Accordingly, the algae are to be divided into 12 Divisions: (1) Cyanophyta, (2) Rhodophyta, (3) Cryptophyta, (4) Dinophyta, (5) Xanthophyta, (6) Chrysophyta, (7) Bacillariophyta, (8) Phaeophyta, (9) Prochlorophyta, (10) Euglenophyta, (11) Chlorophyta and (12) Charophyta. In 1984, for the convenience in practical work, phycologists agreed that the FLORA could be written separately into two parts, the FLORA of Marine Algae and that of the freshwater forms. Because the achievements of researches of the different algal groups are not at the same level, so the work could not be done according to the taxonomic sequence of the algal groups. We may try to publish first the group we have gotten more information and better results about it.

---

1) Liu Guoxiang and Bi Liejue reported *Oedocladium prescottii* Islam from Wuhan in 1993, so all the 3 genera of the Oedogoniales (-aceae) have been reported in China since then.

And, at the same time, the numbers of the sequence of the volumes of the FLORA are also arranged not basing upon the taxonomic series but upon the priority of publications. Thus one volume may be separated into two or more parts if necessary.

In 1988, the first volume of the *Flora Algarum Sinicarum Aquadulcis* "Zygnemataceae" edited by Jao Chinchih was published. In it, 347 species of 9 genera were described, and the types of 219 species were all collected from China. Up to 1999, six volumes of the FLORA had been published, from those we may know it may be concluded that the specimens collected and used are at least 80% and at most 100% from the provinces, municipalities and autonomous regions in China. The descriptions and drawings with very few exceptions are all based on Chinese materials. The taxonomic systems of Chroococophyceae, Charophyta and Euglenophyta had been more or less modified by the editors. The percentage of the number of species in each volume, including the Oedogoniales, to that of the world records is remarkably as large as over 40%. The extreme one is 80% in Chroococophyceae. The number of endemic species is also distinct, for example, in Oedogoniales and Zygnemataceae, they are both over 50%.

The flora of Chinese freshwater algae are plentiful, and the floral composition is evidently peculiar. However, there were still quite a lot of problems to be solved in the editing of the FLORA.

First, in some examples the record of provincial distribution of the country is insufficient. It is unreasonable for a big province to have recorded only a single species. In a number of old literatures, the species description is usually either too simple or lacking, and the drawings are also wanting. For many species, it is very hard to check up with more information because it was reported only once for a very long time. And, an unconquerable difficulty is that the majority of the types, especially in the earlier publications, could not hope some improvements can be made in the successive volumes.

Second, except the Prochlorophyta, freshwater algae could be found in each of the 12 Divisions of algae. Unfortunately, there are a number of subgroups under the Divisions which have not yet been studied especially in the Xanthophyta, Chrysophyta and Cryptophyta. Many dinophytes are investigated by zoologists. In addition, some genera with reputation as "big" taxa, such as the *Navicula*, *Cosmarium*, and *Scenedesmus*, etc., have yet not been collected and studied enough in China.

Third, the taxonomy of algae is a science both old and young. In the past hundreds of years, numerous and valuable information was accumulated. New conceptions in taxonomy and systematics are arising in proceedings of the additions of new taxa, and particularly new facts and ideas are appearing from the new means such as the electron microscopy, culture and molecular biology. The suitable way may be making comprehensive studies in these fields. Unfortunately, this is at present nearly a blank in the phycology research of freshwater algae in China. The combination of traditional and modern methodology is of course necessary and urgent. It is universally hope that more improvements could be achieved in the following volumes.

For the flaws and mistakes in both of the volumes ever published and those to follow, any suggestions and corrections are welcomed by the authors.

Bi Leijue (Hubei University, Wuhan, 430062)
Hu Zhengyu (Institute of Hydrobiology, CAS, Wuhan, 430072)
August 18, 1997

前　言

　　硅藻是一个庞大的类群，目前有描述的种类 2 万多种。管状壳缝作为硅藻一个重要的分类特征，出现在较为进化的类群中，已记录的种有 4000 多个。

　　有关管壳缝目（Aulonoraphidinales）的用法，在国外的著作和期刊上都未查到。在我国最早见于胡鸿钧等（1980）编著的《中国淡水藻类》一书，该目包括窗纹藻科（Epithemiaceae）、菱形藻科（Nitzschiaceae）和双菱藻科（Surirellaceae）3 个科。在我国，也有学者使用双菱藻目（Surirellales）这个名称，其范围与管壳缝目一致（朱惠忠和陈嘉佑，1989b，2000），但在国外文献中 Surirellales 这个目所包含的内容差异较大，常常只是包含本卷册中的双菱藻科（Surirellaceae）种类。

　　近 30 年来，硅藻门管壳缝类群分类和系统变化很大，其中影响较大的是 Round 等（1990）出版的 The Diatoms 一书，在他们的著作中，建立新目：棒杆藻目（Rhopalodiales D. G. Mann. 1990），下设 1 科（棒杆藻科 Rhopalodiaceae）3 属（*Epithemia*、*Rhopalodia* 和 *Protokeelia*）；采用杆状藻目（Bacillariales Hendcy 1937），包括 1 科（杆状藻科 Bacillariaceae）15 属（*Bacillaria*、*Hantzschia*、*Psammodictyon*、*Tryblionella*、*Cymbellonitzschia*、*Gomphonitzschia*、*Gomphotheca*、*Nitzschia*、*Denticula*、*Denticulopsis*、*Fragilariopsis*、*Cylindrotheca*、*Simonsenia*、*Cymatonitzschia* 和 *Perrya*）；建立了双菱藻目（Surirellales）作为一个新目，包括 3 科：Entomoneidaceae（含 1 属：*Entomoneis*）、Auriculaceae（含 1 属：*Auricula*）和 Surirellaceae（含 7 属：*Hydrosilicon*、*Petrodictyon*、*Plagiodiscus*、*Stenopterobia*、*Surirella*、*Campylodiscus* 和 *Cymatopleura*），可以看出，Round 等（1990）所使用的"Surirellales"，已经与朱惠忠和陈嘉佑（1989b，2000）所使用的"Surirellales"有很大差异，所包含的类群也不相同。

　　在 Round 等（1990）所使用的"Surirellales"中，Entomoneidaceae 只有 *Entomoneis* 一属，在《中国淡水藻志》第十四卷中，作为茧形藻属（*Amphiprora*）把它放在舟形藻科（Neviculaceae）；而 Auriculaceae 只有 *Auricula* 一属，我国没有报道。因此，我国淡水 Surirellales 只有 Surirellaceae 一科。

　　学者们不但在形态学上认为棒杆藻科（窗纹藻科）、杆状藻科（菱形藻科）和双菱藻科这三个科没有直接的亲缘关系，近年来分子系统学方面的证据也表明具有管状壳缝的这三个科不是单起源。Ruck 和 Theriot（2011）利用分子生物学手段，重建了具管状壳缝硅藻的系统发生树，提出三个类群的管状壳缝并非同一起源的假设：他们认为管状壳缝经过两次进化，其中 Bacillariaceae 种类具有共同的祖先，Rhopalodiaceae 和 Surirellaceae 具有共同的祖先。

　　尽管具有管状壳缝的这些硅藻类群不是一个自然的类群，放在一个"目"里可能不十分合适。但考虑到它们都具有管状壳缝这一共同特征，可明显和其他类群区分；管壳缝目这一名称在我国使用已久，被许多人所熟知，在《中国淡水藻志》中硅藻系统排列上，作为羽纹纲的5个目之一（见《中国淡水藻志》第十卷）。为保证《中国淡水藻志》的内容统一，方便读者阅读，本卷册仍然沿用了"管壳缝目"这一用法。

本卷册在科属的名称与排列上，基本采用了Round等（1990）的分类系统，使用了杆状藻科（Bacillariaceae）、棒杆藻科（Rhopalodiaceae）和双菱藻科（Surirellaceae）等名称，各科所包含的属也参照了Round等的分类系统。

本卷册于2005年列入《中国淡水藻志》编写计划，但相关研究工作很早就已开始。早在2003年，上海师范大学硕士研究生李海玲就完成了《新疆管壳缝目硅藻初报》的毕业论文；范亚文（2004）出版了《黑龙江省管壳缝目植物研究》一书。本卷册立项以后，尤庆敏博士（2009）完成了《中国淡水管壳缝目硅藻的分类学研究》博士学位论文，为本卷册的编写奠定了基础。之后，又经过6年的补充、修改，完成了本卷册撰写任务。

参加本卷册工作的有上海师范大学的王全喜、尤庆敏、李海玲，哈尔滨师范大学的范亚文、刘妍，中国科学院水生生物研究所的施之新等。

本卷册包括了管壳缝目硅藻的3科13属189个分类单位，其中149种37变种3变型，采集地涵盖了我国31个省、自治区和直辖市。本卷册图版以光学显微镜和扫描电镜照片为主，在这189个分类单位中，有175个分类单位我们采集到了标本，提供了1132张照片，其中光镜照片859张，扫描电镜照片259张；有14个种没有观察到标本或照片不理想，引用了朱惠忠和陈嘉佑（2000）《中国西藏硅藻》一书的绘图。在编研过程中，得到暨南大学齐雨藻教授、中国地质科学院李家英研究员、华东师范大学王幼芳教授的指导和帮助，哈尔滨师范大学包文美教授、厦门大学高亚辉教授等提供资料，上海师范大学的刘琪、李博、刘浩、于潘等参与部分工作，在此一并致谢。

由于我们水平有限，书中定有疏漏和欠妥之处，敬请读者批评指正。

王全喜
2016年12月于上海

目 录

序
中国孢子植物志总序
《中国淡水藻志》序
前言
绪论 ··· 1
  一、管壳缝目硅藻的细胞结构及形态特征 ·· 1
  二、管壳缝目硅藻的生殖 ·· 7
  三、管壳缝目硅藻的生态及在我国的分布 ·· 7
管壳缝目 AULONORAPHIDINALES ··· 10
  杆状藻科 Bacillariaceae ·· 10
    杆状藻属 *Bacillaria* J. F. Gmelin ··· 10
    菱形藻属 *Nitzschia* A. H. Hassall ·· 12
    菱板藻属 *Hantzschia* A. Grunow ··· 49
    盘杆藻属 *Tryblionella* W. Smith ··· 61
    细齿藻属 *Denticula* F. T. Kützing ·· 68
    西蒙森藻属 *Simonsenia* H. Lange-Bertalot ··· 73
    筒柱藻属 *Cylindrotheca* L. Rabenhorst ··· 74
  棒杆藻科 Rhopalodiaceae ·· 75
    窗纹藻属 *Epithemia* F. T. Kützing ··· 75
    棒杆藻属 *Rhopalodia* O. Müller ·· 88
  双菱藻科 Surirellaceae ··· 97
    波缘藻属 *Cymatopleura* W. Smith ··· 98
    双菱藻属 *Surirella* P. J. F. Turpin ··· 104
    长羽藻属 *Stenopterobia* A. de Brébisson ex H. Van Heurck ·················· 124
    马鞍藻属 *Campylodiscus* C. G. Ehrenberg ex F. T. Kützing ··················· 127
附录 I　科、属、种的检索表（英文） ··· 131
附录 II　汉英术语对照表 ·· 142
参考文献 ·· 143
中名索引 ·· 161
学名索引 ·· 164
图版

·xvii·

绪 论

一、管壳缝目硅藻的细胞结构及形态特征

该类群均为单细胞，细胞壁与其他类群硅藻一样，变化多样且充满硅质($SiO_2 \cdot nH_2O$)。细胞壁除了硅质还含一些果胶质，位于壳面和环带周围，环绕形成不连续分布的薄膜。细胞壁(或称壳体，frustule)由上、下两个半壳套合而成。细胞核单个。色素体数量不一，通常1个或2个(偶尔4个)，有些种类具蛋白核。同化产物主要为脂肪，在细胞内呈反光较强的小球体。在 Rhopalodiaceae 硅藻内，常含有体积较小的内共生蓝藻。

该类群主要是单个生活，也可形成链状或星状群体，多以底栖、附着为主。壳体的形态与构造特征是硅藻分类的主要依据，本类群硅藻的特征具体如下所述。

1. 壳体形态和对称性

硅藻细胞壁的结构有两种类型：中心硅藻一般具圆形的细胞壁；羽纹硅藻一般具杆形、舟形或弓形的细胞壁，壳面上的线纹呈羽状排列。管壳缝目硅藻属于羽纹硅藻类型。

在描述硅藻细胞时，有三个轴和三个面是非常重要的(图1)：顶轴(apical axis)连接细胞两端且与壳表面平行，中心纲硅藻的顶轴是无限多的；贯穿上下壳面形态中心的轴是贯壳轴(pervalvar axis)；与顶轴和贯壳轴都成直角，且与壳面平行的轴是横轴(transapical axis)。

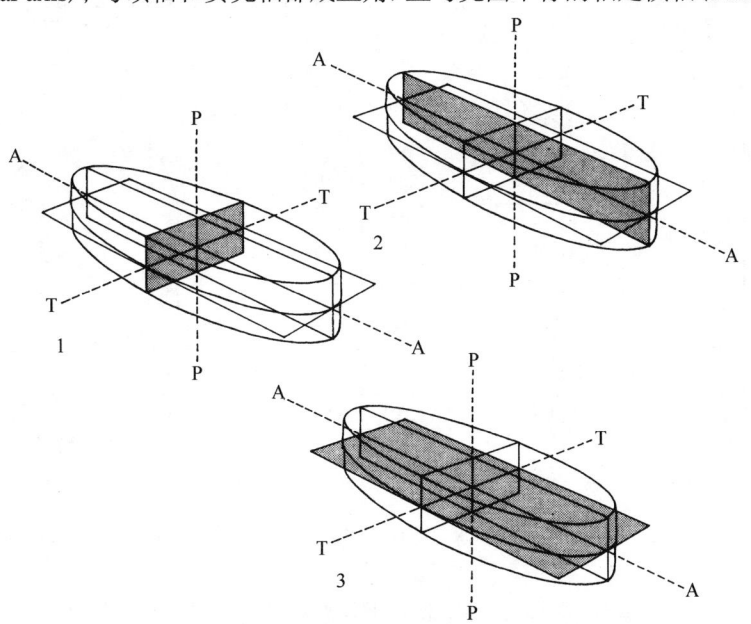

图1 硅藻细胞轴和面的示意图(引自 Krammer and Lange-Bertalot，1986)
A. 顶轴；P. 贯壳轴；T. 横轴。1. 阴影部分是中央横切面；2. 阴影部分是顶面；3. 阴影部分是中央壳面
Fig. 1 Axises and planes of diatom frustule
A. Apical axis；P. Pervalvar axis；T. Transapical axis. 1. Median transapical plane；2. Apical plane；3. Median valve plane

有了这三个轴，就可以定义相关的三个面。顶轴和横轴位于中央壳面(median valve plane)，顶轴和贯壳轴形成顶面(apical plane)，横轴和贯壳轴形成中央横切面(median transapical plane)。

O. Müller 在 1895 年提出的这些术语与数学上的相关术语是有差别的，在几何学里没有弯曲的轴和穹窿形的面。而在硅藻描述中，对轴和面的界定做了修改，轴可以是直或弯曲的线，弯曲或扭曲的面把壳体分成相等或不等的两半。当壳面被分成两个不等部分时，对应的轴穿过发育不相同的两极，称为异极。这三个主要的面并不都是对称的，面的外形和对称性只有在三个轴都是等极的情况下才是相同的。

管壳缝目硅藻包括三个类群的种类，属种较多，形态特征多样，对称性差别大。根据不同的科、属，具体形态和对称性如下所述。

杆状藻科(Bacillariaceae)种类，一般为线形、披针形，关于顶面两侧对称，关于中央横切面两端对称，因为壳缝龙骨系统一般是离心的，所以壳面结构并不是绝对的左右对称(图2A、图2C、图3c)。少数种类弓形或"S"形，如 *Hantzschia* 种类，具背腹之分，关于顶面不对称，关于中央横切面两端对称(图2B、图3b)。

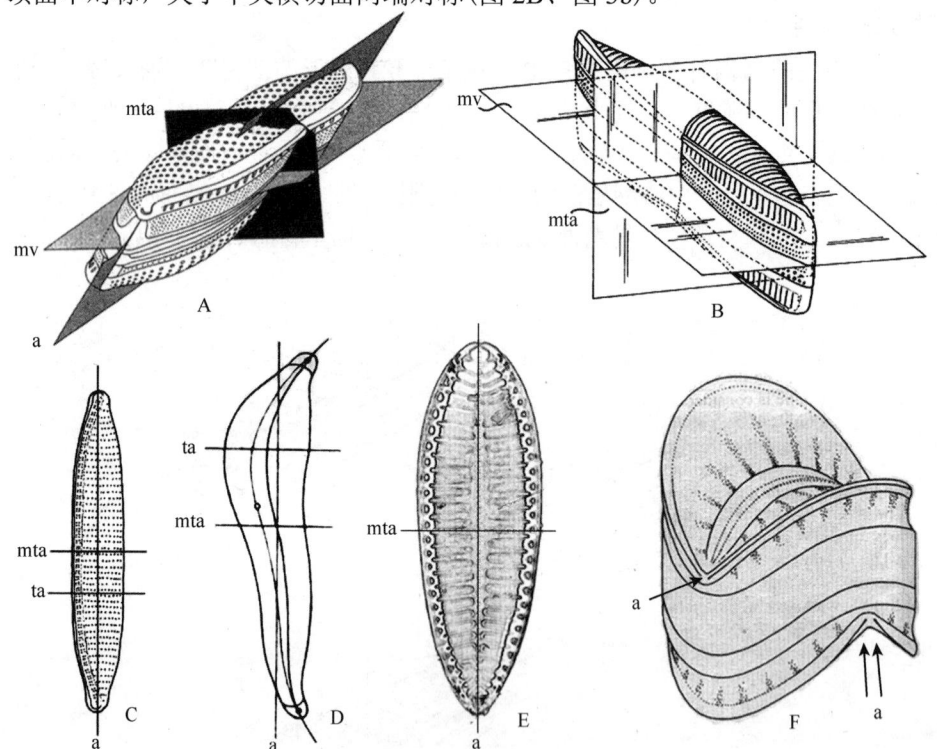

图 2　管壳缝目硅藻细胞对称性(引自 Ruck and Kociolek, 2004; Krammer and Lange-Bertalot, 1986; Mann, 1978; Round *et al.*, 1990)

A、C. 菱形藻属; B. 菱板藻属; D. 棒杆藻属; E. 双菱藻属; F. 马鞍藻属

a. 顶轴; mv. 中央壳面; mta. 中央横轴; ta. 横轴

Fig. 2　Symmetry of canal raphe diatoms

A, C. *Nitzschia*; B. *Hantzschia*; D. *Rhopalodia*; E. *Surirella*; F. *Campylodiscus*

a. Apical axis; mv. Median valve plane; mta. Median transapical axis; ta. Transapical axis

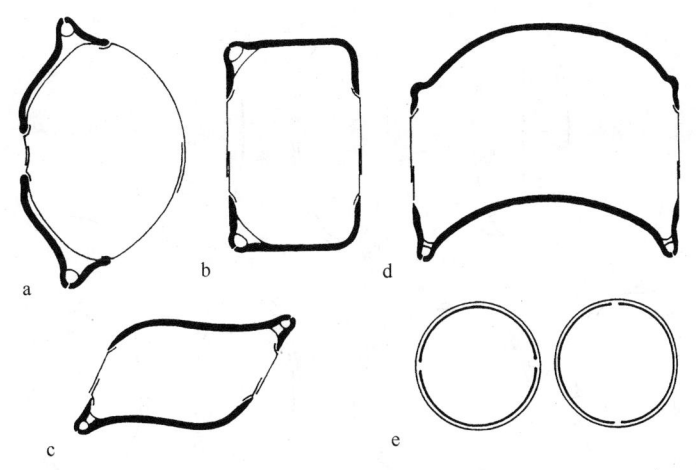

图 3　管壳缝目硅藻的横切面示意图（引自 Round et al., 1990）

a. 棒杆藻属：壳面和环带均具明显的背腹性；b. 菱板藻属：带面略微或不具背腹性，但是壳面明显不对称；c. 菱形藻属：带面不具背腹性，但是壳面明显不对称；d、e. 马鞍藻属：细胞为马鞍形，每个壳面为圆形，e. 示上下两个壳面壳缝末端的相对位置

Fig. 3　Transapical sections of canal raphe diatoms

a. *Rhopalodia*: very marked dorsiventrality of girdle and valves; b. *Hantzschia*: slight or virtually no dorsiventrality of girdle but valves strongly asymmetrical; c. *Nitzschia*: no dorsiventrality of girdle but valves strongly asymmetrical; d, e. *Campylodiscus*: each valve has a circular outline and is saddle-shaped, e. show the relative position of the raphe endings in the 2 valves

棒杆藻科（Rhopalodiaceae）种类，一般为弓形，具背腹之分，关于顶面两侧不对称，在形态和结构上关于中央横切面两端对称或不对称（图 2D、图 3a）。

双菱藻科（Surirellaceae）种类形态比较多样，*Stenopterobia* 种类直或弯曲呈"S"形，直线型种类关于顶面两侧对称，关于中央横切面两端对称，"S"形种类关于壳面两侧两端均不对称；*Cymatopleura* 种类基本关于顶面和中央横切面两侧两端对称；*Surirella* 种类一般关于顶面两侧对称，关于中央横切面两端对称或不对称，该属在外形上两端对称的种类，在结构上不一定对称，因为壳缝在两端的结构可能不相同（图 2E）；*Campylodiscus* 种类由于上下壳面扭曲 90°，因此壳面在外形和结构上关于顶面和中央横切面两侧两端对称（图 2F、图 3d、图 3e）。

2. 壳缝及相关结构

在许多羽纹硅藻的壳面上，常具 1 条或 2 条纵向裂缝，称为壳缝（raphe）（图 4）。壳缝是一个较进化的细胞结构，非常复杂，控制硅藻细胞的运动。壳缝在中部常被一硅质结构分开，比壳面其他部分稍厚，称为中央节（central nodule），壳缝在中央节处的断开部分，称为中缝端（central raphe ending）或近缝端（proximal raphe end），也有一些种类，如 Bacillariaceae 和 Rhopalodiaceae 的一些类群，壳缝在中部不断开，不具中央节和中缝端（图 4c）。壳缝在两极断开处，称为极缝端（polar raphe ending）或远缝端（distal raphe end），常位于一个简单的螺旋舌（helictoglossa）结构上。壳缝不是一个简单的裂缝，横断面呈一个"V"形，据推测这是为了防止在自身膨胀或是外界压力下，细胞壳面纵向裂开而形成的（Round et al., 1990）。

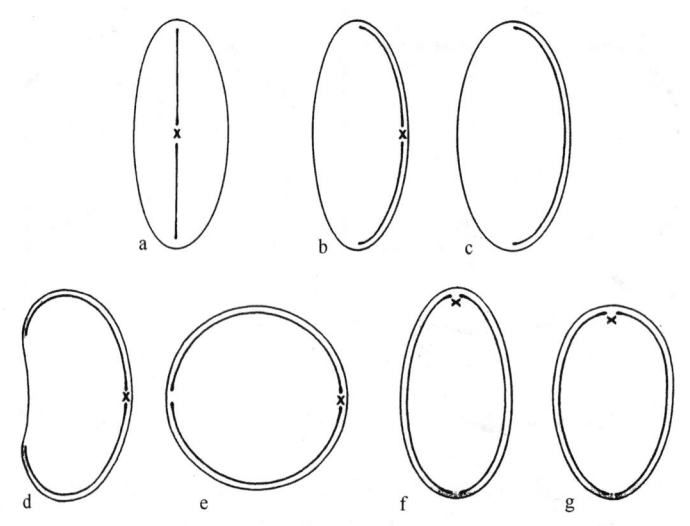

图 4　硅藻壳缝的进化趋势(引自 Round et al., 1990)

a. 舟形类硅藻的壳缝系统从一极至另一极，穿过壳面中部；b. 在离心壳缝系统中开始具龙骨突；c—g 壳缝裂缝的长短和位置开始发生一系列变化。在 b—g 中，龙骨突在壳面内部，在外部看不到；"x"表示中缝端(central raphe ending)或个体发育的等值处

a. 舟形藻属；b、c. 菱形藻属；d. *Auricula*；e. *Campylodiscus*；f. *Cymatopleura*；g. *Surirella*

Fig. 4　Evolutionary trends in advanced raphid diatoms

a. *Navicula*；b、c. *Nitzschia*；d. *Auricula*；e. *Campylodiscus*；f. *Cymatopleura*；g. *Surirella*

(1) 壳缝(raphe)的位置。壳缝位置变化较大(图 4)，可位于壳面中部，或偏离中部(离心)，或壳面边缘，或围绕整个壳面一周。Bacillariaceae 种类的壳缝一般离心(图 4b、图 4c)；Rhopalodiaceae 种类的壳缝一般位于壳面背缘或腹缘，并在壳面中部不同程度伸入壳面，形成"V"形；Surirellaceae 的壳缝一般围绕整个壳面一周(图 4d—g)。

(2) 中缝端(central raphe ending)形状。中缝端形状是指壳缝近中央节末端的形状。管壳缝目硅藻的中缝端在外壳面因科、属不同而异，形状多样，直向而简单，或稍弯曲，或膨大。在内表面，一般直向而简单，常被龙骨突包住，不易观察到。

(3) 极缝端(polar raphe ending)形状。极缝端形状是指壳缝近极节末端的形状。管壳缝目硅藻的极缝端在外壳面略弯曲或弯转呈钩状。在内壳面，形状较为单一，直向，终止于一个小而简单的螺旋舌上。

(4) 相关结构。一般认为，管状壳缝是最完善的壳缝类型，这类壳缝及相关结构一般位于壳缘上特化的龙骨(keel)上，形态结构多样：不同程度凸出于壳面(图 5a—c)；或壳缝下的龙骨融合在一起，甚至部分消失(图 5d、图 5e)，仅留下一排管状结构连接壳缝管和细胞其余部分，称为拟窗龙骨(fenestrated keel)，存在于 *Entomoneis*、*Surirella* 和 *Campylodiscus* 中。管壳缝在内壳面被一组桥状硅质结构包住，称为龙骨突(fibulae)(图 6a、图 6b)。龙骨突的形态多样，从窄肋状到平片状或更复杂的结构，连接龙骨突上面的壳体，在壳缝下形成一个槽，称为管壳缝(canal raphe)(图 6a、图 6b)。龙骨突之间为各种形状的小孔，是细胞内外物质交流的重要通道。龙骨突横向延伸，形成一增厚的硅质结构，称为肋纹(costae)(图 6c)，有时肋纹在内表面明显高于壳面其他部分，称为隔片(partition)。

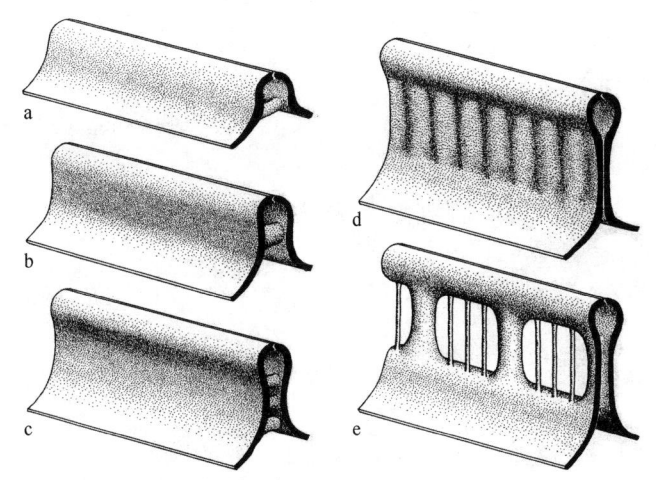

图 5　各种形态的龙骨壳缝系统(引自 Round et al., 1990)

a. 龙骨简单，位置低(如：多数菱形藻属种类)；b. 龙骨较深，龙骨突一排(如：沙网藻属种类)；c. 龙骨深，龙骨突数排(如：膜翼茧形藻)；d. 龙骨两边融合，形成一排窄管状，连接近壳缝和细胞腔(如：部分茧形藻属种类)；e. 具拟窗龙骨(如：部分双菱藻属、长羽藻属和马鞍藻属种类)

Fig. 5　Keeled raphe systems in genera possessing fibulae

a. Simple shallow (e.g. many *Nitzschia*); b. Deeper keel with fibulae at one level (e.g. *Psammodictyon*); c. Deep keel with fibulae at several levels (e.g. *Entomoneis alata*); d. Fusion of the two sides of the keel to create narrow tubes connecting subraphe canal to cell lumen (e.g. some *Entomoneis* spp.); e. Creation of spaces between the connecting tubes by further fusion and simplification of the keel walls (*Surirella*, *Stenopterobia* and *Campylodiscus* spp.)

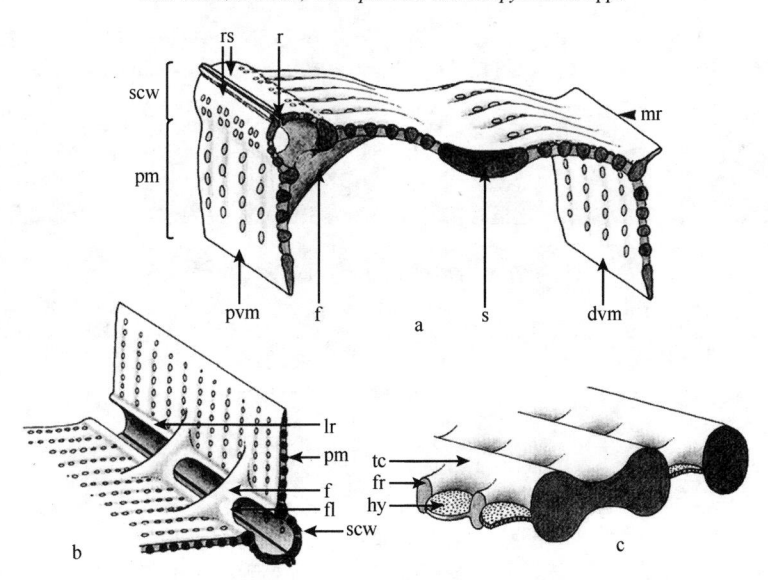

图 6　盘杆藻属壳面结构示意图(引自 Mann，1978)

a. 示壳面整体结构；b. 示管壳缝内部；c. 示具膜的壳面结构

dvm. 远壳缘；f. 龙骨突；fl. 龙骨突边缘延伸结构；fr. 间质片层；hy. 膜；lr. 纵向脊；mr. 边缘脊；pm. 近壳套；pvm. 近壳缘；r. 壳缝裂缝；rs. 壳缝骨；s. 胸骨；scw. 近壳缝管壁；tc. 横肋纹

Fig. 6　Valve construction of *Tryblionella* sp.

a. valve construction; b. interior of raphe canal; c. valve with hymen

dvm. distal valve margin; f. fibula; fl. flange; fr. frets; hy. hymen; lr. longitudinal ridge; mr. marginal ridge; pm. proximal mantle; pvm. proximal valve margin; r. raphe-slit; rs. raphe-sternum; s. sternum; scw. subrahpe canal wall; tc. transapical costae

3. 线纹

1) 线纹的组成形式

线纹由点孔纹（puncta）、拟孔纹（poroid）、"C"形纹（"C"-shaped）或窝孔纹（alveolate）等组成。组成形式有三种类型。

（1）单列型：线纹由单列孔纹组成，如 Bacillariceae、Rhopalodiaceae 和 Surirellaceae 内 *Cymatopleura* 的一些种类。

（2）双列型：线纹由双列孔纹组成，如 *Hantzschia*、*Tryblionella*、*Denticula*、*Rhopalodia* 的一些种类。

（3）多列型：线纹或肋纹由多排孔纹组成，如 *Stenopterobia*、*Surirella* 和 *Campylodiscus* 的一些种类。

2) 组成线纹的孔纹形状

（1）点孔纹：是组成管状壳缝硅藻线纹的主要形式，常见有圆形、椭圆形、裂缝形等形状（图 7a—i、图 7n—p、图 7r—t）。

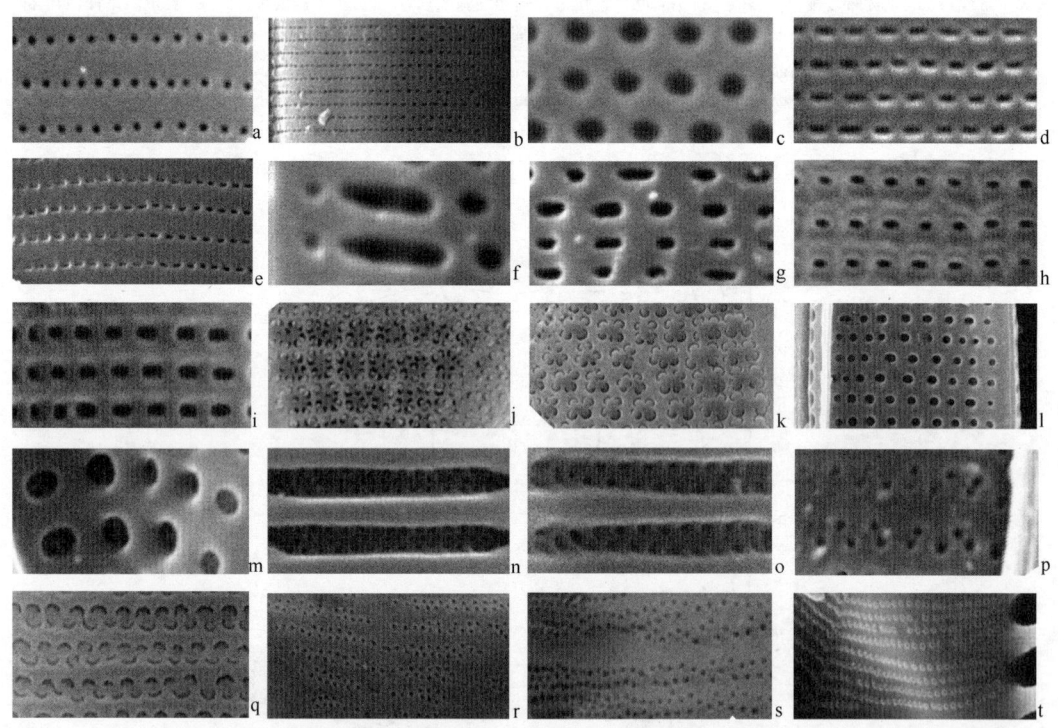

图 7 管壳缝目硅藻的线纹组成和孔纹形状（SEM）

a—i. 单列点孔纹组成形式和孔纹形状：a—c. 圆形，d—f. 裂缝形，g—i. 椭圆形；j，k. 单列"C"形孔纹组成形式和孔纹形状："C"形；l，m. 单列拟孔纹组成形式和孔纹形状：圆形；n—q. 双列点孔纹组成形式和孔纹形状：圆形；q. "C"形；多列点孔纹组成形式和孔纹形状：r—t. 圆形，其中 t 的孔纹具缘

Fig. 7 Shapes of puncta and consisting of striae in canal raphe diatoms（SEM）

a—i. The shapes of puncta and consisting of uniseriata striae: a—c. rounded, d—f. slit-like, g—i. elliptical; j, k. The shapes of "C"-shaped puncta and consisting of uniseriata striae: "C"-shaped; l, m. The shapes of poroid and consisting of uniseriata striae: rounded; n—q. The shapes of puncta and consisting of biseriata striae: rounded; q. "C"-shaped; The shapes of puncta and consisting of multiseriata striae: r—t. rounded, the puncta of "t" with rim

(2) 拟孔纹：在外壳面呈圆形或椭圆形，孔内结构较为复杂，多由分支孔板(vola)组成，如 *Denticula* 一些种具此孔纹(图 7l、图 7m)。

(3) "C"形纹：裂缝在外壳面呈"C"形，在 *Epithemia* 中，一般 4—5(8)个"C"形裂缝围成一圈，似花瓣，也称为窝孔纹(alveolate)。孔内结构较为复杂，多由分支孔板组成(图 7j、图 7k)，在 *Rhopalodia* 中，也具"C"形纹，但是排列方式与 *Epithemia* 种类不同(图 7q)。

二、管壳缝目硅藻的生殖

管壳缝目硅藻的繁殖可分为无性和有性两种生殖方式。无性生殖包括细胞分裂和形成休眠孢子。一般情况下，1 个细胞体进行一分为二的分裂，分裂后期出现 2 个新的子细胞下壳。分裂完成后，其中一个子细胞与母细胞等大，另一个子细胞比母细胞小。而当生活环境发生变化，不适宜细胞正常繁殖时，就会形成休眠孢子。休眠孢子在适合的环境中可以萌发形成一个新细胞，可进行生长繁殖。有性生殖是通过形成增大的结合子或复大孢子来完成的(图 8)，可分为同配和异配两种形式，如在 *Nitzschia amphibia* 中，每个配子囊产生同一类型的配子，称为同配，而在 *Nitzschia palea*、*Surirella peisonis*、*Cymatopleura solea* 等种类中，每个配子囊产生两种不同类型的配子，称为异配。

一些种类进行自体融合或自体受精，如在 *Nitzschia frustulum* var. *perpusilla* 中，发现自体融合现象，即形成合子的两个配子来自同一配子囊；在 *Denticula tenuis* 中，发现自体受精现象，即配子在细胞核融合时，配子囊原生质体不发生分裂。

三、管壳缝目硅藻的生态及在我国的分布

1. 管壳缝目硅藻的生态习性

管壳缝目硅藻多为附着生种类，常附着于水中的沙石、污泥、水草或其他基质上。除 *Epithemia*、*Stenopterobia* 和 *Cymatopleura* 种类仅分布于淡水中，其他大部分种类广泛分布于淡水、咸水或海水中。在内陆水体中，无论是江河、湖泊、水库等开阔的大水体，还是沼泽、水塘、水坑、溪流、泉水、临时性积水、井水以及稻田等小水体都有管壳缝目硅藻的踪迹，甚至在潮湿的土壤、树皮或岩壁等气生、半气生环境中都能生存。管壳缝目硅藻可以生活在不同流速、不同营养程度、不同污染程度等多样化的水体中，其中 *Stenopterobia* 对水体中的 pH 非常敏感，嗜酸性环境。

2. 管壳缝目硅藻在我国的分布

管壳缝目硅藻同其他藻类一样，分布极为广泛。由于生活在不同类型的水体中，决定了硅藻不能像地球上高等植物那样具有明显的地理分布，大量的文献资料表明，小环境对其种类组成有着更大的影响。近来，由于世界各地区硅藻的种类和分布如雨后春笋般地报道出来，科研工作者开始认识到硅藻地理分布特征的重要性。然而，我们要对管壳缝目硅藻的全球地理分布进行系统的总结，还需要更全面的数据。

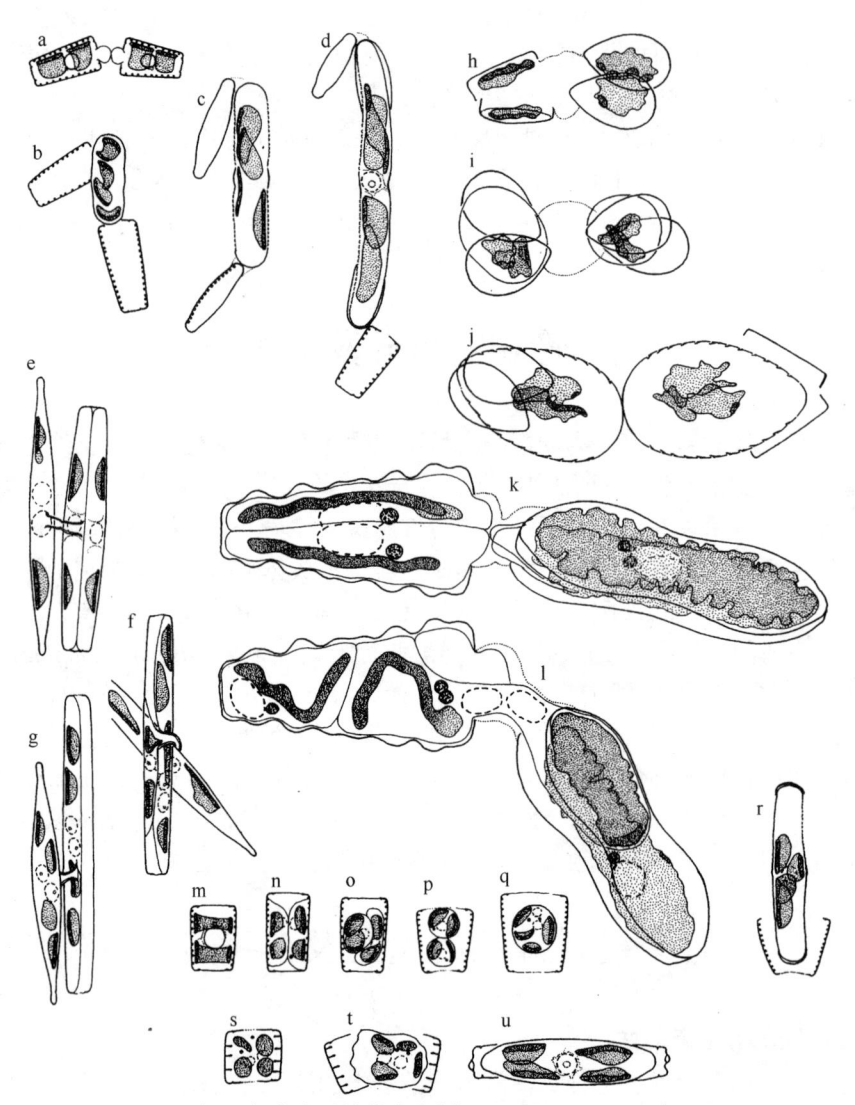

图 8 复大孢子的发育(引自 Round et al., 1990)

a—d. 两栖菱形藻：a. 细胞顶部产生乳突，b—d. 同型孢子融合后复大孢子开始膨大；e—g. 谷皮菱形藻：e. 配子囊内包含两种配子，在中部由一个结合管连接，f. 生理性异配：一对配子已发生融合并移入配子体，另一对还没行动；g. 合子位于配子囊中；h—j. 派松双菱藻：h. 减数分裂 I 后形成成对的配子囊，i. 幼复大孢子，发生在生理性异配融合之后：每一配子囊中具一合子，j. 正在膨大的合子；k，l. 草鞋形波缘藻：k. 配子囊内的成对配子，l. 配子通过结合管移动；m—r. 小片菱形藻 perpusilla 变种：m. 减数分裂前期未成对的配子囊；n—p. 配子形成前，重排期间和重排之后，q. 幼体结合导致配子体内单个合子的形成，r. 膨大的复大孢子；s—u. 小型细齿藻复大孢子自体受精的发育

Fig. 8 Auxospore development

a—d. *Nitzschia amphibia*: a. production of papillae from the cell apices, b—d. expansion of the auxospore following isogamous fusion; e—g. *Nitzschia palea*: e. gametangia containing two gametes apiece, connected by a central copulation tube, f. physiological anisogamy: a pair of gametes has fused and migrated into one gametangium, the other pair has yet to move; g. zygotes lying within the gametangia frustules; h—j. *Surirella peisonis*: h. paired gametangia after meiosis I, i. young auxospores, following physiologically anisogamous fusion: one zygote lies within each gametangium, j. expanding zygotes; k, l. *Cymatopleura solea*: k. paired gametes in each gametangium, l. migration of gametes through copulation canal; m—r. *Nitzschia frustulum* var. *perpusilla*: m. meiotic prophase in unpaired gametangium, n—p. gametes before, during and after rearrangement, q. paedogamy results in the formation of a single zygote lying within the gametangium, r. expanded auxospore; s—u. Autogamous development of the auxospore in a race of *Denticula tenuis*

早期对我国管壳缝目硅藻的报道，大多是采集和鉴定标本的调查研究。所采集的地点主要是在云南、四川、青海、新疆、西藏等边远地区。据不完全统计，Mereschkowsky(1906)、Hustedt(1922a)、Skuja(1937)报道了在我国西藏北部、新疆东南部、青海、云南西北、四川西南、帕米尔东部等地区管壳缝目硅藻共 3 科 6 属近 30 个分类单位。金德祥(1951)搜集了 1847~1946 年 100 年间在中国及蒙古国的硅藻，记录我国淡水管壳缝目硅藻 3 科 9 属 108 个分类单位，其中大部分种类是 Skvortzow 报道的，标本主要采自福建、黑龙江、山东、浙江、江苏、北京、天津等地。

20 世纪 70 年代以后，我国开始有较多学者报道我国各地的管壳缝目硅藻，其中报道种类较多的如下：海南岛菱形藻属 14 种 4 变种(Zhang, 1986)；四川九寨沟自然保护区的管壳缝目硅藻共 3 科 6 属 11 种 2 变种(Bao et al., 1986)；湖南索溪峪的管壳缝目硅藻共 3 科 8 属 49 种 18 变种 2 变型(Zhu and Chen, 1989b)；吉林长白山地区的管壳缝目硅藻共 3 科 7 属 25 种 6 变种 2 变型(Bao et al., 1992)；湖南西北部和贵州东北部的武陵山区(沅江流域、乌江流域、梵净山区)管壳缝目硅藻共 3 科 9 属 64 种 26 变种 3 变型(Zhu and Chen, 1994)；安徽的管壳缝目硅藻共 4 种 3 变种，包括 2 个新种(Yang, 1995, 1999)；西藏的管壳缝目硅藻共 3 科 9 属 80 种 53 变种 3 变型(Zhu and Chen, 2000)；黑龙江、辽宁及内蒙古东部区域的管壳缝目硅藻 3 科 8 属 68 种 27 变种 2 变型，共 97 个分类单位(Fan et al., 2001)；新疆地区的管壳缝目硅藻 63 种 30 变种 1 变型，共 94 个分类单位(Li, 2003)。

目前，我国管壳缝目硅藻共记录有 3 科 10 属 200 多个分类单位，大部分种类仅以名录形式出现，其中仅西藏、新疆、东北等少数地区有较为系统的研究。因此，对我国管壳缝目硅藻地理分布特征的分析，还需要进一步的研究工作。

管壳缝目 AULONORAPHIDINALES

该目主要特征是：上下壳面均具管状壳缝。

该目有 3 科，我国有 3 科。

管壳缝目分科检索表

1. 上下壳面均具一发育良好的管状壳缝系统，围绕整个壳缘，存在龙骨结构 ···· 双菱藻科 Surirellaceae
1. 上下壳面均具一发育良好的管状壳缝系统，纵向贯穿壳面，一般近壳面边缘，龙骨结构存在或缺失 ··· 2
   2. 壳面不具龙骨结构，明显背腹之分，龙骨突延伸形成横隔片 ············· 棒杆藻科 Rhopalodiaceae
   2. 壳面存在龙骨结构，不具或稍具背腹之分，龙骨突不延伸，或稍微延伸一段距离（细齿藻属例外） ··· 杆状藻科 Bacillariaceae

杆状藻科 Bacillariaceae

壳面线形、披针形、弓形或"S"形，罕为椭圆形；壳面一侧具龙骨，龙骨上具壳缝和壳缝管，壳缝管内壁有许多小孔，为龙骨突间距，与细胞内部相联系。常无间生带和隔膜。

目前本科有 16 属，我国淡水中发现有 7 属。

杆状藻科分属检索表

1. 不具龙骨突，在翼状结构上具翼状壳缝管 ·· 6. 西蒙森藻属 Simonsenia
1. 具龙骨突，不具翼状结构和翼状壳缝管 ·· 2
   2. 壳面围绕顶轴扭曲 2—3 圈，像一个螺纹 ··· 7. 筒柱藻属 Cylindrotheca
   2. 壳面不扭曲或稍扭曲 ··· 3
3. 壳面左右两侧不对称，具背腹之分 ·· 3. 菱板藻属 Hantzschia
3. 壳面左右两侧对称，或稍微不对称 ·· 4
   4. 龙骨突横向延伸穿过整个壳面，形成隔片 ······································· 5. 细齿藻属 Denticula
   4. 龙骨突不延伸或稍延伸，不形成隔片 ··· 5
5. 细胞结合成带状，可通过壳表面的联锁结构彼此滑动 ·························· 1. 杆状藻属 Bacillaria
5. 不具上述结构 ··· 6
   6. 壳面直或"S"形，窄；线形、披针形或椭圆形 ································· 2. 菱形藻属 Nitzschia
   6. 壳面直，宽大；椭圆形、线形或提琴形，具明显的纵向褶曲 ············ 4. 盘杆藻属 Tryblionella

杆状藻属 Bacillaria J. F. Gmelin

J. F. Gmelin, Syst. Nat. Ed. 13. 1(6): 3903, 1791.

细胞形成一类特有的运动群体，细胞之间通过壳缝-腹板(raphe-sterna)上的脊(ridge)和槽(groove)相互联锁，彼此往复滑动。群体延伸可呈线性排列（细胞间仅通过末端相连），缩回可呈板状排列(细胞间通过整个壳面相连)。一般色素体 2 个，分别位于细胞两端，与菱形藻属(*Nitzschia*)种类相似。

壳面(valve)线形或线形-披针形，末端喙状或头状。壳表面较平，边缘弯曲嵌入极窄的壳套(mantle)，偶尔可见靠近壳面和壳套的接合处有不规则至网状的硅质脊。线纹单列(偶尔双列)，由膜封闭的小圆孔组成。壳缝系统位于中轴或近中轴，龙骨(keel)细小，具龙骨突(fibulae)。外壳面壳缝-腹板上具一个极明显的脊，位于壳缝一侧；该结构参与细胞之间的联锁作用。壳缝在两极之间是连续的，外壳面极缝端呈孔状、T 形或具钩形的末端裂缝。龙骨突肋状，呈弓形与细胞相连。带面(girdle)由数条断开的环带组成，外面常有极小的瘤状物。

多分布于海水、半咸水，少见于高导电率的淡水中，附着于沉积物或污泥中。

本属种类少，特征明显，是硅藻中最早被描述的属。在杆状藻科其他类群中也存在细胞相连形成群体的现象，如拟脆杆藻属(*Fragilariopsis*)以及一些菱形藻属种类 *Nitzschia* spp.，但都没有明显类似本属的运动方式(Round *et al.*，1990)。

模式种：奇异杆状藻(*Bacillaria paradoxa* Gmelin)。

本志收编 1 种 1 变种。

1. 奇异杆状藻

Bacillaria paradoxa Gmelin，Carolia Linne Systema Naturae per regna tria naturae secundum classes，ordines，genera species cum characteribus，differentiis，synonymis，locis. Ed. 13，Tomus I. Pars VI. p. 3903，1791；Hustedt，Bacillariophyta，*in* Pascher，Süßwass -Fl. Mitteleur. Heft 10，p. 396，fig. 755，1930；Cleve-Euler，Diatomeen schw. Finnland，Teil. V，p. 69，fig. 1457，1952；Krammer & Lange-Bertalot，Bacillariophyceae，2. Teil，*in* Ettl *et al.*，Süßwasserfl. Mitteleur. p. 8，figs. 87：4—7，1988，nachdr. 1997；Xin *et al.*，Journal of Shanxi University(Nat. Sci. Ed.)20(1)：105，1997.

Vibrio paxillifer O. F. Müller，Animalcula infusoria fluviatilia et marina quæ detexit，systematice descripsit et ad vivum delineari curavit Otho Fridericus Müller；sistet opus hoc posthumum，quod cum tabulis Aeneis L. in lucem tradit vidua ejus nobilissima cura Othonis Fabricii. p. 54，figs. VII：3—7，1786.

1a. 原变种　图版 I：1—4，6—7；图版 II：3—4

var. paradoxa

壳面线形或线形-披针形，顶端喙状或头状；长 60—125 μm，宽 5—10 μm；壳缝系统位于中线或稍离心，龙骨突清晰，5—8 个/10 μm；横线纹有 21—24 条/10 μm。

扫描电镜下观察：壳缝管位于龙骨上，壳缝裂缝窄而直，龙骨纵向窄线形，位于壳面中线或稍离心，在壳面中部高于壳面，在两端逐渐与壳面平行。壳面具穿顶线纹，线纹单排，由孔纹组成，55—65 个/10 μm，孔纹内外侧都不具闭塞结构。在内壳面，壳缝管由龙骨突桥接，龙骨突为单一的柱状结构，相应的位于壳面中线或稍离心的位置，排列不均匀。相邻的龙骨突之间形成大小不一椭圆形的孔，成为壳缝管与细胞内部交流的通道。在外壳面，细胞裂缝在壳面中部连续，内壳面，细胞裂缝的形状在壳面中部处未知。外壳面的壳缝末端呈"T"形。在内壳面，可见壳缝末端裂缝位于螺旋舌上。带面窄，具体结构不清楚。

生境：生于湖泊、湖边渗出水、小水渠、沼泽、路边积水中。

国内分布：山西(太原晋祠)，辽宁(辽河)，黑龙江(鸡西兴凯湖)，上海(青浦)，江苏(昆山)，福建(金门)，山东(泰山)，河南(南阳)，湖南(沅江流域)，广东(惠州)，广西(灵川)，海南(崖县)，贵州(赫章、水城、镇宁、沅江流域)，新疆(博乐、赛里木湖、察布查尔、伊宁、布尔津、福海、北屯、阿勒泰、盐湖)，台湾。

国外分布：亚洲(伊朗、新加坡)，非洲(东部)，北美洲(美国、加拿大)，南美洲(阿根廷)，欧洲(奥地利、德国、马其顿、罗马尼亚、西班牙)，大洋洲(澳大利亚、新西兰)。

1b. 肿大变种　　图版 I：5

var. **turmidula** (Grunow) De Toni, Sylloge algarum omnium hucusque cognitarum. Vol. II. Sylloge Bacillariearum. Sectio II. Pseudoraphideae. p. 494, 1892; Hustedt, Bacillariophyta, *in* Pascher, Süßwass -Fl. Mitteleur. Heft 10, p. 849, fig. 756, 1930; Zhu & Chen, The Diatoms of the Suoxiyu Nature Reserve Area, Hunan, China, *in* Li *et al.*, The algal flora and aquatic fauna of the Wulingyuan Nature Reserve Area, Hunan, China. p. 55, 1989b; Fan, Studied on Aulonoraphidinales (Surirellales) from Heilongjiang province. p. 22, fig. 4：2, 2004.

Nitzschia paradoxa var. *tumida* Grunow, *in* Cleve and Grunow, Beiträge zur Kenntniss der arctischen Diatomeen. Kongl. Svenska Vetensk.-Akad. Handl. Ser. 4, 17(2)：68.

与原变种的区别在于：壳面中部明显膨大；长 65—145 μm，宽 4—5 μm。

生境：生于湖泊中。

国内分布：黑龙江(鸡西兴凯湖)；湖南(索溪峪)。

国外分布：北美洲(美国)，欧洲(罗马尼亚、西班牙)。

菱形藻属 Nitzschia A. H. Hassall

A. H. Hassall, Hist. Brit. Freshw. Algae 1：435, 1845.

单细胞或连接成链状、星状群体，或生活在胶质管中。细胞直，针状，有时"S"形，示壳面或带面观。一般色素体 2 个，分别位于细胞两端，简单或复杂呈浅裂叶状，具 1 至多个杆状蛋白核。

壳面直或"S"形，窄，线形、披针形或椭圆形，有时中部膨大，在外形上，基本关于顶轴左右对称，但结构上极不对称。末端形状多样，一般喙状或头状。线纹单列，连续，由膜封闭的小圆孔组成，有时可见筛状孔。在体积较大的"S"形种类中，筛状孔偶尔会分成若干小室。壳缝系统位置变化较大，从中轴至近壳缘，具龙骨突。上下壳面的壳缝关于壳面呈镜面对称 ('hantzschioid' symmetry) 或对角线对称 ('nitzschioid' symmetry)。中缝端 (central raphe ending) 有或无。具中缝端的种类，内壳面结构简单，位于一个双螺旋舌(helictoglossa)结构上，外壳面简单或偏向壳面远缘。一般具末端裂缝(terminal fissure)，朝任意一边弯转或呈钩状。龙骨突形状多样，有时可延伸横穿壳面。带面由数量不等的断开的环带组成，一般有1至多排横向小孔。

在淡水和海水中均可生长，附着于沉积物、污泥中或浮游。

本属种类多，形态多样，难度大。Smith(1853)、Cleve和Grunow(1880)分别对本属进行了分类，后者的分类系统沿用至今，在此基础上，许多学者对其进行过修订，如Hustedt(1930)、Lange-Bertalot和Simonsen(1978)、Mann(1984)和Round等(1990)等。形态学和生殖学方面的研究都表明本属不是一个自然分类群体，同时，基于大亚基rDNA的系统发育学研究也支持这一观点，表明这个属是多源发生的(Lundholm *et al.*, 2002)。Round等(1990)从本属分离出拟脆杆藻属(*Fragilariopsis*)(海水种)、沙网藻属(*Psammodictyon*)(海水种)和盘杆藻属(*Tryblionella*)，本书采用了这一观点。

模式种：类"S"状菱形藻[*Nitzschia sigmoidea* (Nitzsch) W. Smith]。

本志收编58种8变种。

菱形藻属分组检索表

1. 壳缝龙骨上具硅质罩结构；壳缝有或无中缝端······9
1. 壳缝龙骨上硅质罩结构不易分辨······2
  2. 带面观或壳面观呈不同程度"S"形弯曲······3
  2. 带面观或壳面观不呈"S"形弯曲······4
3. 中缝端明显，壳缝裂缝在中部伸进壳面一段距离······钝端组 Sect. *Obtusae*
3. 中缝端不明显······弯型组 Sect. *Sigmata*
  4. 龙骨突延长，横肋纹延伸至壳面中部······类附生组 Sect. *Epithemioideae*
  4. 不具上述结构······5
5. 壳面纺锤状，具明显延长的喙状末端······鸭嘴端组 Sect. *Nitzschiellae*
5. 壳面具短喙状末端······6
  6. 龙骨突延伸形成肋纹，肋纹不延伸到壳面边缘······古鲁维亚组 Sect. *Grunowia*
  6. 龙骨突基本不延伸进壳面······7
7. 环带数量多，带面较宽；壳面具壳缝龙骨的一侧呈不同程度缢缩，一般存在中缝端············可疑和双叶型组 Sect. *Dubiae and Bilobatae*
7. 不具上述特征或不能确定有上述特征······8
  8. 壳体较小，披针形；壳缝龙骨强烈离心，位于壳面边缘······披针型组 Sect. *Lanceolatae*
  8. 壳体较大，线形；此组包括一些很难定义的种类······线型组 Sect. *Lineares*
9. 壳体较大，带面观呈不同程度"S"形弯曲，壳面观弯曲度较小······"S"型组 Sect. *Sigmoideae*
9. 壳体较小，带面观或壳面观都不弯曲······细端组 Sect. *Dissipatae*

"S"型组 Section *Sigmoideae*

"S"型组分种检索表

1. 横线纹约25条/10 μm······4
1. 横线纹约30条/10 μm或密度更大······2

2. 带面稍呈"S"形弯曲，线纹约 30 条/10 μm ················· 1. 蠕虫状菱形藻 *N. vermicularis*
2. 线纹约 40 条/10 μm 或更多，光镜下不宜观察到 ····································· 3
3. 带面稍呈"S"形弯曲，体积较宽大 ·· 2. 尖端菱形藻 *N. acula*
3. 带面明显呈"S"形弯曲 ··· 3. 折曲菱形藻 *N. flexa*
4. 壳面宽度大于 7 μm，龙骨突小于 10 个/10 μm ·············· 4. 类 S 状菱形藻 *N. sigmoidea*
4. 壳面宽度小于 7 μm，龙骨突纤细，一般大于 10 个/10 μm ········· 5. 额雷菱形藻 *N. eglei*

1. 蠕虫状菱形藻　图版 IV: 1—4

Nitzschia vermicularis (Kützing) Hantzsch, *in* Rabenhorst, Die Algen Sachsens. Resp. Mittel-Europa's Gesammelt und herausgegeben von Dr. L. Rabenhorst, p. 60, 1860; Krammer & Lange-Bertalot, Bacillariophyceae. 2. Teil, *in* Ettl *et al.*, Süßwasserfl. Mitteleur., p. 14, pl. 4, figs. 4—5, pl. 7, figs. 1—7, 1988, nachdr. 1997; Zhu & Chen, Bacill. Xiz. Plat., p. 269, no fig., 2000.

Frustulia vermicularis Kützing, *Linnaea* 8: 555, pl. 14, fig. 34, 1833.

带面"S"形，壳面线形、线形-披针形，向两端逐渐呈锥形，小头形末端；长 110—180 μm，宽 6—7 μm；壳缝龙骨离心，龙骨突 8—13 个/10 μm，中间两个龙骨突间距增大或不增大；横线纹细密，光镜下很难分辨。

生境：生于湖边渗出水、河边渗出水、路边积水、沼泽中。

国内分布：山西(绵山清水河)，黑龙江(鸡西兴凯湖、五大连池)，贵州沅江流域，西藏(白地)，宁夏(贺兰山)，新疆(阿克苏、博乐、赛里木湖、喀纳斯、阿勒泰、乌市盐湖)。

国外分布：亚洲(俄罗斯、巴基斯坦、蒙古国、土耳其、新加坡)，北美洲(美国、加拿大、夏威夷群岛)，南美洲(巴西)，欧洲(波罗的海、黑海、英国、德国、丹麦、芬兰、马其顿、波兰、罗马尼亚、冰岛、西班牙)，大洋洲(澳大利亚、新西兰)。

本种与 Krammer 和 Lange-Bertalot(1988)报道的 *Nitzschia vermicularis* 在细胞形状、大小、龙骨突密度等方面较为相似，不同之处：我们的样本壳面中部不凸出，有的种类中间两个龙骨突之间的距离增大。

2. 尖端菱形藻　图版 IV: 9

Nitzschia acula (Kützing) Hantzsch, *in* Rabenhorst, Die Algen Europas, Fortsetzung der Algen Sachsens, resp. Mittel-Europa's. no. 1104d, 1861; Krammer & Lange-Bertalot, Bacillariophyceae, 2. Teil, *in* Ettl *et al.*, Süßwasserfl. Mitteleur. p. 16, figs. 8: 5—8A, 1988, nachdr. 1997.

Synedra acula Kützing, Die Kieselschaligen Bacillarien oder Diatomeen, p. 65, pl. 14, fig. 20, 1844.

Nitzschia acuta (orthogr. statt *acula*) sensu Peragallo & Peragallo, Diatomées marines de France et des districts maritimes voisins. p. 281, pl.72, fig.19, 1899; Zhu & Chen, The Diatoms of the Suoxiyu Nature Reserve Area, Hunan, China, *in* Li *et al.*, The algal flora and aquatic fauna of the Wulingyuan Nature Reserve Area, Hunan, China. p. 55, 1989b;

Fan et al., Bull. Bot. Res. 21: 239—244, 2001.

带面"S"形，壳面线形，两端逐渐变窄，末端尖，略呈小头状；长149 μm，宽8 μm；龙骨突清楚，5—6个/10 μm；横线纹细密，光镜下难分辨。

生境：生于沼泽中。

国内分布：黑龙江(鸡西兴凯湖、五大连池)，福建(金门)，湖南(索溪峪、沅江流域)，四川(九寨沟)，贵州(沅江流域)，西藏(加查、措美、察隅、波密)，新疆(喀纳斯)。

国外分布：亚洲(伊朗、蒙古国)，北美洲(美国)，欧洲(波罗的海、英国、德国、爱尔兰、马其顿、西班牙)，大洋洲(澳大利亚、新西兰)。

本种与 N. dissipata 在细胞形态上非常相似，特别是在壳面中部，不同之处：本种体积明显较大。

3. 折曲菱形藻　图版 IV：5—8

Nitzschia flexa Schumann, Preussische Diatomeen, p. 186, 1862; Krammer & Lange-Bertalot, Bacillariophyceae, 2. Teil, in Ettl et al., Süßwasserfl. Mitteleur. p. 16, fig. 4: 6, figs. 9: 1—4, 1988, nachdr. 1997.

带面呈明显"S"形弯曲，壳面披针形至线形-披针形，小头形末端；长 70—130 μm，宽 4—6 μm；龙骨突较纤细，龙骨突间距比龙骨突宽，6—8 个/10 μm；线纹细密，光镜下很难分辨。

生境：生于河边渗出水、沼泽及稻田中。

国内分布：江苏(昆山)，广东(龙川)，新疆(伊宁、阿克苏、博乐、乌鲁木齐)。

国外分布：亚洲(俄罗斯、伊朗)，北美洲(美国)，欧洲(英国、德国、罗马尼亚)，大洋洲(澳大利亚、新西兰)。

4. 类 S 状菱形藻　图版 II：1—2；图版 III：1—7

Nitzschia sigmoidea(Nitzsch) W. Smith, A synopsis of the British Diatomaceae, p. 38, pl. 13: fig. 104, 1853; Schmidt et al., Atlas Diat.-Kunde, pl. 332, figs. 1—4, 1904; Hustedt, Bacillariophyta, in Pascher, Süßwass -Fl. Mitteleur. Heft 10, p. 419, fig. 810, 1930; Cleve-Euler, Diatomeen schw. Finnland. Teil. V, p. 72, fig. 1467, 1952; Krammer & Lange-Bertalot, Bacillariophyceae, 2. Teil, in Ettl et al., Süßwasserfl. Mitteleur., p. 12, pl. 5, figs. 1—5, 1988, nachdr. 1997; Xin et al., Journal of Shanxi University(Nat. Sci. Ed.)20(1): 105, 1997; Zhu & Chen, Bacill. Xiz. Plat., p. 268, pl. 53, figs. 18—19, 2000; Metzeltin et al., Iconogr. Diatomol.15, pl. 195, figs. 22—23, 2005.

Bacillaria sigmoidea Nitzsch, Neue Schriften der Naturforschenden Gesellschaft zu Halle. 3(1), p. 104, pl. 6, figs. 4—6, 1817.

细胞较大，一般带面比壳面宽，所以常示带面观；带面"S"形弯曲，壳面观直线形，倾斜观察时，壳面就稍呈"S"形弯曲；长 180—450 μm，宽 7—14 μm；龙骨和龙骨突均明显，龙骨突有 6—8 个/10 μm；横线纹细密，光镜下很难分辨。

扫描电镜下观察：壳面的龙骨上覆盖一冠层结构，冠层下的龙骨突纤细，管状，不规则地排列在龙骨上。线纹平行排列，与壳缘垂直，24—27条/10 μm，线纹由单排点纹均匀排列而成，点纹为圆形穿孔；每个龙骨突与2—5条点线纹相连接；中间两个龙骨突之间的距离不增大。壳缝离心，位于龙骨上；靠近壳缝有一缢缩纵向的脊。壳缝管外壁有点纹，均匀成排，每排6—7个点，为壳面线纹的延伸。没有观察到壳缝末端以及环带的结构。

生境：生于湖泊、溪流、小水渠、草地渗出水、沼泽中。

国内分布：河北(昌黎)，山西(太原晋阳湖、太原晋祠、绵山清水河)，内蒙古(阿尔山)，黑龙江(五大连池)，江苏(常熟)，湖南(沅江流域)，贵州(沅江流域、乌江流域)，西藏(定结、浪卡子、林芝、错那、芒康、江达、贡觉、申扎、措勤、普兰、日土)，甘肃(苏干湖)，新疆(赛里木湖、察布查尔、布尔津、喀纳斯、哈巴河、福海)。

国外分布：亚洲(俄罗斯、蒙古国、伊朗、土耳其、新加坡)，北美洲(美国、加拿大、夏威夷群岛)，南美洲(巴西、哥伦比亚、乌拉圭)，欧洲(波罗的海、黑海、英国、德国、爱尔兰、马其顿、波兰、罗马尼亚、西班牙)，大洋洲(澳大利亚、新西兰)。

Knattrup等(2007)对纯培养的 *N. sigmoidea* 进行了详细的显微观察，观察到的标本大小、形状、龙骨突密度及线纹密度与我们的标本相似，除此之外，Knattrup等(2007)观察到本种"hantzschioid"和"nitzschioid"两种类型的细胞。壳面点纹上具膜(TEM)，为六边形的孔纹结构，每个小孔的直径90—100 nm(n=11)。壳缝在中央连续，在外壳面的末端延伸呈钩状裂缝，在内壳面终止于螺旋舌。环带有5—7条，均为开带，在开口的末端呈尖锥形，同时对环带内部的结构也做了详细的描述。

5. 额雷菱形藻　图版 V：1—9

Nitzschia eglei Lange-Bertalot, *in* Lange-Bertalot and Krammer, Bibl. Diatomol. 15, p. 15, figs. 28：1—3, 1987; Krammer & Lange-Bertalot, Bacillariophyceae, 2. Teil, *in* Ettl *et al.*, Süßwasserfl. Mitteleur. p. 18, figs. 10：6—9, 1988, nachdr. 1997.

带面观呈"S"形；壳面直，线形-披针形，小头形末端，由于观察角度不同，壳面会稍呈"S"形；长95—140 μm，宽6—7 μm；壳缝龙骨离心，龙骨突均匀排列，11—13个/10 μm，中间两个龙骨突间距不增大，每个龙骨突与一条横肋纹相连；横线纹24—25条/10 μm。

生境：生于湖泊、溪流、池塘、小水渠中，岩石上附着。

国内分布：海南(崖县)，新疆(阜康、奎屯、察布查尔、哈巴河、布尔津、福海、阿勒泰)。

国外分布：北美洲(美国)，欧洲(英国、德国)。

细端组 Section *Dissipatae*

细端组分种检索表

1. 壳缝龙骨可见，位于壳面中部或稍离心 ··· **6. 细端菱形藻 *N. dissipata***

1. 壳缝龙骨不清楚，离心 ·· **7. 直菱形藻** *N. recta*

6. 细端菱形藻

Nitzschia dissipata (Kützing) Rabenhorst, Die Algen Sachsens. Resp. Mittel-Europa's Gesammelt und herausgegeben von Dr. L. Rabenhorst, no. 968, 1860; Schmidt *et al.*, Atlas Diat.-Kunde, pl. 332, figs. 22—24, 1904; Hustedt, Bacillariophyta, *in* Pascher, Süßwass -Fl. Mitteleur. Heft 10, p. 412, fig. 789, 1930; Cleve-Euler, Diatomeen schw. Finnland, Teil. V, p. 71, fig. 1463, 1952; Zhu & Chen, The Diatoms of the Suoxiyu Nature Reserve Area, Hunan, China, *in* Li *et al.*, The algal flora and aquatic fauna of the Wulingyuan Nature Reserve Area, Hunan, China. p. 56, 1989b; Krammer & Lange-Bertalot, Bacillariophyceae 2. Teil, *in* Ettl *et al.*, Süßwasserfl. Mitteleur. p. 19, pl. 11, figs. 1—7, 1988, nachdr. 1997; Zhu & Chen, Bacill. Xiz. Plat., p. 261, pl. 52, fig. 4, 2000.

Synedra dissipata Kützing, Die Kieselschaligen Bacillarien oder Diatomeen. p. 64, pl. 14, fig. 3, 1844.

6a. 原变种　图版 VI：1—4；图版 VII：1—4

var. **dissipata**

　　壳面披针形，偶尔线形-披针形，末端喙状，很少为尖形或圆形；长 24—55 μm，宽 4—7 μm；壳缝龙骨稍离心，龙骨突排列不均匀，中间两个龙骨突的距离不增大，7—10 个/10 μm；横线纹极细，光镜下很难分辨。

　　生境：生于小河、沼泽，岩石附着。

　　国内分布：辽宁(辽河、本溪)，吉林(长白山)，黑龙江(鸡西兴凯湖、五大连池)，湖南(索溪峪、沅江流域)，广东(东源)，贵州(赫章、沅江流域、乌江流域)，西藏(吉隆、工布江达、墨脱、米林、错那、措美、昌都、芒康、察雅、洛隆、江达、贡觉、类乌齐、札达、噶尔、革吉)，陕西(华山)，新疆(温宿、阿克陶、哈巴河、福海)。

　　国外分布：亚洲(俄罗斯、蒙古国、伊朗、土耳其)，北美洲(美国、加拿大、夏威夷群岛)，南美洲(巴西、哥伦比亚)，欧洲(波罗的海、黑海、英国、德国、丹麦、爱尔兰、马其顿、罗马尼亚、波兰、冰岛、西班牙)，大洋洲(澳大利亚、新西兰)。

6b. 中等变种　图版 VI：9—13

var. **media** (Hantzsch) Grunow, *in* van Heurck, Synopsis des Diatomées de Belgique Atlas, p. 178, pl. 63, figs. 2—3, 1881; Krammer & Lange-Bertalot, Bacillariophyceae. 2. Teil, *in* Ettl *et al.*, Süßwasserfl. Mitteleur., p. 19, pl. 11, figs. 8—14, 1988, nachdr. 1997; Antoniades *et al.*, Iconogr. Diatomol. 17, p. 214, pl. 77, figs. 6—10, 2008; Liu *et al.*, Wuhan Bot. Res. 27(3)：275, figs. 5—6, 2009.

Nitzschia media Hantzsch, Hedwigia 2(7)：40, pl. 6, fig. 9, 1860.

与原变种的区别在于：壳面线形-披针形，壳缝龙骨离心程度大。

生境：生于湖泊、河流、溪流、小池塘、路边积水中。

国内分布：内蒙古(达尔滨湖)，江苏(昆山)，贵州(赫章)，新疆(阿克苏、阜康、察布查尔、布尔津、哈巴河)。

国外分布：亚洲(俄罗斯、土耳其)，北美洲(美国、加拿大)，南美洲(哥伦比亚)，欧洲(波罗的海、英国、德国、马其顿、罗马尼亚、波兰、西班牙)，大洋洲(新西兰)。

7. 直菱形藻 图版 VI：5—8

Nitzschia recta Hantzsch ex Rabenhorst, *in* Rabenhorst, Alg. Eur. p. 1283, 1862; Zhu & Chen, The Diatoms of the Suoxiyu Nature Reserve Area, Hunan, China, *in* Li *et al.*, The algal flora and aquatic fauna of the Wulingyuan Nature Reserve Area, Hunan, China. p. 56, 1989b; Xie *et al.*, Journal of Shanxi University (Nat. Sci. Ed.) 14(4): 417, 1991; Krammer & Lange-Bertalot, Bacillariophyceae. 2. Teil, *in* Ettl *et al.*, Süßwasserfl. Mitteleur., p. 20, pl. 12, figs. 2—6, 1988, nachdr. 1997; Zhu & Chen, Bacill. Xiz. Plat., p. 267, pl. 53, fig. 16, 2000.

壳面线形、线形-披针形至披针形，末端逐渐楔形减小，呈尖头或近圆头状；长 100—130 μm，宽 6—7 μm；壳缝龙骨稍离心，龙骨突纤细，排列不均匀，7—10 个/10 μm；横线纹极细，光镜下很难分辨。

生境：生于河流、湖泊、溪流、小水渠、沼泽、路边积水中。

国内分布：山西(太原晋阳湖)，黑龙江(鸡西兴凯湖、五大连池)，湖南(索溪峪、沅江流域)，贵州(赫章、镇宁、沅江流域)，西藏(波密、江达)，新疆(阿克苏、天池、博乐、察布查尔、布尔津、喀纳斯、福海)。

国外分布：亚洲(俄罗斯、蒙古国)，北美洲(美国)，南美洲(巴西)，欧洲(波罗的海、黑海、英国、德国、捷克、爱尔兰、马其顿、波兰、罗马尼亚、西班牙)，大洋洲(澳大利亚、新西兰)。

根据 Krammer 和 Lange-Bertalot(1988)的描述，扫描电镜下，本种的壳缝连续，没有中央节，龙骨突基部与若干条横肋纹联系。

钝端组 Section *Obtusae*

钝端组分种检索表

1. 体积较大，一般壳面长度 110 μm 以上，宽 7 μm 以上 ················· **8. 钝端菱形藻 *N. obtusa***
1. 体积较小 ··· 2
   2. 当壳面两端不对称时，呈"S"形弯曲，末端不呈短喙状或头状 ······················· 3
   2. 特征与上述不同 ·· 4
3. 壳面接近末端时变窄，明显不对称，呈显著的刀形 ············ **9. 刀形菱形藻 *N. scalpelliformis***
3. 壳面朝两端渐窄，稍不对称性收缩 ······························ **10. 丝状菱形藻 *N. filiformis***
   4. 壳面"S"形弯曲，在末端较宽(不收缩) ······················ **11. 微型菱形藻 *N. nana***

  4. 特征与上述不同 ··· 5
5. 壳面末端喙状，稍朝相反方向弯曲 ·· **12. 克劳斯菱形藻** *N. clausii*
5. 特征与上述不同 ·· 6
  6. 壳面在中部不凹入，末端逐渐变窄 ······· **13. 近粘连菱形藻斯科舍变种** *N. subcohaerens* var. *scotica*
  6. 壳面在中部不同程度凹入，窄，末端喙状 ··· 7
7. 壳面长度变化范围小 ·· **14. 短形菱形藻** *N. brevissima*
7. 壳面长度变化范围大 ··· **15. 土栖菱形藻** *N. terrestris*

8. 钝端菱形藻　图版 X：1-3；图版 XI：1-6

Nitzschia obtusa W. Smith，A synopsis of the British Diatomaceae，p. 39，pl. 13，fig. 109，1853；Schmidt *et al.*，Atlas Diat.-Kunde，figs. 336：20，21，figs. 352：6，7，1924；Hustedt，Bacillariophyta，*in* Pascher，Süßwass -Fl. Mitteleur. Heft 10，p. 422，fig. 817a，1930；Cleve-Euler，Diatomeen schw. Finnland，Teil. V，p. 77 fig. 1476，1952；Mann，Studies in the family Nitzschiaceae（Bacillariophyta），figs. 21：289—292，figs. 21：861—866，1978；Krammer & Lange-Bertalot，Bacillariophyceae，2. Teil，*in* Ettl *et al.*，Süßwasserfl. Mitteleur.，p. 26，figs. 17：1，2，fig. 18：1，1988，nachdr. 1997；Zhu & Chen，Bacill. Xiz. Plat.，p. 266，figs. 53：9—11，2000；Liu *et al.*，Wuhan Bot. Res. 24：41，fig. III：113，2006.

  带面线形，末端稍呈"S"形弯曲；壳面观线形，末端呈不同程度的"S"形弯曲，末端钝圆；长 120—211 μm，宽 7—11 μm；壳缝龙骨在极节处离心程度大，中央节处离心程度小，龙骨突 5—6 个/10 μm，中间两个距离较大；横线纹 24—30 条/10 μm。

  扫描电镜下观察：壳面线形，壳缝管位于龙骨上，龙骨在壳面的一侧边缘，在壳面中部，龙骨稍微伸进壳面。壳面具横线纹，线纹单排，由孔纹组成，圆形或近圆形，直径可达 0.2 μm，未见封闭结构。近壳缝的两排孔纹比较特殊，第一排孔纹较小（有时有两排），圆形，只有壳面上孔纹直径的 1/3，第二排孔纹较大，明显的长圆形，宽度与壳面孔纹的直径相当，长度为其 3—5 倍。壳缝在中部断开，壳缝裂缝伸进壳面，一般穿过近壳缝的两排孔纹，壳缝在壳面末端简单，在螺旋舌上弯曲，弯向壳面的任意一侧，并延伸一小段距离。在内壳面，龙骨突桥接壳缝管和内壳套。龙骨突较大、块状，5—6 个/10 μm。每个龙骨突与 3—6 条横线纹相连，相邻龙骨突之间为一个近圆形或椭圆形的孔，中间两个龙骨突之间的孔最大。内壳面具线纹，单排，由孔纹组成，孔纹类型只有一种，密度与外壳面线纹一样。内壳面，壳缝裂缝在中央节处断开，膨大，在末端中止于一小的螺旋舌。

  生境：生于湖泊、湖边渗出水、小水渠、池塘、路边积水、沼泽中，岩石上附生。

  国内分布：河北（唐山），山西（运城），黑龙江（尚志亚布力），福建（金门），广东（惠州），海南（崖县），西藏（日土），甘肃（月牙泉），新疆（巴楚、天池、奎屯、博乐、赛里木湖、察布查尔、伊宁、布尔津、哈巴河、北屯）。

  国外分布：亚洲（俄罗斯、巴基斯坦），北美洲（美国），南美洲（巴西），欧洲（亚得里亚海、波罗的海、黑海、英国、德国、罗马尼亚、西班牙），大洋洲（澳大利亚、新西兰）。

本种为普生性种类，嗜盐碱的水环境(Krammer and Lange-Bertalot，1997)。在扫描电镜下观察到的超微结构与 Mann(1978)描述的相一致，并发现在外壳面末端，壳缝裂缝可偏向壳面任意一侧。

9. 刀形菱形藻　图版 X：4-7

Nitzschia scalpelliformis Grunow, *in* Cleve & Grunow, Beiträge zur Kenntniss der arctischen Diatomeen, p. 92, 1880; Krammer & Lange-Bertalot, Bacillariophyceae, 2. Teil, *in* Ettl *et al.*, Süßwasserfl. Mitteleur., p. 26, figs. 18：2—5, 1988, nachdr. 1997; Metzeltin *et al.*, Iconogr. Diatomol. 15, figs. 201：3—5, 2005.

Nitzschia obtusa var. *scalpelliformis* (Grunow) Grunow, *in* van Heurck, Synopsis des Diatomées de Belgique Atlas, pl. 67：fig. 2, 1880; Schmidt *et al.*, Atlas Diat.-Kunde, figs. 336：22—24, 1921; Zhu & Chen, Bacill. Xiz. Plat., p. 266, pl. 53, fig. 12, 2000.

壳面线形，中部有时稍微凹入，末端刀形；壳面长 40—110 μm，宽 5—9 μm；壳缝龙骨在极节处离心程度大，中央节处离心程度小，龙骨突 7—10 个/10 μm，中间两个距离较大；横线纹 26—36 条/10 μm。

生境：生于湖泊、湖边渗出水、小水渠、路边积水、沼泽或岩石上附生。

国内分布：山西(运城)，辽宁(辽河)，江苏(昆山)，海南(崖县)，贵州(水城)，西藏(林芝)，新疆(天池、博乐、赛里木湖、察布查尔、伊宁、福海、北屯)。

国外分布：亚洲(俄罗斯)，北美洲(美国)，南美洲(巴西、哥伦比亚、乌拉圭、安第斯山脉)，欧洲(波罗的海、德国、马其顿、罗马尼亚、西班牙)，大洋洲(澳大利亚、新西兰)。

该种与 *Nitzschia obtusa* 在形态上最为相似，主要区别在于：壳面体积小，末端明显呈刀形，龙骨突和横线纹密度大。

10. 丝状菱形藻　图版 VIII：1-7

Nitzschia filiformis (W. Smith) Van Heurck, 1896; Krammer & Lange-Bertalot, Bacillariophyceae. 2. Teil, *in* Ettl *et al.*, Süßwasserfl. Mitteleur., p. 27, pl. 19, figs. 7—13, pl. 20, figs. 1—7, 1988, nachdr. 1997; Zhu & Chen, Bacill. Xiz. Plat., p. 261, pl. 52, fig. 7, 2000.

Homoecladia filiformis W. Smith, A synopsis of the British Diatomaceae, pl. LV, fig. 348, 1856.

带面线形或披针形，略呈"S"形弯曲；壳面线形-披针形，稍呈"S"形弯曲，末端略窄于中部，呈钝圆形；长 40—95 μm，宽 5—7 μm；龙骨突有 8—11 个/10 μm；横线纹在光镜下看不清楚。

生境：生于湖泊，水草附生。

国内分布：辽宁(辽河)，吉林(长白山)，黑龙江(鸡西兴凯湖、五大连池)，江苏(昆山)，广东(紫金、惠州)，海南(三亚)，贵州(赫章)，云南(楚雄、下关)，西藏(错那、芒康)，新疆(博湖、福海)。

国外分布：亚洲（俄罗斯、蒙古国、以色列），北美洲（加拿大、美国），南美洲（阿根廷、哥伦比亚、安第斯山脉），欧洲（波罗的海、英国、德国、爱尔兰、马其顿、罗马尼亚、西班牙），大洋洲（澳大利亚、新西兰）。

11. 微型菱形藻　图版 IX：1—2；图版 XII：1—7

Nitzschia nana Grunow, *in* Van Heurck, Synopsis des Diatomées de Belgique Atlas, pl. 67, fig. 3, 1881; Krammer & Lange-Bertalot, Bacillariophyceae, 2. Teil, *in* Ettl *et al.*, Süßwasserfl. Mitteleur., p. 26, figs. 17：4—8, 1988, nachdr. 1997; Metzeltin *et al.*, Iconogr. Diatomol. 15, figs. 214：10—11, 2005.

Nitzschia obtusa var. *nana* (Grunow) Van Heurck, Synopsis des Diatomées de Belgique, p. 180, 1885; Schmidt *et al.*, Atlas Diat.-Kunde, figs. 347：20—23, 1924.

Nitzschia ignorata Krasske, Botanisches Archiv 27：355, fig. 23, 1929; Fan *et al.*, Bull. Bot. Res. 21：242, fig. II：8, 2001; Zhang, Journal of Jinan University 3：89, fig. I：9, 1986.

带面窄线形，不同程度的"S"形弯曲；壳面中部略呈"S"形弯曲，两端弯曲程度较大，末端钝圆；长 45—165 μm，宽 4—6 μm；龙骨突有 7—11 个/10 μm，中间两个距离明显增大；横线纹密集，在光镜下不容易看清。

生境：生于河流、小水渠、路边积水、沼泽，水草附生、岩石上附着、泉水井边草丛附生。

国内分布：河北（唐山），黑龙江（鸡西兴凯湖、牡丹江镜泊湖、五大连池），江苏（昆山），福建（金门），湖南（沅江流域），海南（崖县、保亭、琼山），贵州（沅江流域、乌江流域），云南（滇池、洱海、下关），新疆（博湖、天池、博乐、察布查尔、布尔津、北屯）。

国外分布：亚洲（俄罗斯、新加坡），北美洲（加拿大、美国、夏威夷群岛），南美洲（巴西、哥伦比亚、乌拉圭），欧洲（波罗的海、英国、德国、爱尔兰、波兰、罗马尼亚、西班牙），大洋洲（澳大利亚、新西兰）。

12. 克劳斯菱形藻　图版 VIII：9—13

Nitzschia clausii Hantzsch, Hedwigia 2(7)：40, pl. 6, fig. 7, 1860; Schmidt *et al.*, Atlas Diat.-Kunde, pl. 336, figs. 7—11, 1921; Krammer & Lange-Bertalot, Bacillariophyceae, 2. Teil, *in* Ettl *et al.*, Süßwasserfl. Mitteleur., p. 31, pl. 19, figs. 1—6A, 1988, nachdr. 1997; Zhu & Chen, Bacill. Xiz. Plat., p. 259, pl. 51, fig. 10, 2000; Metzeltin *et al.*, Iconogr. Diatomol. 15, pl. 200, figs. 5—8, pl. 201, figs. 10—13, 2005; Antoniades *et al.*, Iconogr. Diatomol. 17：211, pl. 75, figs. 6—9, 2008.

Nitzschia sigma var. *clausii* (Hantzsch) Grunow, Algen und Diatomaceen aus dem Kaspischen Meere, p. 119, 1878.

壳面线形，略呈"H"状，末端略延长，圆头状；长 26—55 μm，宽 4—6 μm；龙骨突有 9—13 个/10 μm；横线纹在光镜下看不清楚。

生境：生于湖泊、河流、泉水、小水渠、沼泽，草丛附生、岩石上附着。

国内分布：辽宁（辽河），黑龙江（哈尔滨、五大连池），福建（鼓岭），湖南（长沙、岳阳），广东（龙川、东源、紫金、博罗、东莞），贵州（水城），西藏（申扎），宁夏（贺兰山），新疆（昌吉、博乐、赛里木湖、察布查尔、伊犁河、福海、岳普湖）。

国外分布：亚洲（俄罗斯、蒙古国），北美洲（加拿大、美国、夏威夷群岛），南美洲（阿根廷、巴西、哥伦比亚、乌拉圭），欧洲（波罗的海、英国、德国、波兰、罗马尼亚、冰岛、西班牙），大洋洲（澳大利亚、新西兰）。

13. 近粘连菱形藻斯科舍变种 图版 VIII：15—18；图版 IX：5—6

Nitzschia subcohaerens var. **scotica** (Grunow) Van Heurck, A treatise on the Diatomaceae., p. 406, fig. 127, 1896; Liu *et al.*, Wuhan Bot. Res. 24: 42, fig. III: 119, 2006; Krammer & Lange-Bertalot, Bacillariophyceae, 2. Teil, *in* Ettl *et al.*, Süßwasserfl. Mitteleur., Band 2/2, p. 28, figs. 20: 8—10, 1988, nachdr. 1997.

Homoeocladia subcohaerens var. *scotica* Grunow, *in* Van Heurck, Synopsis des Diatomées de Belgique Atlas, pl. 66, fig. 14, 1881.

细胞较小，线形弯刀形，壳面"S"形弯曲，末端渐尖；壳面长 35—60 μm，宽 3—6 μm；龙骨突有 8—12 个/10 μm；横线纹在光镜下看不清楚。

生境：生于湖泊、池塘、水渠、沼泽，岩石上附生、水草上附着。

国内分布：江苏（昆山），福建（金门），贵州（水城），新疆（天池、奎屯、伊宁、布尔津、福海）。

国外分布：北美洲（美国），欧洲（中部）。

14. 短形菱形藻 图版 IX：3—4；图版 XII：8—15

Nitzschia brevissima Grunow, *in* Van Heurck, Synopsis des Diatomées de Belgique Atlas. pl. LXVII, fig. 4, 1880; Krammer & Lange-Bertalot, Bacillariophyceae, 2. Teil, *in* Ettl *et al.*, Süßwasserfl. Mitteleur., p. 30, pl. 22, figs. 1—6, 1988, nachdr. 1997; Metzeltin *et al.*, Iconogr. Diatomol. 15, pl. 201, figs. 6—9, 2005.

Nitzschia parvula Lewis 1862 non W. Smith 1853; Fan *et al.*, Bull. Bot. Res. 21: 242, 2001.

Nitzschia obtusa var. *brevissima* (Grunow) Van Heurck, Synopsis des Diatomées de Belgique, p. 180, 1885.

带面宽线形，末端略呈"S"形弯曲；壳面以宽线形为基础，稍呈"S"形弯曲，壳面两侧中部凹入，靠近末端处凸出，末端楔形变窄，呈喙状或头状；长 25—46 μm，宽 5—6.5 μm；龙骨突有 5—7 个/10 μm，中间两个距离较大；横线纹在光镜下看不清楚。

生境：生于湖泊、河流、沼泽中。

国内分布：内蒙古（阿尔山），吉林（长白山），黑龙江（鸡西兴凯湖、五大连池），广东（河源、惠东、紫金），贵州（梵净山），新疆（喀纳斯）。

国外分布：亚洲（俄罗斯、新加坡），北美洲（美国、加拿大），南美洲（阿根廷、巴西、哥伦比亚、乌拉圭），欧洲（英国、德国、罗马尼亚、波兰、西班牙），大洋洲（澳大利亚、新西兰)。

15. 土栖菱形藻　图版 VIII：14；图版 XIII：1—8

Nitzschia terrestris (Petersen) Hustedt, Die Diatomeenflora von Poggenpohls Moor bei Dötlingen in Oldenburg, p. 396, 1934; Krammer & Lange-Bertalot, Bacillariophyceae, 2. Teil, *in* Ettl *et al.*, Süßwasserfl. Mitteleur., p. 30, figs. 22: 7—11, 1988, nachdr. 1997; Metzeltin *et al.*, Iconogr. Diatomol. 15, figs. 208: 5—10, 2005; Liu *et al.*, Journal of Shanghai Normal University (Nat. Sci.) 43(3): 270, figs. 8—10, 2014.

Nitzschia vermicularis var. *terrestris* Petersen, The aërial algae of Iceland, *in* Rosenvinge & Warming, The botany of Iceland. Vol. II. Part II, 1928.

带面线形，末端稍呈"S"形弯曲；壳面线形，两侧中部凹入，两端楔形变窄，末端短喙状；长 37—100 μm，宽 4—5 μm；龙骨突有 5—8 个/10 μm；横线纹在光镜下看不清楚。

扫描电镜下观察：壳面线形，壳缝管位于龙骨上，龙骨在壳面的一侧边缘，在壳面中部，龙骨稍微伸进壳面。壳面具横线纹，线纹单排，由孔纹组成，圆形或近圆形，未见封闭结构。壳缝在中部断开，壳缝裂缝伸进壳面，一般穿过近壳缝的两排孔纹，壳缝在壳面末端简单，在螺旋舌上弯曲，弯向壳面的任意一侧，并延伸一小段距离。

生境：浮游、水草附生。

国内分布：山西(运城)，内蒙古(阿尔山)，新疆(布尔津)。

国外分布：亚洲(俄罗斯)，北美洲(美国、加拿大)，南美洲(巴西、乌拉圭)，欧洲(英国、德国、爱尔兰、波兰)，大洋洲(新西兰)。

弯型组 Section *Sigmata*

弯型组分种检索表

1. 龙骨突基本不延伸···**16. 弯菱形藻 *N. sigma***
1. 龙骨突明显延伸进壳面···**17. 簇生菱形藻 *N. fasciculata***

16. 弯菱形藻　图版 XIV：1—9；图版 XV：1—2

Nitzschia sigma (Kützing) W. Smith, A synopsis of the British Diatomaceae, p. 39, pl. 13, fig. 108, 1853; Schmidt *et al.*, Atlas Diat.-Kunde, figs. 336: 1—6, 1921; Hustedt, Bacillariophyta, *in* Pascher, Süßwass -Fl. Mitteleur. Heft 10, p. 420, fig. 813, 1930; Krammer & Lange-Bertalot, Bacillariophyceae, 2. Teil, *in* Ettl *et al.*, Süßwasserfl. Mitteleur., p. 32, figs. 23: 1—9, 1988, nachdr. 1997; Metzeltin *et al.*, Iconogr. Diatomol. 15, figs. 205: 5—7, figs. 207: 1—4, 2005.

Synedra sigma Kützing, Die Kieselschaligen Bacillarien oder Diatomeen, p. 67, pl. 30, fig. 14, 1844.

带面明显"S"形弯曲，壳面稍微至明显的"S"形弯曲，中部线形至线形-披针形，两端长楔形，末端朝反方向弯曲；长 55—285 μm，宽 5—10 μm；龙骨突位于壳面边缘，肋状，排列整齐，有 8—11 个/10 μm，中间两个距离不增大；横线纹有 17—26 条/10 μm。

扫描电镜下观察：外壳面，龙骨发育良好，明显高于壳面，组成线纹的小孔非常不规则，大小不一，圆形至长椭圆形，在壳面形成不规则的纵向"曲线"，末端的无线纹区域分散着一些小孔状凹陷；内壳面平坦，龙骨突桥接壳套和壳面，形态变化不大，肋状，与 1—3 条肋纹相连，龙骨突间距为近圆形的孔，为细胞物质交流的通道，壳面上肋纹的宽度明显大于肋纹间孔纹的宽度，孔纹为狭长线状，不容易看清楚。

生境：生于河流、河流渗出水、小水渠、浅水滩、路边积水或沼泽中。

国内分布：天津，山西(太原晋阳湖)，辽宁(辽河)，吉林(长白山)，黑龙江(五大连池)，上海(松江)，江苏(昆山、苏州)，浙江(宁波)，湖南(长沙)，海南(崖县)，贵州(水城)，甘肃(苏干湖)，青海(西宁)，新疆(阿克苏、博乐、赛里木湖、察布查尔、伊宁、布尔津、哈巴河)。

国外分布：亚洲(俄罗斯、蒙古国、伊朗、土耳其、新加坡)，北美洲(美国、加拿大)，南美洲(阿根廷、巴西、哥伦比亚、乌拉圭)，欧洲(波罗的海、亚得里亚海、黑海、英国、法国、德国、罗马尼亚、西班牙)，大洋洲(澳大利亚、新西兰)。

17. 簇生菱形藻　图版 VIII：8

Nitzschia fasciculata (Grunow) Grunow, *in* Van Heurck, Synopsis des Diatomées de Belgique Atlas. p. 179, pl. 66, figs. 11—13, 1881; Krammer & Lange-Bertalot, Bacillariophyceae. 2. Teil, *in* Ettl *et al.*, Süßwasserfl. Mitteleur., p. 33, pl. 22, figs. 12—13, 1988, nachdr. 1997.

Nitzschia sigma var. *fasciculata* Grunow, p. 119, 1878.

壳面略呈"S"形，长 67 μm，宽 5 μm；龙骨突细，延伸一段距离，6—7 个/10 μm，横线纹极细，光镜下难以分辨。

生境：生于田间清澈水坑。

国内分布：江苏(昆山)。

国外分布：北美洲(美国)，南美洲(巴西)，欧洲(波罗的海、英国、法国、德国、罗马尼亚、西班牙)，大洋洲(澳大利亚、新西兰)。

古鲁维亚组 Section *Grunowia*

18. 弯曲菱形藻

Nitzschia sinuata (Thwaites) Grunow, *in* Cleve & Grunow, Beiträge zur Kenntniss der arctischen Diatomeen, p. 82, 1880; Zhu & Chen, The Diatoms of the Suoxiyu Nature Reserve Area, Hunan, China, *in* Li *et al.*, The algal flora and aquatic fauna of the Wulingyuan Nature Reserve Area, Hunan, China. p. 56, 1989b; Krammer & Lange-Bertalot, Bacillariophyceae, 2. Teil, *in* Ettl *et al.*, Süßwasserfl. Mitteleur., p. 52, figs. 40：1—3, 1988, nachdr. 1997; Zhu & Chen, Bacill. Xiz. Plat., p. 268, fig. 54：1, 2000; Antoniades *et al.*, Iconogr. Diatomol.：17：224, fig. 77：5, 2008.

Denticula sinuata Thwaites *in* W. Smith, A synopsis of the British Diatomaceae, p. 21, pl.

XXXIV, fig. 295, 1856.

18a. 原变种 图版 IX: 9—10
var. sinuata

壳面窄披针形至菱形，两侧波曲，中间膨大呈球茎状，末端小圆头状；长 30—50 μm，宽 6—9 μm；龙骨突窄肋状，延伸至壳面一段距离，3—6 个/10 μm；横线纹由粗糙的孔纹组成，18—23 条/10 μm。

生境：生于河流中。

国内分布：吉林（长白山），湖南（索溪峪），西藏（扎达），新疆（阿克陶）。

国外分布：北美洲（加拿大、美国），欧洲（英国、德国、爱尔兰、罗马尼亚、波兰、西班牙）。

18b. 平片变种 图版 IX: 7, 11—16
var. **tabellaria** (Grunow) Grunow, *in* Van Heurck, Synopsis des Diatomées de Belgique Atlas, p. 176, pl. 60, figs. 12—13, 1881; Schmidt *et al.*, Atlas Diat.-Kunde, p. 331, figs. 331: 28—30, 1921; Hustedt, Bacillariophyta, *in* Pascher, Süßwass -Fl. Mitteleur. Heft 10, p. 409, fig. 782, 1930; Cleve-Euler, Diatomeen schw. Finnland, Teil. V, p. 67, fig. 1452, 1952; Zhu & Chen, The Diatoms of the Suoxiyu Nature Reserve Area, Hunan, China, *in* Li *et al.*, The algal flora and aquatic fauna of the Wulingyuan Nature Reserve Area, Hunan, China. p. 56, 1989b; Krammer & Lange-Bertalot, Bacillariophyceae, 2. Teil, *in* Ettl *et al.*, Süßwasserfl. Mitteleur., p. 53, figs. 39: 10—13, 1988, nachdr. 1997; Xin *et al.*, Journal of Shanxi University (Nat. Sci. Ed.) 20(1): 105, 1997; Zhu & Chen, Bacill. Xiz. Plat., p. 268, fig. 54: 2, 2000.

Denticula tabellaria Grunow, Die Österreichischen Diatomaceen nebst Anschluss einiger neuen Arten von andern Lokalitäten und einer kritischen Uebersicht der bisher bekannten Gattungen und Arten, p. 548, pl. 28/12, fig. 26, 1862.

与原变种的区别在于：壳面较短，菱形，两侧边缘直，只在中部膨大；长 16—22 μm，宽 5—8 μm；龙骨突有 5—7 个/10 μm；横线纹有 19—23 条/10 μm。

生境：生于河边沼泽中。

国内分布：山西（太原晋祠、绵山清水河），辽宁（辽河、本溪），江苏（昆山），安徽（黄山），福建（金门），湖北（武汉植物园），湖南（长沙、索溪峪），广西（灵川），贵州（贵阳），云南（香格里拉），西藏（亚东、工布江达、米林、林芝、加查、察隅、波密、昌都、芒康、江达），陕西（华山），宁夏（贺兰山），新疆（伊宁）。

国外分布：亚洲（土耳其），北美洲（加拿大、美国），欧洲（英国、德国、爱尔兰、罗马尼亚、西班牙），大洋洲（澳大利亚）。

18c. 德洛变种 图版 IX: 8；图版 X: 8—18
var. **delognei** (Grunow) Lange-Bertalot, *Bacillaria* 3: 54—55, figs. 77—86, 155, 156, 1980;

Krammer & Lange-Bertalot, Bacillariophyceae, 2. Teil, *in* Ettl *et al.*, Süßwasserfl. Mitteleur., p. 53, figs. 40: 7, 8, 1988, nachdr. 1997; Metzeltin *et al.*, Iconogr. Diatomol. 15, fig. 207: 13, 2005.

Nitzschia denticula var. *delognei* Grunow, *in* Van Heurck, Synopsis des Diatomées de Belgique Atlas, p. 176, pl. 60, fig. 9, 1881.

Nitzschia interrupta (Reichelt in Kuntze) Hustedt, Archiv für Hydrobiologie 18: 168, 1927; Zhu & Chen, *in* Shi *et al.*, Compilation of Reports on the Survey of Algal Resources in South-Western China, p. 114, 1994.

Nitzschia heidenii var. *pamirensis* Petersen, 1930; Qi & Xie, Journal of Jinan University (1): 102, 1985; Zhu & Chen, Bacill. Xiz. Plat., p. 264, pl. 52, fig. 19, 2000.

Nitzschia solgensis Cleve-Euler, p. 67, fig. 1451c—d, 1952.

与原变种的区别在于：壳面窄披针形，两侧边缘不波曲；其他特征与原变种相似。

生境：生于草地渗出水、小溪中。

国内分布：辽宁（辽河），江苏（常熟），浙江（杭州），安徽（琅琊山、黄山），湖北（武汉植物园、神农架），湖南（沅江流域），广西（灵川），贵州（赫章、镇宁、贵阳、沅江流域、乌江流域、梵净山），云南（丽江），西藏（吉隆、芒康、墨竹工卡、边坝）。

国外分布：北美洲（美国），南美洲（乌拉圭），欧洲（英国、德国、波兰）。

18d. 缢缩变种　图版 IX：17

var. **constricta** Chen & Zhu, New Species and Varieties of the Diatoms from Suoxiyu, Hunan, China, *in* Li *et al.*, The algal flora and aquatic fauna of the Wulingyuan Nature Reserve Area, Hunan, China, p. 36, fig. I: 8, 1989a.

与原变种的区别在于：壳面中部明显收缢；壳面长约 19 μm，最宽处约 5 μm，中间缢缩部宽约 4 μm；龙骨突延长，宽于壳面的 1/2，5—6 个/10 μm；横线纹由明显的孔纹组成，20 条/10 μm。

生境：生于溪流中的岩石上。

国内分布：山西（平定），湖南（索溪峪、沅江流域），贵州（沅江流域、乌江流域）。

可疑和双叶型组 Section *Dubiae* and *Bilobatae*

可疑和双叶型组分种检索表

1. 龙骨突窄，稍延伸至壳面 ·· **19. 毡帽菱形藻** *N. homburgiensis*
1. 龙骨突较宽，基本不延伸至壳面 ·· 2
　2. 壳体大，长度超过 100 μm，龙骨突宽，厚，与 3 条或更多条线纹相连 ··· **20. 凯特菱形藻** *N. kittlii*
　2. 壳体较小，短于 100 μm，龙骨突较窄，一般与 1 条或 2 条线纹相连 ································· 3
　　3. 壳面中部强烈缢缩凹入，壳面舟形或括号形，带面具弯曲，喙状，不对称末端，龙骨突窄，一般与 1 条线纹相连 ·· **21. 杂种菱形藻** *N. hybrida*
　　3. 不具上述特征或特征很难辨认 ·· 4
　　　4. 壳缝龙骨强烈离心，在中缝端处缢缩不明显 ····················· **22. 可疑菱形藻** *N. dubia*

4. 不具上述特征 ··· 5
5. 龙骨突短，与 1 条以上的线纹相连，线纹 24—30 条/10 μm ············ **23. 脐形菱形藻** *N. umbonata*
5. 不具上述特征 ··· 6
　　6. 壳面具中缝端，但很难辨认，中间龙骨突窄，只与 1 条线纹相连，线纹约 20 条/10 μm ············
　　　··· **24. 吉塞拉菱形藻** *N. gisela*
　　6. 壳面具中缝端，容易辨认，壳缝龙骨在中间强烈缢缩，中间两个龙骨突明显分离 ···················
　　　··· **25. 多变菱形藻** *N. commutata*

19. 毡帽菱形藻　图版 XXXI：11—17

Nitzschia homburgiensis Lange-Bertalot，Nova Hedwigia 30：650，pl. 3，figs. 27—30，pl. 4，figs. 40—41，1978；Krammer & Lange-Bertalot，Bacillariophyceae. 2. Teil，*in* Ettl *et al.*，Süßwasserfl. Mitteleur.，p. 64，pl. 49，fig. 6，pl. 50，figs. 4—9，1988，nachdr. 1997.

壳面线形，中部略缢缩，两端呈小头状；壳面长 25—42 μm，宽 4—5 μm；龙骨突 10—14 个/10 μm，横线纹细密，在光镜下难以分辨。

生境：生于湖泊、河流、溪流和水泡中，水草或石头上附生。

国内分布：四川(若尔盖)。

国外分布：北美洲(加拿大、美国)，南美洲(哥伦比亚)，欧洲(波罗的海、英国、德国、爱尔兰、波兰、罗马尼亚)，大洋洲(新西兰)。

20. 凯特菱形藻　图版 XVII：1—3；图版 XVIII：1—2

Nitzschia kittlii Grunow，1882；Schmidt *et al.*，Atlas Diat.-Kunde，p. 347，fig. 347：17，1922；Hustedt，Bacillariophyta，*in* Pascher，Süßwass -Fl. Mitteleur. Heft 10，p. 406，fig. 776；Krammer & Lange-Bertalot，Bacillariophyceae，2. Teil，*in* Ettl *et al.*，Süßwasserfl. Mitteleur.，p. 66，figs. 52：1—2，1988，nachdr. 1997；Zhu & Chen，Bacill. Xiz. Plat.，p. 265，fig. 53：1，2000.

壳面长 132—170 μm，宽 16—20 μm；龙骨突有 2—3.5 个/10 μm；横线纹有 17—24 条/10 μm。

扫描电镜下观察：外壳面，壳缝管位于龙骨上，龙骨发育良好，位于壳面一侧边缘。组成线纹的小孔为单排，规则，短棒状，点纹 38—42 个/10 μm。壳缝在中部断开，壳缝裂缝简单。环带数量较多，可见 1—2 排孔纹。

生境：生于沼泽中。

国内分布：西藏(聂拉木、吉隆、墨脱、错那、措美、察雅、班戈、申扎、措勤、噶尔、革吉)，新疆(阿克陶、博乐、盐湖)。

国外分布：欧洲(英国、德国、罗马尼亚)。

21. 杂种菱形藻　图版 XXI：1—8

Nitzschia hybrida Grunow，*in* Cleve & Grunow，Beiträge zur Kenntniss der arctischen Diatomeen，p. 79，fig. 5：95，1880；Zhu & Chen，The Diatoms of the Suoxiyu Nature

Reserve Area, Hunan, China, *in* Li *et al.*, The algal flora and aquatic fauna of the Wulingyuan Nature Reserve Area, Hunan, China. p. 56, 1989b; Krammer & Lange-Bertalot, Bacillariophyceae, 2. Teil, *in* Ettl *et al.*, Süßwasserfl. Mitteleur., p. 61, figs. 46: 3—6, figs. 47: 1—3, 1988, nachdr. 1997; Zhu & Chen, Bacill. Xiz. Plat., p. 264, fig. 52: 23, 2000.

壳面长 38—91 μm, 宽 5—12 μm, 龙骨突有 7—11 个/10 μm, 横线纹有 16—22 条/10 μm。

生境: 生于路边沼泽、高山草甸沼泽中。

国内分布: 湖南(索溪峪、沅江流域), 海南(三亚), 贵州(沅江流域), 云南(丽江), 西藏(亚东、康马、吉隆、萨嘎、仲巴、昂仁、工布江达、墨脱、林芝、乃东、加查、错那、隆子、措美、昌都、芒康、察雅、洛隆、江达、贡觉、类乌齐、班戈、申扎、改则、札达、普兰、噶尔、革吉、日土), 新疆(布尔津、阿克陶)。

国外分布: 亚洲(俄罗斯), 北美洲(美国), 欧洲(波罗的海、黑海、英国、德国、法国、罗马尼亚、西班牙), 大洋洲(澳大利亚), 南极洲(乔治王岛)。

22. 可疑菱形藻 图版 XVIII: 3—5

Nitzschia dubia W. Smith, p. 41, pl. 13, fig. 112, 1853; Schmidt *et al.*, Atlas Diat.-Kunde, p. 346, figs. 346: 6—7, 1922; Krammer & Lange-Bertalot, Bacillariophyceae, 2. Teil, *in* Ettl *et al.*, Süßwasserfl. Mitteleur., p. 55, figs. 41: 1—2, 1988, nachdr. 1997; Zhu & Chen, Bacill. Xiz. Plat., p. 261, fig. 52: 5, 2000.

壳面长 92—150 μm, 宽 11—14 μm; 龙骨突有 8—11 个/10 μm; 横线纹有 18—24 条/10 μm。

生境: 生于河流、沼泽中。

国内分布: 黑龙江(鸡西兴凯湖), 西藏(昂仁、墨脱、错那、措美、普兰、革吉), 新疆(博湖、博乐、察布查尔)。

国外分布: 亚洲(俄罗斯), 北美洲(加拿大、美国), 欧洲(波罗的海、英国、德国、爱尔兰、波兰、罗马尼亚、西班牙), 大洋洲(澳大利亚、新西兰)。

23. 脐形菱形藻 图版 XVI: 10—16

Nitzschia umbonata (Ehrenberg) Lange-Bertalot, Nova Hedwigia 30: 648—650, Taf. 1, 2, 4, 1978; Krammer & Lange-Bertalot, Bacillariophyceae, 2. Teil, *in* Ettl *et al.*, Süßwasserfl. Mitteleur., p. 65, figs. 51: 1—6A, 1988, nachdr. 1997; Metzeltin *et al.*, Iconogr. Diatomol. 15, figs. 206: 6, 7, 2005.

Navicula umbonata Ehrenberg, Über das Massenverhältnifs der jetzt lebenden Kiesel-Infusorien und über ein neues Infusorien-Conglomerat als Polirschiefer von Jastraba in Ungarn, p. 32, 1837.

Nitzschia thermalis (Ehrenberg) Auerswald, *in* Rabenhorst, Die Algen Europas, Fortsetzung der Algen Sachsens, resp. Mittel-Europas, no. 1064a, 1861; Zhang, Journal of Jinan University 3: 91, fig. II: 5, 1986; Zhu & Chen, *in* Shi *et al.*, Compilation of Reports on the Survey of Algal Resources in South-Western China. p. 114，1994；Xin *et al.*，Journal of Shanxi University (Nat. Sci. Ed.) 20(1)：105，1997；Zhu & Chen，Bacill. Xiz. Plat.，p. 268，fig. 54：5，2000.

Nitzschia stagnorum Rabenhorst，Die Algen Sachsens. Resp. Mittel-Europa's Gesammelt und herausgegeben von Dr. L. Rabenhorst，no. 625，1860；Fan *et al.*，Bull. Bot. Res. 21：240，fig. I：9，2001.

壳面线形，两侧中部平行（较大体积）或稍凹入（较小体积），向两端楔形变窄，末端尖圆或钝圆或短喙状。壳面长30—85 μm，宽6—9 μm，龙骨突有7—10个/10 μm，横线纹有20—30条/10 μm。

生境：生于湖泊、路边积水、稻田、鱼池中。

国内分布：山西（太原晋阳湖、太原晋祠、绵山清水河），辽宁（沈阳），黑龙江（鸡西兴凯湖、五大连池），江苏（苏州），湖南（沅江流域），海南（崖县），贵州（沅江流域），西藏（墨脱、申扎），宁夏（贺兰山），新疆（皮山、阿克苏、莎车、赛里木湖、察布查尔）。

国外分布：亚洲（俄罗斯、土耳其），北美洲（美国），南美洲（阿根廷、哥伦比亚、乌拉圭），欧洲（波罗的海、英国、德国、丹麦、意大利、波兰、爱尔兰、马其顿、罗马尼亚、西班牙），大洋洲（澳大利亚、新西兰）。

24. 吉塞拉菱形藻　图版 XXIV：1—7

Nitzschia gisela Lange-Bertalot，*in* Lange-Bertalot & Krammer，Bibl. Diatomol. 15，p. 21，figs. 20：1—6，1987；Krammer & Lange-Bertalot，Bacillariophyceae，2. Teil，*in* Ettl *et al.*，Süßwasserfl. Mitteleur.，p. 57，figs. 42：7—8，1988，nachdr. 1997.

Nitzschia heufleriana Grunow，Die Österreichischen Diatomaceen nebst Anschluss einiger neuen Arten von andern Lokalitäten und einer kritischen Uebersicht der bisher bekannten Gattungen und Arten，p. 575，1862；Gao，Wuhan Bot. Res. 5(4)：336，1987；Zhu & Chen，The Diatoms of the Suoxiyu Nature Reserve Area，Hunan，China，*in* Li *et al.*，The algal flora and aquatic fauna of the Wulingyuan Nature Reserve Area，Hunan，China. p. 56，1989b；Zhu & Chen，Bacill. Xiz. Plat.，p. 264，pl. 52，fig. 20，2000.

壳面长53—88 μm，宽5—8 μm；龙骨突有8—12个/10 μm；横线纹有18—22条/10 μm。

生境：生于路边积水、河边渗出水、湖边渗出水、小水沟、泉水井边草丛中。

国内分布：黑龙江（鸡西兴凯湖），湖南（索溪峪、沅江流域），贵州（沅江流域），陕西（华山），西藏（芒康、察雅、革吉），新疆（阿克苏、奎屯、赛里木湖、察布查尔、布尔津、喀纳斯、阿勒泰）。

国外分布：亚洲（土耳其），北美洲（美国），欧洲（英国、德国、爱尔兰、马其顿、波兰、罗马尼亚、西班牙）。

25. 多变菱形藻　图版 XV：3—4；图版 XVI：1—9

Nitzschia commutata Grunow，*in* Cleve & Grunow，Beiträge zur Kenntniss der arctischen

Diatomeen, p. 79, 1880; Schmidt et al., Atlas Diat.-Kunde, p. 346, figs. 346: 17—20, 1922; Zhu & Chen, The Diatoms of the Suoxiyu Nature Reserve Area, Hunan, China, in Li et al., The algal flora and aquatic fauna of the Wulingyuan Nature Reserve Area, Hunan, China. p. 56, 1989b; Krammer & Lange-Bertalot, Bacillariophyceae, 2. Teil, in Ettl et al., Süßwasserfl. Mitteleur., p. 56, figs. 42: 1—6, 1988, nachdr. 1997; Zhu & Chen, Bacill. Xiz. Plat., p. 259, fig. 51: 11, 2000; Metzeltin et al., Iconogr. Diatomol. 15, figs. 201: 19—23, 2005; Antoniades et al., Iconogr. Diatomol. 17: 212, figs. 78: 1—3, 7, 8, 2008.

Nitzschia dubia W. Smith pro parte (excl. Typus), A synopsis of the British Diatomaceae, p. 41, pl. 13, fig. 112; Fan & Hu, Acta Hydrobiologica Sinca 28(4): 421—425, 2004.

壳面长35—78 μm，宽4—8 μm；龙骨突有8—12个/10 μm；横线纹有18—22条/10 μm。扫描电镜下观察：外壳面，壳缝管位于龙骨上，龙骨发育良好，位于壳面一侧边缘。组成线纹的小孔为单排，规则，近圆形，点纹34—36个/10 μm。壳缝在中部断开，壳缝裂缝简单。内壳面，单排孔纹，规则，近圆形，与外壳面孔纹类似，龙骨突短棒状。

生境：生于湖泊、湖边渗出水、路边积水、沼泽中。

国内分布：黑龙江(鸡西兴凯湖、五大连池)，湖南(索溪峪)，贵州(乌江流域)，西藏(亚东、康马、吉隆、昂仁、错那、措美、洛隆、类乌齐、班戈、申扎、改则、措勤、札达、革吉、日土)，新疆(阿克苏、赛里木湖、布尔津、哈巴河、福海)。

国外分布：亚洲(俄罗斯、新加坡)，北美洲(加拿大、美国)，南美洲(哥伦比亚、乌拉圭)，欧洲(波罗的海、黑海、英国、德国、爱尔兰、波兰、罗马尼亚、西班牙)，大洋洲(澳大利亚、新西兰)。

线型组 Section *Lineares*

线型组分种检索表

1. 具窄龙骨突，与1条线纹相连，具中央节，中间两个龙骨突间距较宽 ·· **26. 线形菱形藻 *N. linearis***
1. 特征与上述不同 ·· 2
  2. 龙骨突宽，与3至更多条线纹相连，4—8个/10 μm ············ **27. 玻璃质菱形藻 *N. vitrea***
  2. 龙骨突非常窄 ·· 3
3. 壳面约100 μm或更长，龙骨突5—7个/10 μm ············ **28. 木那菱形藻 *N. monachorum***
3. 壳面较短，少于100 μm，龙骨突7—15个/10 μm ············ **29. 近线形菱形藻 *N. sublinearis***

26. 线形菱形藻

Nitzschia linearis W. Smith, A synopsis of the British Diatomaceae, p. 39, pl. XIII, fig. 110, 1853; Schmidt et al., Atlas Diat.-Kunde, p. 334, figs. 334: 22—24, 1921; Hustedt, Bacillariophyta, in Pascher, Süßwass -Fl. Mitteleur. Heft 10, p. 409, fig. 784, 1930; Cleve-Euler, Diatomeen schw. Finnland, Teil. V, p. 80, fig. 1480, 1952; Zhu & Chen,

The Diatoms of the Suoxiyu Nature Reserve Area, Hunan, China, *in* Li *et al.*, The algal flora and aquatic fauna of the Wulingyuan Nature Reserve Area, Hunan, China. p. 56, 1989b; Krammer & Lange-Bertalot, Bacillariophyceae, 2. Teil, *in* Ettl *et al.*, Süßwasserfl. Mitteleur., p. 69, figs. 55: 1—4, 1988, nachdr. 1997; Xin *et al.*, Journal of Shanxi University(Nat. Sci. Ed.) 20(1): 105, 1997; Zhu & Chen, Bacill. Xiz. Plat., p. 265, fig. 53: 3, 2000; Metzeltin *et al.*, Iconogr. Diatomol. 15, figs. 205: 1—3, 206: 2—4, 208: 2, 3, 2005.

26a. 原变种 图版 XIX: 1—8

var. **linearis**

带面观宽至窄线形，边缘中部稍凹入；壳面观线形、线形-披针形至窄披针形，末端楔形减小，圆头，一侧稍凹入，另一侧弧形凸出；壳面长 65—172 μm，宽 4—6 μm；壳缝龙骨强烈离心，龙骨突窄肋状，稍延伸，中间两个距离明显增宽，8—14 个/10 μm；横线纹密集，在光镜下看不清楚。

扫描电镜下观察：壳缝管位于龙骨上，龙骨发育良好，高于壳面，位于壳面的一侧边缘。壳面具横线纹，线纹单排，由孔纹组成，近圆形，未见封闭结构。近壳缝的第一排孔纹较小(有时有两排)，圆形，壳缝在中部断开，中缝端简单，略膨大。壳缝在壳面末端简单，在螺旋舌上弯曲。

生境：生于湖泊、河流、河边渗出水、溪流、路边积水、小水渠、沼泽中，水草附生、泉水井边附生。

国内分布：山西(太原晋祠)，内蒙古(阿尔山)，辽宁(辽河、本溪、沈阳)，吉林(长白山)，黑龙江(鸡西兴凯湖、五大连池)，安徽(黄山、宁国)，福建(金门)，湖南(长沙、索溪峪、沅江流域)，广东(紫金)，广西(灵川)，海南(崖县、屯昌、琼山)，贵州(赫章、水城、贵阳、沅江流域、乌江流域、梵净山)，云南(下关、维西、丽江)，西藏(墨竹工卡、林芝、当雄、聂拉木、定日、亚东、吉隆、萨噶、墨脱、米林、乃东、加查、错那、措美、察隅、芒康、江达、类乌齐、申扎、措勤、札达、革吉)，陕西(华山)，宁夏(贺兰山)，新疆(阿克苏、博湖、天池、博乐、察布查尔、布尔津、喀纳斯、哈巴河、乌鲁木齐盐湖)。

国外分布：亚洲(俄罗斯、蒙古国、土耳其)，非洲(东部)，北美洲(美国、加拿大、夏威夷群岛)，南美洲(巴西、哥伦比亚、乌拉圭)，欧洲(波罗的海、黑海、英国、德国、爱尔兰、马其顿、波兰、罗马尼亚、西班牙)，大洋洲(澳大利亚、新西兰)。

26b. 细变种 图版 XX: 1—7

var. **tenuis** (W. Smith) Grunow, *in* Cleve & Grunow, Beiträge zur Kenntniss der arctischen Diatomeen, p. 93, 1880; Krammer & Lange-Bertalot, Bacillariophyceae, 2. Teil, *in* Ettl *et al.*, Süßwasserfl. Mitteleur., p. 70, figs. 55: 5—6, 1988, nachdr. 1997.

Nitzschia tenuis W. Smith, A synopsis of the British Diatomaceae, p. 40, pl. 13, fig. 111, 1853.

与原变种的区别在于：壳面线形-披针形，末端尖喙状；壳面较窄，4—6 μm。

生境：生于沼泽、湖泊、水坑中，浮游或附生。

国内分布：四川(若尔盖)，新疆(喀纳斯)。

国外分布：亚洲(俄罗斯、蒙古国)，北美洲(美国)，南美洲(哥伦比亚)，欧洲(英国、德国、爱尔兰、罗马尼亚、黑海、西班牙)。

27. 玻璃质菱形藻　图版 XXII：1—3；图版 XXIII：1—12

Nitzschia vitrea Norman，On some undescribed species of Diatomaceae，p. 7，pl. 2，fig. 4，1861；Schmidt *et al.*，Atlas Diat.-Kunde，p. 334，figs. 334：15—18，1921；Hustedt，Bacillariophyta，*in* Pascher，Süßwass -Fl. Mitteleur. Heft 10，p. 411，fig. 787，1930；Cleve-Euler，Diatomeen schw. Finnland，Teil. V，p. 80，figs. 1483 a—c，1952；Zhu & Chen，The Diatoms of the Suoxiyu Nature Reserve Area，Hunan，China，*in* Li *et al.*，The algal flora and aquatic fauna of the Wulingyuan Nature Reserve Area，Hunan，China. p. 56，1989b；Krammer & Lange-Bertalot，Bacillariophyceae，2. Teil，*in* Ettl *et al.*，Süßwasserfl. Mitteleur.，p. 72，figs. 56：1—2，1988，nachdr. 1997；Zhu & Chen，Bacill. Xiz. Plat.，p. 270，fig. 54：11，2000；Metzeltin *et al.*，Iconogr. Diatomol. 15，figs. 217：1，2，2005；Antoniades *et al.*，Iconogr. Diatomol. 17，p. 227，figs. 77：1—3，2008.

带面观线形，有时边缘稍凹入，壳面线形，小体积个体为披针形，朝两端楔形减小，末端喙状，弯向远离壳缝系统一侧；壳面长 55—132 μm，宽 7—12 μm；龙骨点大而清晰，圆矩形，中间两个距离不增宽，4—9 个/10 μm；横线纹由点纹组成，17—25 条/10 μm。

扫描电镜下观察：壳缝管位于龙骨上，龙骨发育良好，位于壳面的一侧边缘。壳面具横线纹，线纹单排，由孔纹组成，近圆形，未见封闭结构，点纹 36—39 个/10 μm。壳缝在壳面末端简单，在螺旋舌上弯曲。

生境：生于湖泊、小水渠、池塘、路边积水、沼泽、岩石上附生。

国内分布：河北(唐山)，辽宁(辽河)，江苏(昆山)，湖北(武汉植物园)，湖南(索溪峪、沅江流域)，贵州(沅江流域、乌江流域)，西藏(康美、察雅、贡觉、申扎、噶尔、革吉)，甘肃(苏干湖)，新疆(巴楚、阿克苏、天池、博乐、赛里木湖、察布查尔、伊宁、布尔津、盐湖)。

国外分布：亚洲(俄罗斯)，非洲(东部)，北美洲(加拿大、美国)，南美洲(巴西、乌拉圭)，欧洲(波罗的海、黑海、英国、德国、罗马尼亚、西班牙)，大洋洲(澳大利亚、新西兰)。

28. 木那菱形藻　图版 XXIV：14—15

Nitzschia monachorum Lange-Bertalot，*in* Lange-Bertalot & Krammer，Bibl. Diatomol. 15，p. 35，figs. 6：1—6，1987；Krammer & Lange-Bertalot，Bacillariophyceae，2. Teil，*in* Ettl *et al.*，Süßwasserfl. Mitteleur.，p. 75，figs. 57：5—8，1988，nachdr. 1997.

带面观线形，偶尔稍呈"S"形弯曲，壳面线形至线形-披针形，向两端稍呈楔形减小，末端呈小圆头状；壳面长 100—130 μm，宽 6—7 μm；壳缝龙骨稍离心，龙骨突窄，稍延长，排列不规则，中间两个间距不增宽，5—7 个/10 μm；横线纹密集，在光镜下看不清楚。

生境：生于沼泽、小水渠中。

国内分布：新疆(阿克苏、博湖)。

国外分布：欧洲(英国、德国)。

29. 近线形菱形藻　图版 XXIV：8—13

Nitzschia sublinearis Hustedt，Bacillariophyta，*in* Pascher，Süßwass -Fl. Mitteleur. Heft 10，p. 411，fig. 786，1930；Cleve-Euler，Diatomeen schw. Finnland. Teil. V，p. 80，fig. 1482，1952；Zhu & Chen，The Diatoms of the Suoxiyu Nature Reserve Area，Hunan，China，*in* Li *et al.*，The algal flora and aquatic fauna of the Wulingyuan Nature Reserve Area，Hunan，China. p. 56，1989b；Krammer & Lange-Bertalot，Bacillariophyceae. 2. Teil，*in* Ettl *et al.*，Süßwasserfl. Mitteleur.，p. 74，figs. 58：10—15，1988，nachdr. 1997；Zhu & Chen，Bacill. Xiz. Plat.，p. 268，fig. 54：4，2000；Antoniades *et al.*，Iconogr. Diatomol.：17：225，figs. 75：1，2，2008.

壳面披针形至线形，向两端楔形减小，末端圆头状；壳面长 55—110 μm，宽 5—7 μm；龙骨突窄，稍延长，中间两个距离不增宽，7—12 个/10 μm；横线纹密集，在光镜下看不清楚。

生境：生于溪流、池塘、稻田、路边积水、沼泽，水草附生。

国内分布：辽宁(沈阳)，黑龙江(五大连池)，湖南(索溪峪、沅江流域)，海南(崖县、琼海)，四川(九寨沟)，贵州(沅江流域、乌江流域、梵净山)，云南(苍山)，西藏(林芝、浪卡子、聂拉木、吉隆、隆子、措美、察隅、芒康、类乌齐、申扎、札达)，宁夏(贺兰山)，新疆(阿克苏、博湖、奎屯、察布查尔、喀纳斯、哈巴河、阿勒泰)。

国外分布：亚洲(俄罗斯、蒙古国、伊朗)，北美洲(加拿大、美国)，南美洲(巴西、哥伦比亚)，欧洲(奥地利、波罗的海、黑海、英国、德国、马其顿、波兰、罗马尼亚、西班牙)，大洋洲(新西兰)。

披针型组 Section *Lanceolatae*

披针型组分种检索表

1. 壳缝在壳面中部不断开(电镜下可见)，中间龙骨突距离不增大 ··· 2
1. 壳缝在壳面中部断开(电镜下可见)，具中缝端，中间两个龙骨突距离增大 ································· 16
  2. 龙骨突小，点状，与 1 条线纹相连，线纹有 24—28 条/10 μm ················ **30. 常见菱形藻 *N. solita***
  2. 龙骨突宽，常与多条线纹相连 ··· 3
3. 线纹间距相对较宽，少于 15—25 条/10 μm(低倍镜下可见)，点状 ·· 4
3. 线纹较密集，多于 25 条/10 μm ·· 9
  4. 壳体小，末端钝圆；线纹粗糙，17—20 条/10 μm ············ **31. 粗肋菱形藻 *N. valdecostata***
  4. 特征与上述不同 ·· 5

5. 壳体小，末端头状，圆或延长 ··· 6
5. 特征与上述不同 ·· 7
   6. 壳面常在中部稍凹入，头状末端窄，线纹有 23—32 条/10 μm ············ **32. 华丽菱形藻 *N. elegantula***
   6. 壳面线形，或在中部稍凸出，末端宽头 ························ **33. 杆状菱形藻 *N. bacilliformis***
7. 壳面一般不长于 40 μm，线纹 21—25 条/10 μm，末端钝圆，常生于低导电率的淡水中 ·················
   ·· **34. 高山菱形藻 *N. alpina***
7. 壳面较长，线形 ··· 8
   8. 壳面宽 4—7 μm ·· **35. 中型菱形藻 *N. intermedia*** (部分)
   8. 壳面宽 2.5—3.5 μm ··· **36. 多样菱形藻 *N. diversa***
9. 线纹密度多于 35 条/10 μm，光镜下不容易看清或根本看不清；壳面较窄，末端呈延长的喙状；龙骨
   突点状，小 ·· **37. 细长菱形藻 *N. gracilis***
9. 线纹密度 25—35 条/10 μm，光镜下不容易看清 ··· 10
   10. 壳面常较小，具显著头状或短喙状末端 ························ **38. 小头菱形藻 *N. microcephala***
   10. 特征与上述不同 ·· 11
11. 末端明显宽圆形，少数稍微延长；宽 4—5.5 μm，线纹约 30 条/10 μm ··· **39. 普通菱形藻 *N. communis***
11. 特征与上述不同 ·· 12
   12. 种群体积较大，线形或披针形 ··· 13
   12. 种群体积明显短，相对较窄 ··· 15
13. 壳面线形，宽大于 4 μm ·· **35. 中型菱形藻 *N. intermedia***
13. 壳面窄披针形，宽小于 4 μm ·· 14
   14. 末端喙状，延长 ·· **40. 近针形菱形藻 *N. subacicularis***
   14. 末端渐窄，不延长 ·· **41. 灌木菱形藻 *N. fruticosa***
15. 较长的壳面为披针形，至少不为窄线形，孔纹和肋纹在同一平面上，因此线纹很难看清 ···············
   ·· **42. 谷皮菱形藻 *N. palea***
15. 较长的壳面为窄线形，线纹 26—32 条/10 μm ······························· **43. 细微菱形藻 *N. perminuta***
   16. 龙骨突较窄，楔形，像牙根，延伸形成一条横肋纹；线纹通常少于 20 条/10 μm ···············
      ··· **44. 两栖菱形藻 *N. amphibia***
   16. 龙骨突较宽，与多条线纹相连，线纹较密集，光镜下看不清楚 ······································ 17
17. 线纹更加细密，超过 20 条/10 μm ··· 18
17. 线纹间距较宽，少于 20 条/10 μm ··· 19
   18. 壳面较长，披针形或线形披针形，末端尖圆或稍呈头状 ········ **45. 泉生菱形藻 *N. fonticola*** (部分)
   18. 壳面呈不同程度披针形，一般生活在导电率中性水体中 ········ **46. 化石菱形藻 *N. fossilis***
19. 壳面小，线纹粗糙，16—19 条/10 μm，小孔呈双排，紧密排列，不容易辨认 ·····························
   ··· **47. 粗条菱形藻 *N. valdestriata***
19. 特征与上述不同 ·· 20
   20. 壳面通常小，椭圆形至线形椭圆形，末端钝圆 ··········· **48. 平庸菱形藻 *N. inconspicua***
   20. 特征与上述不同 ·· 21
21. 较长的壳面呈线形，不呈披针形，壳面末端短，楔形 ············ **49. 吉斯纳菱形藻 *N. gessneri***
21. 较长的壳体呈披针形或线形-披针形，两端为延长的楔形 ··· 22
   22. 壳体趋向线形-披针形，两端喙状 ···························· **50. 管毛菱形藻 *N. tubicola***
   22. 壳面明显披针形 ·· 23
23. 长度少于 40 μm，生活在淡水中，导电率适中，壳缝龙骨上为双排孔 ····· **45. 泉生菱形藻 *N. fonticola***
23. 特征与上述不同，特别是孔纹较简单 ··· 25

24. 壳面披针形至线形-披针形，通常较短，很少到 30 μm，宽度很少达 3 μm，末端不呈圆头状 ··········
·· **51.** 小片菱形藻 *N. frustulum*
24. 特征与上述不同 ··· 26
25. 壳面窄披针形，一般两侧中部不凹入，宽 2.5—3 μm ············ **52.** 辐射菱形藻 *N. radicula*
25. 壳面较宽，一般两侧中部稍凹入，3—6 μm ··················· **53.** 小头端菱形藻 *N. capitellata*

30. 常见菱形藻　图版 XXVIII：30—37

Nitzschia solita Hustedt, Diatomeen aus der Oase Gafsa in Südtunesien, ein Beitrag zur Kenntnis der Vegetation afrikanischer Oasen, p. 152, figs. 3, 4, 1953b; Krammer & Lange-Bertalot, Bacillariophyceae, 2. Teil, *in* Ettl *et al.*, Süßwasserfl. Mitteleur., p. 99, figs. 71：1—12, 1988, nachdr. 1997.

形态变化较大，宽披针形至窄披针形，朝两端楔形减小，末端尖喙状；长 18—50 μm，宽 4—6 μm；龙骨突点状，等距排列，中间两个距离不增大，11—15 个/10 μm；横线纹 24—28 条/10 μm。

扫描电镜下观察：可见组成横线纹的单排孔纹，近圆形，38—42 个/10 μm。

生境：生于鱼池、湖泊、山溪、沼泽中。

国内分布：湖北(武汉植物园)，新疆(阿克陶、奎屯、博乐五一水库、赛里木湖、察布查尔)。

国外分布：亚洲(以色列)，北美洲(美国)，欧洲(英国、德国、马其顿、波兰、罗马尼亚、西班牙)，大洋洲(澳大利亚)。

31. 粗肋菱形藻　图版 XXVIII：28—29

Nitzschia valdecostata Lange-Bertalot & Simonsen, Bacillaria 1, p. 58, figs. 260—263, 269, 270, 1978; Krammer & Lange-Bertalot, Bacillariophyceae, 2. Teil, *in* Ettl *et al.*, Süßwasserfl. Mitteleur., p. 121, figs. 84：1—8, 1988, nachdr. 1997.

壳面线形椭圆形，末端钝圆，长 16—20 μm，宽 3—5 μm。龙骨突宽，6—8 个/10 μm，横线纹粗糙，17—20 条/10 μm。

扫描电镜下观察：龙骨突一般与多条线纹相连，中间龙骨突间距不增大，壳缝在中央节处不断开，孔纹五点梅花状，或呈双排，排列紧密(Krammer and Lange-Bertalot, 1997)。

生境：生于小水渠、路边沼泽中。

国内分布：新疆(阿克苏)。

国外分布：北美洲(美国)，欧洲(英国、德国)。

32. 华丽菱形藻　图版 XXV：29—42

Nitzschia elegantula Grunow, *in* Van Heurck, Synopsis des Diatomées de Belgique Atlas, pl. 69, fig. 22a, 1881; Krammer & Lange-Bertalot, Bacillariophyceae. 2. Teil, *in* Ettl *et al.*, Süßwasserfl. Mitteleur., p. 120, figs. 83：20—24, 1988, nachdr. 1997.

Nitzschia microcephala var. *elegantula* Van Heurck, Synopsis des Diatomées de Belgique, p.

183，1885；Zhu & Chen，Bacill. Xiz. Plat.，p. 266，fig. 53：8，2000.

壳体常较小，壳面两侧中部稍凹入，头状末端窄，延长；长 10—25 μm，宽 2.5—4 μm；龙骨突宽，10—14 个/10 μm，中间两个龙骨突间距不增大；横线纹密集，23—26 条/10 μm。

生境：生于水渠中，多生活在导电率高的水体。

国内分布：西藏(申扎)，新疆(岳普湖、沙雅、尉犁)。

国外分布：非洲(东部)，北美洲(加拿大、美国、夏威夷群岛)，南美洲(智利、哥伦比亚)，欧洲(波罗的海、英国、德国、罗马尼亚、西班牙)，大洋洲(澳大利亚)。

33. 杆状菱形藻　图版 XXV：1—16

Nitzschia bacilliformis Hustedt，Bacillariales aus Innerasien. Gesammelt von Dr. Sven Hedin. *in* Hedin，Southern Tibet，discoveries in former times compared with my own researches in 1906-1908，p. 148；pl. 10，figs. 62—64，1922；Krammer & Lange-Bertalot，Bacillariophyceae，2. Teil，*in* Ettl *et al.*，Süßwasserfl. Mitteleur.，p. 102，figs. 74：18—26，1988，nachdr. 1997；Zhu & Chen，Bacill. Xiz. Plat.，p. 258，fig. 52：1，2000.

壳体通常较小，壳面线形，两侧平行或在中部稍凹入，末端宽头状，有时延长；壳面长 12—35 μm，宽 3—4 μm；龙骨突有 10—12 个/10 μm，中间两个距离不增大；线纹密集，20—25 条/10 μm，通常在光镜下可见。

生境：生于沼泽、泉水中。

国内分布：西藏(定日、吉隆、班戈、申扎)；新疆(阿克陶、和静)。

国外分布：北美洲(美国)，欧洲(英国、德国)。

本种与 *N. alpina* 在形态上最为相似，主要区别在于末端的形状。本书主要依据 Krammer 和 Lange-Bertalot(1997) 的分类，与 Zhu 和 Chen(2000) 中的 fig. 52：1 在外形上有些差别：末端钝圆，不延长。

34. 高山菱形藻　图版 XXV：17—23

Nitzschia alpina Hustedt，Internationale Revue der gesamten Hydrobiologie und Hydrographie 43：232，figs. 60—65，1943；Krammer & Lange-Bertalot，Bacillariophyceae，2. Teil，*in* Ettl *et al.*，Süßwasserfl. Mitteleur.，p. 101，figs. 74：1—10，1988，nachdr. 1997；Antoniades *et al.*，Iconogr. Diatomol. 17：207，figs. 73：11—14，2008.

壳体短小，壳面线形-披针形；长 18—38 μm，宽 3—5 μm；龙骨突宽，8—12 个/10 μm，中间两个距离不增大；线纹密集，21—25 条/10 μm，通常在光镜下可见。

扫描电镜下观察：每条龙骨突常与多条横线纹相连。

生境：生于湖泊、沼泽、稻田、小水沟，常生活在内陆水体中。

国内分布：内蒙古(阿尔山)，广东(河源、紫金)，贵州(水城)，新疆(博乐、赛里木湖、阿克陶、莎车、温宿)。

国外分布：北美洲(加拿大、美国)，南美洲(哥伦比亚)，欧洲(英国、德国、马其顿、波兰)，大洋洲(新西兰)。

本种与 *N. hantzschiana* 在形态结构上较相似，不同在于本种壳面上的中间两个龙骨突距离不增大，但具中央节。

35. 中型菱形藻　图版 XXXI：1—5

Nitzschia intermedia Hantzsch ex Cleve & Grunow，Beiträge zur Kenntniss der arctischen Diatomeen，p. 95，1880；Krammer & Lange-Bertalot，Bacillariophyceae，2. Teil，*in* Ettl *et al.*，Süßwasserfl. Mitteleur.，p. 87，figs. 61：1—10，1988，nachdr. 1997；Xie *et al.*，Journal of Shanxi University (Nat. Sci. Ed.) 14(4)：417，1991；Metzeltin *et al.*，Iconogr. Diatomol. 15，figs. 201：17—18，206：8—10，2005.

壳面较长，线形，长 76—113 μm，宽 4—5 μm；龙骨突 9—11 个/10 μm，中间两个距离不增大；线纹 23—28 条/10 μm，通常光镜下可见。

生境：生于湖泊、河流、泉水、山溪、稻田、路边积水、沼泽中。

国内分布：山西(太原晋阳湖)，辽宁(辽河)，黑龙江(鸡西兴凯湖)，江苏(昆山)，湖北(武汉植物园)，湖南(长沙)，广东(东源、惠城、东莞)，广西(灵川)，贵州(水城、梵净山)，西藏(墨竹工卡)，新疆(阿克苏、阿克陶、和静、博湖、阜康天池、察布查尔、布尔津)。

国外分布：亚洲(俄罗斯)，非洲(东部)，北美洲(加拿大、美国)，南美洲(巴西、哥伦比亚、乌拉圭)，欧洲(波罗的海、黑海、英国、德国、马其顿、波兰、罗马尼亚、西班牙)，大洋洲(澳大利亚、新西兰)。

36. 多样菱形藻　图版 XXVII：15—25

Nitzschia diversa Hustedt，Die Diatomeenflora des Salzlackengebietes im österreichischen Burgenland，p. 436，figs. 14—17，1959；Krammer & Lange-Bertalot，Bacillariophyceae，2. Teil，*in* Ettl *et al.*，Süßwasserfl. Mitteleur.，p. 107，fig. 77：15，1988，nachdr. 1997.

壳面窄线形，朝两端楔形变窄，末端小头状；长 36—80 μm，宽 2.5—3.5 μm；龙骨突有 10—13 个/10 μm，中间两个距离不增大；横线纹密集，24—27 条/10 μm，光镜下不容易看清。

扫描电镜下观察：龙骨发育良好，位于壳面一侧边缘，龙骨上具两排小圆点纹。组成横线纹的单排孔纹，椭圆形，大小不一，28—32 个/10 μm。

生境：生于山溪、香蒲池塘、沼泽中。

国内分布：海南(琼中)，新疆(皮山、阿克陶)。

国外分布：北美洲(美国)，欧洲(英国、德国)。

37. 细长菱形藻　图版 XXIX：1—11

Nitzschia gracilis Hantzsch，Hedwigia 2(7)：40，pl. 6，fig. 8，1860；Schmidt *et al.*，Atlas Diat.-Kunde，p. 349，figs. 349：34—37，1924；Hustedt，Bacillariophyta，*in* Pascher，Süßwass -Fl. Mitteleur. Heft 10，p. 417，fig. 794，1930；Cleve-Euler，Diatomeen schw. Finnland，Teil. V，p. 85，fig. 1493，1952；Krammer & Lange-Bertalot，Bacillariophyceae，2. Teil，*in* Ettl *et al.*，Süßwasserfl. Mitteleur.，p. 93，figs. 66：1—11，1988，nachdr. 1997；

Xin *et al.*, Journal of Shanxi University (Nat. Sci. Ed.) 20(1): 105, 1997; Zhu & Chen, Bacill. Xiz. Plat., p. 263, fig. 52: 14, 2000.

壳面窄线形-披针形, 末端延长呈长喙状; 壳面长 40—110 μm, 宽 3—4 μm; 龙骨突点状, 清晰, 11—13 个/10 μm; 横线纹排列紧密, 不易看清。

扫描电镜下观察: 内壳面, 可见组成线纹的单排孔纹, 近矩形, 25—28 个/10 μm。龙骨突短棒状, 末端具简单螺旋舌。

生境: 生于路边积水、稻田、沼泽中, 多为富含有机质的水体。

国内分布: 山西(太原晋祠、绵山清水河、运城), 黑龙江(鸡西兴凯湖、五大连池), 江苏(昆山), 湖南(沅江流域), 广东(龙川、紫金、博罗、东莞), 海南(崖县、保亭), 贵州(威宁草海、沅江流域、乌江流域), 西藏(当雄、聂拉木、亚东、吉隆、昂仁、乃东、错那、措美、芒康、江达、类乌齐、申扎、革吉), 新疆(疏勒、莎车、尉犁、麦盖提、阿克苏)。

国外分布: 亚洲(俄罗斯、伊朗、土耳其), 非洲(东部), 北美洲(加拿大、美国), 南美洲(巴西、哥伦比亚), 欧洲(波罗的海、黑海、英国、德国、爱尔兰、马其顿、波兰、罗马尼亚、西班牙), 大洋洲(澳大利亚、新西兰)。

38. 小头菱形藻 图版 XXVI: 13—14; 图版 XXVIII: 21—26

Nitzschia microcephala Grunow, *in* Cleve & Möller, Diatoms, 1878; Zhu & Chen, The Diatoms of the Suoxiyu Nature Reserve Area, Hunan, China, *in* Li *et al.*, The algal flora and aquatic fauna of the Wulingyuan Nature Reserve Area, Hunan, China. p. 56, 1989b; Krammer & Lange-Bertalot, Bacillariophyceae. 2. Teil, *in* Ettl *et al.*, Süßwasserfl. Mitteleur., p. 120, figs. 83: 10—18, 1988, nachdr. 1997; Zhu & Chen, Bacill. Xiz. Plat., p. 266, fig. 53: 7, 2000; Metzeltin *et al.*, Iconogr. Diatomol. 15, figs. 214: 12—19, 2005.

壳面常较小, 两侧中部凹入, 具明显头端或短喙状末端, 长 10—15 μm, 宽 2—4 μm, 龙骨突宽, 10—13 个/10 μm, 线纹非常密集, 光镜下不容易看清。

生境: 生于河流、湖泊、沼泽、小水渠中。

国内分布: 黑龙江(五大连池、宁安), 湖南(索溪峪), 广东(惠州), 贵州(施秉舞阳河), 西藏(聂拉木、申扎), 新疆(阿克苏)。

国外分布: 亚洲(俄罗斯、蒙古国), 非洲(东部、南非), 北美洲(美国、夏威夷群岛), 南美洲(巴西、哥伦比亚、乌拉圭), 欧洲(亚得里亚海、波罗的海、英国、丹麦、芬兰、德国、波兰、罗马尼亚、西班牙、瑞典、瑞士、冰岛), 大洋洲(澳大利亚、新西兰)。

39. 普通菱形藻 图版 XXV: 24—28; 图版 XXVI: 15

Nitzschia communis Rabenhorst, Die Algen Sachsens. Resp. Mittel-Europa's Gesammelt und herausgegeben von Dr. L. Rabenhorst, no. 949, 1860; Zhu & Chen, The Diatoms of the Suoxiyu Nature Reserve Area, Hunan, China, *in* Li *et al.*, The algal flora and aquatic fauna of the Wulingyuan Nature Reserve Area, Hunan, China. p. 56, 1989b; Krammer & Lange-Bertalot, Bacillariophyceae, 2. Teil, *in* Ettl *et al.*, Süßwasserfl. Mitteleur., p. 110, figs. 79: 1—6, 1988, nachdr. 1997; Xin *et al.*, Journal of Shanxi University (Nat. Sci. Ed.) 20(1): 105, 1997; Zhu & Chen, Bacill. Xiz. Plat., p. 259, fig. 51: 11, 2000.

壳面线形椭圆形，末端宽圆，少数稍微延长呈钝圆；长 20—55 μm，宽 4—5.5 μm；龙骨突宽，10—14 个/10 μm，中间两个距离不增大；横线纹密集，30—36 条/10 μm，光镜下不容易看清。

扫描电镜下观察：内壳面，可见组成线纹的单排孔纹，近圆形，46—50 个/10 μm。龙骨突块状，末端具简单螺旋舌。

生境：生于湖泊、河流、沼泽中。

国内分布：北京，山西（太原晋祠），黑龙江（鸡西兴凯湖），福建（厦门、鼓岭），湖南（岳阳、索溪峪），云南（维西），西藏（昂仁、乃东、昌都、芒康、类乌齐、革吉），新疆（阿克陶、叶城）。

国外分布：亚洲（俄罗斯），非洲（东部），北美洲（加拿大、美国、夏威夷群岛），南美洲（哥伦比亚），欧洲（波罗的海、黑海、英国、德国、丹麦、爱尔兰、马其顿、波兰、罗马尼亚、西班牙、冰岛），大洋洲（澳大利亚、新西兰）。

40. 近针形菱形藻　图版 XXVII：12—14

Nitzschia subacicularis Hustedt, *in* Schmidt *et al.*, Atlas Diat.-Kunde, p. 348, fig. 348: 76, 1922; Krammer & Lange-Bertalot, Bacillariophyceae, 2. Teil, *in* Ettl *et al.*, Süßwasserfl. Mitteleur., p. 118, figs. 67: 4—10, 1988, nachdr. 1997.

壳面细长，窄披针形至窄线形-披针形，末端明显延长，呈长喙状；壳面长 50—70 μm，宽 2—3 μm；龙骨突点状，13—15 个/10 μm；横线纹在光镜下不容易分辨。

生境：生于鱼池、湖泊、沼泽中。

国内分布：辽宁（辽河），新疆（奎屯、博乐五一水库、赛里木湖）。

国外分布：亚洲（俄罗斯），非洲（东部），北美洲（美国），欧洲（英国、德国、马其顿、波兰、罗马尼亚），大洋洲（澳大利亚、新西兰）。

41. 灌木菱形藻　图版 XXVI：1—5

Nitzschia fruticosa Hustedt, Abhandlungen des Naturwissenschaftlichen Verein zu Bremen 34(3): 349, figs. 81—82, 1957; Krammer & Lange-Bertalot, Bacillariophyceae. 2. Teil, *in* Ettl *et al.*, Süßwasserfl. Mitteleur., p. 86, pl. 60, figs. 8—12, 1988, nachdr. 1997.
Nitzschia actinastroides (Lemmermann) van Goor, Recueil des travaux botaniques néerlandais 22: 320, text figs. 1—2, 1925; Zhu & Chen, Bacill. Xiz. Plat., p. 257, pl. 51, fig. 4, 2000.

壳面线形-披针形；长 45—76 μm，宽 2—4 μm；龙骨突 12—18 个/10 μm，横线纹细密，光镜下难以分辨。

生境：生于河流中，水草附着。

国内分布：湖北(东湖)，四川(若尔盖)，西藏(浪子卡、芒康)。

国外分布：亚洲(俄罗斯)，北美洲(美国)，欧洲(黑海、英国、德国、罗马尼亚、西班牙)，大洋洲(澳大利亚)。

42. 谷皮菱形藻

Nitzschia palea (Kützing) W. Smith, A synopsis of the British Diatomaceae, p. 89, 1856; Schmidt *et al.*, Atlas Diat.-Kunde, p. 349, figs. 349: 1—10, 1924; Hustedt, Bacillariophyta, *in* Pascher, Süßwass -Fl. Mitteleur. Heft 10, p. 416, fig. 801, 1930; Cleve-Euler, Diatomeen schw. Finnland, Teil. V, p. 90, fig. 1504, 1952; Zhu & Chen, The Diatoms of the Suoxiyu Nature Reserve Area, Hunan, China, *in* Li *et al.*, The algal flora and aquatic fauna of the Wulingyuan Nature Reserve Area, Hunan, China. p. 56, 1989b; Krammer & Lange-Bertalot, Bacillariophyceae, 2. Teil, *in* Ettl *et al.*, Süßwasserfl. Mitteleur., p. 85, figs. 59: 1—24, 1988, nachdr. 1997; Xin *et al.*, Journal of Shanxi University(Nat. Sci. Ed.) 20(1): 105, 1997; Zhu & Chen, Bacill. Xiz. Plat., p. 267, fig. 53: 14, 2000; Metzeltin *et al.*, Iconogr. Diatomol. 15, figs. 207: 5—7, 208: 11, 12, 218: 6, 7, 2005; Antoniades *et al.*, Iconogr. Diatomol. 17, p. 218, figs. 74: 1—6, figs. 75: 3—5, 2008.

Synedra palea Kützing, Die Kieselschaligen Bacillarien oder Diatomeen, p. 63, pl. 3, fig. 27, 1844.

42a. 原变种　图版 XXVII: 1—11

var. **palea**

壳面线形-披针形，朝两端楔形减小，末端尖或圆头；长 20—78 μm，宽 3—5 μm；龙骨突清晰，10—15 个/10 μm，横线纹紧密，光镜下不容易看清。

生境：生于河流、湖泊、湖边渗出水、小水渠、池塘、沼泽、路边积水、稻田中。

国内分布：河北(唐山)，山西(太原晋阳湖、太原晋祠、绵山清水河、运城)，内蒙古(阿尔山)，辽宁(辽河、沈阳)，吉林(长白山)，黑龙江(宝清七星河、宁安、鸡西兴凯湖、五大连池)，江苏(昆山、苏州)，浙江(杭州西湖)，安徽(黄山、宁国)，福建(厦门、金门)，江西(鄱阳湖)，山东(泰山)，湖北(武汉植物园)，湖南(长沙、岳阳、索溪峪、沅江流域)，广东(龙川、紫金、惠州、博罗、增城)，广西(灵川)，海南(琼山、琼中、崖县)，重庆(北碚缙云山)，贵州(赫章、水城、沅江流域、乌江流域、梵净山)，西藏(墨竹工卡、林芝、边坝、聂拉木、定日、亚东、康马、吉隆、萨噶、昂仁、工布江达、墨脱、林芝、乃东、朗县、加查、错那、隆子、措美、察隅、八宿、昌都、芒康、察雅、洛隆、江达、贡觉、类乌齐、班戈、申扎、札达、普兰、噶尔、革吉、日土)，陕西(华山)，甘肃(苏干湖)，新疆(巴楚、阿克苏、奎屯、博乐、赛里木湖、察布查尔、布尔津、喀纳斯、哈巴河、福海、阿勒泰)。

国外分布：亚洲(俄罗斯、蒙古国、伊拉克、土耳其、以色列)，非洲(东部、南非)，北美洲(加拿大、美国、夏威夷群岛)，欧洲(波罗的海、英国、德国、比利时、爱尔兰、马其顿、罗马尼亚、波兰、西班牙、冰岛)，大洋洲(澳大利亚、新西兰)。

本种与 *N. gandersheimiensis* 在细胞的形态、大小、线纹和龙骨点密度等方面都比较相似，不同之处在于：本种不存在中央节，中央两个龙骨点和其余龙骨点之间的距离相等；而 *N. gandersheimiensis* 存在中央节，中央两个龙骨点之间的距离最宽(Wendker and Geissler, 1986)。

42b. 微小变种 图版 XXVI: 6—9
var. **minuta** (Bleisch) Grunow, *in* Van Heurck, Syn. Diat. Belgique, pl. 69, fig. 23, 1881; Krammer & Lange-Bertalot, Bacillariophyceae, 2. Teil, *in* Ettl *et al.*, Süßwasserfl. Mitteleur., p. 85, figs. 59: 13—17, 1988, nachdr. 1997.
Nitzschia minuta Bleisch, *in* Rabenhorst, Die Algen Sachsens, p. 950, 1860.

壳面线形-披针形；长 35—42 μm，宽 5 μm；龙骨突 11—13 个/10 μm，横线纹 23—25 条/10 μm。与原变种的区别在于：壳面横线纹稀疏，光镜下可见。

生境：生于水库边沼泽、高山草地沼泽中。

国内分布：新疆(哈密、阿克陶)。

国外分布：亚洲(俄罗斯)，北美洲(美国)，欧洲(黑海、英国、德国、罗马尼亚)。

43. 细微菱形藻 图版 XXV: 43—45
Nitzschia perminuta (Grunow) M. Peragallo, Le catalogue général des diatomées, p. 672, 1903; Krammer & Lange-Bertalot, Bacillariophyceae, 2. Teil, *in* Ettl *et al.*, Süßwasserfl. Mitteleur., p. 99, figs. 72: 1—23A, 1988, nachdr. 1997; Antoniades *et al.*, Iconogr. Diatomol. 17, p. 220—221, figs. 73: 1—9, 2008; Liu *et al.*, Wuhan Bot. Res. 27(3): 275, figs. 7—9, 2009.
Nitzschia palea var. *perminuta* Grunow, *in* van Heurck, Synopsis des Diatomées de Belgique Atlas, pl. LXVIII, fig. 31, 1881.

壳体较小，壳面线形或线形-披针形，两侧近平行或略凸出，末端略延长，尖圆或呈小头状；长 18—32 μm，宽 2.5—3.5 μm；龙骨突明显，有 11—14 个/10 μm；横线纹在光镜下不清晰。

生境：生于泥炭藓沼泽、路边水坑中。

国内分布：内蒙古(阿尔山)，江苏(昆山)。

国外分布：亚洲(俄罗斯、蒙古国、以色列)，非洲(南非)，北美洲(加拿大、美国、夏威夷群岛)，南美洲(哥伦比亚)，欧洲(波罗的海、比利时、英国、法国、德国、爱尔兰、马其顿、波兰、罗马尼亚、冰岛)，大洋洲(澳大利亚、新西兰)，南极洲(詹姆斯罗斯岛、乔治王岛、利文斯顿岛)。

本种形态变化较大，尤其是壳面末端的形状，略延长呈尖喙状，或呈小头状，或不延长略突起呈不明显的小头状。Grunow 认为中部龙骨突的位置是鉴别 *N. perminuta*、*N. acidoclinata*、*N. frustulum* 和 *N. leistikowii* 的关键，但在光学显微镜下很难区分，对这几个小体积种类的生态分布特点也缺乏足够的研究，因此包括 *N. hantzschiana* 在内的这几个 *Nitzschia* "Lanceolatae" 种类很难准确鉴定（Krammer and Lange-Bertalot，1997）。

44. 两栖菱形藻　图版 XXVIII：1—10

Nitzschia amphibia Grunow，Verhandlungen der Kaiserlich-Königlichen Zoologisch-Botanischen Gesellschaft in Wien 12: 574, pl. 28/12, fig. 23, 1862; Zhu & Chen, The Diatoms of the Suoxiyu Nature Reserve Area, Hunan, China, *in* Li *et al.*, The algal flora and aquatic fauna of the Wulingyuan Nature Reserve Area, Hunan, China. p. 55, 1989b; Krammer & Lange-Bertalot, Bacillariophyceae, 2. Teil, *in* Ettl *et al.*, Süßwasserfl. Mitteleur., p. 108, figs. 78: 13—21, 1988, nachdr. 1997; Xin *et al.*, Journal of Shanxi University (Nat. Sci. Ed.) 20(1): 105, 1997; Zhu & Chen, Bacill. Xiz. Plat., p. 258, fig. 51: 6, 2000; Metzeltin *et al.*, Iconogr. Diatomol. 15, figs. 206: 16—19, 2005; Antoniades *et al.*, Iconogr. Diatomol. 17: 209, fig. 73: 16, 2008.

壳体较小，壳面椭圆形、披针形至线形-披针形；长 13—38 μm，宽 4—6 μm；龙骨突稍窄，楔形，像牙根，7—9 个/10 μm，中间两个距离较宽；横线纹粗糙，14—17 条/10 μm。

生境：生于湖边渗出水、小水渠、路边积水、沼泽中。

国内分布：北京，河北（唐山），山西（太原晋祠、运城），辽宁（辽河、本溪、沈阳），吉林（长白山），黑龙江（鸡西兴凯湖、五大连池、宝清七星河、牡丹江镜泊湖），上海（松江），江苏（昆山、苏州），安徽（黄山、宁国），福建（金门），江西（鄱阳湖），山东（泰山），湖北（武汉植物园），湖南（长沙、索溪峪、沅江流域），广西（灵川），海南（崖县、陵水、五指山），贵州（水城、施秉舞阳河、沅江流域、乌江流域），云南（滇池、翠湖、洱海），西藏（边坝、当雄、聂拉木、定日、亚东、吉隆、萨噶、昂仁、拉萨、工布江达、墨脱、林芝、错那、隆子、措美、察隅、芒康、察雅、江达、贡觉、类乌齐、班戈、申扎、札达、普兰、革吉、日土），陕西（华山），宁夏（贺兰山），新疆（博乐、赛里木湖、察布查尔、布尔津、喀纳斯、阿勒泰）。

国外分布：亚洲（俄罗斯、蒙古国、伊朗、土耳其、新加坡），非洲（东部、南非），北美洲（加拿大、美国、夏威夷群岛），南美洲（阿根廷、巴西、哥伦比亚、乌拉圭），欧洲（波罗的海、黑海、英国、德国、爱尔兰、意大利、马其顿、波兰、罗马尼亚、西班牙、瑞典、冰岛），大洋洲（澳大利亚、新西兰）。

45. 泉生菱形藻　图版 XXVIII：11—20

Nitzschia fonticola (Grunow) Grunow, *in* Van Heurck, pl. 69, figs. 15—20, 1881; Schmidt *et al.*, Atlas Diat.-Kunde, p. 348, figs. 348: 61—65, 1922; Hustedt, Bacillariophyta, *in* Pascher, Süßwass -Fl. Mitteleur. Heft 10, p. 415, fig. 800, 1930; Cleve-Euler, Diatomeen schw. Finnland. Teil. V, p. 88, fig. 1500, 1952; Krammer & Lange-Bertalot, Bacillariophyceae, 2. Teil, *in* Ettl *et al.*, Süßwasserfl. Mitteleur., p. 103, figs. 75: 1—22,

1988, nachdr. 1997; Zhu & Chen, Bacill. Xiz. Plat., p. 262, fig. 52: 8, 2000.

Nitzschia palea var. *fonticola* Grunow, *in* Cleve & Möller, Diatoms. Part IV, no. 174, 1879.

壳面明显披针形, 末端尖圆; 长 13—40 μm, 宽 2.5—4 μm; 龙骨突点状, 不延伸, 中间两个距离增大, 10—14 个/10 μm; 横线纹密集, 23—28 条/10 μm。

国内分布: 山西(太原晋祠), 吉林(长白山), 黑龙江(牡丹江镜泊湖、五大连池), 山东(泰山), 湖北(武汉植物园), 湖南(索溪峪、沅江流域), 广东(惠州), 海南(琼中), 贵州(沅江流域、乌江流域、梵净山), 云南(滇池、翠湖、洱海), 西藏(定日、浪卡子、亚东、吉隆、墨脱、措美、察隅、波密、芒康、洛隆、贡觉、班戈、申扎、措勤、札达、噶尔、革吉), 陕西(华山), 新疆(阿克陶、叶城)。

国外分布: 亚洲(俄罗斯、蒙古国、伊拉克、土耳其、以色列), 非洲(东部、南非), 北美洲(加拿大、美国、夏威夷群岛), 欧洲(波罗的海、英国、德国、比利时、爱尔兰、马其顿、罗马尼亚、波兰、西班牙、冰岛), 大洋洲(澳大利亚、新西兰)。

46. 化石菱形藻 图版 XXIX: 18—24

Nitzschia fossilis (Grunow) Grunow, *in* Van Heurk, Synopsis des Diatomées de Belgique Atlas, pl. 68, fig. 24, 1881; Krammer & Lange-Bertalot, Bacillariophyceae, 2. Teil, *in* Ettl *et al.*, Süßwasserfl. Mitteleur., p. 105, figs. 76: 8—16, 1988, nachdr. 1997.

Nitzschia amphibia var. *fossilis* Grunow, *in* Cleve & Grunow, Beiträge zur Kenntniss der arctischen Diatomeen, p. 98, 1880.

壳面披针形至线形, 末端逐渐变窄呈尖圆形; 长 30—50 μm, 宽 3—4 μm; 龙骨突较宽, 8—12 个/10 μm; 横线纹较密集, 19—23 条/10 μm。

生境: 生于小水渠、路边积水、稻田、沼泽中。

国内分布: 新疆(阿克苏、布尔津、喀纳斯、莎车)。

国外分布: 北美洲(美国), 欧洲(英国、德国、波兰、罗马尼亚), 大洋洲(澳大利亚、新西兰)。

本种与 Krammer 和 Lange-Bertalot(1997)中的描述在形态结构上最为相似: 长 30—85 μm, 宽 3.5—5 μm, 龙骨突 7—9 个/10 μm, 横线纹 18—21 条/10 μm。但也存在一些差别: 本书中的种类宽度稍窄, 龙骨突和线纹的密度稍大一些。

47. 粗条菱形藻 图版 XXVIII: 27

Nitzschia valdestriata Aleem & Hustedt, Einige neue Diatomeen von der Südkuste Englands, p. 19, figs. 5 a—b, 1951; Krammer & Lange-Bertalot, Bacillariophyceae, 2. Teil, *in* Ettl *et al.*, Süßwasserfl. Mitteleur., p. 121, figs. 84: 9—12, 1988, nachdr. 1997.

壳体小, 壳面椭圆形至线形椭圆形, 末端钝圆, 长 12.5 μm, 宽 3.5 μm。龙骨突较宽, 9 个/10 μm, 横线纹粗糙, 18 条/10 μm。

生境: 生于小水渠中。

国内分布: 海南(三亚), 新疆(布尔津)。

国外分布: 北美洲(美国), 欧洲(波罗的海、英国、德国、波兰、西班牙), 大洋洲(澳大利亚)。

48. 平庸菱形藻 图版 XXV：46—47

Nitzschia inconspicua Grunow, Die Österreichischen Diatomaceen nebst Anschluss einiger neuen Arten von andern Lokalitäten und einer kritischen Uebersicht der bisher bekannten Gattungen und Arten, p. 579; pl. 28/12, fig. 25, 1862; Krammer & Lange-Bertalot, Bacillariophyceae, 2. Teil, *in* Ettl *et al.*, Süßwasserfl. Mitteleur., p. 95, figs. 69：1—13, 1988, nachdr. 1997; Metzeltin *et al.*, Iconogr. Diatomol. 15, figs. 214：20—26, 2005; You & Wang. Acta Botanica Boreali-Occidentalia Sinica 31(2)：419, figs. 1：24—25, 2011.

壳体小，壳面线形椭圆形，末端钝圆；壳面长 12—15 μm，宽 3—4 μm；龙骨突有 10—13 个/10 μm；横线纹在光镜下看不清楚。

生境：生于沼泽、湖边渗出水中。

国内分布：内蒙古(阿尔山)，广东(紫金、惠州)，贵州(水城)，新疆(博乐五一水库、赛里木湖)。

国外分布：亚洲(俄罗斯)，北美洲(加拿大、美国、夏威夷群岛)，南美洲(哥伦比亚、乌拉圭)，欧洲(波罗的海、英国、芬兰、德国、爱尔兰、马其顿、罗马尼亚、西班牙、瑞典)，大洋洲(澳大利亚、新西兰)。

49. 吉斯纳菱形藻 图版 XXX：1—16

Nitzschia gessneri Hustedt, Diatomeen aus dem Naturschutzgebiet Seeon, p. 632, figs. 3—7, 1953; Krammer & Lange-Bertalot, Bacillariophyceae, 2. Teil, *in* Ettl *et al.*, Süßwasserfl. Mitteleur., p. 106, figs. 77：11—14, 1988, nachdr. 1997.

壳面线形，很少线形-披针形，在壳面中部靠近龙骨突的一侧稍凹入，两端逐渐楔形变窄，呈小头状，一般不延伸；壳面长 35—85 μm，宽 3—4 μm；龙骨突点状，中间两个间距增大，10—13 个/10 μm；横线纹密集，25—28 条/10 μm。

扫描电镜下观察：外壳面，龙骨发育良好，高于壳面，龙骨上具两排点纹；组成线纹的单排孔纹近圆形，36—40 个/10 μm；壳缝在中部断开，中缝端稍膨大。内壳面，龙骨突短棒形，与 1—2 条肋纹相连，略延伸至壳面；末端具简单螺旋舌。

生境：生于沼泽中。

国内分布：新疆(阿克陶、布尔津)。

国外分布：北美洲(美国)，欧洲(英国、德国、罗马尼亚)。

50. 管毛菱形藻 图版 XXXI：6—10

Nitzschia tubicola Grunow, *in* Cleve & Grunow, Kungl. Svenska Vetenskapsakademiens Handlingar. Ser. 4, 17(2)：97, 1880; Krammer & Lange-Bertalot, Bacillariophyceae. 2. Teil, *in* Ettl *et al.*, Süßwasserfl. Mitteleur., p. 90, pl. 63, figs. 8—12, pl. 64, figs. 1—16, 1988, nachdr. 1997.

壳面线形-披针形,两端小头状;长 40—50 μm,宽 4—5 μm;龙骨突 11—13 个/10 μm,横线纹细密,光镜下难以分辨。

生境:生于河流、沼泽和水泡中,丝状藻附着。

国内分布:四川(若尔盖)。

国外分布:亚洲(俄罗斯),北美洲(加拿大、美国),南美洲(巴西),欧洲(英国、芬兰、德国、波兰、罗马尼亚、西班牙)。

51. 小片菱形藻　图版 XXV:48—54

Nitzschia frustulum (Kützing) Grunow, in Cleve & Grunow, Beiträge zur Kenntniss der arctischen Diatomeen, p. 98, 1880; Zhu & Chen, The Diatoms of the Suoxiyu Nature Reserve Area, Hunan, China, in Li et al., The algal flora and aquatic fauna of the Wulingyuan Nature Reserve Area, Hunan, China. p. 55, 1989b; Krammer & Lange-Bertalot, Bacillariophyceae, 2. Teil, in Ettl et al., Süßwasserfl. Mitteleur., p. 94, figs. 68:1—8, 1988, nachdr. 1997; Xin et al., Journal of Shanxi University (Nat. Sci. Ed.) 20(1):105, 1997; Zhu & Chen, Bacill. Xiz. Plat., p. 262, fig. 52:9, 2000; Metzeltin et al., Iconogr. Diatomol. 15, figs. 207:14, 15, 2005.

Synedra frustulum Kützing, Die Kieselschaligen Bacillarien oder Diatomeen, p. 63, pl. 30, fig. 77.

壳体较小,壳面披针形至线形-披针形,很小的个体呈长椭圆形,末端钝圆;长 15—46 μm,宽 2—4 μm;龙骨突 8—13 个/10 μm,中央两个相距较宽,可见中央节;横线纹 18—25 条/10 μm。

生境:生于湖泊、河流、沼泽中。

国内分布:山西(太原晋阳湖、太原晋祠、绵山清水河、运城),辽宁(沈阳),吉林(长白山),黑龙江(宝清七星河、五大连池),福建(厦门、金门),山东(泰山),湖北(武汉植物园),湖南(索溪峪),广东(龙川、东源、紫金、惠城),海南(崖县、陵水),贵州,云南(黑龙潭、翠湖、下关),西藏(白地、亚东、康马、吉隆、昂仁、工布江达、墨脱、错那、措美、察隅、八宿、芒康、察雅、洛隆、江达、贡觉、类乌齐、申扎、革吉),陕西(华山),新疆(博乐、察布查尔、喀纳斯、莎车、焉耆、和静)。

国外分布:亚洲(俄罗斯、蒙古国、巴基斯坦),北美洲(加拿大、美国、夏威夷群岛),南美洲(巴西、阿根廷、哥伦比亚、乌拉圭),欧洲(亚得里亚海、波罗的海、英国、比利时、芬兰、德国、爱尔兰、马其顿、波兰、葡萄牙、罗马尼亚、西班牙、瑞典、冰岛、克罗泽群岛),大洋洲(澳大利亚、新西兰)。

52. 辐射菱形藻　图版 XXVI:10—12

Nitzschia radicula Hustedt, Süßwasser-Diatomeen des indomalayischen Archipels und der Hawaii-Inslen, p. 209; figs. 2—3, 1942; Krammer & Lange-Bertalot, Bacillariophyceae, 2. Teil, in Ettl et al., Süßwasserfl. Mitteleur., p. 105, figs. 77:7—10, 1988, nachdr. 1997.

壳面窄披针形，较长的壳体趋向线形-披针形，末端为延长的楔形；长33—70 μm，宽2.5—3 μm，龙骨突点状或更宽，10—13个/10 μm；横线纹在光镜下不清晰。

生境：生于沼泽中。

国内分布：新疆(焉耆、喀纳斯)。

国外分布：北美洲(美国)，欧洲(德国、马其顿、波兰)，大洋洲(澳大利亚)。

53. 小头端菱形藻 图版XXIX：12—17

Nitzschia capitellata Hustedt，*in* Schmidt *et al.*，Atlas Diat.-Kunde，pl. 348，figs. 57，58，1922；Zhu & Chen，The Diatoms of the Suoxiyu Nature Reserve Area，Hunan，China，*in* Li *et al.*，The algal flora and aquatic fauna of the Wulingyuan Nature Reserve Area，Hunan，China. p. 55，1989b；Krammer & Lange-Bertalot，Bacillariophyceae，2. Teil，*in* Ettl *et al.*，Süßwasserfl. Mitteleur.，p. 88，figs. 62：1—12，1988，nachdr. 1997；Zhu & Chen，Bacill. Xiz. Plat.，p. 259，fig. 51：9，2000；Metzeltin *et al.*，Iconogr. Diatomol. 15，fig. 218：5，2005.

壳面线形至线形-披针形，两侧中部通常略凹入，朝两端楔形变窄，末端圆头状；长35—48 μm，宽3—6 μm；龙骨突有10—15个/10 μm，中间两个距离明显增大；横线纹在光镜下不清楚。

生境：生于沼泽、路边积水，泉水井边附生。

国内分布：黑龙江(五大连池、鸡西兴凯湖、宁安、五大连池)，福建(金门)，江西(鄱阳湖)，湖南(索溪峪、沅江流域)，海南(海口、琼山)，贵州(沅江流域、乌江流域、梵净山)，西藏(吉隆、仲巴、工布江达、林芝、措美、波密、芒康、江达、贡觉)，新疆(博乐、察布查尔、布尔津)。

国外分布：亚洲(俄罗斯)，非洲(南非)，北美洲(加拿大、美国)，南美洲(阿根廷、哥伦比亚、乌拉圭)，欧洲(波罗的海、黑海、英国、德国、爱尔兰、马其顿、波兰、罗马尼亚、西班牙)，大洋洲(澳大利亚、新西兰)。

鸭嘴端组 Section *Nitzschiellae*

鸭嘴端组分种检索表

1. 顶轴直，壳缝具近缝端··2
1. 顶轴"S"形··3
  2. 龙骨突等距··54. 针形菱形藻 *N. acicularis*
  2. 中间两个龙骨突增宽···55. 爪维兰斯菱形藻 *N. draveillensis*
 3. 横线纹粗糙，常少于20条/10 μm···56. 洛伦菱形藻 *N. lorenziana*
 3. 横线纹密集，光镜下不可分辨···57. 反曲菱形藻 *N. reversa*

54. 针形菱形藻 图版XXXIII：4—6

Nitzschia acicularis(Kützing) W. Smith，A synopsis of the British Diatomaceae，p. 43，pl. 15，fig. 122；Schmidt *et al.*，Atlas Diat.-Kunde，p. 335，figs. 335：15—17，1921；Hustedt，

Bacillariophyta, *in* Pascher, Süßwass -Fl. Mitteleur. Heft 10, p. 423, figs. 821, 822, 1930; Cleve-Euler, Diatomeen schw. Finnland, Teil. V, p. 92, fig. 1509, 1952; Zhu & Chen, The Diatoms of the Suoxiyu Nature Reserve Area, Hunan, China, *in* Li *et al.*, The algal flora and aquatic fauna of the Wulingyuan Nature Reserve Area, Hunan, China. p. 55, 1989b; Krammer & Lange-Bertalot, Bacillariophyceae, 2. Teil, *in* Ettl *et al.*, Süßwasserfl. Mitteleur., p. 123, figs. 85: 1—4, 1988, nachdr. 1997; Zhu & Chen, Bacill. Xiz. Plat., p. 257, fig. 51: 3, 2000; Metzeltin *et al.*, Iconogr. Diatomol. 15, figs. 204: 6, 12, 13, 2005.

Synedra acicularis Kützing, Die Kieselschaligen Bacillarien oder Diatomeen, p. 63, pl. 4, fig. 3, 1844.

壳体轻微硅质化，纺锤形，末端急剧变窄，延长呈喙状；壳面长 43—100 μm，宽 3—5 μm；龙骨突点状，中间两个距离不增大，17—20 个/10 μm；横线纹极细，光学显微镜下很难分辨。

生境：生于沼泽、池塘中。

国内分布：山西(太原晋阳湖)，吉林(长白山)，黑龙江(鸡西兴凯湖、五大连池)，江西(鄱阳湖)，湖北(武汉植物园)，湖南(索溪峪、沅江流域)，广东(龙川)，贵州(威宁草海、沅江流域)，西藏(错那、贡觉)，新疆(奎屯、博乐、布尔津、喀纳斯)。

国外分布：亚洲(俄罗斯、蒙古国、伊朗、新加坡、土耳其)，非洲(东部)，北美洲(加拿大、美国、夏威夷群岛)，南美洲(阿根廷、巴西、哥伦比亚、乌拉圭、安第斯山脉)，欧洲(波罗的海、黑海、英国、法国、德国、爱尔兰、马其顿、波兰、罗马尼亚、西班牙、)，大洋洲(澳大利亚、新西兰)，南极洲(亚南极群岛)。

55. 爪维兰斯菱形藻　图版 XXXIII：2—3, 7

Nitzschia draveillensis Coste & Ricard, p. 190, pl. 2, fig. 32, pl. 10, figs. 70, 73—74, 1980; Krammer & Lange-Bertalot, Bacillariophyceae. 2. Teil, *in* Ettl *et al.*, Süßwasserfl. Mitteleur., p. 123, pl. 85, figs. 5—6, 1988, nachdr. 1997.

壳面线形-披针形，末端急剧变窄，延长呈喙状；壳面长 62—100 μm，宽 3—5 μm；龙骨突 9—15 个/10 μm，横线纹细密，在光镜下难以分辨。

生境：生于湖泊、河流、沼泽和水泡中，水草或丝状藻附生。

国内分布：四川(若尔盖)。

国外分布：亚洲(伊朗、蒙古国)，北美洲(美国)，南美洲(哥伦比亚)，欧洲(英国、德国、西班牙)，大洋洲(新西兰)。

56. 洛伦菱形藻

Nitzschia lorenziana Grunow, *in* Cleve & Möller, Diatoms, Nos 208—210, 1879; Krammer & Lange-Bertalot, Bacillariophyceae, 2. Teil, *in* Ettl *et al.*, Süßwasserfl. Mitteleur., p. 125, figs. 86: 6—10, 1988, nachdr. 1997; Fan, Studied on Aulonoraphidinales (Surirellales) from Heilongjiang province. p. 39, fig. 5: 5, 2004; Metzeltin *et al.*, Iconogr.

Diatomol. 15，figs. 204：1—3，2005.

56a. 原变种　图版 XXXIII：8—11，13

var. **lorenziana**

　　带面观和壳面观均呈不同程度"S"形弯曲；壳面窄披针形，朝两端逐渐呈长喙状延伸，末端尖圆；长 100—170 μm，宽 4—7 μm；龙骨突有 6—10 个/10 μm，中间两个距离增大；横线纹清晰可见，13—18 条/10 μm。

　　生境：生于路边小水渠、池塘、沼泽、深沟积水、路边积水中。

　　国内分布：黑龙江(鸡西兴凯湖)，安徽(黄山)，湖北(武汉植物园)，湖南(岳阳)，广东(广州)、海南(保亭、通什、琼山)，新疆(巴楚、阿克苏、奎屯、博乐、察布查尔、布尔津)。

　　国外分布：亚洲(俄罗斯、新加坡)，北美洲(美国)，南美洲(巴西、哥伦比亚、乌拉圭)，欧洲(亚得里亚海、波罗的海、黑海、英国、克罗地亚、德国、马其顿、罗马尼亚、西班牙)，大洋洲(澳大利亚、新西兰)。

56b. 细弱变种　图版 XXXIII：12

var. **subtilis** Grunow，*in* Cleve & Grunow，Beiträge zur Kenntniss der arctischen Diatomeen，
　　p. 102，1880；Zhu & Chen, Bacill. Xiz. Plat.，p. 266，fig. 53：6，2000.

　　壳面长 92 μm，宽 4.5 μm；龙骨突有 8 个/10 μm，横线纹清晰可见，16 条/10 μm。

　　生境：生于湖边草地中水坑、静水。

　　国内分布：西藏(八宿)。

　　国外分布：北美洲(美国)，欧洲(英国、罗马尼亚、西班牙)，大洋洲(澳大利亚、新西兰)。

57. 反曲菱形藻　图版 XXXIII：1

Nitzschia reversa W. Smith，A synopsis of the British Diatomaceae，p. 43，pl. 15，fig. 121
　　1853；Krammer & Lange-Bertalot，Bacillariophyceae，2. Teil，*in* Ettl *et al.*，Süßwasserfl.
　　Mitteleur.，p. 124，figs. 85：7—10，1988，nachdr. 1997；Metzeltin *et al.*，Iconogr. Diatomol.
　　15，fig. 204：4，5，2005.

Nitzschia longissima var. *reversa* Grunow，*in* Cleve & Grunow，Beiträge zur Kenntniss der
　　arctischen Diatomeen，p. 100，1880；Liu *et al.*，Wuhan Bot. Res. 24：41，fig. III：112，
　　2006.

　　带面观和壳面观均呈不同程度"S"形弯曲；壳面纺锤形，末端急剧变窄，延长呈喙状；长 60—90 μm，宽 5—7 μm；龙骨突有 8—13 个/10 μm；横线纹密集，光学显微镜下很难分辨。

　　生境：生于湖泊、河流中。

　　国内分布：福建(金门)。

　　国外分布：亚洲(俄罗斯)，北美洲(美国、古巴)，南美洲(乌拉圭)，欧洲(英国、德

国、罗马尼亚、西班牙)，大洋洲(澳大利亚、新西兰)。

类附生组 Section *Epithemioideae*

58. 类附生菱形藻　图版XXXI：18

Nitzschia epithemoides Grunow，*in* Cleve & Grunow，Beiträge zur Kenntniss der arctischen Diatomeen，p. 82，no fig.，1880；Krammer & Lange-Bertalot，Bacillariophyceae. 2. Teil，*in* Ettl *et al*.，Süßwasserfl. Mitteleur.，p. 51，pl. 40，figs. 9—10，1988，nachdr. 1997；Zhu & Chen，Bacill. Xiz. Plat.，p. 261，pl. 52，fig. 6，2000.

壳面长 45.5—51 μm，宽 6.6—10 μm；龙骨突 7—9 个/10 μm，横线纹 21—35 条/10 μm。

生境：生于泉水小溪、农田边静水塘、采盐井区中。

国内分布：西藏(察隅、波密、芒康、贡觉)。

国外分布：北美洲(加拿大、美国)，南美洲(巴西)，欧洲(英国、德国、西班牙、罗马尼亚)，大洋洲(澳大利亚、新西兰)。

未命名组 Unnamed Section

59. 瓜德罗普菱形藻　图版XXXII：1—4

Nitzschia guadalupensis Manguin，*in* Bourrelly et Manguin，Algues d'eau douce de la Guadeloupe et dépendances，p. 104，pl. 9，fig. 193，1952；Metzeltin & Lange-Bertalot，Iconogr. Diatomol 18，pl. 282，figs. 1—5，2007.

壳面长 130—143 μm，宽 11—13 μm；龙骨突 5—7 个/10 μm，中间两个龙骨突略增宽，横线纹 16—18 条/10 μm。

扫描电镜下观察：内壳面，可见组成线纹的单排孔纹，圆形，31—34 个/10 μm；龙骨突短棒状，排列不均匀，不延伸至壳面；末端具明显的螺旋舌结构。

生境：生于稻田边水沟中。

国内分布：新疆(莎车)。

国外分布：南美洲(哥斯达黎加、瓜德罗普岛、玛丽-加朗特岛)。

菱板藻属　Hantzschia A. Grunow

A. Grunow，Mon. Microsc. J. 18：174，1877.

单细胞，直或"S"形，常示带面观。色素体 2 个(偶尔 4 个)，分别位于中央横切面(median transapical plane)两侧，简单或复杂呈裂叶状，紧靠细胞腹侧；或呈 2 个板状，与带面相对，由一明显的中央蛋白核连接。

壳面具背腹侧之分，腹侧略凹入、直或略凸出，背侧弧形凸出，末端喙状或小头状。壳缝靠近腹侧，上下壳面的壳缝关于壳面(valvar plane)呈镜面对称('hantzschioid' symmetry)。壳面结构简单，具单排或双排小孔组成的线纹，小孔圆形或肾形，由膜或筛状物封闭。有时存在边缘脊。壳缝在两极之间连续或在中部断开，一般为双弧形(壳缝在壳面中部更靠近腹缘),具中缝端时,在外壳面结构简单,或部分被硅质的片层结构覆盖,在内壳面壳缝裂缝连续或相反方向弯转。龙骨突大块状或窄肋状,在内壳面包住壳缝。极缝端(polar raphe ending)简单,或弯向远缘。带面复杂,由断开或闭合的环带组成,有一些环带具 2 至多排孔纹。

广泛分布于海水和淡水中,附着于沉积物或污泥中,特别是在潮间沙滩,也会扩散到气生环境中,如 *H. amphioxys*,广泛存在于土壤中。

本属种类共同特征较少,因此很难定义。目前所知,仅具同一细胞分裂类型,镜面对称的细胞一直产生镜面对称的两个子细胞。这一点与菱形藻属 *Nitzschia* 的细胞分裂类型不同。Round(1970)和 Mann(1977,1980a,1980b,1981)曾经对本属种类进行过详细论述。

模式种:两尖菱板藻[*Hantzschia amphioxys*(Ehrenberg)Grunow]。

本志收编 19 种 1 变种 1 变型。

菱板藻属分种检索表

1. 壳缝不具中缝端,中间两个龙骨突等距或稍宽 ·· 2
1. 壳缝具中缝端,中间两个龙骨突距离增大 ·· 4
  2. 壳体略呈"S"形弯曲,壳面不具 *Hantzschia* 典型的背腹性 ·· 3
  2. 壳体不呈"S"形弯曲,壳面具 *Hantzschia* 典型的背腹性 ················ **1. 长命菱板藻** *H. vivax*
3. 壳体较大,末端略延长呈喙状 ·· **2. 美丽菱板藻** *H. spectabilis*
3. 壳体较小,末端明显楔形,呈小头状 ··· **3. 菱形菱板藻** *H. nitzschioides*
  4. 龙骨突窄,一般只与 1 条横肋纹相连,主要为海水种类 ··· 5
  4. 部分龙骨突较宽,与 1 条以上的横肋纹相连 ·· 6
5. 龙骨突基本不延伸 ·· **4. 显点菱板藻** *H. distinctepunctata*
5. 龙骨突延伸一段距离 ··· **5. 直菱板藻** *H. virgata*
  6. 线纹由粗糙孔纹组成,线纹数目少于 2 倍的龙骨突数目 ··· 7
  6. 线纹由细小孔纹组成,线纹数目多于 2 倍的龙骨突数目 ··· 13
7. 壳面不具 *Hantzschia* 属典型的背腹性,两侧几乎平行 ································ **6. 中华菱板藻** *H. sinensis*
7. 壳面具 *Hantzschia* 属典型的背腹性 ·· 8
  8. 壳面线形,长一般超过 200 μm ··· 9
  8. 壳面不呈明显的线形,长宽比例小 ··· 10
9. 壳面呈膝状弯曲,长宽比例较大 ··· **7. 长菱板藻** *H. elongata*
9. 壳面不呈膝状弯曲,长宽比例较小 ··· **8. 较长菱板藻** *H. longa*
  10. 壳面末端明显变窄,略延长,呈尖锥形 ·· 11
  10. 壳面较宽,一般大于 8 μm,末端呈喙状 ·· 12
11. 背腹性明显,壳体长,为 100—118 μm ··· **9. 巴克豪森菱板藻** *H. barckhausenii*
11. 背腹性不明显,壳体短,为 54—89 μm ·· **10. 伊犁菱板藻** *H. yili*
  12. 壳面腹缘凹入不明显,宽 8—11μm ··· **11. 盖斯纳菱板藻** *H. giessiana*
  12. 壳面腹缘凹入明显,宽 14—16 μm ·· **12. 密集菱板藻** *H. compacta*
13. 壳面基本没有"肩部"或肩部不明显 ··· 14
13. 壳面有"肩部",末端钝圆或小头状 ·· 15
  14. 体积较大,末端窄喙状 ··· **13. 近强壮菱板藻** *H. subrobusta*
  14. 体积较小,末端小头状 ··· **14. 嫌钙菱板藻** *H. calcifuga*

15. 壳体较大，长宽比例大，壳面在两端呈楔形减小，末端小头状 ·· 16
15. 壳体较小，长宽比例小，壳面在两端急剧减小，末端钝圆 ·· 17
　　16. 壳面背侧弧形凸出，横线纹密度 13—15 条/10 μm ············· **15. 活跃菱板藻** *H. vivacior*
　　16. 壳面背侧弧形不明显，横线纹密度 16—18 条/10 μm ········ **16. 拟巴德菱板藻** *H. pseudobardii*
17. 壳体长宽比大，宽小于 10 μm ··· 18
17. 壳体长宽比小，宽大于 10 μm ··· **17. 近石生菱板藻** *H. subrupestris*
　　18. 壳体较长，背腹性明显 ·· **18. 丰富菱板藻** *H. abundans*
　　18. 壳体较小，背腹性不明显 ·· **19. 两尖菱板藻** *H. amphioxys*

1. 长命菱板藻　图版 XXXVIII：1—3

Hantzschia vivax (W. Smith) M. Peragallo *in* Tempere & Peragallo, Diatomées du Monde Entier, p. 56, No. 103—104, 1908; Krammer & Lange-Bertalot, Bacillariophyceae. 2. Teil, *in* Süßwasserfl. Mitteleur., p. 129, fig. 91：5, 1988, nachdr. 1997.

Nitzschia vivax W. Smith, Synopsis of British Diatomaceae, p. 41, pl. 31, fig. 267, 1853.

带面观矩形；壳面腹侧边缘平直，背侧弧形凸出，中部平直或略凹入，末端略延长呈小头端；长 60—180 μm，宽 6—12 μm；龙骨突位于腹侧边缘，纤细，整齐排列，5—6 个/10 μm；横线纹有 11—14 条/10 μm。

生境：生于小水渠中。

国内分布：内蒙古（阿尔山、牙克石、图里河），黑龙江（漠河），四川（若尔盖），新疆（阿克苏）。

国外分布：亚洲（俄罗斯），北美洲（美国），欧洲（波罗的海、英国、德国）。

2. 美丽菱板藻　图版 XL：5—7

Hantzschia spectabilis (Ehrenberg) Hustedt, Die Diatomeenflora des Salzlackengebietes im österreichischen Burgenland, Band 7/2, p. 431, 1959; Krammer & Lange-Bertalot, Bacillariophyceae, *in* Süßwasserfl. Mitteleur. Band 2/2, p. 132, figs. 91：1—4, 1988, nachdr. 1997; You *et al.*, Bulletin of Botanical Research 31(2)：131, figs. 9—11, 2011.

Synedra spectabilis Ehrenberg, Verbreitung und Einfluss des mikroskopischen Lebens in Süd-und Nord-Amerika, p. 389(101), pl. 1/2, fig. 19, pl. 2/3, fig. 4, 1843.

带面稍呈"S"形或不规则的弯曲，两侧近平行；壳面线形，稍弯曲，朝两端楔形减小，末端圆头；长 150—250 μm，宽 10—15 μm；龙骨突有 4—6 个/10 μm；横线纹 9—12 条/10 μm。

生境：生于河边渗出水、小水渠中。

国内分布：新疆（阿克苏、察布查尔）。

国外分布：亚洲（俄罗斯），北美洲（美国），欧洲（波罗的海、英国、德国、西班牙），大洋洲（新西兰）。

3. 菱形菱板藻　图版 XLVI：1—9

Hantzschia nitzschioides Lange-Bertalot, Cavacini, Tagliaventi & Alfinito. Iconogr. Diatomol.

12, p. 62, pl. 115, figs. 1—8, pl. 116, figs. 1—6, 2003; You *et al.*, Phytotaxa 197(1): 10, figs. 58—66, 2015.

壳面常不同程度扭曲或呈"S"形，稍具背腹性。腹缘平直或稍凸出，背缘平直或呈不同程度的弯曲，壳面两端逐渐变窄呈锥形，末端延长，呈小头状；长 84.5—97.5 μm，宽 7.0—9 μm；龙骨突纤细，6—8 个/10 μm，中间两个龙骨突稍宽，一般每个龙骨突与 1 条横肋纹相连；线纹有 15—17 条/10 μm。

扫描电镜下观察：在外壳面，可见组成线纹的小圆孔，41—42 个/10 μm，基本平行排列，孔纹位置明显低于两侧的肋纹。

生境：生于稻田水坑中。

国内分布：新疆(莎车)。

国外分布：欧洲(意大利)。

采自新疆莎车县迪汉其乡的标本，在壳面形态大小、龙骨突形态和密度上，与 Lange-Bertalot 等(2003)报道的 *H. nitzschioides* 吻合，差别在于线纹密度较小，16—17 条/10 μm，而 Lange-Bertalot 等(2003)的标本线纹密度 21 条/10 μm。

4. 显点菱板藻　图版 XXXV：1

Hantzschia distinctepunctata Hustedt, *in* Schmidt *et al.*, Atlas Diat.-Kunde, p. 329, fig. 329: 21, 1921; Krammer & Lange-Bertalot, Bacillariophyceae, 2. Teil, *in* Ettl *et al.*, Süßwasserfl. Mitteleur., p. 131, figs. 88: 8—10, 1988, nachdr. 1997; Metzeltin *et al.*, Iconogr. Diatomol. 15, figs. 217: 3—7, 2005.

带面观一般为矩形，圆角；壳面具背腹性，形态变化大；长 56 μm，宽 6.5 μm；龙骨突纤细，不容易辨认，8—10 个/10 μm；线纹由粗糙的孔纹组成，11—13 条/10 μm。

生境：生于池塘中。

国内分布：新疆(巴楚)。

国外分布：国外分布：亚洲(俄罗斯)，北美洲(美国)，南美洲(乌拉圭)，欧洲(巴西)，大洋洲(澳大利亚)。

本种与 *H. virgata* 在形态结构上相似，主要区别在于龙骨突的密度较大。

5. 直菱板藻　图版 XXXV：9—10

Hantzschia virgata (Roper) Grunow *in* Cleve & Grunow, Beiträge zur Kenntniss der arctischen Diatomeen, p. 104, 1880; Krammer & Lange-Bertalot, Bacillariophyceae, *in* Süßwasserfl. Mitteleur. Band 2/2, p. 130, figs. 90: 1—8, 1988, nachdr. 1997; Metzeltin *et al.*, Iconogr. Diatomol. 15, fig. 215: 15, 2005.

Nitzschia virgata Roper, Notes on some new species and varieties of British marine Diatomaceae, p. 23, pl. 3, fig. 6, 1858.

带面观和壳面观的形态变化较大，带面观一般为矩形，圆角；壳面具背腹性，末端形态变化大；长 40—60 μm，宽 6—8 μm；具中央节，龙骨突纤细，稍延长，4—7 条/10 μm；线纹有粗糙的孔纹组成，10—14 条/10 μm。

生境：生于湖泊中。

国内分布：新疆(博湖)。

国外分布：亚洲(俄罗斯、蒙古国、新加坡)，北美洲(美国、五大湖)，南美洲(巴西、哥伦比亚、乌拉圭)，欧洲(波罗的海、英国、德国、罗马尼亚、西班牙)，大洋洲(澳大利亚、新西兰)。

6. 中华菱板藻　图版 XLIX：10—17；图版 LI：1—4

Hantzschia sinensis Q-M You & J. P. Kociolek，The diatom genus *Hantzschia* (Bacillariophyta) in Xinjiang province，China. Phytotaxa 197(1)：4，figs. 14—25，2015.

壳面的背腹性不明显，背腹两侧几乎平行，中部稍凹入，靠近两端变窄，略延伸，呈小头状，不弯向任何一侧；长 50—90 μm，宽 8—10 μm；龙骨突纤细，8—9 条/10 μm，中间两个龙骨突略增宽；横线纹有 16—18 条/10 μm。

扫描电镜下观察：在外壳面，组成线纹的小圆孔清晰可见，20—36 个/10 μm，基本平行排列，壳缝在壳面中部连续，无中缝端，在末端呈钩状，弯向背侧。在内壳面，孔纹的形态及排列与外壳面相似，龙骨突窄，每个龙骨突与 1 条(偶见 2—4 条)肋纹相连，壳缝具中缝端，裂缝简单，略弯向背侧，壳面末端具明显的螺旋舌结构。

生境：生于稻田边水坑、路边小水坑及河边湿地中。

国内分布：新疆(阿克苏、莎车、伊犁河)。

本种与 Zidarova 等(2010)报道的 *Hantzschia acuticapitata* 在形态和大小上最相似，区别在于 *H. acuticapitata* 的体积略小，壳体长 57—67 μm，宽 7.8—9.1 μm，且 *H. acuticapitata* 的线纹和孔纹均明显密集，线纹 22—23 条/10 μm，孔纹约 50 个/10 μm。

7. 长菱板藻　图版 XL：1

Hantzschia elongata (Hantzsch) Grunow，New diatoms from Honduras，p. 174，1877；*in* Schmidt *et al.*，Atlas Diat.-Kunde，p. 329，fig. 329：1，1921；Hustedt，Bacillariophyta，*in* Pascher，Süßwass -Fl. Mitteleur. Heft 10，p. 395，fig. 751，1930；Krammer & Lange-Bertalot，Bacillariophyceae，2. Teil，*in* Ettl *et al.*，Süßwasserfl. Mitteleur.，p. 128，figs. 89：1—2A，1988，nachdr. 1997；Zhu & Chen，Bacill. Xiz. Plat.，p. 257，fig. 51：1，2000.

Nitzschia vivax var. *elongata* Hantzsch，Hedwigia 2(7)：35，pl. 6，fig. 5，1860.

Hantzschia amphioxys var. *elongata* (Hantzsch). Protic，Wissenschaftliche Mittheilungen aus Bosnien und der Hercegovina，6：717，1899.

壳体具背腹性，狭长，稍呈膝状弯曲，壳面朝两端逐渐变窄呈锥形，末端延长呈尖喙状或近头状；长 180—405 μm，宽 11—13 μm；龙骨突纤细，6—8 个/10 μm，中间两个龙骨突距离较宽，每个龙骨突与 1—3 条肋纹相连；横线纹有 17—18 条/10 μm。

生境：生于湖泊、河流、沼泽中。

国内分布：内蒙古(阿尔山)，辽宁(辽河)，黑龙江(鸡西兴凯湖)，江苏(苏州)，福建(鼓岭)，湖南(索溪峪、沅江流域)，四川(九寨沟)，贵州(沅江流域)，西藏(亚东、吉隆、仲巴、察隅、芒康、申扎、札达、普兰)。

国外分布：亚洲(俄罗斯)，北美洲(美国)，欧洲(英国、德国、爱尔兰、马其顿、罗马尼亚)，大洋洲(澳大利亚、新西兰)。

采自内蒙古的标本较大，长度可达 405 μm，而西藏的标本长度在 200 μm 左右。

8. 较长菱板藻　图版 XL：2—4

Hantzschia longa Lange-Bertalot, Diatoms of Sardinia, Iconogr. Diatomol. 12：60，figs. 100：1—4，figs. 123：8，9，2003；You *et al.*，Bulletin of Botanical Research 31(2)：130，figs. 14—16，2011.

带面呈不同程度扭曲或是"S"形；壳面背腹性不明显，腹缘稍微凹入，中部几乎平直，背缘稍微弧形凸出，不呈膝状弯曲，壳面朝两端逐渐变窄呈锥形，末端近头状；长 200—270 μm，宽 10—13 μm；龙骨突纤细，6—7 个/10 μm，中间两个龙骨突间距稍宽，每个龙骨突与 1—3 条肋纹相连；线纹 17—18 条/10 μm。

生境：生于沼泽、水坑中。

国内分布：内蒙古(阿尔山)，新疆(喀纳斯)。

国外分布：欧洲(意大利)。

Lange-Bertalot(1993) 报道本种，长 200—250 μm，宽 10—12 μm，而本书中来自新疆的标本体积要稍大一些，其他特征都相同。大兴安岭的标本体积要稍小些，稍微呈膝状弯曲，与模式种有点差别，因此是否为 *H. longa* 还有待确定。

本种与 *H. elongata* 最为相似，在光镜下，仍然能较容易将此种与 *H. elongata* 区分开来，*H. elongata* 具膝状弯曲的壳面，中间两个龙骨突距离较宽。本种与 *H. alkaliphila* 也有些相似，区别在于 *H. alkaliphila* 壳面更宽，线纹粗糙，龙骨突间距大。其他体积较大的种类，如 *H. vivacior*、*H. graciosa*、*H. sardiniensis*、*H. rhaetica* 长度一般不会超过 200 μm，或具其他外形及特征。

9. 巴克豪森菱板藻　图版 XXXIX：1—6

Hantzschia barckhausenii Lange-Bertalot & Metzeltin, Indicators of Oligotrophy, Iconogr. Diatomol. 2：75，figs. 66：16—18，figs. 67：1—2，1996；Lange-Bertalot, Iconogr. Diatomol. 12：53，figs. 118：1—6，fig. 123：1，2003；Metzeltin *et al.*, Iconogr. Diatomol. 15，p. 98，no fig.，2005；You *et al.*，Bulletin of Botanical Research 31(2)：130，figs. 20—22，2011.

壳面雅致，背腹之分不明显，腹缘稍微凹入，中部几乎平直，背缘弧形凸出，不呈现膝状弯曲，壳面朝两端逐渐变窄呈长喙形，末端头状。长 100—118 μm，宽 7—8 μm；龙骨突有 7—10 个/10 μm，与 1—3 条横肋纹相连，中间两个距离增大；横线纹有 20—23 条/10 μm。

生境：生于湖边沼泽中。

国内分布：内蒙古(阿尔山)，新疆(伊犁河)。

国外分布：南美洲(乌拉圭)，欧洲(德国)。

Lange-Bertalot(1996a)首次报道本种,壳面两端逐渐变窄呈锥形至长喙状,长 90—130 μm,宽 7.5—9.5μm,线纹 20—25 条/10 μm。采自大兴安岭的标本与之稍有差别: 壳面稍窄,两端的形态更尖细,其他大部分特征是相符合的。

10. 伊犁菱板藻 图版 XLIX:1—9;图版 L:1—5

Hantzschia yili Q-M You & J. P. Kociolek, The diatom genus *Hantzschia*(Bacillariophyta) in Xinjiang province, China. Phytotaxa 197(1):6-8, figs. 26—39, 2015.

壳面具明显背腹性,腹缘直或略凹入,背缘凸出,末端延长,呈小头端,直,不弯向任何一侧;长 54—89 μm,宽 7—8 μm;龙骨突纤细,6—9 个/10 μm,中间两个龙骨突间距明显增宽,横线纹有 17—18 条/10 μm。

扫描电镜下观察:在外壳面,线纹由小圆孔组成,36—44 个/10 μm,平行排列,壳缝在壳面中部连续,无中缝端,在末端呈钩状,弯向背侧。在内壳面,孔纹形态及排列与外壳面相似,龙骨突窄,每个龙骨突与(少 3 条)1—2 条肋纹相连,壳缝具中缝端,裂缝简单,略弯向背侧,两端均具明显的螺旋舌结构。

生境:生于河边湿地中。

国内分布:新疆(伊犁河)。

本种形态与 *H. angusta* Lange-Bertalot(2003)相似,光镜下难以区分。不同之处在于后者较窄(宽 6—6.5 μm),线纹密集(19—23 条/10 μm)。

11. 盖斯纳菱板藻 图版 XLI:1—5

Hantzschia giessiana Lange-Bertalot & Rumrich, *in* Lange-Bertalot, 85 neue taxa und über 100 weitere neu definierte Taxa ergänzend zur Süsswasserflora von Mitteleuropa, Vol. 2/1—4. Bibl. Diatomol. 27, p. 80-81, figs. 94:1—6, 95:1—6, 1993;Metzeltin *et al.*, Iconogr. Diatomol. 20, p. 606, 592, pl. 237, figs. 5—6, 2009.

壳面具有明显的背腹性,背侧边缘平直,腹侧边缘略凹入,两端延长呈小头状;壳面长 48—75 μm,宽 8—11 μm;龙骨突 4—7 个/10 μm,横线纹 16—20 条/10 μm。

生境:生于水泡、沼泽、湖泊、水草、丝状藻附着,底栖。

国内分布:四川(若尔盖),新疆(莎车)。

国外分布:非洲(纳米比亚)。

采自新疆莎车的标本,与 Lange-Bertalot(1993)报道的 *H. giessiana* 形态最为相似,模式种体积稍大,长 110 μm,宽 12 μm,横线纹密度略稀疏,12—13 条/10 μm。

12. 密集菱板藻 图版 XXXVI:1—4

Hantzschia compacta(Hustedt)Lange-Bertalot, *in* Lange-Bertalot & Genkal, Diatomeen aus Sibirien, I: Inselln im Arktischen Ozean(Yugorsky-Shar Strait), Iconogr. Diatomol. 6, p. 62, fig. 73:6. 1999.

Hantzschia amphioxys var. *compacta* Hustedt, *in* Hedin, Southern Tibet, Southern Tibet 6(3)p. 145, pl. 10, fig. 42, 1922;Cleve-Euler, Diatomeen schw. Finnland, Teil. V, p. 48, fig. 1419g, 1952; Zhu & Chen, Bacill. Xiz. Plat., p. 256, fig. 50: 9, 2000; Metzeltin *et al.*, Iconogr. Diatomol. 15, figs. 214: 1—5, 2005.

壳面背侧弧形, 腹侧凹入, 两端逐渐狭窄, 末端略呈头状; 长 65—96 μm, 宽 14—16 μm; 龙骨突有 4—6 个/10 μm, 中间两个距离明显增大; 横线纹有 13—15 条/10 μm。

扫描电镜下观察: 龙骨突形态变化较大, 每个龙骨突可与 1—5 条肋纹相连, 横线纹上的点纹延伸入龙骨间距内, 密度多于 30 个/10 μm。壳环带上也具点纹组成的线纹, 密度与壳面上的相同。壳面内部的末端具一小的螺旋舌结构。

生境: 生于湖泊、溪流、小水沟、路边积水、沼泽中。

国内分布: 内蒙古(阿尔山), 辽宁(辽河), 贵州(乌江流域、梵净山), 西藏(墨竹工卡、聂拉木、定日、康马、吉隆、仲巴、昂仁、乃东、八宿), 新疆(赛里木湖、喀纳斯)。

国外分布: 北美洲(加拿大), 南美洲(乌拉圭), 欧洲(中部)。

采自新疆的标本和 *H. amphiosys* 相似, 不同之处在于: 壳面背腹性明显, 末端明显变窄, 线纹密度稀疏, 12—14 条/10 μm。扫描电镜下, 组成横线纹的点纹数目多于 30 个/10 μm。

13. 近强壮菱板藻 图版 XLII: 1—3; 图版 XLIII: 1—6

Hantzschia subrobusta Q-M You & J. P. Kociolek, The diatom genus *Hantzschia* (Bacillariophyta) in Xinjiang province, China. Phytotaxa 197(1): 8—9, figs. 40—46, 2015.

壳面具背腹性, 腹缘中部明显凹入, 端部凸出, 背缘凸出, 末端延长, 呈窄喙状, 稍微弯向背侧; 长 165—205 μm, 宽 12—14 μm; 龙骨突纤细, 4—6 个/10 μm, 中间两个龙骨突间距明显增宽, 横线纹有 14—17 条/10 μm。

扫描电镜下观察: 在外壳面, 线纹由小圆孔组成, 30—38 个/10 μm, 平行排列, 壳缝在壳面中部连续, 无中缝端, 在末端, 壳缝弯向背侧。在内壳面, 孔纹形态及排列与外壳面相似, 龙骨突窄, 每个龙骨突与(少见 3—5 条)1—2 条肋纹相连, 两端均具明显的螺旋舌结构。

生境: 生于河边湿地中。

国内分布: 内蒙古(阿尔山), 新疆(伊犁河)。

本种与 *H. amphioxys* var. *robusta* Østrup(1910) 相似, 不同在于壳面的体积及肋纹形态, 新疆的标本明显窄, Cleve-Euler(1952) 报道的 *H. amphioxys* var. *robusta* 壳体长 176—297 μm, 宽 15—25.5 μm。本种与 Lange-Bertalot 等(2003) 报道的 *Hantzschia alkaliphila* 较相似, 区别在于后者的壳面末端不延伸, 壳体较宽, 14—16 μm, 与龙骨突连接的肋纹数较多, 一般与 3—7 条肋纹相连。

图 XLII: 1—3 为内蒙古阿尔山的标本, 壳体稍窄, 宽 10—12 μm; 横线纹密度较小, 13—15 条/10 μm。

14. 嫌钙菱板藻 图版 XLII: 4—8

Hantzschia calcifuga Reichardt & Lange-Bertalot, *in* Werum & Lange-Bertalot, Diatoms in

Springs from Central Europe and elsewhele under the influence of hydrogeology and anthropogenic impacts, Iconogr. Diatomol. 13, p. 163, figs. 96: 1—6, 97: 1—4, 2004; Liu et al., Journal of Shanghai Normal University(Natural Sciences)43(3): 269, figs. 1—3, 2014.

壳面具背腹性，弓形，腹缘中部略凹入，背缘略凸出，两端逐渐狭窄，末端延长呈小头状，略向背缘弯曲；长 60—145 μm，宽 8.5—12 μm。龙骨突有 4—6 个/10 μm，横线纹有 13—18 个/10 μm。

扫描电镜下观察：在外壳面，线纹由小圆孔纹组成，单列。壳面靠近龙骨处具一排圆形点纹，同壳面点纹不相连。在内壳面，组成线纹的孔纹明显低于两侧肋纹，龙骨突宽窄不一，每个龙骨突与 1—4 条肋纹相连，两端均具明显的螺旋舌结构。

生境：生于湖泊沿岸带、石塘、溪流、沼泽中。

国内分布：内蒙古(阿尔山、牙克石)。

国外分布：欧洲(波兰)。

Werum 和 Lange-Bertalot(2004)描述的 *H. calcifuga*，体积比我国内蒙古的标本小：长 53.3—92.5 μm，宽 6.5—8 μm。但在壳面形态方面非常相似。

15. 活跃菱板藻　图版 XXXVIII：4—7；图版 XLIV：1—7；图版 XLV：1—4

Hantzschia vivacior Lange-Bertalot, 85 neue taxa und über 100 weitere neu definierte Taxa ergänzend zur Süsswasserflora von Mitteleuropa, Vol. 2/1—4. Bibl. Diatomol. 27, p. 86-87, figs. 104: 1—6, 105: 1—7, 106:1-6, 1993; You et al., Phytotaxa 197(1): 9, figs. 47—57, 2015.

壳面背腹性明显，腹缘凹入，背缘弧形凸出，不呈现膝状弯曲，壳面朝两端逐渐变窄呈锥形，末端近头状；长 70—130 μm，宽 9—12.5 μm；龙骨突纤细，中间两个距离增宽，4—6 个/10 μm，线纹 13—16 条/10 μm。

扫描电镜下观察：在外壳面，可见线纹由小圆孔组成，34—38 个/10 μm，基本平行排列，壳缝在壳面中部连续，无中缝端，在末端呈钩状，弯向背侧。在内壳面，线纹由小圆孔组成，形态及排列与外壳面相似，龙骨突窄，每个龙骨突与 1—2 条肋纹相连，两端均具明显的螺旋舌结构。

生境：生于稻田、沼泽中，水草、苔藓、草叶附着。

国内分布：河北(唐山)，内蒙古(阿尔山)，四川(若尔盖)，新疆(莎车、阿克苏)。

国外分布：北美洲(美国)，南美洲(哥伦比亚)，欧洲(德国、波兰)，大洋洲(新西兰)。

16. 拟巴德菱板藻　图版 XLVII：1—7；图版 XLVIII：1—5

Hantzschia pseudobardii Q-M You & J. P. Kociolek, The diatom genus Hantzschia (Bacillariophyta)in Xinjiang province, China. Phytotaxa 197(1): 3, figs. 1—13, 2015.

壳面具背腹性，腹缘中部略凹入，端部略凸出，背缘略凸出，近端处具明显的"肩部",末端延长,呈窄喙状或小头状,稍微弯向背侧;长 105—125 μm,宽 9—11 μm;龙骨突纤细,5—7 个/10 μm,中间两个龙骨突间距较宽,横线纹有 16—18 条/10 μm。

扫描电镜下观察:在外壳面,可见线纹由小圆孔组成,26—32 个/10 μm,基本平行排列,仅在中部稍呈辐射状,壳缝在壳面中部连续,无中缝端,在末端呈钩状,弯向背侧。在内壳面,线纹由小圆孔组成,形态及排列与外壳面相似,龙骨突窄,每个龙骨突与 1 条肋纹相连,两端均具明显的螺旋舌结构。

生境:生于路边小水渠、稻田及河边湿地中。

国内分布:新疆(阿克苏、莎车、伊犁河)。

本种与 Lange-Bertalot 等(2003)报道的 *Hantzschia bardii* 在形态上最为相似,差别也比较明显:本种体积较大,与 *H. bardii* 的壳体长度(40—80 μm)没有重叠,本种背缘靠近末端的弧度(肩部)更明显,线纹和孔纹的密度更稀疏(*H. bardii* 的线纹 20—24 条/10 μm,孔纹约 40 个/10 μm)。

17. 近石生菱板藻 图版 XXXVII:1—3

Hantzschia subrupestris Lange-Bertalot,85 neue taxa und über 100 weitere neu definierte Taxa ergänzend zur Süsswasserflora von Mitteleuropa,Vol. 2/1—4. Bibl. Diatomol. 27,p. 85,figs. 91:1—6,92:2—7,93:1—6,1993;Metzeltin *et al.*,Iconogr. Diatomol. 20,p. 602,604,pl. 235,figs. 4—7,pl. 236,2—4,2009.

壳面具有明显的背腹性,背侧边缘凸出,中部略凹入,腹侧边缘略凸起,两端延长呈小头状;壳面长 70—88 μm,宽 11—12 μm;龙骨突 4—5 个/10 μm,横线纹 15—16 条/10 μm。

生境:生于湖泊、水泡中,水草、丝状藻附着。

国内分布:四川(若尔盖)。

国外分布:北美洲(美国),欧洲(德国)。

18. 丰富菱板藻 图版 XXXVI:5—10;图版 XXXVII:4—6

Hantzschia abundans Lange-Bertalot,85 neue taxa und über 100 weitere neu definierte Taxa ergänzend zur Süsswasserflora von Mitteleuropa,Vol. 2/1—4. Bibl. Diatomol. 27,p. 75,figs. 85:12—17,figs. 89:1—6,figs. 90:1—6,1993;Lange-Bertalot & Metzeltin,Iconogr. Diatomol. 2,fig. 66:13,1996;Metzeltin *et al.*,Iconogr. Diatomol. 15,figs. 210:6—8,figs. 212:1—3,fig. 213:6,2005;Antoniades *et al.*,Iconogr. Diatomol. 17:155,figs. 72:11—16,2008;You *et al.*,Bulletin of Botanical Research 31(2):129,figs. 1—4,2011.

壳面弓形,背侧略凸出,腹侧凹入,两端逐渐狭窄,末端呈小头状;长 30—80 μm,宽 6—10 μm;龙骨突有 5—8 个/10 μm;横线纹有 13—20 条/10 μm。

扫描电镜下观察:壳面靠近龙骨处有一排孔,中缝端为圆孔,极缝端弯曲,远离龙骨,靠近壳套;线纹单排,中间辐射状,两端平行;龙骨突发育良好,沿着壳面边缘均匀排列;中间两个龙骨突典型增厚,每个龙骨突与多个指状延伸分离的肋纹相连。

生境：生于沼泽、小水渠、路边积水中。

国内分布：江苏（昆山、常熟），安徽（黄山、宁国），四川（若尔盖），贵州（赫章、水城），西藏（墨竹工卡、林芝、边坝、浪卡子），新疆（温宿、巴楚、博乐五一水库、察布查尔、布尔津、喀纳斯）。

国外分布：北美洲（加拿大、美国），南美洲（乌拉圭），欧洲（德国、马其顿、波兰），南极洲。

本种与 *H. amphioxys* 的形态较为相似，主要区别在于本种体积较大，线纹密度较稀疏。Lange-Bertalot(1993)报道 *H. amphioxys*：长 15—50 μm，宽 5—7 μm，横线纹有 20—29 条/10 μm。

19. 两尖菱板藻

Hantzschia amphioxys (Ehrenberg) Grunow, *in* Cleve & Grunow, Beiträge zur Kenntniss der arctischen Diatomeen, p. 103, 1880; Schmidt *et al.*, Atlas Diat.-Kunde, p. 329, figs. 329: 15—20, 1921; Hustedt, Bacillariophyta, *in* Pascher, Süßwass -Fl. Mitteleur. Heft 10, p. 394, fig. 747, 1930; Cleve-Euler, Diatomeen schw. Finnland, Teil. V, p. 46, figs. 1419a—c, 1952; Gasse, Bibl. Diatomol. 11, p. 70, fig. XXXI: 1, 1986; Zhu & Chen, The Diatoms of the Suoxiyu Nature Reserve Area, Hunan, China, *in* Li *et al.*, The algal flora and aquatic fauna of the Wulingyuan Nature Reserve Area, Hunan, China. p. 55, 1989b; Krammer & Lange-Bertalot, Bacillariophyceae, 2. Teil, *in* Ettl *et al.*, Süßwasserfl. Mitteleur., p. 128, figs. 88: 1—7, 1988, nachdr. 1997; Xin *et al.*, Journal of Shanxi University (Nat. Sci. Ed.) 20(1): 105, 1997; Zhu & Chen, Bacill. Xiz. Plat., p. 254, fig. 50: 3, 2000; Lange-Bertalot *et al.*, Iconogr. Diatomol. 12, figs. 101: 1—4; figs. 102: 1—5, 2003; Metzeltin *et al.*, Iconogr. Diatomol. 15, figs. 214: 4, 10—17, 2005.

Eunotia amphioxys Ehrenberg, Verbreitung und Einfluss des mikroskopischen Lebens in Süd-und Nord-Amerika, p. 413(125), pl. 1/1, fig. 26, pl. 1/3, fig. 6, 1843.

19a. 原变种　图版 XXXIV：1—12
var. amphioxys

壳面弓形，背侧略凸出，腹侧凹入，两端显著逐渐狭窄，末端钝尖，呈喙状，长 24—60 μm，宽 5—10 μm，龙骨突有 5—10 个/10 μm，横线纹有 15—24 条/10 μm。

扫描电镜下观察：壳缝管位于龙骨上，龙骨位于壳面一侧，龙骨与壳面齐平或稍高于壳面；壳面具横线纹，单排，由孔纹组成；在内壳面，龙骨突为短柱状，桥接壳缝管，一般不延伸到壳面内部，1 个或 2 个（或更多）融合在一起，5—10 个/10 μm；中间两个龙骨突距离增大，其余龙骨突间距会稍小于龙骨突本身的宽度；内壳面也具由点纹组成的线纹，肋纹之间的硅质片层低于肋纹，使得点纹深陷，不易观察到；在外壳面，壳缝在壳面中部断开，内壳面，壳缝在中部的结构不易观察到；外壳面远缝端的壳缝裂缝，在螺旋舌的上方突然弯向远缘，形成一个钩状，最终与远缘平行，内壳面远缝端的壳缝裂缝位于螺旋舌上；没有观察到带面结构。

生境：生于湖泊、湖边渗出水、路边积水、沼泽中。

国内分布：北京，河北(唐山)，山西(太原晋祠)，内蒙古(满洲里扎赉诺尔、阿尔山)，吉林(长白山)，黑龙江(鸡西兴凯湖、五大连池)，上海(松江)，江苏(苏州)，浙江(杭州西湖)，安徽(黄山、宁国)，福建(厦门)，山东(泰山)，湖北(神农架)，湖南(长沙、岳阳、索溪峪、沅江流域)，广州(惠州)，广西(灵川)，四川(九寨沟、若尔盖)，贵州(沅江流域、乌江流域、梵净山)，云南(大理)，西藏(聂拉木、定日、亚东、康马、吉隆、萨嘎、仲巴、昂仁、墨脱、米林、林芝、乃东、朗县、加查、错那、隆子、措美、察隅、八宿、波密、昌都、芒康、察雅、洛隆、江达、贡觉、类乌齐、班戈、申扎、措勤、札达、普兰、噶尔、革吉、日土)，陕西(华山)，宁夏(贺兰山)，新疆(博乐、赛里木湖、布尔津、喀纳斯、哈巴河、福海、阿勒泰)。

国外分布：亚洲(俄罗斯、蒙古国、土耳其、尼泊尔)，非洲(东部)，北美洲(加拿大、美国、夏威夷群岛)，南美洲(巴西、哥伦比亚、乌拉圭)，欧洲(波罗的海、黑海、英国、比利时、德国、爱尔兰、马其顿、波兰、罗马尼亚、西班牙)，大洋洲(澳大利亚、新西兰)，南极洲。

本种的形态变化比较大。Mann(1978)报道了本属种类中，组成线纹的孔都有膜封闭，而且膜的位置也不一样。Lange-Bertalot(1993)指定了 *H. amphioxys* 的模式标本，这样大大缩小了本种的界限。*H. amphioxys* 的新界限也改变了 *H. abundans* 的分类。

19b. 相等变种　图版 XXXV：3—8

var. **aequalis** Cleve-Euler, Diatomeen schw. Finnland, Teil. V, p. 51, figs. 1420t—u, 1952; Zhu & Chen, Bacill. Xiz. Plat., p. 255, fig. 50: 5, 2000.

光镜下，与原变种的区别在于：壳面背腹性不明显，两侧几乎平行；长 20—48 μm，宽 3—8 μm；龙骨突有 7—10 个/10 μm；横线纹有 15—30 条/10 μm。

生境：生于湖边渗出水、草地渗出水、小水渠、沼泽中。

国内分布：内蒙古(阿尔山)，四川(若尔盖)，福建(金门)，西藏(聂拉木、定日、吉隆、萨嘎、昂仁、乃东、芒康、江达)，新疆(赛里木湖、察布查尔、喀纳斯、阿勒泰)。

国外分布：欧洲(中部)。

19c. 头端变型　图版 XXXV：2

f. **capitata** O. Müller, Bacillariaceen aus SüdPatagonien, p. 34, pl. II, fig. 26, 1909; Hustedt, Bacillariophyta, in Pascher, Süßwass -Fl. Mitteleur. Heft 10, p. 394, fig. 748, 1930; Cleve-Euler, Diatomeen schw. Finnland, Teil. V, p. 49, fig. 1419t, 1952; Zhu & Chen, The Diatoms of the Suoxiyu Nature Reserve Area, Hunan, China, in Li et al., The algal flora and aquatic fauna of the Wulingyuan Nature Reserve Area, Hunan, China. p. 55, 1989b; Zhu & Chen, Bacill. Xiz. Plat., p. 254, fig. 50: 4, 2000.

光镜下，与原变种的区别在于：壳面弓形，细胞末端渐狭呈头状；长 24—81 μm，宽

4—9 μm；龙骨突有 6—11 个/10 μm；横线纹有 15—24 条/10 μm。

生境：生于小水渠、路边积水、沼泽中。

国内分布：吉林(长白山)，黑龙江(五大连池、鸡西兴凯湖)，福建(金门)，湖北(神农架)，湖南(索溪峪、沅江流域)，贵州(威宁草海、沅江流域、乌江流域)，云南(大理)，西藏(聂拉木、定日、吉隆、亚东、昂仁、工布江达、林芝、乃东、措美、波密、芒康、江达、贡觉、申扎、改则、札达、革吉)，新疆(巴楚、博乐、察布查尔)。

国外分布：亚洲(蒙古国、新加坡)，北美洲(美国)，南美洲(巴西)，欧洲(波罗的海、黑海、英国、马其顿)，大洋洲(澳大利亚、新西兰)。

盘杆藻属 Tryblionella W. Smith

W. Smith，Syn. Brit. Diat. 1：35，1853.

单细胞，常示壳面观，色素体 2 个，分别位于中央横切面的两侧。

壳面宽大，椭圆形、线形或提琴形，末端钝圆或尖形。外壳面常具瘤或脊，波状，一侧具龙骨壳缝系统(keeled raphe system)，另一侧边缘常具脊(ridge)，与非常浅的壳套相连。线纹单排至多排，通常被一至多条腹板(sterna)断开，线纹由小圆孔组成，孔外侧多由膜(hymen)封闭，罕见蜂窝状(alveolate)圆孔。壳缝系统靠近壳面边缘，上下壳面的壳缝关于壳面呈对角线对称('nitzschioid' symmetry)。具龙骨和龙骨突。外壳面中缝端非常近，稍微膨大或偏转；偶尔中缝端缺失。内壳面中缝端位于双螺旋舌上。极缝端裂缝短，偏转。龙骨突扁块状，顶轴方向常比横轴方向宽。带面窄，平滑或具稀疏的孔，由断开的环带组成。

本属种类体积较大，分布广泛，常见于高电导率的淡水中，咸水和海水中不常见。附着于沉积物或污泥中。

本属与沙网藻属(*Psammodictyon*)和菱形藻属(*Nitzschia*)关系比较近，目前不能确定是否为单起源，许多种类是从菱形藻属(*Nitzschia*)内的 *Tryblionella* 组、*Circumsutae* 组、*Apiculatae* 组和 *Pseudotryblionella* 组中分离出来的。

模式种：尖锥盘杆藻(*Tryblionella acuminata* W. Smith)。

本志收编 10 种 1 变种。

盘杆藻属分种检索表

1. 壳缝不连续，具中缝端，但不易观察到 ··· 3
1. 壳缝连续，不具中缝端，龙骨突和横肋纹数目一致，因此两者很难区分 ·············· 2
  2. 壳面较长，线形，宽一般大于 7 μm ································· **1. 渐窄盘杆藻 *T. angustata***
  2. 壳面线形-披针形、披针形，宽 6—7 μm ······················· **2. 狭窄盘杆藻 *T. angustatula***
3. 龙骨突容易分辨 ··· 5
3. 龙骨突和横肋纹数目一致，因此两者很难区分 ··· 4
  4. 壳面宽度大于 10 μm ··· **3. 尖锥盘杆藻 *T. acuminata***
  4. 壳面宽度小于 9 μm ··· **4. 细尖盘杆藻 *T. apiculata***
5. 线纹一般由 2 排孔纹组成，在壳面中部被纵向褶曲断开 ·············· **5. 匈牙利盘杆藻 *T. hungarica***

5. 组成线纹的孔纹较精细，偶尔会断开。有时线纹是非常难辨认的 ··· 6
    6. 壳面较大，横肋纹隐约可见，或只在一侧或两侧边缘可见 ············ **6. 岸边盘杆藻 *T. littoralis***
    6. 壳面较小，横肋纹呈阶梯状，连续，或被纵向的褶曲打断，或彼此移位 ··· 7
7. 壳面较窄(相对于长度)，8—11 μm，横肋纹较窄，密度大 ··
·· **7. 暖温盘杆藻 *T. calida***
7. 壳面较宽，一般大于 11 μm，横肋纹较宽，密度小 ··· 8
    8. 壳体较大，横肋纹不明显 ···································· **8. 细长盘杆藻 *T. gracilis***
    8. 壳体较小，横肋纹粗糙，明显 ··· 9
9. 体积较小，横肋纹密度较大 ·· **9. 莱维迪盘杆藻 *T. levidensis***
9. 体积较大，横肋纹密度较小 ·· **10. 维多利亚盘杆藻 *T. victoriae***

1. 渐窄盘杆藻

Tryblionella angustata W. Smith, A synopsis of the British Diatomaceae, p. 36, pl. 30, fig. 262, 1853.

Nitzschia angustata (W. Smith) Grunow, Bemerkungen zu den Diatomeen von Finnmark, dem Karischen Meere und vom Jenissey nebst Vorarbeiten für Monographie der Gattungen *Nitzschia*, *Achnanthes*, *Pleurosigma*, *Amphiprora*, *Plagiotropis*, *Hyalodiscus*, *Podosira* und einiger *Navicula*-Gruppen, p. 70, 1880; Schmidt *et al.*, Atlas Diat.-Kunde, p. 331, pl. 331, figs. 41—43, 1921; Hustedt, Bacillariophyta, *in* Pascher, Süßwass -Fl. Mitteleur. Heft 10, p. 402, fig. 767, 1930; Cleve-Euler, Diatomeen schw. Finnland. Teil. V, p. 59, figs. 1432a—d, 1952; Krammer & Lange-Bertalot, Bacillariophyceae. 2. Teil, *in* Ettl *et al.*, Süßwasserfl. Mitteleur., p. 48, figs. 36: 1—5, 1988, nachdr. 1997; Zhu & Chen, Bacill. Xiz. Plat., p. 258, fig. 51: 7, 2000.

1a. 原变种　　图版 LII：1—10

var. angustata

    壳面线形至线形-披针形，两侧中部平直或稍凹入，朝两端楔形变窄，末端钝圆。壳面长 54—152 μm，宽 6—11 μm，龙骨突不明显，密度和横线纹的相同，11—14 个(条)/10 μm。纵向线形的褶曲明显或不可见，有横线纹穿过。

    扫描电镜下观察：外壳面稍呈波曲，横肋纹窄片状，明显高于两侧的线纹，线纹由粗糙的孔纹组成，孔纹 23—27 个/10 μm。

    生境：生于湖泊、湖边渗出水、路边积水、沼泽中。

    国内分布：山西(太原晋阳湖)，辽宁(辽河)，江苏(苏州)，海南(崖县、琼山、陵水)，西藏(亚东、吉隆、仲巴、错那、措美、昌都、洛隆、江达、班戈、申扎)，陕西(华山)，新疆(阜康天池、赛里木湖、伊宁、察布查尔、喀纳斯)。

    国外分布：亚洲(俄罗斯、蒙古国、伊朗、巴基斯坦、土耳其)，非洲(东部)，北美洲(加拿大、美国、夏威夷群岛)，南美洲(阿根廷)，欧洲(波罗的海、黑海、英国、德国、爱尔兰、马其顿、罗马尼亚、波兰、西班牙)，大洋洲(澳大利亚、新西兰)。

1b. 尖变种　图版 LIII：9

var. **acuta** (Grunow) Bukhtiyarova, Algologia 5(4)：422，1995.

Nitzschia angustata var. *acuta* Grunow, *in* Cleve & Grunow, Beiträge zur Kenntniss der arctischen Diatomeen, p. 70, 1880; Schmidt *et al.*, Atlas Diat.-Kunde, p. 331, figs. 331：44—45, 1904; Hustedt, Bacillariophyta, Pascher, Süßwass -Fl. Mitteleur. Heft 10, p. 402, fig. 769, 1930; Cleve-Euler, Diatomeen schw. Finnland, Teil. V, p. 59, figs. 1432 e—h, 1952; Zhu & Chen, Bacill. Xiz. Plat., p. 258, fig. 51：8, 2000; Antoniades *et al.*, Iconogr. Diatomol. 17, p. 210, figs. 73：17—18, 2008.

　　光镜下与原变种的主要区别：壳面披针形，末端尖圆。

　　生境：生于路边小水沟中。

　　国内分布：湖南(沅江流域)，贵州(沅江流域)，西藏(萨嘎、林芝、芒康、洛隆、察雅、革吉、噶尔)，新疆(阿克陶)。

　　国外分布：亚洲(蒙古国)，北美洲(加拿大、美国)，欧洲(马其顿)。

2. 狭窄盘杆藻　图版 LIII：1—6

Tryblionella angustatula (Lange-Bertalot) Q-M You & Q-X Wang nov. comb.

Nitzschia angustatula Lange-Bertalot, p. 6, pl. 18, figs. 1—4 (as "*angustulata*"), 1987; Krammer & Lange-Bertalot, Bacillariophyceae. 2. Teil, *in* Ettl *et al.*, Süßwasserfl. Mitteleur., p. 48, figs. 36：6—10, fig. 3：6, 1988, nachdr. 1997.

　　壳面线形至线形-披针形，朝两端呈喙状延伸，末端尖圆。壳面长 20—65 μm，宽 6—7 μm，龙骨突不明显，密度和横线纹的相同，14—17 个(条)/10 μm。

　　扫描电镜下观察：外壳面稍呈波曲状，横肋纹明显高于两侧的线纹，线纹由粗糙的孔纹组成，孔纹 13—15 个/10 μm。

　　生境：生于湖泊、河流、沼泽、小水渠、路边积水中。

　　国内分布：上海(松江)，西藏(林芝)、新疆(皮山、阿克苏、新和、岳普湖、阿克陶、和静、尉犁、博湖、阜康天池、博乐、伊宁、察布查尔、布尔津)。

　　国外分布：北美洲(美国)，欧洲(中部)。

　　本种与 *Tryblionella angustata* var. *acuta* 在形态和结构上比较相似，主要区别在于：本种的体积较小，扫描电镜下，肋纹和孔纹的结构差别较大。

3. 尖锥盘杆藻　图版 LIII：7—8

Tryblionella acuminata W. Smith, A synopsis of the British Diatomaceae, p. 36, fig. 10：77, 1853; Round *et al.*, The Diatoms, p. 679, 1990; Antoniades *et al.*, Iconogr. Diatomol. 17, p. 301, fig. 76：7, 2008.

Nitzschia acuminata (W. Smith) Grunow, Algen und Diatomaceen aus dem Kaspischen Meere, p. 118, 1878; Krammer & Lange-Bertalot, Bacillariophyceae. 2. Teil, *in* Ettl *et al.*, Süßwasserfl. Mitteleur. p. 44, figs. 34：4—6, 1988, nachdr. 1997; Xie *et al.*, Journal of Shanxi University (Nat. Sci. Ed.) 14(4)：416, 1991; Metzeltin *et al.*, Iconogr.

Diatomol. 15, figs. 194: 3, 4, fig. 195: 1, 2005.

壳面宽线形，两侧中部稍凹入，朝两端楔形变窄，末端尖圆。壳面长56—68 μm，宽10—11 μm，龙骨突和横线纹密度相等，13—16个(条)/10 μm。纵向线形的褶曲部分呈透明状或是隐约可见有横线纹穿过。

生境：生于湖泊中。

国内分布：天津，山西(太原晋阳湖)，贵州(水城)，新疆(博湖)。

国外分布：亚洲(俄罗斯、以色列)，非洲(东部)，北美洲(加拿大、美国)，南美洲(巴西、哥伦比亚、乌拉圭)，欧洲(亚得里亚海、波罗的海、黑海、英国、爱尔兰、罗马尼亚、西班牙)，大洋洲(澳大利亚、新西兰)。

4. 细尖盘杆藻 图版 LIV：1—14

Tryblionella apiculata Gregory, On the post-Tertiary diatomaceous sand of Glenshira, p. 79, pl. 1, fig. 43, 1857; Round et al., The Diatoms, p. 679, 1990.

Nitzschia apiculata (Gregory) Grunow, Algen und Diatomaceen aus dem Kaspischen Meere, p. 118, 1878.

Synedra constricta Kützing, 1844;

Nitzschia constricta (Kützing) Ralfs, in Pritchard, p. 780, 1861; Krammer & Lange-Bertalot, Bacillariophyceae. 2. Teil, in Ettl et al., Süßwasserfl. Mitteleur., p. 43, figs. 35: 1—6, 1988, nachdr. 1997; Zhu & Chen, Bacill. Xiz. Plat., p. 260, fig. 51: 14, 2000; Metzeltin et al., Iconogr. Diatomol. 15, fig. 194: 15, figs. 195: 3, 4, figs. 196: 1—5, 2005.

壳面线形，两侧中部稍凹入，朝两端楔形变窄，末端轻微喙状。壳面长30—55 μm，宽5—8 μm，龙骨突与横肋纹相连，不容易分辨，两者密度相同，15—18个(条)/10 μm，横线纹看不清楚，壳面中部具一条纵向线形的褶曲(腹板)，呈透明状或是隐约可见有横线纹穿过。

扫描电镜下观察：外壳面稍呈波曲状，壳缝在中部断开。横肋纹表面平整，稍高于两侧的线纹，肋纹之间一般具3排线纹，横线纹有32—36条/10 μm，由点纹组成，点纹密集，约60个/10 μm。壳面中部具纵向褶曲(腹板)，表面常具小颗粒，此处线纹被断开或隐约穿过。

生境：生于湖边渗出水、路边积水、沼泽，水草附生。

国内分布：山西(太原晋阳湖、运城)，黑龙江(五大连池)，广东(紫金)，西藏(曲水、定结、吉隆、错那、芒康、察雅、贡觉、措勤、噶尔、革吉)，新疆(乌鲁木齐盐湖、阿克苏、博湖、博乐、赛里木湖、察布查尔、布尔津、喀纳斯、哈巴河)。

国外分布：亚洲(俄罗斯、蒙古国、土耳其、新加坡)，非洲(东部)，北美洲(美国)，南美洲(乌拉圭)，欧洲(波罗的海、黑海、英国、马其顿、波兰、罗马尼亚、西班牙)，大洋洲(澳大利亚、新西兰)。

Metzeltin 等(2005)提供了本种壳面较为清晰的扫描电镜照片，本书观察的结果与其相吻合，特别是线纹的密度和肋纹之间线纹的密度。

5. 匈牙利盘杆藻　图版 LV：1—8；图版 LVI：1—2

Tryblionella hungarica (Grunow) Frenguelli, Rev. Mus. La Plata, Nueva Serie, Sección Botánica, 5(20)：178，pl. 8，fig. 12，1942：，Round *et al.*, The Diatoms, p. 678, 1990.

Nitzschia hungarica Grunow, Verh. zool. bot. Ges. Wien, p. 568, pl. 28/12, fig. 31, 1862; Schmidt *et al.*, Atlas Diat.-Kunde, p. 331, figs. 331：6—13，1904；Hustedt, Bacillariophyta, *in* Pascher, Süßwass -Fl. Mitteleur. Heft 10, p. 401, fig. 766, 1930; Cleve-Euler, Diatomeen schw. Finnland, Teil. V, p. 61, fig. 1435, 1952; Krammer & Lange-Bertalot, Bacillariophyceae. 2. Teil, *in* Ettl *et al.*, Süßwasserfl. Mitteleur., p. 42, figs. 34：1—3，1988，nachdr. 1997；Xin *et al.*, Journal of Shanxi University (Nat. Sci. Ed.) 20(1)：105，1997；Zhu & Chen, Bacill. Xiz. Plat., p. 264, fig. 52：21，2000.

壳面线形，两侧中部稍凹入，朝两端楔形变窄，大部分末端轻微喙状。壳面长37—78 μm，宽6—10 μm，龙骨突8—10 个/10 μm，横肋纹15—20 条/10 μm，横线纹看不清楚，壳面中部具一条纵向宽线形的褶曲(腹板)，呈透明状或是隐约可见有横线纹穿过。

扫描电镜下观察：外壳面呈波曲状，壳缝在中部断开。横肋纹表面平整，稍高于两侧的线纹，肋纹之间一般具2排线纹，横线纹有32—36 条/10 μm，由点纹组成，点纹密集，约60 个/10 μm。壳面中部具纵向褶曲(腹板)，表面具小颗粒或光滑，此处线纹被断开或隐约穿过。

生境：生于湖泊、小水渠、路边积水、沼泽、稻田、泉水井边草丛。

国内分布：北京，山西(太原晋祠、运城)，内蒙古(牙克石)，吉林(长白山)，黑龙江(五大连池)，上海(松江)，江苏(昆山)，福建(金门)，湖北(武汉植物园)，湖南(沅江流域)，海南(陵水、海口)，贵州(水城、沅江流域、乌江流域)，西藏(当雄、亚东、康马、吉隆、昂仁、错那、措美、洛隆、贡觉、班戈、申扎、改则、措勤、普兰)，新疆(乌鲁木齐盐湖、哈密、阿克苏、库车、岳普湖、莎车、博湖、阜康天池、博乐、赛里木湖、伊宁、察布查尔、布尔津、喀纳斯、哈巴河、北屯)。

国外分布：亚洲(俄罗斯、蒙古国、伊朗、土耳其、以色列)，非洲(东部)，北美洲(加拿大、美国、夏威夷群岛)，南美洲(巴西、哥伦比亚、乌拉圭)，欧洲(波罗的海、黑海、英国、德国、马其顿、爱尔兰、波兰、罗马尼亚、西班牙)，大洋洲(澳大利亚、新西兰)。

本种与 *Tryblionella apiculata* 在形态结构上比较相似，主要区别在于：本种的体积较大，扫描电镜下，可见肋纹间2排线纹，而 *T. apiculata* 体积较小，扫描电镜下，可见肋纹间3排线纹。Metzeltin 等(2005)报道了本种壳面较为清晰的扫描电镜照片，本书观察的结果与其大部分相吻合，不同之处在于：Metzeltin 等(2005)的标本，壳面肋纹之间的线纹一般由1排点纹组成，偶尔2排，而我们观察的结果，多为2排点纹，偶尔为1排。

6. 岸边盘杆藻　图版 LVII：7—9

Tryblionella littoralis (Grunow) D. G. Mann, Round *et al.*, The Diatoms, p. 678, 1990;

Antoniades *et al.*，Iconogr. Diatomol. 17：303，fig. 76：4，2008.

Nitzschia littoralis Grunow，*in* Cleve & Grunow，Beiträge zur Kenntniss der arctischen Diatomeen，p. 75，1880；Cleve-Euler，Diatomeen schw. Finnland，Teil. V，p. 62，figs. 1438 a，b，d，e，1952；Krammer & Lange-Bertalot，Bacillariophyceae. 2. Teil，*in* Ettl *et al.*，Süßwasserfl. Mitteleur.，p. 41，figs. 31：1—5，1988，nachdr. 1997；Zhu & Chen，Bacill. Xiz. Plat.，p. 265，fig. 53：4，2000.

壳面宽椭圆-披针形至线形-椭圆形，一侧或两侧的中部凹入，朝两端楔形变窄，末端钝圆。壳面长 48—70 μm，宽 15—23 μm，龙骨突 7—10 个/10 μm，横肋纹 11—13 条/10 μm。

生境：生于湖边沼泽、芦苇沟、路边积水，水草附生、泉水井边草丛。

国内分布：海南（崖县），西藏（错那、措美），新疆（乌鲁木齐盐湖、博湖、察布查尔、哈巴河）。

国外分布：亚洲（俄罗斯），北美洲（加拿大、美国），南美洲（巴西、哥伦比亚），欧洲（波罗的海、英国、德国、罗马尼亚）。

7. 暖温盘杆藻 图版 LVII：1—6

Tryblionella calida (Grunow) D. G. Mann，Round *et al.*，The Diatoms，p. 678，1990.

Nitzschia calida Grunow，*in* Cleve & Grunow，Beiträge zur Kenntniss der arctischen Diatomeen，17(2)：75，1880；Krammer & Lange-Bertalot，Bacillariophyceae. 2. Teil，*in* Ettl *et al.*，Süßwasserfl. Mitteleur.，p. 40，figs. 30：1—5，1988，nachdr. 1997；Metzeltin *et al.*，Iconogr. Diatomol. 15，figs. 194：5，6，195：5，199：1，1'，2005.

Nitzschia tryblionella var. *calida* (Grunow) Van Heurck，Synopsis des Diatomées de Belgique，p. 171，1885.

壳面线形，两侧中部平直或稍凹入，朝两端楔形变窄，末端短喙状。壳面长 28—56 μm，宽 8—11 μm，龙骨突不明显，有 7—10 个/10 μm，横肋纹（线纹）连续或是在壳缘被纵向线形区域打断，横肋纹 16—20 条/10 μm，横线纹光镜下看不清楚。

扫描电镜下观察：外壳面稍呈波曲状，壳缝在中部断开，且稍微伸入壳面，使壳缘呈现一小的缺刻。横肋纹较窄，表面平整，稍高于两侧的线纹，肋纹之间的距离不均等，一般具 2 排线纹，线纹紧贴肋纹边缘，有时两条肋纹靠近，只通过中间的 1 排点纹彼此区分开。横线纹有 32—36 条/10 μm，由点纹组成，点纹密集，约 60 个/10 μm。纵向褶曲位于壳面远离壳缝的一侧，稍低于壳面，呈透明状或有横线纹穿过。

生境：生于河流、深沟积水、路边积水、小水渠、沼泽、泉水井边附生、水草附生。

国内分布：辽宁（辽河），黑龙江（哈尔滨），上海（松江），江苏（昆山），湖北（武汉植物园），广西（灵川），新疆（阿克苏、博湖、博乐、察布查尔、布尔津）。

国外分布：亚洲（俄罗斯、蒙古国），北美洲（美国），南美洲（乌拉圭），欧洲（英国、德国、波兰、马其顿、罗马尼亚、西班牙），大洋洲（澳大利亚）。

8. 细长盘杆藻 图版 LVIII：1—4

Tryblionella gracilis W. Smith，A synopsis of the British Diatomaceae，p. 35，pl. 10，fig. 75，

1853; Round *et al.*, The Diatoms, p. 679, 1990.

Nitzschia tryblionella Hantzsch, *in* Rabenhorst, Die Algen Sachsens. Resp. Mittel-Europa's Gesammelt und herausgegeben von Dr. L. Rabenhorst, No. 984, 1848-1860; Schmidt *et al.*, Atlas Diat.-Kunde, p. 332, fig. 332: 14, 1921; Hustedt, Bacillariophyta, *in* Pascher, Süßwass -Fl. Mitteleur. Heft 10, p. 399, fig. 757, 1930; Cleve-Euler, Diatomeen schw. Finnland, Teil. V, p. 57, figs. 1430 a, b, 1952; Krammer & Lange-Bertalot, Bacillariophyceae. 2. Teil, *in* Ettl *et al.*, Süßwasserfl. Mitteleur., p. 37, figs. 27: 1—4, 1988, nachdr. 1997; Metzeltin *et al.*, Iconogr. Diatomol. 15, figs. 632: 1, 2, fig. 203: 1, 2005.

壳面线形椭圆形至线形-披针形，朝两端楔形变窄，末端呈钝圆形，长 57—155 μm，宽 18—25 μm。龙骨突有 5—8 个/10 μm，中间两个龙骨突间距较大，壳缝具中缝端，具清楚的中央节。横线纹看不清楚。壳面具一较大的纵向褶曲（腹板）。

生境：生于湖泊、河流、池塘、路边积水、沼泽、泉水中。

国内分布：云南（洱海、滇池），新疆（哈密、皮山、阿克苏、叶城、岳普湖、疏勒、阿克陶、博湖、博乐、赛里木湖、察布查尔、布尔津、哈巴河、北屯）。

国外分布：亚洲（俄罗斯、蒙古国），非洲（东部）；北美洲（美国、夏威夷群岛）；南美洲（乌拉圭）；欧洲（波罗的海、黑海、英国、芬兰、西班牙）；大洋洲（澳大利亚）。

9. 莱维迪盘杆藻　图版 LIII：10—15

Tryblionella levidensis W. Smith, A synopsis of the British Diatomaceae, p. 89, 1856; Round *et al.*, The Diatoms, p. 679, 1990.

Nitzschia levidensis (W. Smith) Grunow, *in* Van Heurck, Synopsis des Diatomées de Belgique Atlas, fig. 57: 15, 1881; Krammer & Lange-Bertalot, Bacillariophyceae. 2. Teil, *in* Ettl *et al.*, Süßwasserfl. Mitteleur., p. 37, figs. 28: 1—4, 1988, nachdr. 1997; Metzeltin *et al.*, Iconogr. Diatomol. 15, figs. 198: 4—8, 199: 2, 2005.

Nitzschia tryblionella var. *levidensis* (W. Smith) Grunow, *in* Cleve & Grunow, Beiträge zur Kenntniss der arctischen Diatomeen, p. 70, 1880; Schmidt *et al.*, Atlas Diat.-Kunde, p. 331, fig. 331: 20, 1904; Hustedt, Bacillariophyta, *in* Pascher, Süßwass -Fl. Mitteleur. Heft 10, p. 399, fig. 760, 1930; Cleve-Euler, Diatomeen schw. Finnland, Teil. V, p. 58, figs. 1430 i—l, 1952; Zhu & Chen, Bacill. Xiz. Plat., p. 264, fig. 52: 21, 2000.

壳面线形-椭圆形，朝两端楔形变窄，末端呈钝圆形，有时一侧或是两侧中部都凹入，壳面长 18—54 μm，宽 9—14 μm，龙骨突与横肋纹的密度一致，9—14 个（条）/10 μm，横线纹看不清楚。

生境：生于湖泊、河流、路边积水、沼泽中。

国内分布：黑龙江（鸡西兴凯湖），江苏（昆山），安徽（黄山），海南（崖县、陵水、南渡江下游、琼山），西藏（亚东、昂仁、措美、申扎、措勤），宁夏（贺兰山），新疆（哈密、博湖、赛里木湖、察布查尔、布尔津、哈巴河）。

国外分布：亚洲(蒙古国、俄罗斯)，非洲(东部)，北美洲(加拿大、美国、夏威夷群岛)，南美洲(巴西、哥伦比亚、乌拉圭、安第斯山脉)；欧洲(波罗的海、黑海、英国、德国、芬兰、爱尔兰、马其顿、罗马尼亚、西班牙)，大洋洲(澳大利亚、新西兰)。

10. 维多利亚盘杆藻　图版 LVI：3—6

Tryblionella victoriae Grunow, Verh. zool. bot. Ges. Wien, p. 553, pl. 28/12, fig. 34, 1862; Round et al., The Diatoms, p. 679, 1990.

Tryblionella tryblionella var. *victoriae* (Grunow) Radzimowsky, Bemerkung Uber das Phytoplankton im Gestrupp des Sudlichen Bugs, p. 96, 1928.

Nitzschia levidensis var. *victoriae* (Grunow) Cholnoky, Österreichische Botanische Zeitschrift 103：57, 1956; Krammer & Lange-Bertalot, Bacillariophyceae. 2. Teil, in Ettl et al., Süßwasserfl. Mitteleur., p. 38, figs. 29：1—5, 1988, nachdr. 1997; Metzeltin et al., Iconogr. Diatomol. 15, figs. 198：2, 3, fig. 199：3, 2005.

Nitzschia tryblionella var. *victoriae* (Grunow) Grunow, in Cleve & Möller, Diatoms, No. 211, 1879; Hustedt, Bacillariophyta, in Pascher, Süßwass -Fl. Mitteleur. Heft 10, p. 399, fig. 758, 1930; Cleve-Euler, Diatomeen schw. Finnland, Teil. V, p. 58, fig. 1430 f, 1952; Xin et al., Journal of Shanxi University(Nat. Sci. Ed.)20(1)：105, 1997; Zhu & Chen, Bacill. Xiz. Plat., p. 269, fig. 54：9, 2000.

壳面宽线形-椭圆形，朝两端楔形变窄，末端呈钝圆形，有时一侧或是两侧中部都凹入，壳面长 30—65 μm，宽 15—26 μm，龙骨突与横肋纹的密度一致，6—9 个(条)/10 μm，横线纹看不清楚。

生境：生于路边积水、沼泽中。

国内分布：天津；山西(太原晋祠)，黑龙江(鸡西兴凯湖)，安徽(黄山)，福建(金门)，湖南(沅江流域)，广西(灵川)，海南(崖县、琼山)，贵州(沅江流域、乌江流域、梵净山)，西藏(康马、昂仁、林芝、错那、措美、洛隆、班戈、申扎、措勤)，新疆(博湖、察布查尔、布尔津)。

国外分布：亚洲(俄罗斯、新加坡)，北美洲(美国、五大湖)，南美洲(巴西、乌拉圭)，欧洲(波罗的海、黑海、英国、爱尔兰、罗马尼亚、西班牙)，大洋洲(澳大利亚、新西兰)。

本种与 *Tryblionella levidensis* 在形态和结构上非常相似，光镜下很难区分，主要区别在于：本种横肋纹密度较小。Metzeltin 等(2005)提供了这两个种类壳面的扫描电镜照片。

细齿藻属 Denticula F. T. Kützing

F. T. Kützing, Kies. Bacill. Diat. 43, 1844.

细胞小，单细胞或以壳面连接成短链状群体(通过非硅质结构相连)，常示带面观，色素体 2 个，简单，对称排列于中央横切面的两侧，细胞质常高度颗粒化，光镜下多呈亮点。

壳面线形或披针形，偶见菱形或椭圆形，外形上基本左右对称，但结构上不对称。末端尖至钝圆形，或轻微延伸呈喙状。线纹单排或双排，由膜封闭的小圆孔组成，有时可见精细的筛状孔。壳缝系统近中轴或稍离心，当位于宽而低的龙骨上时，壳体稍弯。上下壳面的壳缝关于壳面呈对角线对称。龙骨突块状，包围着壳缝系统，并横向延伸贯穿整个壳面形成隔片(partitions)，隔片之间是由数排点纹组成的横线纹。龙骨突基部增宽，使相邻龙骨突间的椭圆形孔隙变小。壳缝有或无中缝端，如果有中缝端，在内、外壳面都简单。极缝端弯成钩状。带面由一些断开的环带或半环带组成。近壳面的环带上有时会有一排横向孔纹。壳套合部(valvocopula)经常会延伸到龙骨突下形成隔膜结构。带面两侧略凸出，呈线形或长方形，末端截形。可见壳内壁横向平行的隔片(即壳面的横肋纹)；有的种类隔片末端呈头状。

分布于淡水和海水中，底栖。

本属因具横向的隔片(横肋纹)，一直与窗纹藻属(*Epithemia*)和棒杆藻属(*Rhopalodia*)一起放在窗纹藻科(Epithemiaceae)中(Hustedt, 1930; Patrick and Reimer, 1975; Krammer and Lange-Bertalot, 1988)。Round等(1990)根据壳缝系统关于壳面呈对角线对称的特征，以及壳面和色素体的形态，认为本属与菱形藻属(*Nitzschia*)的一些种类，如 *N. sinuata*、*N. denticula* 等关系更近。

模式种：华美细齿藻(*Denticula elegans* Kützing)。

本志收编 5 种 2 变种。

细齿藻属分种检索表

1. 隔片位置较低，需仔细聚焦才能观察到 ················· **1. 库津细齿藻 *D. kuetzingii***
1. 隔片位置高，基本与壳套齐平 ··· 2
　2. 壳体较小，长度一般不超过 20 μm，横肋纹密度大于 8 条/10 μm ······ **2. 科瑞提细齿藻 *D. creticola***
　2. 壳体较大，横肋纹密度小于 6 条/10 μm ··· 3
3. 横线纹密集，25—30 条/10 μm，光镜下不容易观察到 ················· **3. 小型细齿藻 *D. tenuis***
3. 横线纹一般少于 20 条/10 μm，光镜下容易观察到 ··· 4
　4. 壳体较小，横肋纹密度一般大于 4 条/10 μm ············· **4. 华美细齿藻 *D. elegans***
　4. 壳体较大，横肋纹密度一般小于 4 条/10 μm ············· **5. 强壮细齿藻 *D. valida***

1. 库津细齿藻

Denticula kuetzingii Grunow, Verh. zool. bot. Ges. Wien, p. 546, 548, pl. 28/12, fig. 27, 1862; Krammer & Lange-Bertalot, Bacillariophyceae. 2. Teil, *in* Ettl *et al.*, Süßwasserfl. Mitteleur., p. 143, figs. 94: 3, 4; figs. 99: 11—23, figs. 100: 1—14, 1988, nachdr. 1997; Antoniades *et al.*, Iconogr. Diatomol. 17: 213, figs. 72: 4—7, 2008.

Nitzschia denticula Grunow, *in* Cleve & Grunow, Beiträge zur Kenntniss der arctischen Diatomeen, p. 82, 1880; Schmidt *et al.*, Atlas Diat.-Kunde, p. 331, figs. 331: 32—39, 1921; Hustedt, Bacillariophyta, *in* Pascher, Süßwass-Fl. Mitteleur. Heft 10, p. 407, fig. 782, 1930; Cleve-Euler, Diatomeen schw. Finnland, Teil. V, p. 66, figs. 1451 a, b，1952；Zhu & Chen，Bacill. Xiz. Plat.，p. 260，fig. 51：16，2000；Metzeltin *et al.*，Iconogr. Diatomol. 15，figs. 195：10—14，2005.

1a. 原变种　图版 LXII：15—17；图版 LXIII：1—28；图版 LXIV：1—5
var. kuetzingii

壳体带面矩形，壳面线形至披针形，或椭圆形(体积小的细胞)，末端圆形或楔形，有时近喙状。壳面长 13—68 μm，宽 2.5—7.5 μm，龙骨突明显，与横肋纹相连，基本延伸至整个壳面，5—8 条/10 μm，不具中缝端，中间一对龙骨突距离不增大。横线纹由粗糙的点纹组成，14—20 条/10 μm。

扫描电镜下观察：壳缝管位于壳面边缘的龙骨上，壳缝裂缝在中部不断开，在两极弯曲呈钩状。有些个体在不具壳缝的壳面边缘有一硅质片状结构，不同个体形态有差异。横线纹，单排，由粗糙的圆孔组成，16—20 个/10 μm，孔内具分支孔板结构。在内壳面，壳缝管由龙骨突桥接，龙骨突与单一肋纹连接，明显高于两侧的横肋纹，基本延伸至整个壳面。带面矩形，表面平滑，环带由开带组成，开口位于两极。

生境：生于湖泊、湖边渗出水、草地渗出水、溪流、小水渠、浅水滩、沼泽、路边积水中。

国内分布：山西(太原晋祠)，辽宁(辽河)，吉林(长白山)，湖南(沅江流域)，四川(九寨沟)，贵州(沅江流域、乌江流域)，西藏(打隆、聂拉木、亚东、吉隆、仲巴、工布江达、墨脱、米林、林芝、乃东、加查、错那、措美、察隅、八宿、波密、昌都、芒康、察雅、洛隆、江达、贡觉、类乌齐、班戈、申扎、改则、措勤、札达、普兰、噶尔、革吉、日土)，甘肃(苏干湖)，宁夏(贺兰山)，新疆(乌鲁木齐盐湖、哈密、皮山、阿克苏、温宿、新和、叶城、莎车、阿克陶、和静、尉犁、博湖、阜康天池、博乐、赛里木湖、伊宁、察布查尔、布尔津、喀纳斯)。

国外分布：亚洲(俄罗斯、蒙古国、土耳其)，非洲(东部)，北美洲(加拿大、美国)，南美洲(阿根廷、哥伦比亚、乌拉圭)，欧洲(英国、德国、马其顿、波兰、爱尔兰、罗马尼亚、西班牙)，大洋洲(澳大利亚、新西兰)。

1b. 汝牧变种　图版 LXI：30—34

var. **rumrichae** Krammer，*in* Lange-Bertalot & Krammer，Bibl. Diatomol. 15，p. 66，figs. 43：11—14，1987；Krammer & Lange-Bertalot，Bacillariophyceae. 2. Teil，*in* Ettl *et al.*，Süßwasserfl. Mitteleur.，p. 143，figs. 18—22，1988，nachdr. 1997；You & Wang. Acta Botanica Boreali-Occidentalia Sinica 31(2)：419，figs. II：12—17，2011a.

光镜下与原变种的主要区别：此变种的群体体积明显大于原变种，长度可达到 120 μm。

生境：生于湖泊、路边水坑、沼泽中。

国内分布：新疆(温宿、阿克陶)。

国外分布：欧洲(英国)。

2. 科瑞提细齿藻　图版 LXI：11—15

Denticula creticola (Østrup) Lange-Bertalot & Krammer, Observations on Simonsenia and some small species of *Denticula* and *Nitzschia*. Nova Hedwigia Beiheft 106, p. 127, fig. 20—36, 1993.

Nitzschia creticola Østrup, 1910, Danske Diatoméer, p. 146, fig. 4：98 a, b.

壳体小，线形椭圆形，末端钝圆。壳面长 10—18 μm，宽 3—4 μm。横肋纹横贯整个壳面，9—11 条/10 μm，横线纹细密，看不清楚。

生境：生于小水渠中。

国内分布：新疆(阿克苏、新和)。

国外分布：北美洲(加拿大)；欧洲(德国)。

3. 小型细齿藻

Denticula tenuis Kützing, Kies. Bacill. Diat., p. 43, fig. 17：8, 1844; Hustedt, Bacillariophyta, *in* Pascher, Süßwass -Fl. Mitteleur. Heft 10, p. 381, fig. 723, 1930; Cleve-Euler, Diatomeen schwed. Finnland, Teil V, p. 33, fig. 1405, 1952; Krammer & Lange-Bertalot, Bacillariophyceae. 2. Teil, *in* Ettl *et al.*, Süßwasserfl. Mitteleur., p. 139, figs. 95：4—25,(?)figs. 100：15—17, 1988, nachdr. 1997; Xin *et al.*, Journal of Shanxi University(Nat. Sci. Ed.)20(1)：105, 1997; Zhu & Chen, Bacill. Xiz. Plat., p. 247, fig. 48：8, 2000.

3a. 原变种　图版 LXI：7—10, 27—28

var. **tenuis**

壳面线形-披针形，末端尖圆至钝圆，长 15—30 μm，宽 4—5 μm，横肋纹 5—7 条/10 μm，横线纹密集，光镜下不容易看清楚，25—30 条/10 μm。

生境：生于湖泊、山溪中。

国内分布：北京，山西(太原晋祠)，辽宁(本溪)，西藏(定日、亚东、察隅、申扎)，新疆(阿克陶、和静)。

国外分布：亚洲(俄罗斯、蒙古国、土耳其)，北美洲(加拿大、美国、五大湖、夏威夷群岛)，南美洲(哥伦比亚、乌拉圭)，欧洲(挪威、英国、芬兰、德国、爱尔兰、意大利、马其顿、波兰、罗马尼亚、西班牙)，大洋洲(澳大利亚、新西兰)。

3b. 粗变种　图版 LXI：29

var. **crassula** (Nägeli) Hustedt, Bacillariophyta, *in* Pascher, Süßwass -Fl. Mitteleur. Heft 10, p. 381, fig. 724, 1930; Zhu & Chen, The Diatoms of the Suoxiyu Nature Reserve Area, Hunan, China, *in* Li *et al.*, The algal flora and aquatic fauna of the Wulingyuan Nature Reserve Area, Hunan, China. p. 55, 1989b; Xin *et al.*, Journal of Shanxi University(Nat. Sci. Ed.)20(1)：105, 1997; Zhu & Chen, Bacill. Xiz. Plat., p. 247, fig. 48：8, 2000.

光镜下与原变种的主要区别：壳面较粗。

生境：生于湖泊、河流、沼泽中。

国内分布：山西(太原晋祠)，山东(泰山)，湖南(索溪峪、沅江流域)，贵州(沅江流域、乌江流域、梵净山)，西藏(聂拉木、亚东、吉隆、墨脱、米林、林芝、朗县、错那、隆子、昌都、芒康、察雅、洛隆、江达、贡觉、类乌齐、班戈、申扎、札达、噶尔)，陕西(华山)，宁夏(贺兰山)。

国外分布：亚洲(蒙古国)，北美洲(美国)，欧洲(英国、爱尔兰、马其顿、罗马尼亚、西班牙、瑞典)。

4. 华美细齿藻　图版 LXI：20—26

Denticula elegans Kützing，Kies. Bacill. Diat.，p. 44，fig. 17：5，1844；Hustedt，Bacillariophyta，*in* Pascher，Süßwass -Fl. Mitteleur. Heft 10，p. 382，fig. 725，1930；Cleve-Euler，Diatomeen schwed. Finnland，Teil V，p. 33，fig. 1404，1952；Krammer & Lange-Bertalot，Bacillariophyceae. 2. Teil，*in* Ettl *et al.*，Süßwasserfl. Mitteleur.，p. 141，figs. 96：10—33，figs. 97：1—5，1988，nachdr. 1997；Zhu & Chen，Bacill. Xiz. Plat.，p. 247，fig. 48：7，2000.

壳面舟形，末端钝圆，长 15—30 μm，宽 4—7 μm，横肋纹有 4—5 条/10 μm，肋纹之间有孔纹 2—4 排，横线纹有 16—20 条/10 μm。

生境：生于泉水渗出水、小水坑中。

国内分布：辽宁(辽河)，贵州(梵净山)，西藏(吉隆、昌都、芒康、贡觉、班戈、申扎、札达)；新疆(莎车、和静)。

国外分布：亚洲(俄罗斯、蒙古国、巴基斯坦)，北美洲(加拿大、美国、墨西哥、夏威夷群岛)，南美洲(巴西、哥伦比亚)，欧洲(挪威、奥地利、英国、德国、马其顿、波兰、罗马尼亚、西班牙、瑞典)，大洋洲(新西兰)。

5. 强壮细齿藻　图版 LXI：16—19；图版 LXII：1—14

Denticula valida(Pedicino)Grunow，*in* Van Heurck，Synopsis des Diatomées de Belgique Atlas，fig. 49：5，1881；Krammer & Lange-Bertalot，Bacillariophyceae. 2. Teil，*in* Ettl *et al.*，Süßwasserfl. Mitteleur.，p. 142，figs. 97：9—17，figs. 98：1—7，1988，nachdr. 1997；Metzeltin *et al.*，Iconogr. Diatomol. 15，figs. 218：9—15，2005；You & Wang. Acta Botanica Boreali-Occidentalia Sinica 31(2)：420，figs. II：18—20，2011a.

Denticula elegans f. *valida* Pedicino，Pochi studi sulle Diatomee viventi presso alcune terme dell'isola d'Ischia，p. 7，pl. 1，figs. 42—45，1867.

壳面线形至线形-披针形，两侧平行或稍凹入，朝两端楔形变窄，末端钝圆。壳面长 28—45 μm，宽 6—8 μm。横肋纹有 2.5—4 条/10 μm，肋纹之间有孔纹 4—7 排，横线纹有 16—19 条/10 μm。

生境：生于湖泊、小水渠中。

国内分布：新疆(乌市盐湖、哈密、温宿、库车、赛里木湖、察布查尔)。

国外分布：北美洲(美国)，南美洲(乌拉圭)，欧洲(英国、德国、西班牙)。

采自新疆的标本，壳体偏小，Krammer 和 Lange-Bertalot(1997)报道的种类宽度为 7—11 μm，其他结构特征吻合。

西蒙森藻属 Simonsenia H. Lange-Bertalot

H. Lange-Bertalot, Bacillaria 2：127-136，1979.

单细胞，带面观矩形，圆形角，壳面披针形，一般具尖而延伸的末端。具纵向的壳缝龙骨系统，对角线对称。壳缝形成管状通道，壳缝管位于翼状管上，壳缝管通过翼状管与细胞内部连通。

本属种类少，淡水和海水中均有分布，可生活在低至高电导率的水体中。

本属的形态结构介于菱形藻属(*Nitzschia*)和双菱藻属(*Surirella*)之间，翼状管是 *Surirella* 典型的结构，而对角线对称的壳缝龙骨系统是与 *Nitzschia* 相符的，因此本属的系统进化位置处于 *Nitzschia* 和 *Surirella* 之间。

模式种：德洛西蒙森藻[*Simonsenia delognei*(Grunow) Lange-Bertalot]。

本志收编 2 种。

西蒙森藻属分种检索表

1. 壳体较小，壳面窄披针形，横肋纹 20—21 条/10 μm ·················· **1. 德洛西蒙森藻 *S. delognei***
1. 壳体较大，壳面披针形至椭圆披针形，横肋纹 13—17 条/10 μm ········ **2. 茂兰西蒙森藻 *S. maolaniana***

1. 德洛西蒙森藻　图版 LXI：3—6

Simonsenia delognei (Grunow) Lange-Bertalot, *Simonsenia*, a new genus with morphology intermediate between *Nitzschia* and *Surirella*. Bacillaria 2：132，1979；Krammer & Lange-Bertalot, Bacillariophyceae. 2. Teil, *in* Ettl *et al.*, Süßwasserfl. Mitteleur., p. 135, pl. 84, figs. 13—19, 1988, nachdr. 1997.

Nitzschia delognei Grunow, *in* Van Heurck, Synopsis des Diatomées de Belgique, p. 184, Suppl. pl. C, fig. 38, 1885.

壳体小，带面观矩形，壳面窄披针形，末端尖、略延长，长 8.5—13 μm，宽 2.0—2.5 μm，横肋纹 20—21 条/10 μm；线纹细密，在光镜下不可见。

生境：附着。

国内分布：贵州(荔波)。

国外分布：北美洲(美国)，欧洲(英国、德国、波兰)。

本种与 Lange-Bertalot(1979)报道的模式种(长 8—15 μm，1.6—2 μm，横肋纹 16—22 条/10 μm)相比，壳面较宽。

2. 茂兰西蒙森藻　图版 LIX：1—13；图版 LX：1—6

Simonsenia maolaniana Q-M You & J.P. Kociolek, *in* Q-M You *et al.*, Diatom Research

31(3)：269—275，2016.

壳体小，带面观矩形，壳面披针形至椭圆披针形，末端窄喙状、不延长，长 9—25.5 μm，宽 2.5—3.5 μm，横肋纹 13—17 条/10 μm；线纹细密，在光镜下不可见。

扫描电镜下观察：外壳面波曲，壳缝位于壳面边缘，壳缝裂缝细、简单，不具中缝端。壳缝位于龙骨上，两侧各有一排肋状硅质结构，支撑管状壳缝，与横肋纹密度相同。线纹 2 至多排，28—36 条/10 μm。内壳面平坦，壳套窄，龙骨突 6—7 个/10 μm，末端具小的螺旋舌结构。

生境：附着。

国内分布：贵州（荔波）。

本种仅发现于贵州荔波县境内的茂兰国家级自然保护区，在同一生境内，还发现德洛西蒙森藻（*S. delognei*）。

筒柱藻属 Cylindrotheca L. Rabenhorst

Rabenhorst，Die Algen Sachsens resp. Mittel-Europa's. Decas 81-82，1859.

单细胞，针状，直或弓形，壳体通常关于顶轴强烈扭曲，色素体 2 至多个，一般板状或圆盘状。

壳面长，窄，仅轻度硅质化或部分壳面具硅质。如果壳面有线纹，一般不规则排列。具壳缝和龙骨突，龙骨突数量大，窄，肋状，硅质相对增厚。中缝端存在或缺失，如具中缝端，直或轻微膨大。极缝端简单。环带窄，数条，由轻度硅质化的简单条带组成。

本属种类少，多分布于世界各地的海岸带，淡水中少见。大多数种类附着于沉积物或污泥中。

本属与菱形藻属（*Nitzschia*）关系比较近，主要区别是壳体关于顶轴螺形扭曲（*C. closterium* 仅在两端扭曲），硅质化不明显，易碎，因此在酸处理过程中容易破损。

模式种：戈斯腾博格筒柱藻（*Cylindrotheca gerstenbergeri* Rabenhorst）。

本志收编 1 种。

细筒柱藻　图版 LXI：1—2

Cylindrotheca gracilis(Brébisson ex Kützing)Grunow，*in* van Heurck，Synopsis des Diatomées de Belgique，p. 186，pl. 80，fig. 2，1882；Krammer & Lange-Bertalot，Bacillariophyceae. 2. Teil，*in* Ettl *et al.*，Süßwasserfl. Mitteleur.，p. 134，pl. 87，fig. 3，1988，nachdr. 1997.

Ceratoneis gracilis Brébisson ex Kützing，*in* Kützing，F.T.，Species algarum. p. 89，1849.

壳体扭曲，壳面线形到线形-披针形，中间较宽，向两端逐渐变细，两端略弯曲；长 66 μm，宽 2.5 μm；壳缝及龙骨相互环绕，扭曲，龙骨突 24—25 个/10 μm；线纹细密，在光镜下不可见。

生境：附着。

国内分布：江苏（连云港），福建（沿海），台湾。

国外分布：北美洲(加拿大、美国)，欧洲(波罗的海、英国、法国、德国、罗马尼亚、葡萄牙、西班牙)，大洋洲(澳大利亚)，北极(斯瓦尔巴特群岛)。

本种与 Krammer 和 Lange-Bertalot(1997)报道的 *Cylindrotheca gracilis*(长 60—240 μm，宽 4—6 μm)相比，壳面较窄，在壳体形态、长度和龙骨突数量方面相似。

棒杆藻科 Rhopalodiaceae

壳面弓形，上、下壳面均具发达的管状壳缝，壳缝常在壳面上呈"V"形曲折或位于背侧边缘的龙骨上。在管壳缝内壁上具通入细胞内的小孔或无，壳面具横肋纹，在横肋纹之间具横线纹或蜂窝状孔纹，中央节或极节退化或完全没有。色素体 1 个，侧生片状。

目前本科有 3 属，我国淡水中发现有 2 属。

棒杆藻科分属检索表

1. 壳缝大部分位于壳面腹侧边缘，特别是接近两极的部分，而在壳面中部会不同程度的延伸进壳面，呈"V"形 ··· 1. 窗纹藻属 *Epithemia*
1. 壳缝常位于背侧或接近背侧边缘的龙骨上 ··· 2. 棒杆藻属 *Rhopalodia*

窗纹藻属 Epithemia F. T. Kützing

F. T. Kützing, Kies. Bacill. Diat.: 33, 1844.

单细胞，偶尔通过壳面形成短链状(通过非硅质结构相连)，常示带面观，线形至椭圆形。色素体 1 个，大，呈深裂叶状，紧靠细胞腹侧。所有种类都含有少数体积较小的内共生(endosymbiotic)蓝藻。

壳面具明显背腹侧之分，呈弓形，末端钝圆至宽圆形。外壳面平坦，偶尔会有瘤状物，有时在壳套背缘的连接处有边缘脊。线纹单排，组成线纹的网眼孔(areolae)非常复杂，很难确定其特征和边界，有时形成隔室(loculate)。网眼孔在外壳面被帽边状结构(半球形顶盖)封闭，仅留下窄的新月形(crescent-shaped)裂缝。横肋纹(transapical costae)粗壮，有些在壳体内部增厚并贯穿壳面背腹两侧，具龙骨突。壳缝系统离心，双弧形(biarcuate)，在内部形成管状结构，通过相邻龙骨突之间小圆形或卵形孔与细胞内部联系。外壳面的中缝端(central raphe endings)和极缝端(polar raphe endings)简单或稍微膨大，不具末端裂缝；内壳面的中缝端连续，极缝端终止于一个小的螺旋舌(helictoglossae)。带面有时在背侧增宽，复杂，由断开和闭合的环带共同组成。近壳面的环带结合形成隔膜(septum)，在完整的细胞膜上，紧密结合周围主要的横肋纹。

分布于淡水中，附着于沉积物或污泥中，喜营养丰富的基质环境。

Sims(1983)仔细研究了本属模式种类 *Epithemia turgida*，认为与棒杆藻属(*Rhopalodia*)的关系比较近，而不是与细齿藻属(*Denticula* senus stricto)。有些种类，如 *D. vanheurckii*，按传统的分类方法放在细齿藻属(*Denticula*)中，然而基于壳面和原生质体的结构，这些种类更接近于窗纹藻属(*Epithemia*)，它们只是外形上像 *Denticula*(Round et al., 1990)。

模式种：膨大窗纹藻[*Epithmia turgida* (Ehrenberg) Kützing]。

本志收编 7 种 12 变种 1 变型。

窗纹藻属分种检索表

1. 带面观头状隔片清楚，壳缝的中缝端位于壳面背侧，隔片上的裂缝靠近壳面腹侧 ··· **1. 光亮窗纹藻 *E. argus***
1. 带面观，头状隔片不清楚 ··· 2
    2. 壳缝位于腹侧边缘，只能在中部看清中缝端，肋纹平行排列 ············ **2. 弗里克窗纹藻 *E. frickei***
    2. 壳缝朝腹侧弯曲，或位于壳面上 ··· 3
3. 一般在两条肋纹间有 3 条或少于 3 条线纹 ··· 5
3. 一般在两条肋纹间有 3 条或多于 3 条线纹 ··· 4
    4. 壳缝分支纵穿整个壳面，弯向背侧中部 ·································· **3. 施密斯窗纹藻 *E. smithii***
    4. 壳缝分支沿着腹侧，几乎纵穿整个壳面，但只在中部可见，很少到达壳面中部 ······················ 6
5. 龙骨突 5 个/10 μm 或更多，线纹多于 10 条/10 μm。背侧强烈凸出，壳缝分支在壳面中部到达背侧边缘 ··· **4. 鼠形窗纹藻 *E. sorex***
5. 龙骨突少于 5 个/10 μm，线纹少于 10 条/10 μm。背侧稍凸出，壳缝分支仅在壳体中部弯向背侧 ····· ··· **5. 膨大窗纹藻 *E. turgida***
    6. 壳面背腹之分明显 ··· **6. 侧生窗纹藻 *E. adnata***
    6. 壳面背腹之分不明显 ··· **7. 光亮型窗纹藻 *E. arguiformis***

1. 光亮窗纹藻

Epithemia argus (Ehrenberg) Kützing, Kies. Bacill. Diat., p. 35, figs. 22: 55, 56, 1844; Schmidt *et al.*, Atlas Diat.-Kunde, p. 251, figs. 251: 11—13, figs. 16—19, 1904; Hustedt, Bacillariophyta, *in* Pascher, Süßwass -Fl. Mitteleur. Heft 10, p. 383, fig. 727a, 1930; Cleve-Euler, Diatomeen schw. Finnland, Teil. V, p. 34, figs. 1406 a—d, 1952; Patrick & Reimer, Diatoms U. S., 2(1): 175, fig. 23: 1, 1975. Gasse, Bibl. Diatomol. 11: 50, no fig., 1986; Krammer & Lange-Bertalot, Bacillariophyceae. 2. Teil, *in* Ettl *et al.*, Süßwasserfl. Mitteleur., p. 147, figs. 102: 1—9, 1988, nachdr. 1997; Shi *et al.*, Compil. Rep. Surv. Alg. Resour. South-West. China, p. 112-113, 1994; Zhu & Chen, Bacill. Xiz. Plat., p. 248, fig. 48: 10, 2000.

Eunotia argus Ehrenberg, Verbreitung und Einfluss des mikroskopischen Lebens in Süd-und Nord-Amerika, 1841: 413, pl. 2/6, fig. 33; pl. 3/4, fig. 7, 1843.

1a. 原变种　　图版 LXVIII: 1—4

var. argus

　　细胞带面和横截面均为矩形，或是在截面的中央有些膨胀；壳面背侧凸出，腹侧微凹入，两端逐渐变窄，末端钝圆，不与壳面主体分开；长 30—80 μm，宽 6—15 μm，长宽比 5—6；一般在整个壳面都能看到管壳缝，壳缝在壳面中部弯向背侧，呈"V"形；中央孔位于壳面的中部或是近背侧的一半；壳缝两分支形成的角度一般小于 *E. adnata* 的壳缝分支角度；肋纹 2—4 条/10 μm；窝孔纹 13—15 条/10 μm；两条肋纹间有窝孔纹 2—6

条；隔片发育良好，带面观横肋纹的末端呈圆头状。

扫描电镜下观察：外壳面壳缝裂缝的两侧没有像 E. adnata 的薄的硅质结构，裂缝在顶端几乎位于中线上，离背腹两侧的距离相等。在裂缝的整个背侧，具一宽的透明带，有时，也存在于腹侧，但一般会窄于背侧的带宽。壳面的外表面能看到顶向和切顶向规则排列的半球形顶盖(domed caps)，4—8 个半球形顶盖组成一个窝孔纹，通常是 4 个。在半球形顶盖的下面，简单的分支孔板填充每个窝孔纹的间隙。在壳体内部，壳面是由规则的切顶向的肋纹和肋间杆组成。内部的壳缝裂缝在中央节(central nodule)处连续。

生境：生于湖泊、湖边渗出水、小水渠、沼泽、芦苇滩中。

国内分布：河北(衡水湖)，湖南(沅江流域)，贵州(施秉舞阳河、沅江流域、梵净山)，西藏(聂拉木、亚东、仲巴、墨脱、米林、波密、芒康、班戈、申扎、改则、札达、噶尔、革吉)，新疆(赛里木湖、察布查尔、北屯、盐湖)。

国外分布：亚洲(蒙古国、伊朗、土耳其)，非洲(东部)，北美洲(加拿大、美国)，南美洲(哥伦比亚)，欧洲(挪威、波罗的海、英国、德国、爱尔兰、马其顿、波兰、罗马尼亚、西班牙)，大洋洲(澳大利亚、新西兰)。

1b. 高山变种　图版 LXVIII：5—7；图版 LXIX：9—10

var. **alpestris** (W. Smith) Grunow, Verh. Zool. - Bot. Ges. Wien, 12：329, fig. 5：28, 1862, illustration, Verh. Zool. - Bot. Ges. Wien, 10, pl. 5 (Grun., pl. 3), fig. 28. 1860; Schmidt *et al.*, Atlas Diat.-Kunde, p. 251, figs. 251：2, 3, 9, 1904; Hustedt, Bacillariophyta, *in* Pascher, Süßwass -Fl. Mitteleur. Heft 10, p. 383, fig. 727b, 1930; Patrick & Reimer, Diatoms U. S., 2(1)：176, fig. 23：4, 1975; Krammer & Lange-Bertalot, Bacillariophyceae. 2. Teil, *in* Ettl *et al.*, Süßwasserfl. Mitteleur., p. 148, figs. 103：1—5, 1988, nachdr. 1997; Zhu & Chen, Bacill. Xiz. Plat., p. 249, fig. 48：11, 2000.

Epithemia alpestrus W. Smith, A synopsis of the British Diatomaceae, vol. 1, p. 13, fig. 1：7, 1853.

Epithemia argus var. *capitata* Fricke *in* A. Schimidt *et al.*, fig. 251：14, 1904.

光镜下与原变种的区别在于：壳面背侧明显凸出，末端明显喙状或头状，有些弯曲；长 30—57.5 μm，宽 9—11.5 μm；横肋纹有 2—4 条/10 μm；窝孔纹有 12—15 条/10 μm；在两条横肋纹之间的窝孔纹有 2—6 条。

生境：生于湖泊、湖边渗出水、芦苇滩、路边积水、小水渠、稻田、沼泽、泉水井边附生、岩石上附生。

国内分布：吉林(长白山)，黑龙江(绥芬河)，西藏(定日、波密、班戈、申扎、札达)，新疆(阿克苏、天池、赛里木湖、察布查尔、布尔津、北屯)。

国外分布：亚洲(俄罗斯、蒙古国)，北美洲(美国)，欧洲(英国、德国、爱尔兰、西班牙)。

1c. 长角变种　图版 LXX：1—5

var. **longicornis** (Ehrenberg) Grunow, Verh. Zool.- Bot. Ges. Wien, 15：329, 1862; Schmidt *et al.*, Atlas Diat.-Kunde, p. 251, figs. 251: 1, 6, 15, 1904; Hustedt, Bacillariophyta, *in* Pascher, Süßwass -Fl. Mitteleur. Heft 10, p. 383, figs. 727 c, d, 1930; Cleve-Euler, Diatomeen schw. Finnland, Teil. V, p. 35, figs. 1406 e, f, 1952; Patrick & Reimer, Diatoms U. S., 2(1): 177, figs. 23: 2, 3, 1975; Zhu & Chen, Bacill. Xiz. Plat., p. 249, fig. 48: 12, 2000.

Eunotia longicornis Ehrenberg, Passatstaub und Blutregen. Ein Grofses organisches unsichtbares Wirken und Leben in der Atmosphäre, p. 272, pl. 1(1), fig. 17; pl. 1(2), figs. 19; pl. 3(1), fig. 8; pl. 4(A), fig. 23; pl. 4(B), fig. 21; pl. 5(2), fig. 19—22; pl. 6(1), fig. 19; pl. 6(2), fig. 11, 1849.

光镜下与原变种的区别在于：壳面明显延长，体积大；长 60—120 μm，宽 12—16 μm，长宽比大，6.0—10.5；横肋纹有 1—3 条/10 μm；窝孔纹有 8—12 条/10 μm；在两条横肋纹之间的窝孔纹有 3—8 条。

生境：生于高山草甸沼泽，水清澈、路边水坑，营养化程度高。

国内分布：西藏(波密、申扎)，新疆(阿克陶、沙雅、尉犁)。

国外分布：亚洲(俄罗斯、菲律宾)，北美洲(美国)，欧洲(爱尔兰、英国、罗马尼亚)。

本变种在我国不常见，分类较为混乱，Schmidt 和 Fricke(1904)报道了本变种(figs. 251: 1, 6, 15)，其中 fig. 251: 1, 6 较为相似，长宽比 5.75—6.5，末端基本不变窄，也不延伸，而 fig. 251: 15 的长宽比约为 9，末端明显变窄，稍有延伸且反曲；Husted(1930)中的种类与 Schmidt 等(1904)报道的 fig. 251: 1, 6 相符；Cleve-Euler(1952)的报道：长 60—120 μm，宽 13—19 μm。Fig. 1406: e 和 f 差别比较大，fig. 1406: e 的长宽比约为 10，末端稍延伸呈钝圆，与 Schmidt 等(1904)中 fig. 251: 15 相符。Fig. 1406: f 的长宽比约为 5.6，末端不延伸，与 Schmidt 等(1904)中 fig. 251: 1, 6 相符，但是不能确定 fig. 1406: f 为本变种。本文中的种类符合 Schmidt 等(1904)中 fig. 251: 15 及 Cleve-Euler(1952)中 fig. 1406: e。本变种的壳面末端与原变种以及其他变种的末端相似，主要区别在于壳面明显延长。

1d. 伸长变种 图版 LXIX：1

var. **protracta** Mayer, Die bayerischen Epithemien, p. 100, fig. 6: 15, figs. 7: 1—4, 1936; Cleve-Euler, Diatomeen schw. Finnland, Teil. V, p. 35, figs. 1406 k, l, n, 1952; Patrick & Reimer, Diatoms U. S., 2(1): 177, fig. 23: 5, 1975; Zhu & Chen, Bacill. Xiz. Plat., p. 249, fig. 48: 13, 2000.

光镜下与原变种的区别在于：背侧较高，末端延长但不呈头状；壳面长 30.5—32 μm，宽 8—11 μm；横肋纹有 3—4 条/10 μm；窝孔纹有 15 条/10 μm；在两条肋纹间有窝孔纹 2—4 条。

生境：生于湖泊、河流、沼泽、岩石表面。

国内分布：西藏(亚东、波密、墨脱)。

国外分布：北美洲(美国)，欧洲(中部)。

1e. 龟形变种　图版 LXIX：2—5

var. **testudo** Fricke, *in* Schmidt *et al.*, Atlas Diat.-Kunde, p. 251, fig. 251：4, 1904.

　　光镜下与原变种的区别在于：壳面明显短小，呈龟背形，背侧弧形凸出，腹侧几乎平直；长 25—35 μm，宽 8—12 μm，长宽比小，2.1—3.5；横肋纹有 2—3 条/10 μm；窝孔纹有 10—12 条/10 μm；在两条横肋纹之间的窝孔纹有 3—8 条。

　　生境：生于高山草甸沼泽，水清澈、路边水坑，营养化程度高。

　　国内分布：新疆（阿克陶、沙雅）。

　　国外分布：欧洲（中部）。

　　Husted(1930) 将本变种并入了 *E. argus* var. *alpestris* 中，而本文依据 Schmidt 等 (1904) 将这个变种独立出来，因为我们观察到的标本形态和体积变异不大，一般体积小于 *E. argus* var. *alpestris*，很容易区分开来。和同体积的 *E. argus* var. *alpestris* 比较，长宽比较小，腹侧平直也是一个区别的特征。

2. 弗里克窗纹藻　图版 LXIX：6—8

Epithemia frickei Krammer, *in* Lange-Bertalot & Krammer, Bibl. Diatomol. 15, p. 71, no fig. 1987；Krammer & Lange-Bertalot, Bacillariophyceae. 2. Teil, *in* Ettl *et al.*, Süßwasserfl. Mitteleur., p. 151, figs. 104：1—7, 1988, nachdr. 1997.

Epithemia intermedia Fricke, *in* Schmidt *et al.*, Atlas Diat.-Kunde, p. 249, figs. 249：14—18, 1904；Hustedt, Bacillariophyta, *in* Pascher, Süßwass -Fl. Mitteleur. Heft 10, p. 387, fig. 732, 1930；Patrick & Reimer, Diatoms U. S., 2(1)：179, fig. 24：2, 1975.

Epithemia zebra var. *intermedia* (Fricke) Hustedt, Die Diatomeenflora von Poggenpohls Moor bei Dötlingen in Oldenburg, p. 394, 1934；Zhu & Chen, *in* Shi *et al.*, Compilation of Reports on the Survey of Algal Resources in South-Western China. p. 112, 1994.

　　壳面背侧凸出，腹侧近于平直，中部略凹入，末端钝圆，末端不与壳面主体分开；壳面长 25—55 μm，宽 9—15 μm；壳缝几乎全部位于腹侧边缘，只在中部稍微弯向背侧；横肋纹有 3—5 条/10 μm；窝孔纹有 10—13 条/10 μm；两条肋纹间有窝孔纹 2—4 条；带面观横肋纹的末端呈圆头状。

　　生境：生于小水沟、河边沼泽中。

　　国内分布：内蒙古（牙克石），黑龙江（五大连池），湖南（沅江流域），贵州（沅江流域），新疆（喀纳斯、阿勒泰）。

　　国外分布：亚洲（俄罗斯），北美洲（美国），欧洲（波罗的海、英国、德国、波兰、爱尔兰、罗马尼亚），大洋洲（澳大利亚、新西兰）。

3. 施密斯窗纹藻　图版 LXXIII：2—3

Epithemia smithii Carruthers, The Diatomaceae, *in* Gray, Handb. British Freshw. Weeds, p. 76, 1864；Patrick & Reimer, Diatoms U. S., 2(1)：187, fig. 27：3a—b, 1975；Krammer & Lange-Bertalot, Bacillariophyceae. 2. Teil, *in* Ettl *et al.*, Süßwasserfl. Mitteleur., p. 152, figs. 105：1—6, 1988, nachdr. 1997.

壳面的背侧强烈弧形凸起，腹侧凹入，末端延长，呈圆形，偶尔平截或稍有反曲；长 50—78 μm，宽 9—15 μm；壳缝大部分位于壳面，靠近两端的部分靠近腹侧边缘，而在中部伸进壳面，中央孔会超过壳面的一半，一般靠近背侧边缘，呈"V"形；横肋纹呈放射状排列，2—3 条/10 μm；窝孔纹 8—10 条/10 μm；两横肋纹间具窝孔纹 3—6 条。

生境：生于水坑中，有高等水生植物附生。

国内分布：甘肃（月牙泉），新疆（尉犁）。

国外分布：北美洲（加拿大、美国），欧洲（英国、德国、爱尔兰、波兰、罗马尼亚），大洋洲（新西兰）。

本种和 *E. sorex* 相近，区别在于细胞的体积一般比 *E. sorex* 大，横肋纹更加清楚，肋纹的间距明显较大，有窝孔纹 3—6 排，而 *E. sorex* 肋纹间窝孔纹只有 2—3 条。

4. 鼠形窗纹藻

Epithemia sorex Kützing, Kies. Bacill. Diat., p. 33, figs. 5: 12, 5 a—c, 1844; Schmidt *et al.*, Atlas Diat.-Kunde, p. 252, figs. 252: 22—28, 1904; Hustedt, Bacillariophyta, *in* Pascher, Süßwass -Fl. Mitteleur. Heft 10, p. 388, fig. 736, 1930; Cleve-Euler, Diatomeen schw. Finnland, Teil. V, p. 41, figs. 1412 a, b, 1952. Patrick & Reimer, Diatoms U. S., 2(1): 188, fig. 27: 4, 1975; Zhu & Chen, The Diatoms of the Suoxiyu Nature Reserve Area, Hunan, China, *in* Li *et al.*, The algal flora and aquatic fauna of the Wulingyuan Nature Reserve Area, Hunan, China. p. 55, 1989b; Krammer & Lange-Bertalot, Bacillariophyceae. 2. Teil, *in* Ettl *et al.*, Süßwasserfl. Mitteleur., p. 154, figs. 106: 1—13, 1988, nachdr. 1997; Zhu & Chen, Bacill. Xiz. Plat., p. 249, fig. 48: 15, 2000; Metzeltin *et al.*, Iconogr. Diatomol. 15, fig. 190: 12, figs. 191: 5—14, 2005.

Eunotia sorex (Kützing) Rabenhorst, Die Süsswasser-Diatomaceen (Bacillarien.), p. 18, pl. 1, fig. 7, 1853.

4a. 原变种　图版 LXX：6—7；图版 LXXI：1—18

var. **sorex**

细胞带面椭圆披针形，环带关于壳面不对称，背侧的环带和壳套比腹侧的宽；壳面具强烈的背腹之分，背侧明显凸起，腹侧凹入，末端明显变窄，呈喙状-头状，朝壳面背侧反曲，末端反曲主要是背侧边缘弯曲造成的；壳面长 15—65 μm，宽 6—15 μm；壳缝双弧形，在中部弯向背侧，大部分位于壳面，中央孔一般位于壳面背侧的一半，有时靠近背侧边缘；横肋纹辐射排列，5—8 条/10 μm；窝孔纹 12—15 条/10 μm；两条横肋纹间有窝孔纹 2—3 条；肋纹清晰，带面观横线纹的末端不呈头状或稍呈头状。

扫描电镜下观察：外壳面壳缝裂缝的两侧围绕一圈薄的硅质结构，绕过中央孔一直到壳面腹侧边缘，此结构稍低于壳面，因此不是很明显。裂缝在中缝端膨大呈圆形，在远缝端几乎位于中线上，离背腹两侧的距离相等。在裂缝的整个背侧，具一宽的透明带，一般裂缝的腹侧没有此透明带。壳面的外表面能看到顶向和切顶向规则排列的半球形顶盖，硅质化程度高，一般很难看到顶盖的半月形裂缝，在顶盖下面，简单的分支孔板填充每个窝孔纹的间隙。内壳面是由规则的切顶向的肋纹和肋间杆组成。肋纹发育良好，贯穿壳面两侧，一般两个肋纹间会有一个小圆孔，是壳缝管与细胞内部相连的通道，内部的壳缝裂缝在中央节(central nodule)处连续。

生境：生于河流、湖边渗出水、小水渠、沼泽、路边积水，水草附生。

国内分布：北京，辽宁(辽河)，吉林(长白山)，黑龙江(鸡西兴凯湖、宝清七星河、牡丹江镜泊湖、五大连池)，江苏(昆山)，安徽(黄山)，湖北(武汉植物园、神农架)，湖南(岳阳、长沙、索溪峪、沅江流域)，广西(灵川)，海南(三亚、崖县)，贵州(赫章、威宁草海、沅江流域、梵净山)，云南(大理、滇池)，西藏(林芝、浪卡子、聂拉木、白地、定日、亚东、康马、吉隆、仲巴、昂仁、工布江达、墨脱、米林、加查、乃东、错那、措美、察隅、八宿、波密、芒康、察雅、申扎、措勤、札达、革吉、日土)，陕西(华山)，甘肃(月牙泉)，新疆(赛里木湖、察布查尔、布尔津、哈巴河、福海、北屯)。

国外分布：亚洲(俄罗斯、菲律宾、蒙古国、土耳其)，非洲(东部、南非)，北美洲(加拿大、美国)，南美洲(哥伦比亚、乌拉圭)，欧洲(波罗的海、黑海、英国、芬兰、德国、爱尔兰、罗马尼亚、马其顿、波兰、西班牙、瑞典)，大洋洲(澳大利亚、新西兰)。

本种的形状变异较大，细胞较长的种类呈线性-披针形，较短的种类呈椭圆形。壳面不仅在外形上与 *E. adnata* 和 *E. argus* 区别明显，而且在壳面结构上差别也很大：主要区别在于本种壳面的硅质化程度较高，半球形顶盖的裂缝不易观察到。

4b. 细长变种　　图版 LXXIII：6

var. **gracilis** Hustedt, Die Bacillariaceen-Vegetation des Lunzer Seengebietes(Nieder-Österreich), p. 237, pl. 3, fig. 4, 1922b; Hustedt, Bacillariophyta, *in* Pascher, Süßwass-Fl. Mitteleur. Heft 10, p. 388, fig. 737, 1930; Cleve-Euler, Diatomeen schw. Finnland, Teil. V, p. 41, figs. 1412 e, f, 1952; Krammer & Lange-Bertalot, Bacillariophyceae. 2. Teil, *in* Ettl *et al.*, Süßwasserfl. Mitteleur., p. 154, fig. 106：14, 1988, nachdr. 1997; Zhu & Chen, Bacill. Xiz. Plat., p. 249, fig. 48：15, 2000.

光镜下与原变种的区别在于：末端略延长钝尖，反曲明显；长 36—84 μm，宽 9.5—12 μm；横肋纹有 2—3 条/10 μm；窝孔纹有 11—14 条/10 μm；在两条肋纹间有窝孔纹 2—4 条。

生境：生于湖泊、河流、沼泽、泉水、路边小水塘中。

国内分布：海南(崖县、三亚)，西藏(浪卡子、康马、措勤、日土)，新疆(博湖、天池、赛里木湖、喀纳斯、北屯)。

国外分布：亚洲(俄罗斯)，欧洲(英国、爱尔兰、罗马尼亚)。

5. 膨大窗纹藻

Epithemia turgida(Ehrenberg)Kützing, Kies. Bacill. Diat., p. 34, pl. 5, fig. 14, 1844; Hustedt, Bacillariophyta, *in* Pascher, Süßwass-Fl. Mitteleur. Heft 10, p. 387, fig. 733, 1930; Schmidt *et al.*, Atlas Diat.-Kunde, p. 250, fig. 250：1—6, 1904; Cleve-Euler,

Diatomeen schw. Finnland, Teil. V, p. 39, fig. 1410a—e, p, 1952; Patrick & Reimer, Diatoms U. S., 2(1): 182, figs. 25: 1a—b, 1975; Sims, Bacillaria, 6: 219, figs. 14—24, 1983; Krammer & Lange-Bertalot, Bacillariophyceae. 2. Teil, *in* Ettl *et al.*, Süßwasserfl. Mitteleur., p. 155, figs. 109: 4—7, 1988, nachdr. 1997; Zhu & Chen, Bacill. Xiz. Plat., p. 250, figs. 48: 17, 2000.

Navicula turgida Ehrenberg, Phys. Abh. Akad. Wiss. Berlin, p. 80, 1832.

Eunotia turgida (Ehrenberg) Ehrenberg, Ber. Akad. Wiss. Berlin, for 1837: 45, 1837b.

5a. 原变种　　图版 LXXIV：1—7
var. **turgida**

细胞带面矩形，中间稍膨大，横截面梯形；壳面背侧弧形凸起，腹侧平直或略凹入，背侧向腹侧逐渐狭窄，末端钝圆，有时稍延伸呈头状，略弯向背侧，末端反曲主要是背侧边缘弯曲造成的；长 38—145 μm，宽 10—19 μm；壳缝大部分位于腹侧边缘，特别是在靠近两端的部分，而在壳面中部会伸进壳面，一般到达壳面的一半或稍微超过一半，呈"V"形；横肋纹呈放射状排列，3—6 条/10 μm；窝孔纹有 7—12 条/10 μm；两条横肋纹间具窝孔纹 2—3 条。

扫描电镜下观察：外壳面壳缝裂缝的两侧围绕一圈薄的硅质结构，绕过中央孔一直到壳面腹侧边缘，裂缝在顶端几乎位于中线上，离背腹两侧的距离相等。在裂缝的整个背侧，具一宽的透明带，一般腹侧不存在，透明带稍高于裂缝，成为壳缝管的一部分。从横截面可以看出，壳缝裂缝是一条蜿蜒曲折的缝隙，管壁的背侧朝外壳面凸出伸进腹侧，同时腹侧伸进背侧，但是在壳缝管内部不明显。壳面的外表面能看到顶向和切顶向规则排列的半球形顶盖(domed caps)，4—8 个半球形顶盖组成一个窝孔纹，通常是 4 个。在半球形顶盖的下面，简单的分支孔板填充每个窝孔纹的间隙。在壳体内部，壳面是由规则的切顶向的肋纹和肋间杆组成，肋纹发育良好，贯穿壳面两侧，壳缝开口于一个结实管状的槽，壳缝管的直径与肋纹龙骨之间的棒状结构等宽，一般两个肋纹间会有两个小圆孔，是壳缝管与细胞内部相连的通道，内部的壳缝裂缝在中央节(central nodule)处连续。

生境：生于河流、湖边渗出水、小水渠、沼泽中。

国内分布：北京，内蒙古(牙克石)，辽宁(铁岭)，黑龙江(五大连池、宝清七星河、加格达奇、鸡西兴凯湖)，上海(淀山湖)，广东(广州)，贵州(赫章)，西藏(定结、亚东、米林、错那、芒康)，新疆(赛里木湖、察布查尔、布尔津、喀纳斯、哈巴河、北屯、阿勒泰)。

国外分布：亚洲(俄罗斯、菲律宾、蒙古国、土耳其)，非洲(南非)，北美洲(加拿大、美国、墨西哥)，南美洲(巴西、哥伦比亚)，欧洲(波罗的海、黑海、英国、德国、爱尔兰、马其顿、波兰、罗马尼亚、西班牙、瑞典)，大洋洲(澳大利亚)。

在标本观察中发现，原变种的分布和数量都不及变种多。这一现象与 Hustedt(1930) 中描述的吻合。

5b. 头端变种　图版 LXXII：3—6

var. **capitata** Fricke, *in* Schmidt *et al.*, Atlas Diat.-Kunde, p. 250, fig. 250: 7, 1904; Hustedt, Bacillariophyta, *in* Pascher, Süßwass -Fl. Mitteleur. Heft 10, p. 387, no fig., 1930; Cleve-Euler, Diatomeen schw. Finnland, Teil. V, p. 39, figs. 1410 f, g, 1952; Zhu & Chen, Bacill. Xiz. Plat., p. 250, fig. 48: 18, 2000.

光镜下与原变种的区别在于：壳面背侧弓形，腹侧平直，末端呈膨大头状；长 48—88 μm，宽 10—14 μm；横肋纹 6—7 条/10 μm；窝孔纹 12—13 条/10 μm；在两条横肋纹之间的窝孔纹有 1—2 条。

生境：生于湖边渗出水、路边积水、池塘、沼泽，水草附生。

国内分布：广西(灵川)，西藏(边坝、聂拉木)，新疆(博湖、奎屯、博乐、赛里木湖、察布查尔、哈巴河)。

国外分布：北美洲(美国)，欧洲(英国、爱尔兰、罗马尼亚)，大洋洲(澳大利亚、新西兰)。

5c. 颗粒变种　图版 LXXII：7—9

var. **granulata** (Ehrenberg) Brun, Diat. Alpes Jura, p. 44, fig. 2: 13. 1880; Schmidt *et al.*, Atlas Diat.-Kunde, p. 250, figs. 250: 10—19, 1904; Hustedt, Bacillariophyta, *in* Pascher, Süßwass -Fl. Mitteleur. Heft 10, p. 387, fig. 734, 1930; Cleve-Euler, Diatomeen schw. Finnland, Teil. V, p. 39, fig. 1410h, 1952; Patrick & Reimer, Diatoms U. S., 2(1): 183, fig. 25: 3, 1975; Krammer & Lange-Bertalot, Bacillariophyceae. 2. Teil, *in* Ettl *et al.*, Süßwasserfl. Mitteleur., p. 156, figs. 108: 4—8, 1988, nachdr. 1997; Zhu & Chen, Bacill. Xiz. Plat., p. 250, fig. 48: 19, 2000.

Eunotia granulata Ehrenberg, Poggendorff's Annalen der Physik und Chemie, p. 220, pl. 4, fig. 2, 1836.

Navicula granulata Ehrenberg, Verbreitung Ann. Phys. u. Chem., 38: 220, pl. 3, fig. 2, 1836b.

Epithemia granulata (Ehrenberg) Kützing, Kies. Bacill. Diat., p. 35, pl. 5, fig. 20, 1844.

光镜下与原变种的区别在于：背侧边缘与腹侧边缘几乎平行；长 65—123 μm，宽 9—17 μm；横肋纹 2—5 条/10 μm；窝孔纹 9—13 条/10 μm；在两条横肋纹之间的窝孔纹有 1—2 条。

生境：生于小水沟、路边积水、沼泽，泉水井边丛生、石上附生。

国内分布：黑龙江(鸡西兴凯湖)，江西(鄱阳湖)，湖南(长沙)，云南(洱海)，西藏(聂拉木、多庆、吉隆、仲巴、拉萨、错那、察隅)，宁夏(贺兰山)，新疆(察布查尔、布尔津、喀纳斯、福海、北屯、阿勒泰)。

国外分布：亚洲(俄罗斯、蒙古国)，北美洲(美国)，欧洲(波罗的海、黑海、英国、爱尔兰、德国、马其顿、罗马尼亚、西班牙)，大洋洲(澳大利亚、新西兰)。

5d. 具褶变种　图版 LXXII：2；图版 LXXIII：1

var. **plicata** Meister, Die Kieselalgen der Schweiz. Beiträge zur Kryptogamenflora der

Schweiz, p. 197, pl. 33, fig. 19, 1912; Cleve-Euler, Diatomeen schw. Finnland, Teil. V, p. 41, figs. 1410 k—m, 1952; Zhu & Chen, Bacill. Xiz. Plat., p. 250, fig. 49: 1, 2000.

光镜下与原变种的区别在于：壳面背缘显著凸出，腹侧凹入明显，末端延长或呈小头状；壳面宽 13—21.5 μm，长 52—96 μm；横肋纹有 4—5 条/10 μm；窝孔纹有 7—10 条/10 μm；在两条肋纹间有窝孔纹 2—3 条。

生境：生于高山泉水、河流、小水沟、积水坑中。

国内分布：西藏(浪卡子、波密、芒康、措勤)。

国外分布：北美洲(美国)，欧洲(中部)。

5e. 韦斯特曼变种　　图版 LXXIII：7—10

var. **westermannii** (Ehrenberg) Grunow, Verh. Zool.-Bot. Ges. Wien, 12: 325, pl. 3 (Grunow, pl. 6), fig. 8, 1862; Cleve-Euler, Diatomeen schw. Finnland, Teil. V, p. 40, figs. 1410 n, o, 1952; Patrick & Reimer, Diatoms U. S., 2(1): 184, figs. 25: 2a—b, 1975; Krammer & Lange-Bertalot, Bacillariophyceae. 2. Teil, in Ettl et al., Süßwasserfl. Mitteleur., p. 156, figs. 109: 1—3, 1988, nachdr. 1997; Zhu & Chen, Bacill. Xiz. Plat., p. 251, fig. 49: 2, 2000.

Navicula westermannii Ehrenberg, Phys. Abh. Akad. Wiss. Berlin, 1833: 261, 266, 1834.

Eunotia westermannii (Ehrenberg) Ehrenberg, Die Infusionsthierchen als vollkommene Organismen, p. 190, fig. 14: 6, 1838.

Epithemia westermannii (Ehrenberg) Kützing, Kies. Bacill. Diat., p. 33, figs. 5: 12 (1—4), fig. 30: 4, 1844.

Epithemia westermannii var. *stricta* Temp. & Paragallo, Diat. Monde Entier, 2nd ed., p. 194, No. 364-366, 1910.

光镜下与原变种的区别在于：细胞体积小，长宽比小；壳面长 44 μm，宽 13 μm；横肋纹 4—5 条/10 μm；窝孔纹 11 条/10 μm；在两条肋纹间有窝孔纹 1—3 条。

生境：生于池塘中，淡水或半咸水。

国内分布：黑龙江(鸡西兴凯湖)、浙江(象山港)、云南(维西)、西藏(波密、芒康)、新疆(布尔津)。

国外分布：亚洲(俄罗斯、菲律宾、蒙古国)，北美洲(美国)，欧洲(波罗的海、英国、德国、马其顿、罗马尼亚、瑞典)，大洋洲(澳大利亚、新西兰)。

6. 侧生窗纹藻

Epithemia adnata (Kützing) Brébisson, Consid. Diat., p. 16, 1838; Patrick & Reimer, Diatoms U. S., 2(1): 179, figs. 24: 3, 4, 1975; Krammer & Lange-Bertalot, Bacillariophyceae. 2. Teil, in Ettl et al., Süßwasserfl. Mitteleur., p. 152, figs. 107: 1—11, 1988, nachdr. 1997. Fan et al., Bull. Bot. Resch., 21(2): 239, fig. I: 1, 2001; Metzeltin et al., Iconogr. Diatomol. 15, figs. 190: 1—11, figs. 191: 2—4, 2005.

Frustulia adnata Kützing, *Linnaea* 8: 544, pl. 13, 1833.

Navicula zebra Ehrenberg, Dritter Beitrag zur Erkenntniss grosser Organisation in der Richtung des kleinsten Raumes, p. 261, 1833.

Epithemia zebra (Ehrenberg) Kützing, Kies. Bacill. Diat., p. 34, fig. 5/12, 6 a—c, 1844; Schmidt *et al.*, Atlas Diat.-Kunde, p. 252, fig. 252: 1, 1904; Hustedt, Bacillariophyta, *in* Pascher, Süßwass -Fl. Mitteleur. Heft 10, p. 384-385, fig. 729, 1930; Cleve-Euler, Diatomeen schw. Finnland, Teil. V, p. 34, figs. 1406 a—d, 1952; Zhu & Chen, The Diatoms of the Suoxiyu Nature Reserve Area, Hunan, China, *in* Li *et al.*, The algal flora and aquatic fauna of the Wulingyuan Nature Reserve Area, Hunan, China. p. 55, 1989b; Zhu & Chen, Bacill. Xiz. Plat., p. 251, fig. 49: 3, 2000.

6a. 原变种　图版 LXV：1—2
var. **adnata**

细胞带面和横截面均为矩形，壳面新月形，稍有弯曲，背侧凸出，腹侧略微凹入，两侧近平行，顶端钝圆，不延长，不与壳面主体分开；长 15—150 μm，宽 7—14 μm；大部分壳缝位于腹侧边缘，中央孔位于壳面近腹侧的一半，一般不超过中线，呈 "V" 形；壳缝两分支常形成钝角，约 120°；横肋纹平行排列或稍有弯曲，(常 3—5 条) 2—8 条/10 μm；窝孔纹有 12—14 条/10 μm；两条肋纹间有窝孔纹 3—7 条；隔片微弱发育，带面观横肋纹的末端圆形但呈不清楚的头状。

扫描电镜下观察：外壳面壳缝裂缝的两侧围绕一圈薄的硅质结构，绕过中央孔一直到壳面腹侧边缘，裂缝在顶端几乎位于中线上，离背腹两侧的距离相等。在裂缝的整个背侧，具一宽的透明带，有时，也存在于腹侧。壳面的外表面能看到顶向和切顶向规则排列的半球形顶盖 (domed caps)，4—8 个半球形顶盖组成一个窝孔纹，通常是 4 个。在半球形顶盖的下面，简单的分支孔板填充每个窝孔纹的间隙。在壳体内部，壳面是由规则的切顶向的肋纹和肋间杆组成。内部的壳缝裂缝在中央节 (central nodule) 处连续。

生境：生于湖边渗出水、溪流、小水沟、池塘、路边沼泽中。

国内分布：北京，河北 (衡水湖)，内蒙古 (牙克石)，吉林 (长白山)，黑龙江 (鸡西兴凯湖、五大连池)，浙江 (西湖)，安徽 (宁国)，福建 (金门)，湖北 (武汉植物园)，湖南 (索溪峪、沅江流域)，贵州 (赫章、沅江流域、乌江流域)，云南 (滇池、翠湖、洱海、楚雄)，西藏 (聂拉木、白地、康马、亚东、浪卡子、仲巴、米林、林芝、措美、八宿、波密、芒康、贡觉、申扎、措勤、札达、普兰)，甘肃 (月牙泉)，新疆 (阿克苏、奎屯、博乐、赛里木湖、阿勒泰、布尔津、喀纳斯)。

国外分布：亚洲 (俄罗斯、菲律宾、蒙古国、伊朗)，非洲 (东部、南非)，北美洲 (加拿大、美国、墨西哥)，南美洲 (巴西、哥伦比亚、乌拉圭)，欧洲 (波罗的海、挪威、比利时、黑海、英国、芬兰、德国、爱尔兰、马其顿、波兰、罗马尼亚、瑞典、冰岛)，大洋洲 (澳大利亚)。

本种与 *E. zebra* (Ehrenberg) Kützing 互为同物异名现象，在国外早期的著作中，Hustedt (1930)、Cleve-Euler (1955)、Schmidt 等 (1874-1959) 等常使用种名 *E. zebra*，而在

后期的著作中，Patrick 和 Reimer(1975)、Krammer 和 Lange-Bertalot(1997)等常使用种名 *E. adnata*。国内的情况也基本一样，早期的著作，如金德祥(1951)、朱蕙忠和陈嘉佑(1989a，1994，2000)等都使用种名 *E. zebra*，而在包文美等(1992)、范亚文等(2001)、范亚文(2004)等文献中使用种名 *E. adnata*。此种下的变种也是同样的情况。鉴于种名 *E. adnata* 的合法发表时间(1838 年)早于 *E. zebra* 的合法发表时间(1844 年)，因此本书使用 *E. adnata*，而将 *E. zebra* 作为其同物异名处理。

本种种内变异较大，变种与原变种之间的差异主要在两端的形状，Krammer 和 Lange-Bertalot(1997)更是将这些变种全部放在原变种下，而这样的处理让这个种更加复杂。本书主要依据 Patrick 和 Reimer(1975)对本种的分类，将变种分开，依据如下所述。

6b. 顶生变种(象鼻变种)　图版 LXVII：1—2，5—7

var. **proboscidea** (Kützing) Hendey, Journal of the Marine Biological Association of the United Kingdom 33：557，1954.

Epithemia proboscidea Kützing, Kies. Bacill. Diat., p. 35, fig. 5：13，1844；

Epithemia zebra var. *proboscidea* (Kützing) Grunow, Verh. Zool. - Bot. Ges. Wien, 12：329, pl. 3 (Grunow, pl. 6), fig. 5, 1862 [as：*E. zebra* var. *proboscoidea*.]；Schmidt *et al.*, Atlas Diat.-Kunde, p. 252, fig. 252：2，1904；Cleve-Euler, Diatomeen schw. Finnland, Teil. V, p. 34, figs. 1409 l—p, 1952.

光镜下与原变种的区别在于：壳面末端变窄，呈喙状或头状，稍有反曲，明显与壳面主体分开；长 40—70 μm，宽 10—13 μm；横肋纹有 3—5 条/10 μm；窝孔纹有 12—14 条/10 μm；两条肋纹间有窝孔纹 3—5 条。

生境：生于河边渗出水、湖边渗出水、小水沟、路边积水、沼泽，水草附生、岩石上附生。

国内分布：黑龙江(鸡西兴凯湖)，西藏(浪卡子、林芝)，新疆(阿克苏、博湖、赛里木湖、察布查尔、布尔津、喀纳斯、哈巴河、福海、阿勒泰、盐湖)。

国外分布：亚洲(俄罗斯、蒙古国)，北美洲(美国、夏威夷群岛)，欧洲(英国、马其顿)，大洋洲(澳大利亚、新西兰)。

本变种与 *E. adnata* var. *porcellus* 在形态上最为相似，详见下述。

6c. 顶生变种(象鼻变种)二齿变型　图版 LXXII：1

var. **proboscidea** f. **bidens** (A. Cleve) Q-M You & Q-X Wang nov. comb.

Epithemia zebra var. *proboscidea* f. *bidens* A. Cleve, Lule Lpm., p. 27 (als v.), 1895；Cleve-Euler, Diatomeen schw. Finnland, Teil. V, p. 38, fig. 1409 p, 1952；Zhu & Chen, Bacill. Xiz. Plat., p. 251, fig. 49：4，2000.

Epithemia zebra var. *porcellus* f. *bidens* A. Cleve, Tåk., p. 36, fig. 74, 1932.

光镜下与象鼻变种的区别在于：壳面背缘中部具两个波状凸起；壳面宽 11.5—12 μm，长 103 μm；横肋纹有 4—5 条/10 μm；窝孔纹有 13 条/10 μm；在两条肋纹间有窝孔纹 1—4 条。

生境：生于高山泉水、净水塘中。

国内分布：西藏(波密)。

国外分布：欧洲(中部)。

本变型曾放在 E. zebra var. proboscidea 下，现在国际上常用 E. zebra var. proboscidea 作为 E. adnata var. proboscidea 的同物异名，大多数学者已经接受将 E. adnata var. proboscidea 作为正名。因此本书将变型 f. bidens 移至 E. adnata var. proboscidea 下作为新组合 E. adnata var. proboscidea f. bidens。

6d. 蛆形变种　图版 LXVI：3—4；图版 LXVII：3—4

var. **porcellus**(Kützing)Patrick, *in* Patrick & Reimer, Diatoms U. S., 2(1)：180, fig. 24：6, 1975.

Epithemia porcellus Kützing, Kies. Bacill. Diat., p. 34, figs. 5：18, 19, 1844.

Epithemia zebra var. *porcellus*(Kützing)Grunow, Verh. Zool.-Bot. Ges. Wien, 12：328, pl. 3/6, figs. 3, 4, 1862; Schmidt *et al.*, Atlas Diat.-Kunde, p. 252, figs. 252：15—21, 1904; Hustedt, Bacillariophyta, *in* Pascher, Süßwass -Fl. Mitteleur. Heft 10, p. 385, fig. 731, 1930; Cleve-Euler, Diatomeen schw. Finnland, Teil. V, p. 34, figs. 1409 q, s, 1952; Zhu & Chen, The Diatoms of the Suoxiyu Nature Reserve Area, Hunan, China, *in* Li *et al.*, The algal flora and aquatic fauna of the Wulingyuan Nature Reserve Area, Hunan, China. p. 55, 1989b; Zhu & Chen, Bacill. Xiz. Plat., p. 251, fig. 49：5, 2000.

光镜下与原变种的区别在于：壳面弯曲程度大，两端变窄，膨大呈头状，稍有反曲，明显与壳面主体分开；长 40—95 μm，宽 8—13 μm；横肋纹有 2—5 条/10 μm；窝孔纹有 11—15 条/10 μm；在两条横肋纹之间的窝孔纹有 2—7 条。

生境：生于小水渠、路边积水、沼泽中。

国内分布：湖南(索溪峪)，西藏(墨竹工卡、聂拉木、白地、多庆、浪卡子、亚东、吉隆、仲巴、昂仁、拉萨、工布江达、墨脱、林芝、错那、措美、芒康、察雅、洛隆、江达、类乌齐、班戈、申扎、措勤、札达、普兰、革吉、日土)，新疆(博乐、察布查尔、布尔津、喀纳斯、北屯、阿勒泰)。

国外分布：亚洲(蒙古国)，非洲(东部)，北美洲(美国)，欧洲(中部)。

本变种与 E. adnata var. proboscidea 在形态上最为相似，Hustedt(1930)把这两个变种一起放在 E. adnata var. porcellus 下，他认为将它们分开是不合适的。Krammer 和 Lange-Bertalot(1997)把 E. adnata 下的所有变种都放在原变种中。在我们观察标本的过程中，发现这两个变种的数量都比较大，而变种内的形态差异不大，较容易分辨。因此本书依据 Patrick 和 Reimer(1975)的分类，将这两个变种分开。

6e. 萨克森变种(索桑变种)　图版 LXV：3—7；图版 LXVI：1—2

var. **saxonica**(Kützing)Patrick, *in* Patrick & Reimer, Diatoms U. S., 2(1)：182, fig. 24：9, 1975; Fan *et al.*, 21(2)：239, fig. I：2, 2001.

Epithemia saxonica Kützing, Kies. Bacill. Diat., p. 35, pl. 5, fig. 15, 1844.

Epithemia zebra var. *saxonica* (Kützing) Grunow, Verh. Zool. -Bot. Ges. Wien, 12: 328, pl. 3, fig. 6, 1862; Schmidt *et al.*, Atlas Diat.-Kunde, p. 252, figs. 252: 3—14, 1904; Hustedt, Bacillariophyta, *in* Pascher, Süßwass -Fl. Mitteleur. Heft 10, p. 385, fig. 730, 1930; Zhu & Chen, The Diatoms of the Suoxiyu Nature Reserve Area, Hunan, China, *in* Li *et al.*, The algal flora and aquatic fauna of the Wulingyuan Nature Reserve Area, Hunan, China. p. 55, 1989b; Zhu & Chen, Bacill. Xiz. Plat., p. 251, fig. 49: 6, 2000.

光镜下与原变种的区别在于：细胞小，壳面腹缘凹入明显；长35—50.5 μm，宽8—10.5 μm，肋纹4—5条/10 μm；窝孔纹有12—14条/10 μm；两条肋纹间有窝孔纹3—6条。

生境：生于河边渗出水、路边积水、沼泽中。

国内分布：黑龙江（五大连池），福建（鼓岭、金门），江西（鄱阳湖），湖南（索溪峪、沅江流域），贵州（沅江流域、梵净山），西藏（聂拉木、白地、多庆、浪卡子、亚东、吉隆、仲巴、昂仁、拉萨、工布江达、墨脱、林芝、错那、措美、芒康、察雅、洛隆、江达、类乌齐、班戈、申扎、措勤、札达、普兰、革吉、日土），陕西（华山），新疆（阿克苏、博湖、喀纳斯）。

国外分布：亚洲（蒙古国），非洲（东部），北美洲（美国），欧洲（中部）。

7. 光亮型窗纹藻　图版LXXIII: 4—5

Epithemia arguiformis Q-M You & Q-X Wang, *in* You *et al.*, Taxonomy and distribution of diatoms in the genera *Epithemia* and *Rhopalodia* from the Xinjiang, China. Nova Hedwigia 89 (3—4): 413, figs. 60—61, 2009.

壳面背腹之分不明显，背侧略凸出，腹侧中部略凹入，两端略凸出。两端宽楔形，不反曲，顶端常位于壳面纵向中线附近；壳面长30—50 μm，宽9—11 μm；一般在整个壳面都能看到管壳缝，壳缝在壳面中部弯向背侧，呈"V"形；中央孔位于壳面中部附近；横肋纹有2—3条/10 μm；窝孔纹有12—16条/10 μm；两条肋纹间有窝孔纹3—8条。

生境：生于泉水流出水中。

国内分布：新疆（皮山）。

本种发现于新疆皮山县稚普泉水库旁的泉水流出水中，壳面形态和壳缝走向与 *E. argus* 最为相近，但本种的背腹之分不明显。

棒杆藻属 Rhopalodia O. Müller

O. Müller, Bot. Jahrb. Syst. 22: 57, 1895

单细胞，常示带面观，线形、披针形或椭圆形，异极或等极。色素体1个，盘形，边缘裂叶状，靠近细胞腹侧。有些种类含少数体积较小的内共生蓝藻。

壳面具背腹侧之分，呈线形或弓形。壳套的边界难辨认。外壳面光滑，有时会有瘤状物。线纹单排至多排，由1个或几个分支孔板（volae）封闭的拟孔（poroid）组成。横肋纹粗壮，有些在壳体内部增厚并贯穿壳面背腹两侧，具龙骨突。壳缝系统离心，近背侧，一般位于龙骨上，在内部形成管状结构，通过相邻龙骨突之间的小圆形或卵形孔与细胞内部连通。外壳面的中缝端膨大，形成明显的中央节，有时稍偏向腹侧，内壳面的中缝端简单。外壳面的极缝端简单。带面的背侧宽于腹侧，复杂，由断开和闭合的具孔的环带组成，有些种类的环带比较简单。与窗纹藻属(*Epithemia*)相比，没有与壳面形成精细的互锁结构。

广泛分布于淡水和海水中，附着于沉积物、污泥或附生在植物上。

本属与关系较近的窗纹藻属(*Epithemia*)相比，结构更加多样，生活环境更加广泛。细胞外形、带面结构和壳缝可将两个属区分开来。本属包含两个组：弯棒杆组(*Gibba*-group)和驼峰棒杆组(*Gibberula*-group)，一直以来，弯棒杆组(*Gibba*-group)存在的分类问题比较少，而对于驼峰棒杆组(*Gibberula*-group)，由于组内种间的超微结构和壳缝构造有很大不同，所以传统的分类方法(如利用壳面外形或壳体带面观)已经不能解决这个组的分类问题，只能通过电镜和分子生物学技术的应用来解决。Kranmmer(1988a，1988b)对驼峰棒杆组(*Gibberula*-group)进行了非常详细的研究，该组在形态上与窗纹藻属(*Epithemia*)更为接近。

模式种：弯棒杆藻[*Rhopalodia gibba*(Ehrenberg)O. Müller]。

本志收编 9 种 3 变种。

棒杆藻属分组检索表

1. 壳体带面观线形，或只在中部横向膨大(主要为淡水种类) ························· 弯棒杆藻组 Sect. *Gibba*
1. 壳体带面观椭圆形或卵形(主要为咸水种类) ························· 驼峰棒杆藻组 Sect. *Gibberula*

弯棒杆藻组 Section *Gibba*

弯棒杆藻组分种检索表

1. 细胞带面观具不同程度膨大的中部，壳面括号形，一般末端弯向腹侧，尖圆形··· **1.** 弯棒杆藻 *R. gibba*
1. 细胞带面观两侧非常平行，至多在中部稍微有点膨大·· 2
    2. 体积较大，朝向两端略微变窄，壳体末端钝圆···················· **2.** 平行棒杆藻 *R. parallela*
    2. 体积较小，朝向两端明显变窄，壳体末端尖圆···················· **3.** 纤细棒杆藻 *R. gracilis*

1. 弯棒杆藻

Rhopalodia gibba(Ehrenberg)O. Müller, Bot. Jahrb., 22：65，figs. 1：15—17. 1895；Schmidt *et al.*，Atlas Diat.-Kunde, p. 253，figs. 253：1—13，1905；Hustedt, Bacillariophyta, *in* Pascher, Süßwass -Fl. Mitteleur. Heft 10, p. 390, fig. 740, 1930；Cleve-Euler, Diatomeen schw. Finnland, Teil. V, p. 44, figs. 1416 a—e, 1952；Patrick & Reimer, Diatoms U. S., 2(1)：189，fig. 28：1，1975；Gasse, *Bibliotheca Diatomologia*, 11：162, figs. XXXIX：8—9, 1986；Krammer & Lange-Bertalot, Bacillariophyceae, 2. Teil, *in* Ettl *et al.*，Süßwasserfl. Mitteleur., p. 159, figs. 111：1—13，1988，nachdr. 1997；Mitsuzo, Flora North-Eastern Prov. China, part. IV, p. 1411, pl. 4, fig. 22, 1970；Zhu & Chen, The Diatoms of the Suoxiyu Nature Reserve Area, Hunan, China, *in* Li *et* *al.*, The algal flora and aquatic fauna of the Wulingyuan Nature Reserve Area, Hunan, China. p. 55, 1989b; Shi *et al.*, Compil. Rep. Surv. Alg. Resour. South-West. China, p 112—113, 1994; Xin *et al.*, Journal of Shanxi University(Nat. Sci. Ed.)20(1): 105, 1997; Zhu & Chen, Bacill. Xiz. Plat., p. 252, figs. 49: 7—8, 2000; Metzeltin *et al.*, Iconogr. Diatomol. 15, figs. 192: 1—5, 193: 10, 2005.

Navicula gibba Ehrenberg, Phys. Abh. Akad. Wiss. Berlin, for 1830: 64, 65, 1832. [described.] Ibid., for 1831: 80, 1832.

Epithemia gibba(Ehrenberg)Kützing, Kies. Bacill. Diat., p. 35, fig. 4: 22, 1844.

1a. 原变种　　图版 LXXV: 1—5
var. **gibba**

细胞单生，壳体等极，常示带面观；壳体线形、线形-披针形，壳面弓形，背侧弧形，中部有一小的缺刻，腹侧平直，两端逐渐狭窄，呈楔形或尖端，向腹侧弯曲；壳体长 49—200 μm，宽 18—30 μm，壳面宽 8—11 μm；背侧具一条龙骨，龙骨上具一条不明显的管壳缝，中央节不清楚。肋纹发育良好，平行排列，4—10 条/10 μm；横线纹 12—16 条/10 μm；在两条横肋纹之间的横线纹有 1—3 条。

扫描电镜下观察：管壳缝位于背侧边缘的龙骨上，龙骨表面没有窝孔纹分布，在背侧边缘形成一条纵向的空白区域。壳缝在壳面中部断开，稍伸进壳面，形成一个小的缺刻，缺刻周围分布着若干凹陷，与窝孔纹的大小相似，也位于龙骨上。壳面上的横线纹由单排窝孔纹组成，窝孔纹外侧具瓣状结构，遮盖了孔内的结构，只留下窄的弓形裂缝。环带之间通过纵向的孔和硅质的小突起紧密地铆合在一起，不容易分开，远离壳面的环带上有时可见 1—2 排纵向的孔纹。

生境：淡水普生性种类。

国内分布：北京，天津，山西(太原晋祠)，辽宁(铁岭)，吉林(长白山)，黑龙江(鸡西兴凯湖、五大连池、宝清七星河、伊春、牡丹江镜泊湖)，江苏(苏州)，浙江(西湖)，福建(厦门、鼓岭、金门)，湖南(岳阳、索溪峪、沅江流域)，广西(灵川)，四川(西南部)，贵州(赫章、威宁草海、沅江流域、乌江流域)，海南(琼中)，云南(西双版纳、下关、南华)，西藏(边坝、浪卡子、白地、多庆、亚东、康马、吉隆、仲巴、昂仁、拉萨、墨脱、米林、林芝、乃东、错那、加查、措美、察隅、八宿、波密、昌都、芒康、察雅、贡觉、类乌齐、班戈、申扎、改则、措勤、札达、普兰、革吉、日土)，甘肃(月牙泉)，新疆(巴楚、阿克苏、博湖、天池、奎屯、博乐、赛里木湖、察布查尔、伊宁、布尔津、喀纳斯、哈巴河、阿勒泰、福海、北屯)。

国外分布：亚洲(俄罗斯、菲律宾、蒙古国、伊拉克、巴基斯坦、土耳其)，非洲(东部、南非)，北美洲(加拿大、美国、墨西哥、夏威夷群岛)，南美洲(巴西、哥伦比亚、乌拉圭)，欧洲(波罗的海、比利时、英国、德国、爱尔兰、马其顿、波兰、罗马尼亚、西班牙、冰岛、瑞典、芬兰、葡萄牙、德国)，大洋洲(澳大利亚、新西兰)。

1b. 偏肿变种　　图版 LXXVI：6—9

var. **ventricosa** (Kützing) H. & M. Paragallo, Diat. Mar. France, p. 302, figs. 77: 3—5, 1900; in Schmidt *et al.*, Atlas Diat.-Kunde, p. 253, figs. 253: 1—17, 1905; Hustedt, Bacillariophyta, *in* Pascher, Süßwass -Fl. Mitteleur. Heft 10, p. 391, fig. 741, 1930; Patrick and Reimer, Diatoms U. S., 2(1), p. 190, figs. 28: 3, 4, 1975; Zhu *& al.*, Diatom. Suoxiyu Nat. Preserv. Area Hunan, China, p. 55, 1989; Shi *et al.*, Compil. Rep. Surv. Alg. Resour. South-West. China, p. 112—113, 1994; Zhu & Chen, Bacill. Xiz. Plat., p. 252—253, fig. 49: 9, 2000; Fan *et al.*, 21(2): 240, fig. I: 5, 2001.

Epithemia ventricosa Kützing, Kies. Bacill. Diat., p. 35, figs. 30: 9 a, b, 1844.

Epithemia gibba var. *ventricosa* (Kützing) Grunow, *in* V. Heurck, Syn. Diat. Belgique, pl. 32, figs. 4, 5, 1881.

Rhopalodia ventricosa Kützing, Revisio Gen, Plant., vol. 2, p. 891, 1891.

Rhopalodia ventricosa (Kützing) O. Müller, Bot. Jahrb., 22: 65, pl. 1, figs. 20, 21, 1895.

　　与原变种的区别在于：光镜下，细胞短小，披针形至披针-椭圆形，壳面观腹侧中部略向背侧弯曲；壳面宽 7—9 μm，长 34.5—72 μm；横肋纹有 8—10 条/10 μm；两条横肋纹之间有横线纹 1—2 条。扫描电镜下，由窝孔纹形成的横线纹之间的距离紧凑，明显比原变种小。

　　生境：生于湖水、池塘、小水沟、稻田、浅水滩、沼泽、路边水坑中。

　　国内分布：吉林(长白山)，黑龙江(鸡西兴凯湖、伊春、五大连池)，江苏(昆山)，江西(鄱阳湖)，湖北(武汉植物园)，湖南(索溪峪、沅江流域)，海南(琼中)，贵州(沅江流域、梵净山)，云南(南华)，西藏(边坝、浪卡子、亚东、吉隆、昂仁、林芝、加查、察隅、波密、芒康、类乌齐、申扎、普兰)，新疆(巴楚、阿克苏、博湖、博乐、赛里木湖、察布查尔、米粮泉、布尔津、阿勒泰、哈巴河、北屯、五指泉、盐湖)。

　　国外分布：亚洲(印度、菲律宾、蒙古国、伊朗、新加坡、土耳其、伊拉克)，非洲(埃及)，北美洲(美国、墨西哥)，南美洲(巴西)，欧洲(黑海、英国、法国、爱尔兰、马其顿、罗马尼亚)，大洋洲(澳大利亚、新西兰)。

2. 平行棒杆藻

Rhopalodia parallela (Grunow) O. Müller, Bot. Jahrb., 22: 64, figs. 1: 13, 14. 1895; Schmidt *et al.*, Atlas Diat.-Kunde, p. 252, figs. 252: 33—36, 1905; Hustedt, Bacillariophyta, *in* Pascher, Süßwass -Fl. Mitteleur. Heft 10, p. 389, fig. 739, 1930; Cleve-Euler, Diatomeen schw. Finnland, Teil. V, p. 44, fig. 1417, 1952; Patrick & Reimer, Diatoms U. S., 2(1): 190, fig. 28: 2, 1975; Qi *et al.*, Journal of Jinan University, 3: 102, 1985; Bao *et al.*, J. Southwest Nor. Univ., 3: 64, 1986; Zhu & Chen, Bacill. Xiz. Plat., p. 253, fig. 50: 2, 2000.

Epithemia gibba var. *parallela* Grunow, Verh. Zool. -Bot. Ges. Wien, 12: 327, pl. 3 (Grun., pl. 6), fig. 7. 1862.

2a. 原变种　图版 LXXVI：1—5；图版 LXXVIII：2—3

var. parallela

带面观矩形；壳面背侧与腹侧几乎平行，背侧中部有一小的缢缩，两端略微变窄，并弯向腹侧；长 45—260 μm，壳体宽 15—38 μm，壳面宽 9—11 μm；横肋纹平行排列，6—9 条/10 μm；横线纹 12—15 条/10 μm；在两条横肋纹之间的横线纹有 1—4 条。

扫描电镜下观察：管壳缝位于背侧边缘的龙骨上，龙骨表面没有窝孔纹分布，在背侧边缘形成一条纵向的空白区域，靠近腹侧的边缘有 1—2 排点状凹陷，凹陷的大小及间距都与窝孔纹相似，壳缝在壳面中部断开，稍伸进壳面，形成一个小的缺刻。壳面上的横线纹由双排窝孔纹组成，窝孔纹外侧不具瓣状结构。

生境：生于湖泊、泉水、小水渠、河边渗出水、浅水滩、沼泽、路边积水中。

国内分布：湖北(神农架)，四川(九寨沟)，西藏(定日、亚东、康马、吉隆、萨噶、察隅、昌都、芒康、江达、札达、噶尔)，宁夏(贺兰山)，新疆(阿克苏、博湖、天池、博乐、赛里木湖、察布查尔、米粮泉、喀纳斯、盐湖)。

国外分布：亚洲(蒙古国)，北美洲(美国)，欧洲(英国、爱尔兰、波兰、罗马尼亚、西班牙、阿尔卑斯山)。

光镜下本种和 *R. gibba* 外形上较相似，均为线形。而 *R. parallela* 的壳面两侧非常平行，或在中部稍有膨大，一般体积较大，长宽比例大。电镜下 *R. parallela* 壳面有双排孔纹组成的横线纹，而 *R. gibba* 的横线纹是由单排孔纹组成，这是两个种在结构上的较大区别，因此，本书将 *R. parallela* 作为独立的种处理。

2b. 扭曲变种　图版 LXXVII：3—6

var. distorta Fricke, *in* Schmidt *et al.*, Atlas Diat.-Kunde, p. 252, figs. 252：29—32，1904.

与原变种的区别在于：龙骨及壳缝绕顶轴扭曲，一般扭曲少于 180°；长 45—200 μm，壳体宽 15—30 μm，壳面宽 8—11 μm；横肋纹平行排列，6—9 条/10 μm；横线纹 12—15 条/10 μm；在两条横肋纹之间的横线纹有 2—3 条。

生境：生于沼泽、河边渗出水中。

国内分布：新疆(阿克苏、阿克陶、察布查尔)。

国外分布：欧洲(中部)。

2c. 巨大变种　图版 LXXVIII：1

var. ingens Fricke, *in* Schmidt *et al.*, Atlas Diat.-Kunde, p. 252, figs. 252：37，38，1904.

与原变种的区别在于：体积明显比原变种大；壳面线形，中部不膨大；长 215 μm，壳面宽 10 μm；横肋纹平行排列，6—8 条/10 μm；横线纹 12—14 条/10 μm；在两条横肋纹之间的横线纹有 2—3 条。

生境：生于小水沟中。

国内分布：新疆(尉犁)。

国外分布：欧洲(中部)。

3. 纤细棒杆藻　图版 LXXVII：1—2

Rhopalodia gracilis O. Müller, Bot. Jahrb., p. 63, figs. 1: 8—12, figs. 2: 5, 6, 1895; Schmidt *et al.*, Atlas Diat.-Kunde, p. 255, figs. 255: 22—27, 1905; Gasse, Bibliotheca Diatomologia, 11: 163, figs. XXXIX: 1—3, 7, 1986.

带面观矩形；壳面背侧与腹侧几乎平行，中部不呈横向膨大，有一小的缢缩，两端逐渐狭窄，并略微弯向腹侧，末端尖圆形；壳面长 55—110 μm，宽 8—12 μm；横肋纹平行排列，8—9 条/10 μm；横线纹有 11—12 条/10 μm；在两条横肋纹之间的横线纹有 1—3 条。

生境：生于路边积水、浅水滩、稻田中。

国内分布：云南（丽江），新疆（阿克苏、博湖、察布查尔）。

国外分布：非洲（东部），欧洲（中部）。

驼峰棒杆藻组 Section *Gibberula*

驼峰棒杆藻组分种检索表

1. 在光镜高倍镜头下，很难看到或看不到孔，一般多于 30 个/10 μm ·········· **4. 具盖棒杆藻 R. operculata**
1. 在光镜高倍镜头下，可看见线纹上的孔，一般少于 30 个/10 μm ·· 2
  2. 壳面宽，镰刀形，线纹上的孔少于 16 个/10 μm，壳体具强烈弯曲的背侧，外形简单 ·················
                                                                      5. 肌状棒杆藻 R. musculus
  2. 结构精细 ··· 3
3. 腹侧凹入，壳面呈镰刀形，龙骨发育弱，壳缝沿着壳体边缘，在壳面难以分辨 ·······················
                                                                     6. 驼峰棒杆藻 R. gibberula
3. 腹侧直或稍微凸出，壳体呈圆形的一部分或披针形 ·· 4
  4. 体积稍小，长度小于 40 μm，腹侧直或凹入，线纹由单排孔组成 ········ **7. 毕列松棒杆藻 R. brebissonii**
  4. 体积较大，长度可大于 40 μm，腹侧直 ·· 5
5. 线纹由双排孔组成，不规则排列（光镜下不易分辨）············· **8. 缢缩棒杆藻 R. constricta**
5. 线纹由单排孔组成，规则排列 ··· **9. 石生棒杆藻 R. rupestris**

4. 具盖棒杆藻　图版 LXXX：1—6

Rhopalodia operculata (Agardh) Håkansson, Beihefte zur Nova Hedwigia 64, p. 166, 167, 1979; Krammer, *Nova Hedwigia*, 47(1—2): 163, figs. 103—116, 1988; Krammer & Lange-Bertalot, Bacillariophyceae. 2. Teil, *in* Ettl *et al.*, Süßwasserfl. Mitteleur., p. 165, figs. 115: 9—12, 1988, nachdr. 1997; Metzeltin *et al.*, Iconogr. Diatomol. 15, figs. 193: 8, 9, 2005.

Frustulia operculata Agardh, Flora 10(40): 627, 1827.

壳面新月形，背侧弧形凸出，中部有一小的缢缩，腹侧平直或略微凹入，两端逐渐狭窄，并略弯向腹侧；壳体长 20—50 μm，宽 15—25 μm，壳面宽 8—10 μm，肋纹清楚，横肋纹 3—4 条/10 μm，横线纹 30—40 条/10 μm，在两条横肋纹之间的横线纹有 3—5 条。

扫描电镜下观察：龙骨高于壳面，在背侧边缘形成一条纵向的空白区域，龙骨表面有点纹分布。管壳缝位于背侧边缘的龙骨上，壳缝在壳面中部断开，稍伸进壳面，形成一个小的缺刻(notch)。整个壳面的线纹均由双排孔组成，排列不规则，对生或互生，孔为细小的圆孔，直径约 0.1 μm，内外开口都简单，没有闭塞物。

生境：广泛分布于河流、小溪、水渠和渗出水中。

国内分布：贵州(赫章)，西藏(墨竹工卡、边坝)，新疆(乌鲁木齐盐湖、博乐、察布查尔、福海)。

国外分布：亚洲(俄罗斯)，北美洲(美国)，南美洲(哥伦比亚、乌拉圭)，欧洲(英国、德国、罗马尼亚)，大洋洲(新西兰)。

扫描电镜下观察的标本，龙骨及与龙骨连接的肋纹要比 Krammer(1988，fig. 116)报道的位置低。

5. 肌状棒杆藻　图版 LXXVII：7—8

Rhopalodia musculus (Kützing) O. Müller, Hedwigia, 38(5, 6)：278，294，1900；Schmidt *et al.*，Atlas Diat.-Kunde，p. 255，figs. 255：1—12，1905；Hustedt, Bacillariophyta, *in* Pascher，Süßwass -Fl. Mitteleur. Heft 10，p. 392，fig. 745，1930；Gasse, Bibliotheca Diatomologia，11：163，figs. XXXIX：1—3，1986；Patrick & Reimer，Diatoms U. S.，2(1)：191，fig. 28：5，1975；Krammer，Nova Hedwigia，47(1-2)：163，figs. 59—75，1988；Krammer & Lange-Bertalot, Bacillariophyceae. 2. Teil, *in* Ettl *et al.*，Süßwasserfl. Mitteleur.，p. 163，figs. 114：1—8，1988，nachdr. 1997；Zhu & Chen，Bacill. Xiz. Plat.，p. 253，fig. 50：1，2000；Metzeltin *et al.*，Iconogr. Diatomol. 15，fig. 192：6，2005.

Epithemia musculus Kützing，Kies. Bacill. Diat.，p. 33，pl. 30，fig. 6，1844.

Rhopalodia gibberula var. *musculus* (Kützing) Cleve-Euler, Diatomeen schw. Finnland，p. 43，figs. 1415 o—r，1952.

Rhopalodia gibberula var. *musculus* (Kützing) Muschler，Enumération des Algues marines et d'eau douce observées jusqu'à ce jour en Egypte，5(3)：141—237，1908.

壳面新月形，背侧弧形，腹侧平直或略微向背侧弯曲，两端逐渐狭窄，并略弯向腹侧，末端钝尖；壳面长 20—34 μm，宽 8—10 μm，横肋纹在 10 μm 内有 6—7 条，横线纹 13—15 条，两条横肋纹之间有横线纹 2—5 条。

生境：生于湖泊、河边渗出水、小水渠、路边积水、沼泽中。

国内分布：山西(运城)，内蒙古(牙克石)，黑龙江(尚志)，江苏(昆山)，贵州(赫章)，西藏(拉萨、加查、申扎)，新疆(博乐、察布查尔、布尔津、盐湖)。

国外分布：亚洲(俄罗斯、蒙古国、新加坡)，非洲(东部)，北美洲(加拿大、美国、夏威夷群岛、关岛)，南美洲(巴西、乌拉圭)，欧洲(亚得里亚海、波罗的海、黑海、英国、德国、希腊、爱尔兰、马其顿、罗马尼亚、西班牙)，大洋洲(澳大利亚、新西兰)。

6. 驼峰棒杆藻　图版 LXXIX：1—6

Rhopalodia gibberula (Ehrenberg) O. Müller, Bot. Jahrb., p. 58, 1895; Schmidt *et al.*, Atlas Diat.-Kunde, p. 253, figs. 253: 23—37, p. 254, figs. 254: 12—21, 1905; Hustedt, Bacillariophyta, *in* Pascher, Süßwass -Fl. Mitteleur. Heft 10, p. 391, fig. 742, 1930; Cleve-Euler, Diatomeen schw. Finnland, Teil. V, p. 42, fig. 1415, 1952; Patrick & Reimer, Diatoms U. S., 2(1): 191, fig. 28: 6, 1975; Gasse, Bibliotheca Diatomologia, 11: 163, figs. XXX: 15—16, figs. XXXIX: 10, 1986; Krammer, Nova Hedwigia, 47(1-2): 160, figs. 1—23, 1988; Krammer & Lange-Bertalot, Bacillariophyceae. 2. Teil, *in* Ettl *et al.*, Süßwasserfl. Mitteleur., p. 160, figs. 112: 1—6, figs. 113: 4—6, 1988, nachdr. 1997; Zhu & Chen, The Diatoms of the Suoxiyu Nature Reserve Area, Hunan, China, *in* Li *et al.*, The algal flora and aquatic fauna of the Wulingyuan Nature Reserve Area, Hunan, China. p. 55, 1989b; Shi *et al.*, Compil. Rep. Surv. Alg. Resour. South-West. China, p. 112—113, 1994; Zhu & Chen, Bacill. Xiz. Plat., p. 253, fig. 49: 10, 2000.

Eunotia gibberula Ehrenberg, Phys. Abh. Akad. Wiss. Berlin, 1841: 414, pl. 3/4, fig. 8, 1843.

Epithemia gibberula (Ehrenberg) Kützing, Kies. Bacill. Diat., p. 35, pl. 29, fig. 54, pl. 30, fig. 3, 1844.

壳面新月形，背侧弧形凸出，中部有一小的缢缩，腹侧平直或略微凹入，两端逐渐狭窄，并略弯向腹侧；长 40—75 μm，宽 12—15 μm。龙骨和窝孔纹均不清楚，横肋纹 3—7 条/10 μm，横线纹 14—18 条/10 μm，在两条横肋纹之间的横线纹有 1—8 条。

扫描电镜下观察：龙骨稍高于壳面或是与壳面齐平，在背侧边缘形成一条纵向的空白区域，龙骨表面没有窝孔纹分布。管壳缝位于背侧边缘的龙骨上，壳缝在壳面中部断开，稍伸进壳面，形成一个小的缺刻(notch)。窝孔纹由双排孔组成，在不倾斜的壳面上，可见窝孔纹外边缘"C"形的孔，而当壳面具一定倾斜角度时，可见非常不规则的浮雕形。

生境：生于河流、草原中流水、小水渠、路边积水、沼泽、泉水井边草丛。

国内分布：天津、山西(运城)、吉林(长白山)、黑龙江(鸡西兴凯湖)、湖南(索溪峪、沅江流域)、广东(紫金)、贵州(沅江流域、乌江流域)、西藏(聂拉木、吉隆、昂仁、墨脱、林芝、措美、察隅、八宿、波密、芒康、察雅、江达、贡觉、札达、革吉、日土)、新疆(博乐、察布查尔、喀纳斯、哈巴河、北屯、阿勒泰)。

国外分布：亚洲(俄罗斯、新加坡、土耳其)、北美洲(美国、墨西哥、夏威夷群岛)、非洲(东部、南非)、南美洲(巴西、哥伦比亚)、欧洲(亚得里亚海、波罗的海、黑海、英国、德国、爱尔兰、罗马尼亚、西班牙、冰岛)、大洋洲(澳大利亚、新西兰)。

Grunow(1862) 和 Müller(1990) 在本种下放了很多变种，包括 var. *protracta*、var. *vanheurckii* 等，这些变种在外形上与 *R. gibberula* 相似，而在扫描电镜下会发现这些种类的壳面结构和壳缝构造有很大区别，因此本书采用 Krammer(1988b) 的观点，将原先的变种划分出去。

7. 毕列松棒杆藻 图版 LXXVII: 10—12

Rhopalodia brebissonii Krammer, *in* Lange-Bertalot & Krammer, p.76, figs. 48: 7—10, 1987; Krammer, Nova Hedwigia, 47(1—2): 165, figs. 76—90, 1988; Krammer & Lange-Bertalot, Bacillariophyceae, 2. Teil, *in* Ettl *et al.*, Süßwasserfl. Mitteleur., p. 164, figs. 113: 7—13, figs. 113A: 7—12, 1988, nachdr. 1997.

细胞体积小，背侧弧形凸出，腹侧近平直，两端逐渐狭窄，弯向腹侧，末端钝尖；壳体长 20—34 μm，宽 14—20 μm，壳面宽 6—8 μm，壳缝清晰可见，主要位于壳面背侧边缘，在中部形成一钝角，横肋纹清晰，稍呈辐射状，3—5 条/10 μm，横线纹 15—18 条/10 μm，两条横肋纹之间有横线纹 2—5 条。

生境：生于池塘、水沟中。

国内分布：新疆（皮山、察布查尔）。

国外分布：亚洲（俄罗斯），北美洲（加拿大、美国），南美洲（哥伦比亚），欧洲（英国、德国、波兰、罗马尼亚），大洋洲（澳大利亚、新西兰）。

在光镜下本种很容易与 *R. musculus*、*R. gibberula* 混淆，而 *R. brebissonii* 的体积相对其他两种要小一些，本书的标本，主要依据 Krammer(1988b) 中的 figs. 85—87，壳缝和肋纹均清晰可见，壳缝在壳面中部形成一个钝角，这些特征使其很容易和其他种类分开。Krammer(1988b) 观察到 *R. brebissonii* 的扫描电镜照片：壳缝管两侧为双排孔，壳面其他部分为单排孔。双排孔的构造与 *R. gibberula* 上孔的构造相似，孔的外侧为圆形、椭圆形或"C"形的结构。

R. brebissonii 与 *R. operculata* 的外形也很相似，但是 *R. brebissonii* 具清楚的线纹，15—18(20) 条/10 μm，而 *R. operculata* 的双排孔不清楚，30—40 条/10 μm。本种在标本中数量比较少，因此没有得到扫描电镜照片。

8. 缢缩棒杆藻 图版 LXXVII: 9；图版 LXXVIII: 4—5

Rhopalodia constricta (W. Smith) Krammer, *in* Lange-Bertalot & Krammer, Bibl. Diatomol. 15, p. 77, fig. 47: 12, figs. 48: 1—6, 1987; Krammer, Nova Hedwigia, 47(1—2): 169, figs. 149—167, 1988; Krammer & Lange-Bertalot, Bacillariophyceae. 2. Teil, *in* Ettl *et al.*, Süßwasserfl. Mitteleur., p. 164, fig. 110: 3, figs. 113A: 1—6, 1988, nachdr. 1997.

Epithemia constricta W. Smith, Syn. British Diat., p. 14, fig. 30: 248, 1853.

Rhopalodia gibberula var. *constricta* (W. Smith) Cleve-Euler, Diatomeen schw. Finnland. Teil. V, p. 43, figs. 1415k—l, 1952.

Rhopalodia gibberula var. *constricta* (W. Smith) Karsten, Die Diatomeen der Kieler Bucht, p. 98, fig. 122, 1899.

Rhopalodia musculus var. *constricta* (W. Smith) H. Peragallo & M. Peragallo, Diat. Mar. France, p. 303, figs. 77: 11—17, 1897—1908.

带面观宽椭圆形，壳面新月形，背侧强烈弧形凸出，中部有一明显的缢缩，腹侧平直或略微凹入，两端逐渐狭窄，并略弯向腹侧；壳面长 25—55 μm，宽 8—15 μm，横肋纹 3.5—6

条/10 μm，横线纹 15—20 条/10 μm，两条横肋纹之间有横线纹 2—4 条。

扫描电镜下观察：壳面不在一个水平面上，中间靠近背侧部分稍低，并有明显缢缩。龙骨与壳面差不多齐平，在背侧边缘形成一条纵向的区域，龙骨表面有肋纹分布。管壳缝位于背侧边缘的龙骨上，壳缝在壳面中部断开，伸进壳面，形成一个小的缺刻。整个壳面的线纹均由双排孔组成，排列不规则，对生或互生，孔为细小的圆孔，直径 0.1—0.2 μm，外侧开口简单，没有闭塞物。

生境：生于小水坑中。

国内分布：广东(连平)，贵州(水城)，新疆(阿克苏)。

国外分布：亚洲(俄罗斯)，北美洲(美国)，欧洲(英国、德国、西班牙、罗马尼亚)，大洋洲(澳大利亚、新西兰)。

本种经常被作为 *R. musculus* 的变种处理。我们在扫描电镜下发现本种与 *R. musculus* 壳面形状和壳面上的结构差别很大。因此，本书采用 Patrick 和 Reimer(1975)的观点，将 *R. constricta* 作为独立的种处理。

9. 石生棒杆藻　图版 LXXX：7—9

Rhopalodia rupestris (W. Smith) Krammer, *in* Lange-Bertalot & Krammer, p. 86, pl. 49, figs. 1—6, 1987; Krammer & Lange-Bertalot, Bacillariophyceae. 2. Teil, *in* Ettl *et al.*, Süßwasserfl. Mitteleur., p. 165, pl. 115, figs. 1—8, 1988, nachdr. 1997.

Epithemia rupestris W. Smith, A synopsis of the British Diatomaceae. p. 14, pl. 1, fig. 12, 1853.

壳面长 30—45 μm，宽 7—10 μm；横肋纹 3—4 条/10 μm，横线纹 18—22 条/10 μm。

生境：生于水库边渗出水、山溪中。

国内分布：贵州(赫章)，海南(五指山)。

国外分布：北美洲(美国)，欧洲(英国、德国、罗马尼亚)，大洋洲(新西兰)。

双菱藻科 Surirellaceae

单细胞，壳面呈波状上下起伏，或平直，或弯曲，龙骨及翼状结构围绕整个壳缘，管壳缝通过翼沟与细胞内部相联系，带面窄，简单。

本科有 7 属，我国淡水中有 4 属。

双菱藻科分属检索表

1. 壳面弯曲呈鞍形，上下壳面的顶轴呈直角 ………………………………… **4. 马鞍藻属 Campylodiscus**
1. 壳面不弯曲呈鞍形，上下壳面的顶轴平行排列 ………………………………………………………… 2
　2. 壳面窄，顶轴"S"形或直；中间区域窄线形，边界清楚 ……… **3. 长羽藻属 Stenopterobia**
　2. 壳面外形和结构与上述不同 …………………………………………………………………… 3
　　3. 横向波纹和线纹在中线区域是断开的 …………………………………… **2. 双菱藻属 Surirella**
　　3. 横向波纹和线纹在中线区域是不断开的 ……………………………… **1. 波缘藻属 Cymatopleura**

波缘藻属 Cymatopleura W. Smith

W. Smith, Ann. Mag. Nat. Hist. Ser. 2. 7: 12, 1851

单细胞，示壳面或带面观。壳体(frustule)等极，偶尔关于顶轴扭曲。色素体 1 个，由两片大而紧贴壳面的盘状结构组成，通过靠近细胞一端的极窄狭部连接(连接部位在异极种类中靠近窄的一端)。边缘呈深裂状。

壳面提琴形，或线形至椭圆形，具一个环绕壳面的壳缝系统。壳面顶轴方向呈横向波纹，一般在壳面中部较浅，上下壳面的波纹准确互补。壳面强烈硅质化，外壳面常具肋纹以及网状(reticulate)增厚结构，内壳面平坦。壳缝系统位于浅的龙骨上，龙骨壁常一同呈波纹状。线纹单排，由小圆孔组成，孔在内壳面具一边缘结构。线纹在壳面中部断开，形成一条纵向假线纹。壳缝简单，外壳面末端直或稍弯曲，内壳面形成一连续的槽沟(groove)。带面由两侧断开的环带组成，表面具少许瘤状物。

本属种类体积较大，分布于淡水中，常附着于沉积物或污泥中，喜高电导率的(碱)水环境。

本属分类较为清楚，种类容易辨认，数量不多。与双菱藻属(*Surirella*)和马鞍藻属(*Campylodiscus*)关系较近。壳面外部看似等极，但内部的结构和功能并非等极。关于模式种 *C. solea* 的异名仍有争论：Mann(1987)认为 *C. librile* 不是 *C. solea* 的异名，而 Krammer 和 Lange-Bertalot(1997)把 *C. librile* 作为 *C. solea* 的异名，本书采用了后者的观点。

模式种：草鞋形波缘藻[*Cymatopleura solea*(Brébisson) W. Smith]。

本志收编 4 种 5 变种。

波缘藻属分种检索表

1. 壳面椭圆形，等极，末端宽圆 ··· **1. 椭圆波缘藻 *C. elliptica***
1. 壳面提琴形，龙骨突延伸形成窄而连续的横向波纹，一直到两极 ····························· 2
   2. 壳面在同一平面上，不扭曲 ··· **2. 草鞋形波缘藻 *C. solea***
   2. 壳面沿顶轴方向稍扭曲 ··· 3
3. 壳面中部缢缩 ··· **3. 扭曲波缘藻 *C. aquastudia***
3. 壳面中部不缢缩 ··· **4. 新疆波缘藻 *C. xinjiangiana***

1. 椭圆波缘藻

Cymatopleura elliptica (Brébisson) W. Smith, Ann. Mag. Nat. Hist. Ser. 2, p. 13, pl. 3, figs. 10—11, 1851; Schmidt *et al.*, Atlas Diat.-Kunde, p. 276, 277, fig. 276: 7, figs. 277: 1—7, 1912; Hustedt, Bacillariophyta, *in* Pascher, Süßwass -Fl. Mitteleur. Heft 10, p. 427, fig. 825, 1930; Gasse, Bibliotheca Diatomologia, 11: 39, fig. XLIV: 1, 1986; Shi *et al.*, Compil. Rep. Surv. Alg. Resour. South-West. China, p. 116—117, 1994; Krammer & Lange-Bertalot, Bacillariophyceae. 2. Teil, *in* Ettl *et al.*, Süßwasserfl. Mitteleur., p. 170, figs. 119: 1—4, figs. 120: 1—6, 1988, nachdr. 1997; Zhu & Chen,

The Diatoms of the Suoxiyu Nature Reserve Area, Hunan, China, *in* Li *et al.*, The algal flora and aquatic fauna of the Wulingyuan Nature Reserve Area, Hunan, China. p. 57, 1989b; Zhu & Chen, Bacill. Xiz. Plat., p. 270, fig. 55: 1, 2000; Metzeltin *et al.*, Iconogr. Diatomol. 15, fig. 226: 3, 2005.

Surirella elliptica Brébisson ex Kützing, Kies. Bacill. Diat., p. 61, pl. 28, fig. 28, 1844.

1a. 原变种　　图版 LXXXI: 1—6; 图版 LXXXII: 2
var. **elliptica**

壳面宽线形、宽椭圆形至菱形-椭圆形，等极，末端宽圆形至楔圆形，壳缝系统位于壳面边缘浅的龙骨上，周生；壳面和带面均可见 4—6 条粗糙的波纹；壳面长 43—150 μm，宽 30—68 μm；龙骨突有 3—6 个/10 μm；横线纹有 17—20 条/10 μm。

生境：生于湖泊、草原中流水。

国内分布：山西(太原晋阳湖)，内蒙古(阿尔山)，辽宁(辽河)，浙江(西湖)，湖南(索溪峪、沅江流域)，贵州(沅江流域、乌江流域)，云南(西双版纳)，西藏(吉隆、错那、措美、察隅、措勤、普兰)，新疆(哈巴河、阿勒泰、福海)。

国外分布：亚洲(俄罗斯、蒙古国、土耳其)，非洲(东部)，北美洲(加拿大、美国)，南美洲(乌拉圭)，欧洲(波罗的海、黑海、英国、德国、爱尔兰、马其顿、波兰、罗马尼亚、西班牙)。

1b. 缢缩变种　　图版 LXXXII: 5
var. **constricta** Grunow 1862, Verh. zool. bot. Ges. Wien, p. 464(150), pl. 11, fig. 13; Schmidt *et al.*, Atlas Diat.-Kunde, p. 279, figs. 279: 4—7, 1912; Hustedt, Bacillariophyta, *in* Pascher, Süßwass -Fl. Mitteleur. Heft 10, p. 427, fig. 826, 1930; Cleve-Euler, Diatomeen schw. Finnland, Teil. V, p. 99, fig. 1520f, 1952; Zhu & Chen, Bacill. Xiz. Plat., p. 271, fig. 55: 2, 2000.

与原变种的区别在于：壳面中部明显缢缩；长 58—102 μm，宽 14—26 μm；龙骨突有 7—10 个/10 μm；横线纹有 24—25 条/10 μm。

生境：生于河流、小水渠、沼泽中。

国内分布：湖南(沅江流域)，四川(九寨沟)，贵州(沅江流域)，西藏(亚东、吉隆、墨脱、错那、措美、察隅、芒康、申扎、改则、日土)，宁夏(贺兰山)。

国外分布：北美洲(美国)，欧洲(英国、爱尔兰、罗马尼亚、西班牙)。

1c. 冬生变种　　图版 LXXXII: 1
var. **hibernica** (W. Smith) Van Heurck, A treatise on the Diatomaceae, p. 367, fig. 31: 863, 1896; Krammer & Lange-Bertalot, Bacillariophyceae. 2. Teil, *in* Ettl *et al.*, Süßwasserfl. Mitteleur., p. 170, figs. 119: 1—4, figs. 120: 1—6, 1988, nachdr. 1997.

Cymatopleura hibernica W. Smith, Ann. Mag. Nat. Hist. Ser. 2, p. 13, pl. 3, fig. 12, 1851.

?*Cymatopleura elliptica* var. *nobilis* (Hantzsch) Hustedt, *in* A. Schmidt *et al.*, Atlas

Diat.-Kunde, pl. 278, figs. 2, 4, 5, 1912; Fan *et al.*, Bull. Bot. Res. 21: 243, 2001.

与原变种的区别在于：壳面粗壮，菱形-椭圆形，末端楔圆形；长 118 μm，宽 67 μm；龙骨突有 3—4 个/10 μm；横线纹有 20—24 条/10 μm。

生境：湖泊中浮游。

国内分布：黑龙江(五大连池)，新疆(福海)。

国外分布：亚洲(俄罗斯、蒙古国)，北美洲(美国)，欧洲(波罗的海、英国、德国、爱尔兰)。

2. 草鞋形波缘藻

Cymatopleura solea (Brébisson) W. Smith, Ann. Mag. Nat. Hist. Ser. 2, p. 12, fig. 3: 9, 1851; Schmidt *et al.*, Atlas Diat.-Kunde, p. 275-276, figs. 275: 3—7, figs. 276: 2—3, 1911; Hustedt, Bacillariophyta, *in* Pascher, Süßwass -Fl. Mitteleur. Heft 10, p. 425—426, fig. 823a, 1930; Cleve-Euler, Diatomeen schw. Finnland, Teil. V, p. 95-96, fig. 1519, 1952; Zhu & Chen, The Diatoms of the Suoxiyu Nature Reserve Area, Hunan, China, *in* Li *et al.*, The algal flora and aquatic fauna of the Wulingyuan Nature Reserve Area, Hunan, China. p. 57, 1989b; Krammer & Lange-Bertalot, Bacillariophyceae. 2. Teil, *in* Ettl *et al.*, Süßwasserfl. Mitteleur., p. 168, figs. 116: 1—4, figs. 117: 1—5, figs. 118: 1—8, 1988, nachdr. 1997; Xin *et al.*, Journal of Shanxi University(Nat. Sci. Ed.)20(1): 105, 1997; Zhu & Chen., Bacill. Xiz. Plat., p. 271, fig. 55: 3, 2000; Ruck & Kociolek, Bibliotheca Diatomologica, 50: 42, figs. 48: 1—6, figs. 49: 7—11, figs. 50: 12—16, 2004.

Cymbella solea Brébisson, in Brébisson & Godey, Algues des environs de Falaise, décrites et dessinées par MM. de Brébisson et Godey, p. 51, fig. 7, 1835.

2a. 原变种　图版 LXXXII: 3—4; 图版 LXXXIII: 1—6; 图版 LXXXIV: 1—8; 图版 LXXXV: 1—4; 图版 LXXXVI: 1—6

var. **solea**

壳面宽线形，等极，中部缢缩，末端呈钝圆-楔形；壳面具粗糙的波纹，一般中部有或没有；长 42—200 μm，宽 20—40 μm；龙骨突有 7—10 个/10 μm；带面两侧具明显的波状皱褶。

扫描电镜下观察：壳面具细小横肋纹，稍高于壳面，肋纹在壳面中部断开，形成一条纵向的假线纹。肋纹之间具硅质的小枝结构，除了近壳面边缘(龙骨)处没有，其他部分都存在，长短不一，在壳面中部的小枝状结构较疏松，不规则，使整个壳面呈现网状结构。外壳面可见由细小孔纹组成的线纹，孔纹边缘不凸出，由于内壳面表面平整，孔纹更易观察到，边缘凸起或不凸起，而龙骨突之间的孔纹边缘不凸起。线纹密度 30—38 条/10 μm，线纹在顶端以及壳面边缘的密度要大于壳面中部。壳面内外的龙骨突上均不具线纹，龙骨突之间具双排线纹，其他部分为单排线纹，壳面中部的线纹近平行排列，两端的线纹斜向极节，壳套内侧也具线纹，单排，外侧情况不清楚。龙骨突在内壳面桥接壳缝管，向两侧扩展，在两个龙骨突之间形成一个圆孔，连接细胞内部和壳缝管，龙骨突单个均匀地分布在壳面内侧边缘，有7—10个/10 μm。壳缝管在外侧不高于壳面，所以不存在翼状突、翼状管和窗栏开孔。在外壳面，壳缝为简单的裂缝，在两极不连续，末端简单，内壳面，壳缝裂缝连续，并且容易观察到。环带由开带组成，在侧面打开，而不是两极。

生境：普生性种类。

国内分布：河北（昌黎），山西（太原晋阳湖、太原晋祠、绵山清水河），内蒙古（阿尔山），辽宁（辽河、沈阳），吉林（长白山），黑龙江（鸡西兴凯湖、五大连池），上海（淀山湖），江苏（昆山、常熟、苏州），浙江（西湖），湖南（岳阳、索溪峪、沅江流域），广西（灵川），贵州（威宁草海、施秉舞阳河、沅江流域、乌江流域），云南（西双版纳、滇池、洱海、维西、丽江），西藏（林芝、多庆、打隆、亚东、康马、吉隆、萨嘎、仲巴、昂仁、拉萨、加查、措美、察隅、波密、昌都、芒康、洛隆、申扎、改则、措勤、普兰、噶尔、革吉、日土），陕西（华山），宁夏（贺兰山），新疆（巴楚、阿克苏、博湖、阜康天池、博乐五一水库、赛里木湖、察布查尔、布尔津、喀纳斯、哈巴河、阿勒泰、福海）。

国外分布：亚洲（俄罗斯、菲律宾、蒙古国、土耳其、伊朗），非洲（埃及），北美洲（美国），欧洲（波罗的海、黑海、德国、爱尔兰、波兰、罗马尼亚、西班牙、挪威），大洋洲（澳大利亚）。

本种是一个世界广布种，在许多地区都有报道，并有清楚的（包括整体和局部）光镜和扫描电镜照片（Schmidt and Hustedt，1911；Round et al.，1990；Krammer and Lange-Bertalot，1997；Ruck and Kociolek，2004）。

本种的外形和壳面结构与其他种类差别较大，因此容易辨认。但是种间变异非常大，VanLandingham（1969）记录了将近40个变种。原变种的形态也比较多，Krammer 和 Lange-Bertalot（1997）报道的种类（fig. 116：1—4，117：1—5，118：1—8），形态差异明显，其中还将体积较大、长宽比大的变种 var. *gracilis* Grunow 归并进来（本书仍将该类型作为变种处理）。我们认为 Zhu 和 Chen（2000）报道的 *C. solea* var. *subconstricta*（fig. 55：5）应归并到原变种中。

本文观察到的大部分特征与文献相符：包括不连续的外壳缝，外部横肋纹，内部的龙骨突，以及壳体的大小，龙骨突的密度。同时，我们也观察到一些文献中没有描述过的特征：①外壳面上的线纹，由孔纹组成，单排（龙骨突之间的线纹为双排），孔纹边缘不凸出。而 Ruck 和 Kociolek（2004）报道：在外壳面没有观察到线纹（p. 43），并指出这与以往文献中描述的一致。②内壳面孔纹边缘凸起或不凸起，龙骨突之间的孔纹边缘不凸起。而在文献中，只报道过内壳面孔纹边缘凸起的特征（Round et al.，1990；Krammer and Lange-Bertalot，1997；Ruck and Kociolek，2004）。本书的显微观察结果将进一步完善对本种超微结构的认识。

2b. 细长变种　　图版 LXXXVII：1—2

var. **gracilis** Grunow，Verh. zool. bot. Ges. Wien，p. 466(152)，1862；Schmidt *et al.*，Atlas

Diat.-Kunde, p. 275, fig. 275: 2, 1911; Cleve-Euler, Diatomeen schw. Finnland, Teil. V, p. 96, fig. 1519 d, 1952.

光镜下与原变种的区别在于：壳面较大，纤细，长宽比明显大；长 180—250 μm，宽 31—35 μm；龙骨突有 7—9 个/10 μm。

生境：生于河边水坑中。

国内分布：内蒙古(阿尔山)，上海(松江)，青海(西宁)，新疆(阜康天池)。

国外分布：亚洲(俄罗斯、菲律宾)，北美洲(美国)，欧洲(罗马尼亚)。

Krammer 和 Lange-Bertalot(1997)将此类群归并到原变种中，本书认为该种的体积大小和形态与原变种差别较大，因此仍将这一类群作为变种处理。

2c. 细尖变种 图版 LXXXVII：3—7

var. **apiculata** (W. Smith) Ralfs, in Pritchard, A history of infusoria, p. 793, 1861; Schmidt et al., Atlas Diat.-Kunde, p. 275, figs. 275: 8—13, 1911; Hustedt, Bacillariophyta, in Pascher, Süßwass -Fl. Mitteleur. Heft 10, p. 426, fig. 823c, 1930; Cleve-Euler, Diatomeen schw. Finnland, Teil. V, p. 96, figs. 1519 f—h, 1952; Krammer & Lange-Bertalot, Bacillariophyceae. 2. Teil, in Ettl et al., Süßwasserfl. Mitteleur., p. 169, figs. 118: 4—8, 1988, nachdr. 1997; Yang, Bull. Bot. Resch., 15(3): 316, pl. I, fig. 3, 1999.

Cymatopleura apiculata W. Smith, Syn. British Diat., p. 37, pl. 10, fig. 79, 1953.

光镜下，与原变种的区别在于：壳面短粗，中部缢缩明显，末端稍延长；壳面中部具粗糙波纹。长 50—120 μm，宽 12—40 μm；龙骨突有 8—10 个/10 μm。

生境：生于河流、小水渠、小溪、沼泽、路边积水，石上附生。

国内分布：安徽(琅琊山)，云南(丽江)，新疆(察布查尔、布尔津、喀纳斯、哈巴河)。

国外分布：亚洲(土耳其)，北美洲(美国)，欧洲(英国、德国、爱尔兰、马其顿、波兰、罗马尼亚)，大洋洲(新西兰)。

2d. 整齐变种 图版 LXXXVI：7—9

var. **regula** (Ehrenberg) Grunow, Verh. zool. bot. Ges. Wien, p. 466(152), 1862; Hustedt, Bacillariophyta, in Pascher, Süßwass -Fl. Mitteleur. Heft 10, p. 426, fig. 823b, 1930; Cleve-Euler, Diatomeen schw. Finnland, Teil. V, p. 97, fig. 1519 s—v, 1952; Zhu et al., Diatom. Suoxiyu Nat. Preserv. Area Hunan, China, p. 57, 1989; Shi et al., Compil. Rep. Surv. Alg. Resour. South-West. China, p. 57, 1994; Zhu & Chen, Bacill. Xiz. Plat., p. 271, fig. 55: 4, 2000.

Surirella regula Ehrenberg, Verbreitung und Einfluss des mikroskopischen Lebens in Süd-und Nord-Amerika, p. 136, pl. 3/5, fig. 3, 1843.

Cymatopleura regula (Ehrenberg) Ralfs., in Schmidt et al., Atlas Diat.-Kunde, p. 276, figs. 276: 10—11, 1911.

光镜下与原变种的区别在于：壳面两侧平直；长 50—85 μm，宽 14—24 μm；龙骨突有 7—9 个/10 μm；带面两侧波状皱褶不清晰。

生境：生于河流、河边沼泽、湖边、小水渠、路边积水中。

国内分布：内蒙古(阿尔山)，黑龙江(五大连池)，湖南(沅江流域)，贵州(沅江流域、乌江流域)，西藏(曲水、康马、吉隆、昂仁)，新疆(察布查尔、伊宁、布尔津)。

国外分布：亚洲(俄罗斯)，北美洲(美国)，欧洲(爱尔兰、罗马尼亚)。

3. 扭曲波缘藻　图版 LXXXVIII：1—6；图版 LXXXIX：1—6；图版 XC：1—4
Cymatopleura aquastudia Q-M You & J.P. Kociolek, *in* You *et al.*, Fottea, 17(2)：297, 2017.

壳面沿顶轴方向稍扭曲。壳面提琴形，上下不对称，一端宽圆形，另一端窄，稍延伸，中部缢缩。壳面具 3—6 个粗糙的横波纹。壳面长 50.5—125 μm，中部缢缩位置的壳面宽 17—25 μm；最宽处的壳面 18—28 μm。在最宽处壳面的长宽比 2.9—5.8。肋纹有 7—9 个/10 μm，壳面边缘发育较良好，一直延伸到轴区；轴区窄线形，横线纹细，光镜下不清晰。

扫描电镜下观察：外壳面具窄的横肋纹，稍高于壳面，肋纹在壳面中部断开，形成一条纵向的窄的轴区。壳表面具一层硅质，覆盖了表面的筛孔纹。壳面边缘具双排线纹，在中部形成单排线纹。筛孔纹 80—95 个/10 μm，近壳缝处较为密集。壳缝位于浅的龙骨上，简单，在两极都不连续，具壳缝末端，末端的壳缝裂缝不增大，稍弯向壳面。

内壳面为单排线纹，线纹由小圆形的筛孔组成，80—95 个/10 μm。在内壳面边缘，龙骨突之间的位置，线纹是双排的。龙骨突肋状，8—9 个/10 μm，均匀分布，连接壳面和壳套。壳缝简单，在两极都连续，不具壳缝末端。

生境：生于路边临时性的水塘中，伴生轮藻。

国内分布：新疆(叶城、阿克苏、博湖、察布查尔)。

本种与 *C. solea* 在形态和结构上最相似，最大区别在于：本种壳面沿纵轴方向稍扭曲，且壳面上下不对称。

4. 新疆波缘藻　图版 XCI：1—6；图版 XCII：1—5；图版 XCIII：1—5
Cymatopleura xinjiangiana Q-M You & J.P. Kociolek, *in* You *et al.*, Fottea, 17(2)：297, 2017.

壳面沿顶轴方向稍扭曲。壳面宽楔形，上下不对称，一端延伸、呈宽圆形，另一端窄，楔形，中部不缢缩。壳面具 3—5 个粗糙的横波纹。壳面长 45—75 μm，宽 18—21 μm。壳面的长宽比 2.4—3.5。肋纹有 7—9 个/10 μm，壳面边缘发育较良好，一直延伸到轴区；轴区窄线形，横线纹细，光镜下不清晰。

扫描电镜下观察：外壳面具窄的横肋纹，稍高于壳面，肋纹在壳面中部断开，形成一条纵向的窄的轴区。肋纹之间具不规则的硅质结构，形成一个网面。壳表面具一层硅质，覆盖了表面的筛孔纹。壳缝位于浅的龙骨上，简单，在两极都不连续，具壳缝末端，末端的壳缝裂缝不增大稍弯向壳面。

内壳面为单排线纹，线纹由小圆形的筛孔组成，75—85 个/10 μm。在内壳面边缘，龙骨突之间的位置，线纹是双排的。龙骨突肋状，9—11 个/10 μm，均匀分布，连接壳面和壳套。壳缝简单，在两极都连续，不具壳缝末端。

生境：生于路边临时性的水塘中，伴生轮藻。

国内分布：新疆（巴楚、阿克苏、察布查尔）。

本种与 *C. cochlea* Brun 和 *C. elliptica* f. *spiralis* Boyer 在形态和结构上相似，最大区别在于：本种体积较小，壳面具圆形的末端，且在较宽的一端稍延伸。

双菱藻属 Surirella P. J. F. Turpin

P. J. F. Turpin, Mém. Mus. Hist. Nat. 16: 363, 1828.

单细胞，示壳面或带面观。壳体等极，或异极呈楔形，偶尔关于顶轴扭曲。色素体1个，由两片大而紧贴壳面的盘状结构组成，通过靠近细胞一端的极窄狭部连接（连接部位在异极种类中靠近窄的一端）；或色素体2个，边缘呈深裂状。

壳面线形至椭圆形，或倒卵形，有时提琴形，具一个环绕壳面的壳缝系统。壳面强烈硅质化，表面平坦，或呈凹面，有时具波纹，与顶轴平行，表面有时具硅质的瘤或脊，偶尔在壳面中线附近具刺。外壳面肋纹不明显，线纹常多排，由具分支孔板(volae)的小圆孔组成，线纹在壳面中部常被一凸出的脊断开。壳缝系统位于龙骨上，龙骨壁也呈波纹状，有时两者融合在一起。龙骨突肋状或盘状，在内壳面包住壳缝。壳缝简单，内外壳面的壳缝末端直，不膨大，有时在内壳面中部连续，壳缝末端位于两极。带面由数条断开的环带组成。

本属种类体积较大，分布于淡水和海水中，常附着于沉积物或污泥中。

Schmidt(1979a)对本属细胞分裂过程研究表明：细胞的一端（异极种类较宽的一端）与舟形类硅藻的中部同源，而另一端则与舟形类硅藻的两端同源。细胞核在有丝分裂周期中移向"中部"一端，此处为子细胞硅质沉积的开端。Krammer 和 Lange-Bertalot(1987) 讨论了 *S. ovalis* 及变种的复杂性。本属可分为三个组：羽纹组(*Pinnatae* group)、粗壮组(*Robustae* group)和华壮组(*Fastuosae* group)。其中 *Fastuosae* group 为海水种类，本书不收录。

模式种：具条纹双菱藻(*Surirella striatula* Turpin)。

本志收编25种3变种1变型。

双菱藻属分组检索表

1. 壳面具翼及翼状管，壳面边缘可见环形结构，外形通常线形-卵圆形·············· 粗壮组 Sect. *Robustae*
1. 体积一般较小，壳面不具翼及翼状管，外形通常宽椭圆形，具假漏斗结构，龙骨突呈管状·············· ·············· 羽纹组 Sect. *Pinnatae*

羽纹组 Section *Pinnatae*

羽纹组分种检索表

1. 细胞两端相等(等极) ··· 2
1. 细胞两端不相等(异极) ··· 5
    2. 体积较大，粗壮线形 ·· **1. 细长双菱藻** *S. gracilis*
    2. 体积较小，精细线形 ·· 3
3. 壳面哑铃形，两端宽，中部窄 ·· **2. 泰特尼斯双菱藻** *S. tientsinensis*
3. 壳面窄线形，朝两端逐渐变窄 ··· 4
    4. 体积较小，带面的长宽比 5∶1，横向波纹到达中线 ·· **3. 窄双菱藻** *S. angusta*
    4. 体积较长，窄，带面的长宽比可达 15∶1 ··· **4. 拉普兰双菱藻** *S. lapponica*
5. 横向波纹平行排列，延伸到中线，长宽比达 5∶1，龙骨突 60—80 个/100 μm
                                                                  5. 微小双菱藻 *S. minuta*
5. 特征与上述不同 ·· 6
    6. 壳面宽阔，梨形 ·· 7
    6. 壳面卵圆形至线形-卵圆形 ·· 8
7. 壳面具略会聚或是不会聚的波纹 ·· **6. 囊形双菱藻** *S. crumena*
7. 壳面具明显会聚的波纹，龙骨突和假漏斗结构共同形成一个窄环形边缘 ············ **7. 派松双菱藻** *S. peisonis*
    8. 壳面两端均为楔形(在体积大的种类中较明显) ··· 9
    8. 在中型至大型的种类中，壳面一端或两端均为宽圆形 ··· 11
9. 壳面呈线形椭圆形 ··· **8. 盘状双菱藻** *S. patella*
9. 壳面呈卵圆形 ··· 10
    10. 假漏斗结构边界不清楚，壳面具稍会聚的波纹 ··· **9. 卵圆双菱藻** *S. ovalis*
    10. 假漏斗结构边界清楚，壳面具强烈会聚的波纹 ··· **10. 布赖韦尔双菱藻** *S. brightwellii*
11. 横向波纹或线纹清晰 ··· 12
11. 横向波纹或线纹不清晰 ··· 13
    12. 壳面一般短于 50 μm ··· **11. 近盐生双菱藻** *S. subsalsa*
    12. 壳面一般长于 70 μm，波纹顶部和底部基本一样宽
                                                         12. 具条纹双菱藻 *S. striatula*
13. 壳面线形-椭圆形 ·· **13. 维苏双菱藻** *S. visurgis*
13. 壳面卵形 ··· **14. 布列双菱藻** *S. brebissonii*

1. 细长双菱藻　图版 XCV：7—12；图版 XCVI：2—3；图版 XCVII：1—3
Surirella gracilis (W.Smith) Grunow, Verh. zool. bot. Ges. Wien, p. 144(458), fig. 7/10：
    11, 1862; Cleve-Euler, Diatomeen schw. Finnland, Teil. V, p. 118, figs. 1556 a, b,
    1952; Zhu & Chen, The Diatoms of the Suoxiyu Nature Reserve Area, Hunan, China,
    in Li *et al.*, The algal flora and aquatic fauna of the Wulingyuan Nature Reserve Area,
    Hunan, China. p. 57, 1989b; Krammer & Lange-Bertalot, Bacillariophyceae. 2. Teil,
    in Ettl *et al.*, Süßwasserfl. Mitteleur., p. 188, figs. 136：1—4, 1988, nachdr. 1997.
Tryblionella gracilis W.Smith, Syn. British Diat., p. 35, pl. 10, fig. 75, 1853.
？*Surirella moelleriana* sensu Hustedt, p. 435, fig. 842, 1930; Zhu & Chen, Bacill. Xiz. Plat.,

p. 274, fig. 55: 7, 2000.

壳体等极, 带面观线形。壳面线形, 两侧平行, 稍微凸出或凹入, 末端楔圆形; 长40—100 μm, 宽11—15 μm; 没有翼状结构, 壳面结构与 S. angusta 相似, 具清楚的假漏斗结构, 可至壳面中部; 龙骨突有70—90个/100 μm; 横线纹在光镜下不清楚。

生境: 生于河流、河边渗出水、水渠、路边积水, 水草附生。

国内分布: 内蒙古(牙克石), 黑龙江(鸡西兴凯湖、漠河), 江苏(昆山), 湖南(长沙、岳阳、索溪峪), 贵州(乌江流域), 西藏(乃东), 新疆(阿克苏、叶城、库尔勒、察布查尔、布尔津)。

国外分布: 亚洲(俄罗斯), 北美洲(美国), 欧洲(波罗的海、英国、德国、爱尔兰、马其顿、罗马尼亚), 大洋洲(澳大利亚、新西兰)。

新疆的标本在形态和大小上与 Krammer 和 Lange-Bertalot (1997) 报道的相似, 但是龙骨突密度较大, 为70—90个/100 μm, 而 Krammer 和 Lange-Bertalot (1997) 报道的龙骨突仅为40—60个/100 μm。西藏的标本在宽度上稍窄(宽8.5—11 μm), 其他特征相似。

2. 泰特尼斯双菱藻　图版 CII: 1

Surirella tientsinensis Skvortzow, Journal of Botany 65 (772): 107, fig. 24, 1927; Fan, Studied on Aulonoraphidinales (Surirellales) from Heilongjiang province. p. 44, fig. 5: 13, 2004.

壳面线形, 两侧明显凹入, 两端宽圆; 长44—60 μm, 宽10—13 μm; 龙骨突有60个/100 μm, 横线纹在光镜下看不清楚。

生境: 生于河流、水坑中。

国内分布: 天津, 内蒙古(阿尔山), 黑龙江(鸡西兴凯湖、五大连池), 江苏(昆山), 广西(灵川)。

国外分布: 亚洲(俄罗斯), 欧洲(中部)。

3. 窄双菱藻　图版 XCIV: 1—11; 图版 XCV: 1—4; 图版 XCVI: 1

Surirella angusta Kützing, Kies. Bacill. Diat., p. 61, fig. 30: 52, 1844; Hustedt, Bacillariophyta, *in* Pascher, Süßwass -Fl. Mitteleur. Heft 10, p. 435, figs. 844—845, 1930; Gasse, Bibliotheca Diatomologia, 11: 171, figs. XXXIX: 11—12, 1986; Zhu & Chen, The Diatoms of the Suoxiyu Nature Reserve Area, Hunan, China, *in* Li *et al.*, The algal flora and aquatic fauna of the Wulingyuan Nature Reserve Area, Hunan, China. p. 57, 1989b; Shi *et al.*, Compil. Rep. Surv. Alg. Resour. South-West. China, p. 116-117, 1994; Krammer & Lange-Bertalot, Bacillariophyceae. 2. Teil, *in* Ettl *et al.*, Süßwasserfl. Mitteleur., p. 187, figs. 133: 6—13, 134: 6—10, 1988, nachdr. 1997; Zhu & Chen, Bacill. Xiz. Plat., p. 272, fig. 54: 12, 2000; Metzeltin *et al.*, Iconogr. Diatomol. 15, figs. 219: 5—9, 221: 1—7, 2005; Antoniades *et al.*, Iconogr. Diatomol.: 17: 296, figs. 79: 1—5? 6—9, 2008.

Surirella ovalis var. *angusta* (Kützing) Van Heurck, Synopsis des Diatomées de Belgique, p. 189, pl. 73, fig. 13, 1885.

Surirella ovata var. *angusta* (Kützing) Cleve-Euler, Diatomeen schw. Finnland. Teil. V, p. 123, figs. 1566 k—l, 1952.

壳体等极或稍异极，带面线形-矩形；壳面线形，等极，末端楔形；长 16—50 μm，宽 6—10 μm，长宽比例为 1：3—1：4，最多可达到 1：5；没有翼状结构，龙骨突有 50—80 个/100 μm；壳缘具假漏斗结构，其上有细密的线纹，光镜下看不清。

扫描电镜下观察：外壳面的肋纹分组明显，两组肋纹均开始于壳面边缘并且延伸至壳面中部(在中部形成一条纵向窄线形的空白区域)，一组肋纹高于壳面，而另一组肋纹几乎与壳面齐平，一般两条高的肋纹之间会有 2—5 条与壳面齐平的肋纹，这样就使壳面呈现轻微的波纹状，光镜下常观察到的肋纹就是这里高于壳面的肋纹。在壳面边缘，两条肋纹(包括两种类型的肋纹)之间会形成一个小孔，小孔的直径约为肋纹宽度的一半。小孔外侧的龙骨上有点纹，围绕壳面一圈，与肋纹密度差不多。内壳面，龙骨突单一且均匀地分布在壳面边缘，圆柱形，从边缘延伸至壳面中部，高度逐渐与内壳面齐平，位置与外壳面高的肋纹相对应，密度也为 50—80 条/100 μm，龙骨突之间有 2—5 条细小的肋纹，壳面边缘，龙骨突之间有一矩形的孔，壳缝管通过此孔与细胞内部交流，细小的肋纹从孔的边缘延伸至壳面中部。壳缝管几乎与壳面齐高，不存在翼状突，翼状管和窗栏开孔。壳套结构关于壳缝裂缝镜面对称。在外壳面，两个壳缝裂缝不连续，末端简单。在内壳面，壳缝裂缝的情况未观察到。

生境：生于湖泊、溪流、河边渗出水、小水沟、池塘、路边积水、沼泽，水草附生。

国内分布：天津，山西(太原晋祠、绵山清水河)，内蒙古(牙克石、阿尔山达尔滨湖)，辽宁(辽河)，吉林(长白山)，黑龙江(鸡西兴凯湖、五大连池、漠河)，江苏(昆山、苏州)，安徽(黄山、宁国)，福建(厦门、金门)，湖北(武汉植物园、神农架)，湖南(长沙、岳阳、索溪峪、沅江流域)，广东(龙川、紫金、惠州)，广西(灵川)，贵州(赫章、沅江流域、乌江流域、梵净山)，云南(西北部)，西藏(墨竹工卡、林芝、浪卡子、聂拉木、亚东、吉隆、萨嘎、昂仁、工布江达、墨脱、米林、乃东、错那、隆子、措美、察隅、八宿、波密、昌都、芒康、察雅、洛隆、江达、贡觉、类乌齐、申扎、改则、措勤、普兰、革吉)，新疆(巴楚、阿克苏、博湖、博乐、察布查尔、布尔津、喀纳斯、哈巴河、福海、阿勒泰)。

国外分布：亚洲(俄罗斯、蒙古国、伊朗、以色列、土耳其)，非洲(东部)，北美洲(加拿大、美国、夏威夷群岛)，南美洲(阿根廷、巴西、哥伦比亚、乌拉圭)，欧洲(奥地利、波罗的海、英国、芬兰、德国、爱尔兰、马其顿、波兰、罗马尼亚、西班牙)，大洋洲(澳大利亚、新西兰)。

根据显微观察，采自我国的标本可分为两种形态类型，形态 I：肋纹密度小，5—6 条/10 μm；形态 II：肋纹密度大，7—8 条/10 μm。形态 I 与 Krammer 和 Lange-Bertalot(1997) 中 fig. 133：6—13 相吻合；形态 II 与 Krammer 和 Lange-Bertalot(1997) 中 fig. 134：1 相吻合。

4. 拉普兰双菱藻　图版 CII：3—6

Surirella lapponica A. Cleve，On recent freshwater Diatoms from Lule Lappmark in Sweden，
p. 25，fig. 1：26，1895；Cleve-Euler，Diatomeen schw. Finnland，Teil. V，p. 118，
figs. 1558 a—d，1952；Krammer & Lange-Bertalot，Bacillariophyceae. 2. Teil，*in* Ettl *et al.*，Süßwasserfl. Mitteleur.，p. 188，figs. 135：15—17，1988，nachdr. 1997；Zhu & Chen，
Bacill. Xiz. Plat.，p. 273，fig. 54：16，2000.

壳体等极，带面线形，壳面窄线形，两侧平行，末端楔圆形；长 50—79 μm，宽 6.5—11 μm，长宽比例可达到 6；没有翼状结构，龙骨突有 30—50 个/100 μm；横线纹有 20—28 条/10 μm。

国内分布：内蒙古(阿尔山达尔滨湖)，江苏(昆山)，贵州(乌江流域)，西藏(定日、墨脱)。

国外分布：亚洲(俄罗斯)，北美洲(加拿大)，欧洲(波罗的海、英国、德国)。

5. 微小双菱藻　图版 XCVIII：1—9；图版 XCIX：1—2

Surirella minuta Brébisson *in* Kützing，Species algarum，p. 38. 1849；Krammer & Lange-Bertalot，Diat. Res.，2(1)：89，figs. 69—87，1987；Krammer & Lange-Bertalot，Bacillariophyceae. 2. Teil，*in* Ettl *et al.*，Süßwasserfl. Mitteleur.，p. 186，figs. 135：1—14，1988，nachdr. 1997；Xin *et al.*，Journal of Shanxi University (Nat. Sci. Ed.) 20(1)：106，1997；Metzeltin *et al.*，Iconogr. Diatomol. 15，fig. 225：4，2005；Antoniades *et al.*，Iconogr. Diatomol. 17：298，figs. 79：10—14，2008.

Surirella ovata Kützing，Kies. Bacill. Diat.，p. 62，pl. 7，figs. 1—4，1844.

Surirella pinnata W. Smith，Syn. British Diat.，p. 34，pl. 9，fig. 72，1853.

Surirella minuta var. *pinnata* (W.Smith) Grunow，Verh. zool. bot. Ges. Wien，p. 146 (460)，1862.

Surirella ovata var. *pinnata* (W.Smith) Brun，Diat. Alpes Jura，p. 98，pl. 2，fig. 5，1880.

Surirella ovalis var. *ovata* (Kützing) Van Heurck，Syn. Diat. Belgique，p. 188，pl. 73，figs. 5—7，1885.

Surirella ovalis var. *pinnata* (W. Smith) Van Heurck，Syn. Diat. Belgique，p. 189，pl. 73，fig. 12，1885.

Surirella ovata var. *pinnata* (W. Smith) Hustedt，Bacillariophyta，*in* Pascher，Süßwass -Fl. Mitteleur. Heft 10，p. 442，fig. 866，1930.

?*Surirella ovata* var. *salina* (W. Smith) Van Heurck，Treat. Diat.，1896；Zhu & Chen，Bacill. Xiz. Plat.，p. 275，fig. 56：5，2000.

壳体异极，带面楔形，壳面线形椭圆形，一端宽圆形，另一端楔形。长 20—45 μm，宽 9—11 μm，没有翼状结构。假漏斗结构可到达中线，龙骨突 50—80 个/100 μm，横线纹在光镜下看不清楚。

扫描电镜下观察：壳面横线纹和横肋纹交替出现。在外壳面，具肋纹和单排线纹，肋纹高于线纹。没有看到内壳面的点纹。外壳面边缘的大部分肋纹与壳缝管相融合，同

时延伸进壳面窄线形的中线；也有一些肋纹以短小、单一或分支的形式终止于壳面边缘。每2—5个肋纹以统一的形式与壳缝管边缘连接，其间有1—2条肋纹与壳缝管较低的地方连接，这样壳面边缘呈现出裙状花边。龙骨突桥接壳面内部和壳套，50—80个/100 μm，龙骨突通常单个均匀地分布在壳面内部，偶尔也会两个连在一起，龙骨突上没有线纹。每一对龙骨突之间有一个孔，细胞内部通过此孔与壳缝管发生联系。在外侧，壳缝管几乎与壳面等高，所以不存在翼状突、翼状管和窗栏开孔。壳缝裂缝在外壳面不连续，末端简单；在内壳面，宽圆形一端的壳缝裂缝连续，楔形一端的壳缝裂缝不连续。

生境：生于溪流、小水沟、沼泽、路边积水中。

国内分布：天津，内蒙古(阿尔山达尔滨湖)，山西(太原晋祠)，辽宁(辽河、本溪)，黑龙江(五大连池)，江苏(昆山)，安徽(宁国)，湖南(长沙、岳阳、索溪峪、沅江流域)，广东(东江流域)，广西(灵川)，海南(琼中)，贵州(赫章、沅江流域、乌江流域、梵净山)，西藏(察隅、芒康、班戈、申扎、日土)，新疆(乌鲁木齐盐湖、阿克苏、察布查尔、布尔津、喀纳斯、哈巴河、阿勒泰)。

国外分布：亚洲(俄罗斯、蒙古国、伊朗、以色列、土耳其)，非洲(东部)，北美洲(加拿大、美国、墨西哥)，南美洲(阿根廷、巴西、哥伦比亚)，欧洲(亚得里亚海、波罗的海、黑海、丹麦、英国、德国、爱尔兰、马其顿、波兰、罗马尼亚、西班牙)，大洋洲(澳大利亚、新西兰)。

S. minuta 的壳面结构与 *S. ovalis* 非常相似，主要不同在于壳面的外形和大小。光镜下 *S. minuta* 小型个体与 *S. brebissonii* var. *kuetzingii* 相似。Krammer和Lange-Bertalot(1987)通过对Brébisson和W. Smith 的模式材料进行对比，得出 *S. minuta* 和 *S. pinnata* 为同一种类，特别是壳面外形、壳面结构以及壳缝管结构都非常相似。假漏斗在壳面中部汇聚，一直到壳面中间窄的轴区。

S. minuta 常与 *S. angusta* Kutzing(1844)一起出现。一般壳面结构和壳缝管的孔都非常相似。壳面外形的不同在于 *S. angusta* 两端是楔形的，而 *S. minuta* 两端是钝圆的。

6. 囊形双菱藻　　图版 CVI：3

Surirella crumena Brébisson ex Kützing, Species Algarum, p. 38, 1849; Krammer & Lange-Bertalot, Bacillariophyceae. 2. Teil, *in* Ettl *et al.*, Süßwasserfl. Mitteleur., p. 182, figs. 129: 1—5, 1988, nachdr. 1997.

Surirella ovata var. *crumena* (Brébisson) Husedt, Bacillariophyta, *in* Pascher, Süßwass -Fl. Mitteleur. Heft 10, p. 445, fig. 867, 1930; Zhu & Chen, Bacill. Xiz. Plat., p. 275, fig. 57: 1, 2000.

Surirella ovalis var. *crumena* (Brébisson ex Kützing) Van Heurck, Synopsis des Diatomées de Belgique, p. 188, pl. 73, fig. 1, 1885; Cleve-Euler, Diatomeen schw. Finnland, Teil. V, p. 121, figs. 1565 g, h, 1952.

壳面很宽，近圆形；壳面长31—85 μm，宽25—71 μm，在不同种群中，龙骨突密度变化较大，其中体积大的类群龙骨突密度大，25—60 个/100 μm，横线纹有12—17 条/10 μm。

生境：生于湖泊，岩石水草上附生。

国内分布：西藏(浪卡子、昂仁、申扎)。

国外分布：北美洲(加拿大、美国)，欧洲(波罗的海、英国、德国、爱尔兰、波兰)。

本种生活史尚不清楚，常被误认为是 *S. ovata* 的变种。然而，本种通常是以一个群体形式存在的，具有连续的细胞分裂变化模式，这一点与其他相似种类不同。单独一个体积较大的种类容易被鉴定成 *S. brebissonii*(如 Krammer and Lange-Bertalot, 1987 中，fig. 34)。

S. crumena 嗜高导电率的水体，没有 *S. brebissonii* 常见。*S. crumena* 和 *S. brightwellii* 都具有圆形壳体，外形相似，因此常出现错误鉴定。W. Smith(1853)认为 *S. crumena* 和他观察到的 *S. brightwellii* 是同一个种。然而，两个类群的超微结构和生活史类型不同，并和 Pinnatae 组中其他种类分开(Krammer and Lange-Bertalot, 1987)。

采自西藏的标本，海拔较高，为 4440—5100 m，水温较低，为 0—10℃，与 Krammer 和 Lange-Bertalot(1997)中的描述(壳面长 30—65 μm，宽 27—31 μm，龙骨突有 35—80 个/100 μm，线纹有 17—25 条/10 μm)相比，壳面宽度和线纹密度差别较大，有待进一步研究。

7. 派松双菱藻　图版 CIV：1—4；图版 CV：1—4；图版 CVI：1

Surirella peisonis Pantocsek, Kieselalgen oder Bacillarien des Balaton, p. 96(123), pl. 12, fig. 288, 1902; Krammer & Lange-Bertalot, Bacillariophyceae. 2. Teil, *in* Ettl *et al.*, Süßwasserfl. Mitteleur., p. 183, figs. 131: 1—3, 1988, nachdr. 1997.

Surirella ovalis var. *maxima* Grunow, 1862; Zhu & Chen, Bacill. Xiz. Plat., p. 274, fig. 56: 3, 2000.

壳体较大，异极，壳面卵圆形至楔圆形，长 60—120 μm，宽 40—70 μm，龙骨突有 30—50 个/100 μm，线纹有 15—18 条/10 μm。

生境：生于湖泊中，偏好半咸水和淡水。

国内分布：黑龙江(牡丹江镜泊湖)，西藏(班戈)，新疆(尉犁、博湖、赛里木湖)。

国外分布：亚洲(蒙古国)，北美洲(美国)，欧洲(罗马尼亚、西班牙)。

本种是一类体积较大的种群，种内变异也比较大。目前，难点在于没有学者看见模式标本。Pantocsek(1902)描绘了此种为长椭圆形以及在"典型的"*S. peisonis* 种群中很少见的结构。*S. peisonis* var. *pyriformis* Pantocsek 1901 在外形和结构上与新歇德勒湖(奥地利)和附近湖泊的种群非常相似，这些水体都是高导电率的(Krammer and Lange-Bertalot, 1987)。

本种与 *S. ovalis* 在外形上较相似，不同之处在于：本种的同心波纹和龙骨突更明显。Schmidt(1979a)发现 *S. peisonis* 与 *S. ovalis* 的初始细胞都是椭圆-披针形；*S. ovalis* 长约 130 μm，这与 Pantocsek 报道的 *S. peisonis*(112—120 μm)长度相似。而这些并不能证明 *S. peisonis* 与 *S. ovalis* 是同一物种，因为同一地点两个种类的变异模式是不同的，此外用初始细胞的相似性来证明它们是同种不够充分，因为初始细胞通常与后期细胞的外形不相同；初始细胞一般很少具有成熟细胞的特征(Krammer and Lange-Bertalot, 1987)。

8. 盘状双菱藻　图版 CIX：3

Surirella patella Kützing, Kies. Bacill. Diat., p. 61, fig. 7: 5, 1844; Hustedt, Bacillariophyta, in Pascher, Süßwass -Fl. Mitteleur. Heft 10, p. 445, fig. 868, 1930; Cleve-Euler, Diatomeen schw. Finnland, Teil. V, p. 120, fig. 1561 c—e, 1952; Krammer & Lange-Bertalot, Bacillariophyceae. 2. Teil, in Ettl et al., Süßwasserfl. Mitteleur., p. 185, figs. 137: 1—9, 1988, nachdr. 1997; Zhu & Chen, Bacill. Xiz. Plat., p. 275, fig. 57: 5, 2000.

壳体异极，带面楔形，壳面线形-卵圆形至线形-披针形，两端均为钝楔圆形；长 47—88 μm，宽 25—34 μm，没有翼状结构；龙骨突粗壮，只在壳面边缘可见，有时小一些的龙骨突会延伸进壳面，30—50 个/100 μm；壳面具微弱的同心波纹，壳缘具假漏斗结构；横线纹有 15—16 条/10 μm。

生境：生于湖泊、浅水滩，泥上附着。

国内分布：西藏（昂仁、班戈、申扎）。

国外分布：欧洲（中部）。

本种与宽椭圆形的 S. ovalis 较为相似，不同之处在于：本种壳面较长圆一些，壳面结构更粗糙。同时，在 S. ovalis 中，大体积到小体积壳体的长宽比例减少速度比 S. patella 快。因此如果能观察到群体，较容易鉴定到种。小而单一的种类只能在扫描电镜下分辨出来。此外，本种壳缝管、孔纹和龙骨突的结构与 S. brebissonii 非常相似 (Krammer and Lange-Bertalot, 1987)。

9. 卵圆双菱藻　图版 XCIX：3—4；图版 C：1—4；图版 CI：1—7

Surirella ovalis Brébisson, Consid. Diat., p. 17, 1838; Schmidt et al., Atlas Diat.-Kunde, p. 24, figs. 24: 1—4, 1885; Cleve-Euler, Diatomeen schw. Finnland, Teil. V, p. 120, figs. 1565 a, b, 1952; Mitsuzo, Flora North-Eastern Prov. China, part. IV, p. 1413, fig. 4: 25, 1970; Gasse, Bibl. Diatomol. 11: 172—173, fig. XLIII: 3, 1986; Krammer & Lange-Bertalot, Diat. Res., 2(1): 82, figs. 1(ov1), 2, 3, 6—8, 10—20, 1987; Krammer & Lange-Bertalot, Bacillariophyceae. 2. Teil, in Ettl et al., Süßwasserfl. Mitteleur., p. 178, figs. 125: 1—7, 1988, nachdr. 1997; Xin et al., Journal of Shanxi University (Nat. Sci. Ed.) 20(1): 106, 1997; Zhu & Chen, Bacill. Xiz. Plat., p. 274, fig. 55: 8, 2000; Ruck & Kociolek, Bibl. Diatomol. 50: 21, figs. 14—17: 1—21, 2004; Metzeltin et al., Iconogr. Diatomol. 15, figs. 219: 1—4, figs. 220: 1, 2, 2005.

Suriraya ovalis (Brébisson) Pfitzer, Untersuchungen uber Bau und Entwickelung der Bacillariaceen (Diatomaceen), p. 112, 1871.

Surirella ovata var. *ovalis* (Brébisson) Kirchner, Algen. in Cohn, Kryptogamen-Flora von Schlesien. Part 1. Vol. 2, p. 201, 1878.

壳面椭圆披针形，异极（大型种类近等极），一端楔形，另一端宽圆形，大型种类两端均为楔形；带面观线形至轻微楔形；壳面具与边缘平行的同心波纹，壳缘具假漏斗结构；壳面长 27—101 μm，宽 20—44.5 μm，长宽比例 3：1—1：1；龙骨突有 35—60 个/100 μm；

横线纹有 13—16 条/10 μm。

扫描电镜下观察：壳面横线纹和横肋纹交替出现，肋纹稍高于线纹。在外壳面，具肋纹（只存在于外壳面），组成线纹的点纹不易观察到。在内壳面，很容易看到双排（偶尔单排或三排）点纹，排列不规则，一般第三排从龙骨突开始一直伸进龙骨突间距，点纹边缘不凸出，50—60 个/10 μm。外壳面边缘的大部分肋纹与壳缝管相融合，同时延伸进壳面同心波曲的凹陷区域；也有一些肋纹以短小、单一或分支的形式终止于壳面边缘。每 2—5 个肋纹以统一的形式与壳缝管边缘连接，其间有 1—2 条肋纹与壳缝管较低的地方连接，这样壳面边缘呈现出裙状花边。壳缝管内侧的龙骨突桥接壳面内部和壳套，3.5—6 个/10 μm，龙骨突通常单个均匀地分布在壳面内部，偶尔也会两个连在一起，龙骨突上没有线纹。每一对龙骨突之间有一个椭圆形或线形椭圆形的孔，细胞内部通过此孔与壳缝管发生联系。在外侧，壳缝管几乎与壳面等高，所以不存在翼状突、翼状管和窗栏开孔。壳缝裂缝在外壳面不连续，末端简单；在内壳面，宽圆形一端的壳缝裂缝连续，楔形一端的壳缝裂缝不连续。壳缝末端不与任何结构联系。环带在两侧打开，中间环带上存在舌片唇舌和反舌片唇舌。

生境：生于小水渠、路边积水、河边渗出水、沼泽、泉水井边草丛，偏好半咸水。

国内分布：黑龙江（鸡西兴凯湖、五大连池），山西（太原晋祠），浙江（西湖），西藏（亚东、吉隆、加查、措美、察雅、班戈、申扎、措勤、札达），新疆（乌鲁木齐盐湖、阿克苏、博乐、察布查尔、布尔津、哈巴河、阿勒泰）。

国外分布：亚洲（俄罗斯、蒙古国），非洲（东部、埃及），北美洲（美国），南美洲（乌拉圭），欧洲（瑞典、芬兰）。

本种与 *S. brebissonii* 在壳面外形和结构上都比较相似，光镜下很难区别。一般认为，*S. ovalis* 是体积大一些的类型，壳面有强烈的同心波纹，而且具严格界限的边缘区域和容易辨认的线纹；*S. ovata*（与 *S. brebissonii* 同物异名）的体积比较小，边缘区域和同心波纹也相对模糊（Hustedt，1930；Cleve-Euler，1952）。通过对 *S. brebissonii* 模式标本以及大量其他种群的调查，显示这些特征的分布规律、边界和尚有争议的"经典"种类之间不是绝对关联的。然而，对整个种群的观察来看，上述特征确实能使 *S. brebissonii* 和 *S. ovalis* 以及 Pinnatae 组其他种类区分开来，特别是都存在体积较大的种类时，从形态上容易辨别：*S. ovalis* 较宽的一端是楔形-圆形，而 *S. brebissonii* 较宽的一端是宽圆形（Krammer and Lange-Bertalot，1987）。

通过扫描电镜观察，发现更大的差别，主要是内壳面龙骨突的结构：*S. ovalis* 壳缝管内壁相邻的龙骨突之间有 3—4 个孔，龙骨突的长度较短，有限制；而 *S. brebissonii* 内壁相邻的龙骨突之间只有 1 个孔，龙骨突伸进同心波纹的凹陷区域。

10. 布赖韦尔双菱藻　图版 CII：11—12

Surirella brightwellii W. Smith, Syn. British Diat., p. 33, fig. 9: 69, 1853; Krammer & Lange-Bertalot, Bacillariophyceae. 2. Teil, *in* Ettl *et al.*, Süßwasserfl. Mitteleur., p. 183, figs. 132: 1—8, 1988, nachdr. 1997.

Surirella ovalis var. *brightwellii* (W. Smith) H. Peragallo & M. Peragallo, Diat. Mar. France,

p. 258, pl. 67, figs. 9—10, 1899; Zhu & Chen, Bacill. Xiz. Plat., p. 274, fig. 55: 9, 2000.

Surirella ovalis var. *brightwelli* (W.Smith) Cleve-Euler, Diatomeen schw. Finnland. Teil. V, p. 121, figs. 1565 e, f, 1952.

壳体异极，壳面近椭圆形，较宽的一端为圆形，另一端楔形至尖圆形；长15—55 μm，宽20—35 μm；不具翼状结构；壳面具非常明显的同心波纹，壳面边缘具假漏斗结构；龙骨突清晰可见，30—45个/100 μm；线纹14—19个/10 μm。

生境：生于湖泊中。

国内分布：西藏(定结、昂仁、班戈、申扎)，新疆(赛里木湖)。

国外分布：亚洲(俄罗斯)，北美洲(美国)，欧洲(波罗的海、英国、德国、罗马尼亚、西班牙)。

11. 近盐生双菱藻 图版 CII：2

Surirella subsalsa W. Smith, Syn. British Diat., p. 34, fig. 31: 259, 1853; Cleve-Euler, Diatomeen schw. Finnland, Teil. V, p. 105, figs. 1526 a—d, f, 1952; Zhu & Chen, Bacill. Xiz. Plat., p. 276, fig. 56: 8, 2000.

壳体很小，异极，带面观线形，略呈楔形；壳面呈宽倒卵形；长20—46 μm，宽9—13 μm；龙骨突有30—50个/100 μm；横线纹在光镜下看不清楚。

生境：生于沼泽、湖边、稻田中，水草及岩石上附生，偏好富含有机质的水体。

国内分布：安徽(天柱山)，西藏(墨脱、八宿、波密)。

国外分布：北美洲(美国)，欧洲(英国、德国、波兰、罗马尼亚、西班牙)。

12. 具条纹双菱藻 图版 CVI：2

Surirella striatula Turpin, Mémoires du Musée d'Histoire Naturelle 16：363, pl. 15, figs. 2—10, 1828; Hustedt, Bacillariophyta, *in* Pascher, Süßwass -Fl. Mitteleur. Heft 10, p. 445, fig. 869, 1930; Cleve-Euler, Diatomeen schw. Finnland, Teil. V, p. 120, fig. 1569 a, 1952; Krammer & Lange-Bertalot, Bacillariophyceae. 2. Teil, *in* Ettl *et al.*, Süßwasserfl. Mitteleur., p. 190, figs. 140: 4—5, 1988, nachdr. 1997; Metzeltin *et al.*, Iconogr. Diatomol. 15, figs. 231: 3—5, 2005.

壳体异极，带面窄楔形，壳面宽卵圆形，一端为宽圆形，另一端稍窄，呈楔形；壳面长90—150 μm，宽50—80 μm，龙骨突有8—13个/100 μm，横线纹有18—20条/10 μm；具横向波纹，一直延伸至壳面中部，形成一条纵向窄线形区域；波纹底部和顶部的宽度几乎相等。

生境：生于静水沟中。

国内分布：山西(运城)，新疆(尉犁)。

国外分布：亚洲(蒙古国)，北美洲(加拿大、美国)，南美洲(乌拉圭)，欧洲(波罗的海、黑海、英国、德国、马其顿、罗马尼亚、西班牙)，大洋洲(澳大利亚、新西兰)。

13. 维苏双菱藻　图版 CII：7—10

Surirella visurgis Hustedt，Abh. Naturw. Ver. Bremen 34，p. 363，figs. 1：8—10，1957；Krammer & Lange-Bertalot，Bacillariophyceae. 2. Teil，*in* Ettl *et al.*，Süßwasserfl. Mitteleur.，p. 184，figs. 134：3—5，1988，nachdr. 1997.

壳体略呈异极，壳面线形至线形-卵圆形，两侧近平行或稍凸出，一端宽圆形，另一端钝楔形；长 20—46 μm，宽 11—13 μm；假漏斗结构不清楚，龙骨突有 30—40 个/100 μm；线纹清楚，可延伸至中部，甚至穿过中部窄线区，15—17 条/10 μm。

生境：生于路边积水、沼泽、岩石上附生。

国内分布：新疆（天池、博乐、察布查尔）。

国外分布：亚洲（俄罗斯），欧洲（英国、德国、波兰）。

14. 布列双菱藻　图版 CIII：1—8

Surirella brebissonii Krammer & Lange-Bertalot，Diat. Res.，2(1)：82，figs. 1(ov2)，4，5，9，21—23，1987；Xie *et al.*，Journal of Shanxi University 14(4)：417，1991；Krammer & Lange-Bertalot，Bacillariophyceae，2. Teil，*in* Ettl *et al.*，Süßwasserfl. Mitteleur.，p. 179，figs. 126：2—11，figs. 127：1—13，1988，nachdr. 1997；Antoniades *et al.*，Iconogr. Diatomol. 17：297，figs. 79：16—18，2008.

Surirella ovata Kützing sensu Hustedt，Bacillariophyta，*in* Pascher，Süßwass.-Fl. Mitteleur. Heft 10，p. 442，figs. 863，864，1930；Cleve-Euler，Diatomeen schw. Finnland，Teil. V，p. 122，fig. 1566，1952；Gasse，East African Diatom.，*in* Cramer，Bibliotheca Diatomologia，Band. 11：173，pl. XL，fig. 7，1986；Mitsuzo，Flora North-Eastern Prov. China，part. IV，p. 1413，pl. 4，fig. 24，1970；Zhu *et al.*，Diatom. Suoxiyu Nat. Preserv. Area Hunan，China，p. 57，1989；Shi *et al.*，Compil. Rep. Surv. Alg. Resour. South-West. China，p. 118-119，1994；Zhu & Chen，Bacill. Xiz. Plat.，p. 275，fig. 56：6，2000；Metzeltin *et al.*，Iconogr. Diatomol. 15，figs. 222：3，4，2005.

壳面卵圆-椭圆形至宽卵圆形，异极，大型种类线形-卵圆形，一端楔形圆形，另一端宽圆形；中型种类卵圆形，两端几乎等圆；小型种类宽椭圆形至近圆形，一端楔形，另一端宽圆形。带面观轻微楔形。具有与壳面边缘平行的同心波纹。壳面长 16—70 μm，宽 16—30 μm，长宽比例 2.4：1—1：1，龙骨突有 30—60 个/100 μm，横线纹有 16—19 条/10 μm。

扫描电镜下观察：壳面上横线纹和横肋纹交替出现，肋纹稍高于线纹（光镜下肋纹和线纹不容易分辨）。外壳面边缘的大部分肋纹与壳缝管相融合，同时延伸至窄线形的中线；也有一些肋纹以短小、单一或分支的形式终止于壳面边缘。每 2—5 条肋纹以统一的形式与壳缝管边缘连接，其间有 1—2 条肋纹与壳缝管较低的地方连接，这样壳面边缘呈现出裙状花边。壳缝管内侧的龙骨突桥接壳面内部和壳套，龙骨突通常单个均匀地分布在壳面内部，每一对龙骨突之间有 3—4 个矩形孔，细胞内部通过这些孔与壳缝管发生联系。外壳面上，没有翼状突、翼状管和窗栏开孔。壳缝裂缝在外壳面不连续，末端简单；在内壳面，宽圆形一端的壳缝裂缝连续，楔形一端的壳缝裂缝不连续。壳缝末端不与任何结构联系。环带在两侧打开，中间环带上存在舌片唇舌和反舌片唇舌。

生境：生于湖边渗出水、小水渠、路边积水、沼泽，水草附生、岩石上附生，淡水及半咸水。

国内分布：山西(太原晋阳湖、运城)，内蒙古(牙克石)，吉林(长白山)，黑龙江(鸡西兴凯湖、牡丹江镜泊湖、五大连池)，江苏(苏州)，湖南(索溪峪、沅江流域)，贵州(沅江流域、乌江流域、梵净山)，云南(西双版纳)，西藏(白地、浪卡子、定结、打隆、康马、吉隆、萨噶、昂仁、米林、乃东、错那、措美、察隅、昌都、芒康、察雅、洛隆、江达、贡觉、类乌齐、班戈、申扎、措勤、札达、日土)，甘肃(苏干湖)，宁夏(贺兰山)，新疆(帕米尔东部、博湖、博乐、赛里木湖、察布查尔、布尔津、哈巴河、喀纳斯、阿勒泰、盐湖、天池)。

国外分布：亚洲(俄罗斯、蒙古国、土耳其)，非洲(东部)，北美洲(加拿大、美国)，南美洲(哥伦比亚、乌拉圭)，欧洲(波罗的海、黑海、英国、芬兰、德国、爱尔兰、马其顿、波兰、罗马尼亚、西班牙、瑞典)，大洋洲(澳大利亚)。

目前很难找到 *Surirella ovata* Kützing 的模式标本，Kützing(1844)的描述比较粗糙，至少涉及三个不同种类，硅藻学者根据现有文献，很难准确定义这个种，因此这个种变成了一个"破布袋"，很多形态相似的种类，都被放进来。鉴于此，Krammer 和 Lange-Bertalot(1978)依据 Hustedt(1930)报道的 *Surirella ovata* Kützing 建立了种 *Surirella brebissonii*。本书采用了这一观点。

本书扫描电镜下观察到的外壳面结构，与 Krammer 和 Lange-Bertalot(1978)报道的一致，内壳面未观察到，主要依据 Krammer 和 Lange-Bertalot(1978)中 fig. 4，5 的描述。本种与 *S. ovalis* 的区别详见 *S. ovalis* 的描述部分。

粗壮组 Section *Robustae*

粗壮组分种检索表

1. 壳体沿顶轴方向强烈扭曲 ·· **15. 螺旋双菱藻 *S. spiralis***
1. 壳体沿顶轴方向不扭曲或是稍微扭曲 ·· 2
   2. 细胞两端相等 ·· 3
   2. 细胞两端不相等 ·· 6
3. 壳面披针形，中间区域窄披针形，横向波纹顶部靠近壳面边缘具明显的小刺 ·· **16. 二额双菱藻 *S. bifrons***
3. 壳面披针形至线形-披针形，横向波纹在中间区域可见 ·· 4
   4. 体积大，粗壮，翼状管少于 20 条/100 μm，在波纹顶部常具多而精细的刺 ·· **17. 二列双菱藻 *S. biseriata***
   4. 体积小，精细，翼状管多于 20 条/100 μm ·· 5
5. 壳面线形至线形-披针形 ·· **18. 线性双菱藻 *S. linearis***
5. 壳面椭圆披针形，在中间区域和肋纹上具大量小刺 ·· **19. 淡黄双菱藻 *S. helvetica***
   6. 壳面在较宽一端的前部有一个粗壮的刺，偶尔也会在较窄一端出现 ······ **20. 卡普龙双菱藻 *S. capronii***
   6. 不存在典型的刺，翼状管少于 35 条/100 μm，横向波纹在壳面中央清楚 ·· 7
7. 翼状突起清楚 ·· 8
7. 翼状突起不太清楚或根本不清楚，体积大 ·· **21. 美丽双菱藻 *S. elegans***

8. 翼状管多于 20 条/100 μm，体积大，中间具清晰的纵向肋纹·················· **22. 柔软双菱藻 S. tenera**
8. 翼状管少于 20 条/100 μm ·· 9
9. 壳面沿顶轴方向稍微扭曲 ··· **23. 阿斯特里双菱藻 S. astridae**
9. 壳面沿顶轴方向不扭曲 ·· 10
10. 翼状管一般 10 条或更少/100 μm，横向波纹非常清楚 ······················· **24. 粗壮双菱藻 S. robusta**
10. 横向波纹较精致，翼状突起小且不清楚，翼状管通常多于 10 条/100 μm ·· **25. 华彩双菱藻 S. splendida**

15. 螺旋双菱藻　图版 CXXII: 1—2; 图版 CXXIII: 1—2

Surirella spiralis Kützing, Kies. Bacill. Diat., p. 60, pl. 3, fig. 64, 1844; Schmidt *et al.*, Atlas Diat.-Kunde, p. 56, figs. 56: 25, 26, 1886; Hustedt, Bacillariophyta, *in* Pascher, Süßwass -Fl. Mitteleur. Heft 10, p. 445, fig. 870, 1930; Cleve-Euler, Diatomeen schw. Finnland, Teil. V, p. 124, fig. 1567, 1952; Krammer & Lange-Bertalot, Bacillariophyceae. 2. Teil, *in* Ettl *et al.*, Süßwasserfl. Mitteleur., p. 206, figs. 168: 1—7, 1988, nachdr. 1997; Zhu & Chen, Bacill. Xiz. Plat., p. 276, figs. 58: 1—3, 2000.

Campylodiscus spiralis(Kützing) W. Smith, Ann. Mag. Nat. Hist. Ser. 2, p. 6, pl. 1, fig. 2a, 1851.

Campylodiscus spiralis(Kützing) R.Gutwinski, Flora Glonów Okolic Lwowa (Flora algarum agri Leopoliensis), p. 102, 1891.

Spirodiscus spiralis(Kützing) Jurilj, Nove Dijatomeje-Surirellaceae-iz Ohridskog Jezera i njihovo filogenetsko znacenje, 186, 1949.

壳体等极，但沿纵轴强烈扭曲，常呈"8"字形，而通常"8"字的一半明显大于另一半，甚至表现出披针形；壳面椭圆形至线形-椭圆形，末端楔形-圆形；长 68—130 μm，宽 25—91 μm；翼状突清晰可见，翼状管比窗栏开孔窄，肋状，15—30 个/100 μm。

生境：生于山涧溪流、草地渗出水、路边静水沟中。

国内分布：辽宁，湖南（沅江流域），四川（九寨沟），贵州（施秉舞阳河、沅江流域、乌江流域），云南（丽江、德钦、维西），西藏（亚东、聂拉木、定日、吉隆、林芝、错那、察隅、八宿、波密、芒康、江达、类乌齐、班戈、札达、革吉），宁夏（贺兰山），新疆（尉犁）。

国外分布：亚洲（蒙古国、新加坡），北美洲（加拿大、美国），南美洲（巴西），欧洲（波罗的海、黑海、英国、德国、爱尔兰、马其顿、波兰、罗马尼亚、西班牙），大洋洲（澳大利亚、新西兰）。

采自云南的标本与 Krammer 和 Lange-Bertalot(1997) 中描述的生境较符合，为山间溪流，贫营养水体；而采自新疆的标本，生活在路边的静水沟中，有机质较丰富。

16. 二额双菱藻　图版 CVIII: 1—3; 图版 CIX: 1—2; 图版 CX: 1—3; 图版 CXI: 1—2; 图版 CXII: 1—3

Surirella bifrons Ehrenberg, Verbreitung und Einfluss des mikroskopischen Lebens in Süd-und Nord-Amerika, p. 388(100), pl. 3/5, fig. 5, pl. 4/3, fig. 1, 1843; Krammer & Lange-Bertalot, Bacillariophyceae. 2. Teil, *in* Ettl *et al.*, Süßwasserfl. Mitteleur., p.

195, figs. 145: 2—4, figs. 147: 1—5, 1988, nachdr. 1997; Metzeltin *et al.*, Iconogr. Diatomol. 15, figs. 228: 1—2, 2005.

Surirella biseriata var. *bifrons* (Ehrenberg) Hustedt, Beiträge zur Algenflora von Bremen IV. Bacillariaceen aus der Wumme, p. 305, 1911; Cleve-Euler, Diatomeen schw. Finnland, Teil. V, p. 105, fig. 1528 f, 1952.

壳面近菱形,等极;长96—150 μm,宽30—60 μm;翼状管有12—22个/100 μm。

扫描电镜下观察:壳缝管高于壳面,位于发达的龙骨或翼状突上,存在翼状管和窗栏开孔。壳面顶面观呈波纹状,形成波纹的肋纹从壳面边缘延伸至线形-披针形的中线,中线区域稍高于壳面但不高于龙骨。在壳面边缘,肋纹与翼状管相连,肋纹之间的凹陷部分与窗栏开孔相连。翼状管单一而均匀地分布在壳面边缘,内部为一个椭圆形开口,是管壳缝和细胞内部交流的通道。窗栏开孔上有2—8条(通常4—6条)细棒状的小窗条,以随意的形式互相融合或分离。在外壳面的肋纹顶部和中线区域具短刺,光镜下呈小点状。整个壳面均具横线纹,线纹一般由单列或双列小孔组成(28—40个/10 μm),在壳面中部平行排列,末端汇聚。

生境:生于沼泽、香蒲水滩中。

国内分布:内蒙古(阿尔山达尔滨湖),黑龙江(宁安),福建(鼓岭),海南(琼中),新疆(阿克苏、喀纳斯)。

国外分布:亚洲(俄罗斯、蒙古国),北美洲(美国),南美洲(乌拉圭),欧洲(波罗的海、英国、德国、爱尔兰、马其顿、波兰、罗马尼亚),大洋洲(新西兰)。

17. 二列双菱藻

Surirella biseriata Brébisson, *in* Brébisson & Godey, Algues des environs de Falaise, p. 53, pl. 7, 1835; Schmidt *et al.*, Atlas Diat.-Kunde, p. 283, fig. 283: 2, 1912; Hustedt, Bacillariophyta, *in* Pascher, Süßwass -Fl. Mitteleur. Heft 10, p. 433, figs. 831—832, 1930; Cleve-Euler, Diatomeen schw. Finnland, Teil. V, p. 105, fig. 1528 a, 1952; Krammer & Lange-Bertalot, Bacillariophyceae. 2. Teil, *in* Ettl *et al.*, Süßwasserfl. Mitteleur., p. 195, figs. 141: 1—3, figs. 142: 1—5, figs. 143: 1—9, 1988, nachdr. 1997; Xin *et al.*, Journal of Shanxi University (Nat. Sci. Ed.) 20(1): 105, 1997; Zhu & Chen, Bacill. Xiz. Plat., p. 272, fig. 54: 13, 2000.

17a. 原变种　图版 CXIII: 1—3; 图版 CXIV: 2

var. **biseriata**

壳面披针形至线形-披针形,等极或异极;长124—200 μm,宽32—43 μm;壳面具横向的波纹,在中间区域可见,中部具一条清晰的纵向肋纹;壳面不具明显的刺,而在波纹靠近壳缘的一端常具多而精细的刺;靠近两端的肋纹偏斜角度大;翼状管有 8—20个/100 μm。

生境:生于河流、湖泊、小水沟,岩石上附生。

国内分布:山西(太原晋祠),河北(昌黎),内蒙古(阿尔山),黑龙江(鸡西兴凯湖、五大连池)，湖南(沅江流域)，海南(琼中)，贵州(沅江流域、乌江流域、梵净山)，西藏(革吉)，新疆(布尔津、哈巴河、福海、北屯、阿勒泰)。

国外分布：亚洲(俄罗斯、蒙古国)，非洲(东部)，北美洲(加拿大、美国)，南美洲(巴西、哥伦比亚)，欧洲(波罗的海、英国、丹麦、德国、爱尔兰、马其顿、波兰、罗马尼亚、西班牙)，大洋洲(澳大利亚、新西兰)。

17b. 缩小变种　图版 CXIV：3—4

var. **diminuta** Cleve-Euler，Diatomeen schw. Finnland. Teil. V，p. 106，figs. 1528 h—I，1952；Zhu & Chen，Bacill. Xiz. Plat.，p. 272，fig. 54：14，2000.

光镜下，与原变种的区别在于：壳体较小；长 38—45 μm，宽 14—16 μm；翼状管有 25—30 个/100 μm。

生境：生于河流中。

国内分布：西藏(林芝)，新疆(察布查尔、北屯)。

国外分布：北美洲(美国)，欧洲(瑞典、芬兰)。

18. 线性双菱藻

Surirella linearis W. Smith，Syn. British Diat.，p. 31，fig. 8：58a，1853；Husedt，Bacillariophyta，*in* Pascher，Süßwass -Fl. Mitteleur. Heft 10，p. 434，figs. 837，838，1930；Cleve-Euler，Diatomeen schw. Finnland，Teil. V，p. 109，figs. 1535 a, b，1952；Zhu & Chen，The Diatoms of the Suoxiyu Nature Reserve Area，Hunan，China，*in* Li *et al.*，The algal flora and aquatic fauna of the Wulingyuan Nature Reserve Area，Hunan，China. p. 57，1989b；Krammer & Lange-Bertalot，Bacillariophyceae. 2. Teil，*in* Ettl *et al.*，Süßwasserfl. Mitteleur.，p. 198，figs. 149：1—9，fig. 150：1，fig. 151：1，1988，nachdr. 1997；Zhu & Chen，Bacill. Xiz. Plat.，p. 273，fig. 56：7，2000；Metzeltin *et al.*，Iconogr. Diatomol. 15，figs. 221：8—11，2005.

18a. 原变种　图版 CXVI：1—5

var. **linearis**

壳面同极或稍异极，线形至线形-披针形，两侧平行、稍凹入或凸出，两端楔形或钝圆；壳面长 36—86 μm，宽 10—25 μm；翼状突在壳面上的角度不同，清楚或是较窄，20—30 个/100 μm；在壳缘具一个波纹状的圆环，波峰和波谷等宽；窗栏开孔一般略宽于翼状管，横肋纹可到达中线，在壳面中部形成一个线形-披针形区域。

生境：生于河流、湖边香蒲滩、沼泽，岩石上附生。

国内分布：内蒙古(牙克石、阿尔山达尔滨湖)，辽宁(辽河、沈阳)，吉林(长白山)，黑龙江(五大连池、漠河)，江苏(苏州)，浙江(安吉、西湖)，福建(厦门)，湖北(武汉植物园)，湖南(长沙、索溪峪、沅江流域)，四川(九寨沟)，贵州(沅江流域、乌江流域、梵净山)，云南(丽江)，西藏(林芝、聂拉木、定日、亚东、吉隆、萨噶、墨脱、米林、乃东、加查、错那、措美、察隅、芒康、江达、类乌齐、申扎、措勤、札达、革吉)，宁夏(贺兰山)，新疆(博湖、察布查尔、布尔津、喀纳斯、福海)。

国外分布：亚洲(俄罗斯、菲律宾、蒙古国、新加坡、尼泊尔、伊朗、土耳其)，北美洲(美国)，南美洲(巴西、哥伦比亚、乌拉圭)，欧洲(波罗的海、英国、德国、爱尔兰、马其顿、波兰、罗马尼亚、西班牙)，大洋洲(澳大利亚、新西兰)。

18b. 缢缩变种　图版 CXVII：1—5

var. **constricta** Grunow Verh. zool. bot. Ges. Wien, p. 141(455), 1862; Husedt, Bacillariophyta, *in* Pascher, Süßwass -Fl. Mitteleur. Heft 10, p. 434, fig. 839, 1930; Cleve-Euler, Diatomeen schw. Finnland, Teil. V, p. 109, figs. 1535 g, h, 1952; Zhu & Chen, Bacill. Xiz. Plat., p. 273, fig. 57: 7, 2000; Metzeltin *et al.*, Iconogr. Diatomol. 15, figs. 221: 12—17, 2005.

光镜下，与原变种的区别在于：壳面线形，中部缢缩；长 74—96 μm，宽 18—20 μm；翼状管有 20—40 个/100 μm。

生境：生于路边积水中。

国内分布：辽宁(沈阳)，福建(厦门)，湖南(索溪峪、沅江流域)，贵州(沅江流域)，西藏(聂拉木、亚东、仲巴、墨脱、措美、察隅、芒康、措勤)；新疆(察布查尔)。

国外分布：亚洲(蒙古国)，非洲(东部)，北美洲(美国)，南美洲(巴西、哥伦比亚、乌拉圭、安第斯山脉)，欧洲(英国、爱尔兰、马其顿、罗马尼亚、西班牙)，大洋洲(澳大利亚、新西兰)。

19. 淡黄双菱藻　图版 CXVIII：1—4；图版 CXIX：1—2

Surirella helvetica Brun, Diat. Alpes Jura, p. 100, pl. 2, fig. 4, pl. 9, fig. 28, 1880; Cleve-Euler, Diatomeen schw. Finnland, Teil. V, p. 107, figs. 1531 a—d, 1952.

Surirella linearis var. *helvetica* (Brun) Meister Die Kieselalgen der Schweiz. Beiträge zur Kryptogamenflora der Schweiz, p. 223, pl. 41, fig. 6, 1912; Zhu & Chen, The Diatoms of the Suoxiyu Nature Reserve Area, Hunan, China, *in* Li *et al.*, The algal flora and aquatic fauna of the Wulingyuan Nature Reserve Area, Hunan, China. p. 57, 1989b; Krammer & Lange-Bertalot, Bacillariophyceae. 2. Teil, *in* Ettl *et al.*, Süßwasserfl. Mitteleur., p. 199, figs. 151: 2—4, 1988, nachdr. 1997; Zhu & Chen, Bacill. Xiz. Plat., p. 273, fig. 57: 4, 2000.

壳面同极或稍异极，椭圆披针形，两侧平行或稍凸出，两端楔形或钝圆；壳面长 65—95 μm，宽 17—22 μm；壳面具大量相对较长的刺，特别是在中线和肋纹上，光镜下呈小突起；翼状管较清楚，20—30 个/100 μm，窗栏开孔一般略宽于翼状管；一般横肋纹可到达中线，在壳面中部形成一个线形-披针形区域。

生境：生于湖泊、湖边渗出水中。

国内分布：辽宁(辽河)，湖南(索溪峪、沅江流域)，贵州(沅江流域、梵净山)，云南(丽江)，西藏(多庆、聂拉木、察隅)，陕西(华山)，新疆(赛里木湖)。

国外分布：北美洲(美国)，南美洲(安第斯山脉)，欧洲(英国、马其顿、波兰)。

本种与 S. linearis 在形态结构上最相似，不同之处在于：本种壳体相对较宽，壳面具大量较长的刺，同时翼状管轮廓更为清晰。

20. 卡普龙双菱藻　图版 CXIV：1

Surirella capronii Brébisson & Kitton, in Brébisson, Annales de la Société Phytologique et Micrographique de Belgique 1: p. 61; figs. 43—44, 1869; Bacillariophyta, in Pascher, Süßwass -Fl. Mitteleur. Heft 10, p. 440, fig. 857, 1930; Cleve-Euler, Diatomeen schw. Finnland, Teil. V, p. 110, fig. 1537, 1952; Zhu & Chen, The Diatoms of the Suoxiyu Nature Reserve Area, Hunan, China, in Li et al., The algal flora and aquatic fauna of the Wulingyuan Nature Reserve Area, Hunan, China. p. 57, 1989b; Krammer & Lange-Bertalot, Bacillariophyceae. 2. Teil, in Ettl et al., Süßwasserfl. Mitteleur., p. 205, figs. 166: 1—4, figs. 167: 1—4, 1988, nachdr. 1997; Zhu & Chen, Bacill. Xiz. Plat., p. 273, fig. 56: 1, 2000.

壳面卵形，两端不等宽；长 125—160 μm，宽 52—56 μm；翼状管有 15—20 个/100 μm；壳面靠近较宽一端有一粗壮的刺；带面广楔形。

生境：生于湖泊、河流、路边积水、小水沟中。

国内分布：山西(太原晋阳湖)，辽宁(沈阳)，黑龙江(鸡西兴凯湖、五大连池)，上海(淀山湖)，福建(鼓岭)，湖南(索溪峪、沅江流域)，贵州(施秉舞阳河、沅江流域、乌江流域、梵净山)，西藏(拉萨)，甘肃(苏干湖)，新疆(布尔津、福海、北屯、阿勒泰)。

国外分布：亚洲(俄罗斯、蒙古国、伊朗)，北美洲(美国)，南美洲(哥伦比亚)，欧洲(波罗的海、黑海、英国、德国、罗马尼亚)，大洋洲(澳大利亚)。

21. 美丽双菱藻　图版 CXV：1—2

Surirella elegans Ehrenburg, Verbreitung und Einfluss des mikroskopischen Lebens in Süd-und Nord-Amerika, p. 424(136), pl. 3/1, fig. 22, 1843; Bacillariophyta, in Pascher, Süßwass -Fl. Mitteleur. Heft 10, p. 440, figs. 858, 859, 1930; Cleve-Euler, Diatomeen schw. Finnland, Teil. V, p. 116, fig. 1552, 1952; Krammer & Lange-Bertalot, Bacillariophyceae. 2. Teil, in Ettl et al., Süßwasserfl. Mitteleur., p. 204, fig. 160: 5, figs. 161: 1, 2, figs. 162: 1—7, figs. 163: 1—4, 1988, nachdr. 1997; Xin et al., Journal of Shanxi University (Nat. Sci. Ed.) 20(1): 106, 1997.

壳面异极至几乎等极，壳面卵圆形、卵圆-披针形至披针形；长 82—258 μm，宽 21—61 μm；翼状突几乎垂直于壳面边缘，翼状管在壳面非常窄，看不清楚，而在带面较宽；翼状管比窗栏开孔宽很多，12—21 个/100 μm；肋纹几乎平行排列，很难到达中线，常被壳面中部宽披针形的轴区断开。

生境：湖边浮游。

国内分布：山西(太原晋祠)，河北(昌黎)，内蒙古(阿尔山达尔滨湖)，黑龙江(鸡西兴凯湖、五大连池、漠河)，浙江(安吉)，河南(南阳)，湖南(沅江流域)，贵州(沅江流域)。

国外分布：亚洲(俄罗斯、蒙古国)，北美洲(加拿大、美国、夏威夷群岛)，欧洲(波罗的海、英国、德国、爱尔兰、波兰、西班牙)，大洋洲(新西兰)。

22. 柔软双菱藻

Surirella tenera Gregory, Quarterly Journal of Microscopical Science, 4: 11, fig. 1: 38, 1856; Hustedt, Bacillariophyta, *in* Pascher, Süßwass -Fl. Mitteleur. Heft 10, p. 439, fig. 853, 1930; Cleve-Euler, Diatomeen schw. Finnland, Teil. V, p. 104, figs. 1525 a, b, 1952; Zhu & Chen, The Diatoms of the Suoxiyu Nature Reserve Area, Hunan, China, *in* Li *et al.*, The algal flora and aquatic fauna of the Wulingyuan Nature Reserve Area, Hunan, China. p. 57, 1989b; Krammer & Lange-Bertalot, Bacillariophyceae. 2. Teil, *in* Ettl *et al.*, Süßwasserfl. Mitteleur., p. 203, figs. 164: 1—4, figs. 165: 1—3, 1988, nachdr. 1997; Zhu & Chen, Bacill. Xiz. Plat., p. 276, fig. 57: 6, 2000.

22a. 原变种　图版 CXXV：1—3；图版 CXXVI：1—3

var. **tenera**

细胞体积大，壳面椭圆-披针形至线形-披针形，异极，一端钝圆，另一端尖圆；壳面长 86—156 μm，宽 23—45 μm；壳面具波纹，形成波纹的肋纹一般从壳缘延伸至壳面中部的线形-披针形透明区域，透明区域中部具清晰的纵向肋纹，肋纹上不具刺；翼状管清晰，20—35 个/100 μm；带面广楔形。

生境：生于河流、溪流、小水沟、池塘、路边积水、沼泽，岩石上附生。

国内分布：内蒙古(阿尔山)，辽宁(辽河)，黑龙江(鸡西兴凯湖、宁安、五大连池)，安徽(黄山)，福建(金门)，河南(南阳)，湖南(岳阳、索溪峪、沅江流域)，广西(灵川)，海南(琼中)，贵州(赫章、沅江流域、梵净山)，西藏(墨脱、米林、林芝、加查、芒康、江达、类乌齐)，新疆(奎屯、博乐、察布查尔、布尔津、喀纳斯、哈巴河、阿勒泰、福海、北屯)。

国外分布：亚洲(俄罗斯、蒙古国)，北美洲(美国、夏威夷群岛)，南美洲(巴西、哥伦比亚)，欧洲(波罗的海、英国、德国、爱尔兰、罗马尼亚、西班牙)，大洋洲(澳大利亚、新西兰)。

区别于本组其他大型种类：结构更加精细(龙骨突密度大于 25 个/100 μm)，体积稍小，光镜下壳面显得柔软(可能是细胞壁硅质成分积累较少)；此外，壳面中间的径向肋纹明显。Krammer 和 Lange-Bertalot(1997)认为 Hustedt(1930)发现于同一材料中的 *S. tenera*(fig. 853)和 *S. tenera* var. *nervosa*(fig. 854)是同一种类，fig. 854 中壳面中部径向肋纹上起伏的齿状结构是不稳定的特征。Krammer 和 Lange-Bertalot(1997)把 *S. tenere* f. *cristata*(fig. 165：3)作为 *S. tenera* 的同物异名处理，认为将 *S. tenera* var. *nervosa* 重新组合成 *S. nervosa* 是不合适的，并在观察中发现不具齿与具齿种类的生活环境并不完全重合，且具齿的种类数量也相当丰富，本书采用了 Hustedt(1930)的观点，将具齿的种类放在 *S. tenera* var. *nervosa* 里。

22b. 具脉变种　图版 CXXVII：1—2

var. **nervosa** A. Schmidt, *in* Schmidt *et al.*, pl. 23, figs. 15—17, 1875; Hustedt, Bacillariophyta, *in* Pascher, Süßwass -Fl. Mitteleur. Heft 10, p. 439, figs. 854—855, 1930; Cleve-Euler, Diatomeen schw. Finnland, Teil. V, p. 104, figs. 1525 c—e, 1952. *Surirella nervosa* (Schmidt) Mayer, Berichte des naturwissenschaftlichen (früher zoologisch-mineralogischen) Vereins zu Regensburg 14：341, pl. 23, fig. 5, pl. 28, figs. 8—9; Schmidt *et al.*, Atlas Diat.-Kunde, p. 366, figs. 366：4—5, 1927; Zhu & Chen, The Diatoms of the Suoxiyu Nature Reserve Area, Hunan, China, *in* Li *et al.*, The algal flora and aquatic fauna of the Wulingyuan Nature Reserve Area, Hunan, China. p. 57, 1989b; Zhu & Chen, Bacill. Xiz. Plat., p. 274, fig. 56：2, 2000.

光镜下与原变种的区别在于：壳面中部纵向肋纹上具起伏的齿状结构，中央无线纹区域为线形-披针形；壳面长 72—145 μm，宽 19.5—50 μm；翼状管有 20—30 个/ 100 μm。

生境：生于河流，泉水井边草丛、岩石上附生。

国内分布：内蒙古（阿尔山达尔滨湖），黑龙江（鸡西兴凯湖、宁安），上海（淀山湖），安徽（琅琊山），湖南（索溪峪、沅江流域），贵州（沅江流域、乌江流域），西藏（亚东、墨脱、林芝、芒康、江达、类乌齐），新疆（察布查尔、布尔津、哈巴河、福海）。

国外分布：亚洲（俄罗斯），北美洲（美国、夏威夷群岛），南美洲（巴西），欧洲（英国、爱尔兰、罗马尼亚、西班牙），大洋洲（新西兰）。

23. 阿斯特里双菱藻　图版 CVII：1—3

Surirella astridae Hustedt, *in* Schmidt *et al.*, Atlas Diat.-Kunde, pl. 363, figs. 8—9, 1925; Krammer & Lange-Bertalot, Bacillariophyceae. 2. Teil, *in* Ettl *et al.*, Süßwasserfl. Mitteleur., p. 202, figs. 160：1—2, 1988, nachdr. 1997.

壳面异极，一端较宽，末端广圆形，另一端渐窄，末端尖圆形，在纵轴方向略扭曲。长 80—120 μm，宽 35—55 μm。翼状突起明显，翼状管 9—13 个/100 μm。

生境：湖中浮游，湖边沼泽、水草附着。

国内分布：内蒙古（阿尔山达尔滨湖），江苏（常熟），湖南（索溪峪）。

国外分布：欧洲（英国）。

24. 粗壮双菱藻

Surirella robusta Ehrenberg, Characteristik von 274 neuen Arten von Infusorien, 1840：215, 1840; Schmidt *et al.*, Atlas Diat.-Kunde, p. 22, fig. 22：3, 1885; Zhu & Chen, The Diatoms of the Suoxiyu Nature Reserve Area, Hunan, China, *in* Li *et al.*, The algal flora and aquatic fauna of the Wulingyuan Nature Reserve Area, Hunan, China. p. 57, 1989b; Krammer & Lange-Bertalot, Bacillariophyceae. 2. Teil, *in* Ettl *et al.*, Süßwasserfl. Mitteleur., p. 201, figs. 156：1—5, figs. 157：1—4, 1988, nachdr. 1997; Xin *et al.*, Journal of Shanxi University (Nat. Sci. Ed.) 20 (1)：106, 1997; Ruck & Kociolek,

Bibliotheca Diatomologica, 50: 25, figs. 21—23: 1—13, 2004; Zhu & Chen, Bacill. Xiz. Plat., p. 276, fig. 57: 2, 2000.

24a. 原变种 图版 CXX: 1—2
var. robusta

壳面卵形，异极，一极宽圆形，另一极楔形；壳面长 158—297 μm，宽 50—81 μm；翼状管有 7—9 个/100 μm；线纹有 40—60 条 /10 μm；壳缝管位于发达的龙骨上，壳面顶面观波纹形，显示颗粒状，这些波纹状肋纹从壳面边缘延伸到中间线形-披针形的透明区域。

生境：生于湖泊、溪流、沼泽中。

国内分布：山西(太原晋阳湖、太原晋祠)，内蒙古(阿尔山)，辽宁(沈阳)，上海(淀山湖)，浙江(西湖)，湖南(索溪峪、沅江流域)，广东(龙川、河源)，贵州(沅江流域、乌江流域)，云南(西双版纳、丽江)，西藏(墨脱、林芝、错那、察隅、芒康)，宁夏(贺兰山)，新疆(喀纳斯、哈巴河、福海)。

国外分布：亚洲(俄罗斯、蒙古国、新加坡)，非洲(东部)，北美洲(美国)，南美洲(巴西、哥伦比亚)，欧洲(波罗的海、英国、德国、爱尔兰、马其顿、罗马尼亚、西班牙)，大洋洲(澳大利亚、新西兰)。

24b. 宽大变型 图版 CXX: 3；图版 CXXI: 1—2
f. lata Hustedt, Archiv für Hydrobiologie 18, p. 170, fig. 1, 1927; Simonsen R, Atlas and Catalogue of the Diatom Types of Friedrich Hustedt, p. 104, pl. 162, fig. 8, 1987.

与原变种的区别在于：壳面宽大，宽可达 100 μm。

生境：生于草地渗出水中。

国内分布：云南(丽江)。

国外分布：大洋洲(澳大利亚)。

25. 华彩双菱藻 图版 CXXIV: 1—3
Surirella splendida (Ehrenberg) Kützing, Kies. Bacill. Diat., p. 62, pl. 7, fig. 9, 1844; Krammer & Lange-Bertalot, Bacillariophyceae. 2. Teil, *in* Ettl *et al.*, Süßwasserfl. Mitteleur., p. 202, figs. 158: 1—3, figs. 159: 1—6, 1988, nachdr. 1997; Ruck & Kociolek, Bibliotheca Diatomologica, 50: 28, figs. 27—29: 1—17, 2004; Metzeltin *et al.*, Iconogr. Diatomol. 15, figs. 223: 1, 2, figs. 226: 1, 2, figs. 227: 1—4, 2005.

Navicula splendida Ehrenberg, Über die Entwicklung und Lebensdauer der Infusionstheiere, nebst ferneren Beiträgen zu einer Vergleichung ihrer organischen Systeme. 1831: 81, 1832.

Surirella robusta var. *splendida* (Ehrenberg) Van Heurck, Syn. Diat. Belgique, p. 187, 1885; Hustedt, Bacillariophyta, *in* Pascher, Süßwass -Fl. Mitteleur. Heft 10, p. 437, figs. 851, 852, 1930; Cleve-Euler, Diatomeen schw. Finnland, Teil. V, p. 104, fig. 1524 h, 1952; Zhu & Chen, The Diatoms of the Suoxiyu Nature Reserve Area, Hunan, China. In: Li *et al.* The algal flora and aquatic fauna of the Wulingyuan Nature Reserve Area, Hunan, China. p. 57, 1989b; Zhu & Chen, Bacill. Xiz. Plat., p. 276, fig. 57: 3, 2000.

壳面椭圆-披针形，异极，一端钝圆，另一端近圆；壳面长75—250 μm，宽49—65 μm，翼状管12—18个/100 μm；壳缝管位于较发达的龙骨上；壳面具波纹，形成波纹的肋纹从壳缘延伸到线形-披针形的透明线形区域，带面广楔形。

扫描电镜下观察：壳缝管高于壳面，位于发达的龙骨或翼状突上，存在翼状管和窗孔。壳面顶面观呈波曲状，形成波曲的肋纹从壳面边缘延伸到线形-披针形的中线，中线区域稍高于壳面但不高于龙骨。在壳面边缘，一条肋纹与一条翼状管相连，肋纹之间的凹陷部分与窗栏开孔相连。翼状管单一而均匀地分布在壳面边缘，内部为一椭圆形开口，为管壳缝和细胞内部交流的通道。窗栏开孔上有2—8条(通常4—6条)细棒状的小窗条，以随意的形式互相融合或分离。整个壳面均具横线纹，线纹一般由单列小孔组成(28—40条/10 μm)，在壳面中部平行排列，末端汇聚。外壳面的孔简单，孔缘不凸出，内壳面的孔边缘凸出。外部壳缝裂缝不连续，顶面观很难看到。在倾斜的标本中，可见外壳面较宽一端的壳缝裂缝弯向壳套，末端简单，另一端壳缝裂缝也弯向壳套，但是末端膨大。在内壳面，较宽一端的壳缝裂缝连续，穿过宽平或凸出的节结区域，另一端壳缝裂缝不连续，末端有两个螺旋舌似的结构。

生境：生于河流、湖泊、小水渠、路边积水、沼泽，水草附生。

国内产地：内蒙古(阿尔山)，吉林(长白山)，黑龙江(鸡西兴凯湖)，上海(淀山湖)，江苏(昆山)，福建(厦门、鼓岭)，湖南(索溪峪、沅江流域)，广东(河源、黄石)，广西(灵川)，贵州(赫章、水城、镇宁、施秉舞阳河、沅江流域、乌江流域)，云南(滇池、洱海、南华、翠湖)，西藏(亚东、吉隆、察隅)，陕西(华山)，新疆(阿克苏、博湖、博乐、察布查尔、布尔津、喀纳斯、福海、阿勒泰)。

国外产地：亚洲(俄罗斯、蒙古国、伊朗、土耳其)，北美洲(加拿大、美国、夏威夷群岛)，南美洲(巴西、哥伦比亚、乌拉圭)，欧洲(波罗的海、黑海、英国、德国、爱尔兰、马其顿、罗马尼亚、西班牙)，大洋洲(澳大利亚、新西兰)。

采自新疆的标本，窗孔上小窗条不具小刺，而Ruck和Kociolek(2004)采自Mountain Lake 的标本，在每个窗孔上，有一个或两个小窗条具刺(fig. 28: 10)，这些小刺通常在小窗条的上部(接近龙骨顶端的部分)，且只出现在壳面小窗条的一侧，壳套一边没有发现。

长羽藻属 Stenopterobia A. de Brébisson ex H. Van Heurck

A. de Brébisson ex H. Van Heurck, Treat. Diat.: 374, 1896.

单细胞，示壳面观或带面观。色素体1个，由两片大而紧贴壳面的盘状结构组成，通过靠近细胞一端的极窄狭部连接。

壳面"S"形或直，窄，线形，表面轻微波曲，具一个环绕壳面的壳缝系统，壳面和壳套部分内折或融合形成龙骨突。线纹多排，由小圆孔组成，孔外没有膜封闭。横肋纹表面具蘑菇状的突起或瘤状物，一直延伸到壳面中线位置，彼此融合形成一条窄的径向肋纹或腹板，有时表面会有一些分散的穿孔。外壳面壳缝末端简单或稍弯曲，内壳面壳缝末端简单。环带结构尚不清楚。

分布于淡水中，附着于沉积物或污泥中，喜生于酸性贫营养湖泊和泥炭沼泽中。

种类较少，一般认为只包含"S"形种类 Sten. sigmatella (=Sten. 'intermedia')。壳面结构和壳缝系统相对简单，与双菱藻属种类关系最接近，因狭长的外形和对酸性水体的生态偏好，将其从双菱藻属中分出来。

模式种：弯曲长羽藻[*Stenopterobia curvula* (W. Smith) Krammer]。

本志收编 3 种 1 变种。

长羽藻属分种检索表

1. 壳面狭长，"S"形 ··· 2
1. 壳面线形，直 ··· **1. 优美长羽藻 *S. delicatissima***
   2. 横线纹密度小于 18 条/10 μm ··· **2. 剑形长羽藻 *S. anceps***
   2. 横肋纹密度大于 18 条/10 μm ··· **3. 中型长羽藻 *S. intermedia***

1. 优美长羽藻　图版 CXXVII：5

Stenopterobia delicatissima (Lewis) Brébisson ex Van Heurck, *in* Van Heurck, Treat. Diat., p. 374, 1896; Krammer & Lange-Bertalot, Bacillariophyceae. 2. Teil, *in* Ettl *et al.*, Süßwasserfl. Mitteleur., p. 210, figs. 170: 5—6, figs. 173: 1—8, figs. 174: 1—12, 1988, nachdr. 1997; Liu *et al.*, Journal of Shanghai Normal University (Natural Sciences) 43(3): 270, fig. 13. 2014.

壳面线形-披针形，直，顶端呈喙状；长 30—100 μm，宽 4—9 μm；横肋纹有 4—5 条/10 μm，横线纹有 18—25 条/10 μm。

扫描电镜下观察：在外壳面，可见横肋纹，肋纹在壳面中部形成纵向窄线形的腹板。壳缝管位于龙骨上，稍高于壳面，具翼状突。

生境：生于沼泽中，喜酸性水体。

国内分布：内蒙古(阿尔山)。

国外分布：北美洲(美国)，欧洲(瑞典、芬兰)，大洋洲(澳大利亚、新西兰)。

2. 剑形长羽藻　图版 CXXVII：3—4；图版 CXXVIII：1—2

Stenopterobia anceps (Lewis) Brébisson ex Van Heurck, *in* Van Heurck, Treat. Diat., p. 374, 1896; Krammer & Lange-Bertalot, Bacillariophyceae. 2. Teil, *in* Ettl *et al.*, Süßwasserfl. Mitteleur., p. 208, figs. 171: 1—4, 1988, nachdr. 1997; Liu *et al.*, Journal of Shanghai Normal University (Natural Sciences) 43(3): 270, figs. 11—12.

Surirella anceps Lewis, Proceedings of the Academy of Natural Sciences of Philadelphia 15: 342, pl. 1, fig. 3, 1864.

壳面狭长，"S"形，顶端圆形；长 130—175 μm，宽 6—7 μm；横肋纹有 4—5 条/10 μm，

横线纹有 17—18 条/10 μm。

扫描电镜下观察：在外壳面，可见横肋纹，肋纹在壳面中部形成纵向窄线形的腹板，大部分肋纹末端与相对应壳缝管下的壳套融合，而其他肋纹延伸至翼状管外壁的壳面和壳套上。壳缝管位于龙骨上，稍高于壳面，具一个大小适中的翼状突，壳缝管通过翼状管与细胞内部相连，翼状管单一均匀地分布在细胞内部，具单一的孔状开口。在外壳面，壳缝裂缝不连续，末端急剧弯向壳套，裂缝末端膨大，内壳面，壳缝裂缝不连续，末端简单。

生境：生于沼泽中，喜酸性水体。

国内分布：内蒙古(阿尔山、根河)。

国外分布：北美洲(美国)，欧洲(瑞典、芬兰)，大洋洲(澳大利亚、新西兰)。

3. 中型长羽藻

Stenopterobia intermedia (Lewis) Brébisson ex Van Heurck, Treat. Diat., p. 374, 1896; Hustedt, Bacillariophyta, *in* Pascher, Süßwass -Fl. Mitteleur. Heft 10, p. 428—429, fig. 830, 1930; Cleve-Euler, Diatomeen schw. Finnland, Teil. V, p. 100, figs. 1522 a—d, 1952; Yang, Bull. Bot. Resch., 15(3): 336, fig. I: 4, 1995; Zhu & Chen, Bacill. Xiz. Plat., p. 271—272, fig. 55: 6, 2000.

Surirella intermedia Lewis, On some new and singular intermediate forms of diatomacea, Proc. Acad. Nat. Sci. Philadelphia, 15: 336—346, 1863.

3a. 原变种　图版 CXXVII：6

var. intermedia

壳面窄，"S"形，中部两侧最宽，朝两端渐狭，顶端朝相反方向弯曲，呈喙状；长 95.6—142 μm，宽 4—7 μm；横肋纹有 3—6 条/10 μm，横线纹有 23—25 条/10 μm。

生境：生于冰渍物中流出的清泉小水坑，缓流、稻田与坡地之间的清水坑。

国内分布：安徽(天柱山)，西藏(亚东、墨脱)。

国外分布：北美洲(美国)，欧洲(瑞典、芬兰)，大洋洲(澳大利亚、新西兰)。

3b. 头端变种　图版 CXXVIII：3—8

var. capitata Fontell, Ark. Bot., p. 46, pl. 2, fig. 46, 1917; Cleve-Euler, Diatomeen schw. Finnland. Teil. V, p. 100, figs. 1522 e, f, 1952.

光镜下，与原变种的区别在于：壳面较长，末端楔形；长 150—180 μm，宽 6—8 μm；横肋纹有 4—5 条/10 μm，横线纹有 19—23 条/10 μm。

生境：生于石塘、溪流、沼泽中。

国内分布：内蒙古(阿尔山、根河、满归)。

国外分布：欧洲(瑞典、芬兰)。

马鞍藻属 Campylodiscus C. G. Ehrenberg ex F. T. Kützing

C. G. Ehrenberg ex F. T. Kützing, Kies. Bacill. Diat.: 59, 1844.

单细胞，马鞍形，常示壳面观。上下壳面相似，但呈 90°扭曲互相连接而成。由于观察的角度不同，细胞可呈现三角形或"V"形。色素体 1 个，由两片大而紧贴壳面的盘状结构组成，通过靠近细胞一端的极窄狭部连接。

壳面圆形或近圆形，壳面沿着顶轴方向凸起，横轴方向凹入。上下壳面的顶轴呈 90°交叉。外壳面上常具瘤状物和脊，以及折叠。线纹双排至多排，被腹板(sterna)断开，线纹由小圆孔组成，有时是较大的筛状孔。具一个环绕壳面的壳缝系统，壳缝位于龙骨上，龙骨突小肋状，或是由两边龙骨融合形成较大结构。内外壳面的壳缝末端简单或稍膨大。带面由数条断开的环带组成。

本属种类较多，多分布于半咸水和海水中，偶见于淡水中，常附着于沉积物或污泥中。

本属种类容易辨认，但可能不是单源属，Paddock (1985)认为双菱藻属(*Surirella*)和马鞍藻属(*Campylodiscus*)是平行进化的，这一观点也说明了 *Campylodiscus* 的近圆形壳面和上下壳面 90°扭曲不是一次进化的结果。

模式种：盾状马鞍藻[*Campylodiscus clypeus* (Ehrenberg) Ehrenberg ex Kützing]。

本志收编 5 种。

马鞍藻属分种检索表

1. 波纹底部非常窄，呈现出窄、辐射状、轮廓清晰的肋纹 ··· 2
1. 波纹底部宽，具不规则辐射状排列的肋纹 ··· 4
  2. 壳面呈明显的马鞍形，沿顶轴扭曲，波纹辐射状排列 ············ **1. 莱温德马鞍藻 *C. levanderi***
  2. 壳面马鞍形，但是不沿着顶轴扭曲 ··· 3
3. 辐射状排列的波纹 20—30 个/100 μm，常延伸到壳面中部 ············ **2. 诺克里马鞍藻 *C. noricus***
3. 辐射状排列的波纹 10—20 个/100 μm，在壳面中部常具大面积的无波纹区域 ·· **3. 冬生马鞍藻 *C. hibernicus***
  4. 体积较小，边缘辐射状排列的波纹不被打断，壳面中央的结构与边缘相同 ·· **4. 二肋纹马鞍藻 *C. bicostatus***
  4. 体积较大，边缘辐射状排列的波纹被一透明同心环状结构打断 ············ **5. 盾状马鞍藻 *C. clypeus***

1. 莱温德马鞍藻 图版 CXXXIII: 7; 图版 CXXXV: 1—4

Campylodiscus levanderi Hustedt, Diatomeen. *in* Järnefelt, Ann. Soc. Zool.-Bot. Fennicae Vanamo Helsinki 2: 328, 1925; Krammer & Lange-Bertalot, Bacillariophyceae. 2. Teil, *in* Ettl *et al.*, Süßwasserfl. Mitteleur., p. 215, figs. 181: 4—6, 1988, nachdr. 1997; Elizabeth & Kociolek, Bibliotheca Diatomologica 50: 28, figs. 42—44: 1—13, 2004.

细胞马鞍形，壳面近圆形至椭圆形，壳面均匀排列着辐射状的波纹，小刺随意排列在波纹顶部。中间无线纹区域呈菱形。壳面具不同程度的扭曲。壳面长 90—110 μm，宽 60—100 μm。漏斗 14—20 个/100 μm。

生境：生于河流、小溪、草地渗出水，石上附生，海拔 2000—2800 m。

国内分布：云南(维西、丽江)。

国外分布：北美洲(美国)，欧洲(英国、德国、马其顿)。

本种在我国首次发现，不常见，在模式产地芬兰和马其顿的数量也比较少。Ruck 和 Kociolek(2004)详细描述了本种光镜和扫描电镜下的壳面结构，本书只在光镜下观察到此种，壳面形态、体积大小和漏斗的密度均与 Ruck 和 Kociolek(2004)中的一致。

2. 诺克里马鞍藻　图版 CXXXIII：8

Campylodiscus noricus Ehrenberg ex Kützing, *in* Kützing, Kies. Bacill. Diat., p. 59, 1844; Schmidt *et al.*, Atlas Diat.-Kunde, p. 55, fig. 55：8, 1886; Hustedt, Bacillariophyta, *in* Pascher, Süßwass -Fl. Mitteleur. Heft 10, p. 446, fig. 871, 1930; Cleve-Euler, Diatomeen schw. Finnland, Teil. V, p. 130, fig. 1583, 1952; Krammer & Lange-Bertalot, Bacillariophyceae. 2. Teil, *in* Ettl *et al.*, Süßwasserfl. Mitteleur., p. 213, figs. 182：1—5, 1988, nachdr. 1997; Zhu & Chen, Bacill. Xiz. Plat., p. 277, fig. 58：5, 2000.

Surirella norica (Ehrenberg) Brun, Diat. Alpes Jura, p. 101; pl. 1, fig. 16; pl. 9, fig. 30, 1880.

Campylodiscus hibernicus var. *noricus* (Ehrenberg; Ehrenberg) van Heurck, Treat. Diat., p. 379, pl. 14, fig. 594, 1896.

细胞马鞍形，壳面圆形，在壳面边缘可见漏斗结构，漏斗斜向顶端排列，龙骨突非常小，一般延伸到漏斗形成的波纹，有时可达壳面中部。波纹明显分成 4 个区域，由纵轴和横轴划分开。波纹起伏较小，在每 1/4 的区域内几乎彼此平行排列，而在 4 个顶端呈辐射排列，此处的波纹最短。有时，壳面具随机排列的小刺。壳面直径 28—95 μm，漏斗 20—30 个/100 μm。

生境：生于水塘、沼泽中。

国内分布：西藏(芒康、申扎、措勤)。

国外分布：亚洲(俄罗斯、蒙古国)，北美洲(美国)，欧洲(波罗的海、英国、德国、爱尔兰、马其顿、罗马尼亚、西班牙)。

本种仅发现于我国的西藏地区,标本采集地海拔比较高,特别是申扎地区,高达 5400 m。采自这里的种类，直径明显比中欧地区的标本小[Cleve-Euler(1952)：直径 90—140 μm; Krammer and Lange-Bertalot(1997)：直径 60—150 μm]。

3. 冬生马鞍藻　图版 CXXXIV：1—4

Campylodiscus hibernicus Ehrenberg, Vorläufige zweite Mettheilung über die weitere Erkenntnifs der Beziehungen des kleinsten organischen Lebens zu den vulkanischen Massen der Erde, p. 154, 1845; Schmidt *et al.*, Atlas Diat.-Kunde, p. 55, figs. 55：14—16, 1886; Cleve-Euler, Diatomeen schw. Finnland, Teil. V, p. 129, fig. 1581, 1952; Krammer & Lange-Bertalot, Bacillariophyceae. 2. Teil, *in* Ettl *et al.*, Süßwasserfl. Mitteleur., p. 214, figs. 179：1—4, figs. 180：1—7, figs. 181：1—3, 1988, nachdr. 1997.

Campylodiscus noricus var. *hibernicus* (Ehrenberg) Grunow, Verh. zool. bot. Ges. Wien, p.

439(125), 1862; Hustedt, Bacillariophyta, in Pascher, Süßwass -Fl. Mitteleur. Heft 10, p. 447, fig. 872, 1930; Zhu & Chen, Bacill. Xiz. Plat., p. 277, figs. 58：6—7, 2000.

细胞马鞍形，壳面近圆形，在壳面边缘可见漏斗结构，漏斗斜向顶端排列，中间区域是一个椭圆形的无线纹区域，上下壳面分别关于顶面观对称。壳缝系统靠近壳面边缘，位于龙骨上，并且围绕整个壳面的边缘。壳面直径 80—130 μm，漏斗 10—20 个/100 μm。

生境：生于湖泊、小水沟、泉水井边草丛中。

国内分布：湖南(沅江流域)，四川(九寨沟)，贵州(沅江流域)，云南(维西)，西藏(亚东、错那、察隅、芒康、班戈、措勤、日土)，新疆(温宿、博湖、察布查尔、福海)。

国外分布：亚洲(俄罗斯、土耳其)，北美洲(美国)，欧洲(波罗的海、英国、德国、爱尔兰、马其顿、波兰、罗马尼亚、西班牙)，大洋洲(澳大利亚)。

4. 二肋纹马鞍藻　图版 CXXIX：1—4；图版 CXXX：1—2；图版 CXXXIII：1—6

Campylodiscus bicostatus W. Smith ex F. C. S. Roper, in Roper, Transactions of the Microscopical Society 2：75, fig. 6：4, 1854; Schmidt et al., Atlas Diat.-Kunde, p. 55, figs. 55：5—6, 1886; Cleve-Euler, Diatomeen schw. Finnland, Teil. V, p. 127—128, fig. 1578, 1952; Krammer & Lange-Bertalot, Bacillariophyceae. 2. Teil, in Ettl et al., Süßwasserfl. Mitteleur., p. 215, figs. 178：1—6, 1988, nachdr. 1997.

Campylodiscus clypeus var. *bicostatus* (W. Smith) Hustedt, Bacillariophyta, in Pascher, Süßwass -Fl. Mitteleur. Heft 10, p. 448—449, fig. 874, 1930; Xie et al., Journal of Shanxi University 6(3)：337, 1993.

细胞马鞍形，壳面近圆形。在壳面边缘可见漏斗结构，不规则辐射状排列，中间区域内具横线纹，漏斗和横线纹之间具一圈无线纹区域，呈现出一个明显的"曰"形。壳面直径 20—80 μm，漏斗 10—30 个/100 μm。

扫描电镜下观察：在外壳面，不规则的分布着一些瘤状物，主要分布在有线纹的区域。外壳面的线纹区域稍高于中间区域，漏斗位于壳面边缘位置较高的区域，这样，呈"曰"形的无孔纹区域位置相对比较低；在内壳面，这些结构的位置高低正好相反。外壳面，每个漏斗上都有线纹，没有观察到孔纹的详细情况，漏斗之间的区域没有孔纹。线纹在内壳面更容易观察到，均为小圆孔，边缘不凸出，横线纹和漏斗上的线纹均由单排孔组成，横线纹的密度 15—25 条/10 μm，漏斗上的线纹要比横线纹的密度大近一倍。在内壳面，壳缝管上具龙骨突，开始于管壳缝边缘延伸一小段至壳面，龙骨突窄而深。每个漏斗内有 1—3 个龙骨突，每对龙骨突之间呈现一个椭圆形开口，与漏斗上的线纹相连，此开口成为壳缝管与细胞内部联系的通道，漏斗之间的无孔纹区域窄，直接与壳面边缘相连，不具孔。壳套窄，几乎关于管壳缝与壳面边缘镜面对称。内外壳面，壳缝裂缝均不连续，没有观察到壳缝末端的详细情况。

生境：生于湖泊、路边清澈缓流水沟、路边静水沟中。

国内分布：山西(运城)，新疆(岳普湖、尉犁、博湖)。

国外分布：亚洲(蒙古国、以色列)，北美洲(美国)，南美洲(哥伦比亚)，欧洲(波罗的海、英国、德国、罗马尼亚、西班牙)，大洋洲(澳大利亚)。

本种与 *C. clypeus* 在外形和壳面结构上比较相似，Hustedt(1930)就将本种放在 *C. clypeus* 下作为变种，此后，Cleve-Euler(1952)、Krammer 和 Lange-Bertalot(1997)又将此种独立出来，本书采用后者的观点。原因如下：①边缘辐射状排列的波纹不被打断，壳面中央的结构与边缘相同，而 *C. clypeus* 边缘波纹被一透明同心环状结构打断。②体积明显比 *C. bicostatus* 小，两者基本没有重叠，此外，采自新疆的标本，最小直径可达 20 μm，且数量比较多，而 Krammer 和 Lange-Bertalot(1997)中描述的壳面直径为 40—85 μm。本书中小体积标本，经扫描电镜观察，壳面结构，包括漏斗、线纹和龙骨突的密度与 *C. bicostatus* 相符合，因此，可以确定是 *C. bicostatus*。③外壳面具一些不规则分布的瘤状物，主要分布在有线纹的区域，而 *C. clypeus* 的外表面光滑，不具瘤状物。④中间区域形状变化较大，从宽线形到菱形-披针形，而 *C. clypeus* 的中间区域一般为宽线形。⑤横线纹的密度 15—25 条/10 μm，漏斗上的线纹要比横线纹密度多近一倍。而 *C. clypeus* 横线纹和漏斗上线纹密度差不多，7—14 条/10 μm。可见，两个种的区别还是相当大的，因此，我们将本种从 *C. clypeus* 中分离出来，提升到种的水平。

5. 盾状马鞍藻　图版 CXXXI：1—7；图版 CXXXII：1—4

Campylodiscus clypeus(Ehrenberg)Ehrenberg ex Kützing, Kies. Bacill. Diat. p. 59, pl. 2. figs. V: 1—6, 1844; Schmidt *et al.*, Atlas Diat.-Kunde, p. 54, figs. 54: 7—8, 1878; Hustedt, Bacillariophyta, *in* Pascher, Süßwass-Fl. Mitteleur. Heft 10, p. 448, fig. 873, 1930; Cleve-Euler, Diatomeen schw. Finnland, Teil. V, p. 128, fig. 1579, 1952; Krammer & Lange-Bertalot, Bacillariophyceae. 2. Teil, *in* Ettl *et al.*, Süßwasserfl. Mitteleur., p. 214, figs. 175: 3, 4, figs. 177: 1—5, 1988, nachdr. 1997; Metzeltin *et al.*, Iconogr. Diatomol. 15, fig. 62: 2, 2005.

Cocconeis clypeus Ehrenberg, Infusionsthierchen, p. 195. 1838.

Surirella clypeus(Ehrenberg)Kützing, Kies. Bacill. Diat., p. 59, 1844.

细胞马鞍形，壳面近圆形。在壳面边缘可见漏斗结构，不规则辐射状排列，被一透明同心环状结构打断，中间区域具横线纹，漏斗和横线纹之间具一圈无线纹区域，呈现出一个明显的"曰"形。壳面直径 70—130 μm，漏斗 10—20 个/100 μm。

扫描电镜下观察：外壳面不具瘤状物，线纹区域稍高于中间区域，漏斗位于壳面边缘位置较高的区域，这样，呈"曰"形的无孔纹区域位置相对比较低；在内壳面，这些结构的位置高低正好相反。外壳面，每个漏斗上都具线纹，没有观察到孔纹的详细情况，漏斗之间的区域没有孔纹。线纹在内壳面更容易观察到，为圆孔状，边缘不凸出，横线纹和漏斗上的线纹均由单排孔组成，线纹密度 7—14 条/10 μm。

生境：生于香蒲沼泽、缓流或静止的水沟中。

国内分布：新疆(岳普湖、疏勒、莎车、尉犁)。

国外分布：亚洲(俄罗斯、蒙古国)，非洲(东部)，北美洲(美国)，南美洲(巴西、乌拉圭)，欧洲(波罗的海、英国、德国、罗马尼亚、西班牙)，大洋洲(澳大利亚)。

附录I 科、属、种的检索表(英文)
Key to families, genera and species of Aulonoraphidinales

Key to the families of Aulonoraphidinales

1. Frustule with well-developed raphe system and a keel on each valve positioned along the periphery of the valve ·· **Surirellaceae**
1. Frustule with well-developed raphe systems running from pole to pole and mostly laterally to the median line, never around the entire periphery of the valve; with or without a keel system ················ 2
   2. Valves without a keel system, obviously dorsiventral, fibulae extended and forming transapical costae ·· **Rhopalodiaceae**
   2. Valves with a keel system, raphe along one margin of each valve, not or slightly dorsiventral, fibulae usually not or slightly extended across the valve face (exception: *Denticula*) ················ **Bacillariaceae**

Key to the genera of Bacillariaceae

1. Fibulae absent, an alar raphe canal lies distally on a wing ·· 6. ***Simonsenia***
1. Fibulae always present, wing and alar canals absent ·· 2
   2. Valves twisted two or three times around the apical axis, like a screw thread ············ 7. ***Cylindrotheca***
   2. Valves usually not twisted ·· 3
3. Apical axis of the valves usually heteropolar, valves dorsiventral ······························ 3. ***Hantzschia***
3. Apical axis of the valves usually isopolar or slightly heteropolar ·· 4
   4. Fibulae extended and forming transapical partitions ·································· 5. ***Denticula***
   4. Fibulae not or slightly extended, not forming partitions ·· 5
5. Frustules combined in table like bands, each individual is able to slide against the others with its central keel ··· 1. ***Bacillaria***
5. Not with the above combination of characteristics ··· 6
   6. Valves straight or sigmoid, narrow; linear, lanceolate or elliptical ·················· 2. ***Nitzschia***
   6. Valves robust, broad; elliptical, linear or panduriform, with more or less obvious longitudinal fold ···· ··· 4. ***Tryblionella***

Key to the species of *Bacillaria*

Only 1 species——*Bacillaria paradoxa* in China.

Key to the Sections of *Nitzschia*

1. Raphe keel on the outside of the valve covered by a canopy; raphe with or never with central raphe endings ·· 9
1. Canopy not positively distinguishable ··· 2
   2. Frustules more or less sigmoidly curved in girdle and/or valve view ························ 3
   2. Frustules not sigmoidly curved in girdle and valve view ······································· 4
3. Raphe with obviously central raphe endings, raphe fissures slightly extended into valve in the center ·········· ·· Sect. ***Obtusae***

3. Not with obviously central raphe endings ···Sect. ***Sigmata***
    4. Fibula elongated, transapical ribs run far into the valve face ······················Sect. ***Epithemioideae***
    4. Not with the above combination of characteristics ··· 5
5. Valves fusiform, with conspicuously elongated rostrate ends ································· Sect. ***Nitzschiellae***
5. Valves ends short, rostrate ·· 6
    6. Fibulae extended forming ribs, not reach the other valve edge ·····················Sect. ***Grunowia***
    6. Fibulae do not extend as far into the surface of the valve ··· 7
7. Frustules in girdle view often quite wide because of numerous bands. The edge with a raphe keel more strongly constricted than the edge without the keel, central raphe endings always present·······························
···Sect. ***Dubiae*** and ***Bilobatae***
7. Not with or uncertain the above combination of characters·· 8
    8. Valves much shorter, lanceolate, raphe keel strongly eccentric, positioned at the edge between valve surface and mantle··Sect. ***Lanceolatae***
    8. Valves much longer, linear, containing many related taxa which are difficult to define·······················
···Sect. ***Lineares***
9. Valves much longer, frustules more or less sigmoid curved in girdle view······················Sect. ***Sigmoideae***
9. Valves much shorter, frustules not curved in girdle or valve view ································ Sect. ***Dissipatae***

Key to the species of Sect. *Sigmoideae*

1. Around 25 striae /10 μm·· 4
1. Around 30 striae /10 μm or more·· 2
    2. Frustules weakly sigmoid in girdle view, striae around 30 /10 μm·················1. ***N. vermicularis***
    2. Striae around 40 /10 μm or more, not able to be resolved without specialized lighting ················ 3
3. Frustules weakly sigmoid in girdle view, valve usually longer, wider···································2. ***N. acula***
3. Frustules very strongly sigmoid in girdle view ··3. ***N. flexa***
    4. Width of the valves more than 7 μm, fibulae less than 10/10 μm ·····················4. ***N. sigmoidea***
    4. Width of the valves less than 7 μm, fibulae very slender, more than 10/10 μm ···············5. ***N. eglei***

Key to the species of Sect. *Dissipatae*

1. Raphe keel visible approximately in the centre of the valve···6. ***N. dissipata***
1. Raphe keel moderately eccentric ···7. ***N. recta***

Key to the species of Sect. *Obtusae*

1. Valves long, length of the valves usually greater than 110 μm, with a width at least 7 μm········8. ***N. obtusa***
1. Valves short·· 2
    2. Sigmoid curve in valve view only suggested by the asymmetric outlines of the poles, poles not short rostrate or capitate ··· 3
    2. Not with the above combination of characteristics ·· 4
3. Ends narrowed very close to the poles, strongly asymmetrical, resulting in a conspicuous scalpel-shape ····
··9. ***N. scalpelliformis***
3. Ends gradually narrowed and less asymmetrically constricted ·····································10. ***N. filiformis***
    4. Sigmoid curve in valve view, ends relatively broad at the poles (not constricted in any way) ················
··11. ***N. nana***

4. Not with the above combination of characteristics ·· 5
5. Valve ends rostrate and more or less bent to opposite sides ·································· 12. *N. clausii*
5. Not with the above combination of characteristics ·· 6
    6. Valves not concave in the centre, ends gradually narrowed ················ 13. *N. subcohaerens* var. *scotica*
    6. Valves usually more or less concave in the centre, narrowed, and ends more or less rostrate ············· 7
7. Valve length within relatively narrow limits ·· 14. *N. brevissima*
7. Valve length less narrowly limited ·· 15. *N. terrestris*

Key to the species of Sect. *Sigmata*

1. Fibulae slightly extended into the valve face ·· 16. *N. sigma*
1. Fibulae obviously extended into the valve face ·· 17. *N. fasciculata*

Key to the species of Sect. *Grunowia*

Only 1 species——*Nitzschia sinuata* in China

Key to the species of Sect. *Dubiae* and *Bilobatae*

1. Fibulae narrow, more or less extended into the valve face ·· 19. *N. homburgiensis*
1. Fibulae wider, slightly extended ·· 2
    2. Frustules very large, length more than 100 μm, fibulae wide, thick, combined with 3 or more striae ··
    ··· 20. *N. kittlii*
    2. Frusltules short, length less than 100 μm, fibulae narrow, combined with 1 or 2 striae ···················· 3
3. Valve in the middle strongly constricted, valves "naviculoid"or"bracket-shaped" in girdle valve with bent,
    rostrate, asymmetric ends, fibulae always narrow and combined with only 1 stria································
    ···21. *N. hybrida*
3. Not with above combination of characteristic or characters difficult to distinguish ································ 4
    4. Raphe keel more strongly eccentric and less strongly constricted at the central raphe endings ················
    ·· 22. *N. dubia*
    4. Not with above combination of characteristic ··· 5
5. Fibulae short, combined with more than one stria, striae 24—30/10 μm ·······················23. *N. umbonata*
5. Not with above combination of characteristic ·· 6
    6. Valves with central raphe endings difficult to distinguish, with the middle fibulae narrow and only
    combined with one stria, striae around 20/10 μm ···24. *N. gisela*
    6. Valves with central raphe endings easy to distinguish, with raphe keel more strongly constricted in the
    centre, the two central ones distinctly separated ··· 25. *N. commutata*

Key to the species of Sect. *Lineares*

1. Fibulae narrow, each combined with one stria ··
··26. *N. linearis*
1. Not with above combination of characteristic ·· 2
    2. Fibulae broad, combined with 3 or more striae, 4—8 fibulae /10 μm······························27. *N. vitrea*
    2. Fibulae very narrow ·· 3
3. Valves around 100 μm or longer, fibulae 5—7/10 μm ··· 28. *N. monachorum*
3. Valves shorter, less than 100 μm, fibulae 5—7/10 μm..29. *N. sublinearis*

Key to the species of Sect. *Lanceolatae*

1. The centre fibulae usually equidistant, combined with the characteristic of a continuous raphe uninterrupted from pole to pole ···················· 2
1. Raphe is usually interrupted in the centre between the poles by central raphe endings, the two fibulae in the middle are usually more widely spaced than the others ···················· 16
   2. Fibulae short and pointed, combined with one stria, striae 24—28/10 μm ···················· 30. *N. solita*
   2. Fibulae wider, combined with more than one stria ···················· 3
3. Striae comparatively more widely spaced, fewer than 15 to 25/10 μm, usually distinguishable as punctate 4
3. Striae more dense, more than 25/10 μm ···················· 9
   4. Fustules short, ends bluntly rounded; Striae coarse, 17—20/10 μm ···················· 31. *N. valdecostata*
   4. Not with above combination of the characteristic ···················· 5
5. Fustules short, ends capitate, round or elongate ···················· 6
5. Not with above combination of the characteristic ···················· 7
   6. Valves usually more or less weakly concave in the centre, capitate ends narrow, striae 23—32/10 μm ···················· 32. *N. elegantula*
   6. Valves linear or weakly concex in the centre, broad capitate ends ···················· 33. *N. bacilliformis*
7. Valves rarely greater than 40 μm long, striae 21—25/10 μm, ends bluntly rounded, tribes in electrolyte poor freshwater ···················· 34. *N. alpina*
7. Valves greater, linear ···················· 8
   8. Valve width 4—7 μm ···················· 35. *N. intermedia* in part
   8. Valve width 2.5—3.5 μm ···················· 36. *N. diversa*
9. The dense of striae more than 35/10 μm, difficult to visible or not visible at all in the light microscope; valves generally narrow, rostrate ends elongate; fibulae short, point-like ···················· 37. *N. gracilis*
9. The dense of striae 25—35/10 μm, difficult to visible in the light microscope ···················· 10
   10. Valves usually short, with conspicuously capitate or short rostrate ends ···················· 38. *N. microcephala*
   10. Not with above combination of the characteristic ···················· 11
11. Ends quite broadly rounded, rarely slightly elongated and bluntly rounded; width 4—5.5 μm, striae 30/10 μm ···················· 39. *N. communis*
11. Not with above combination of the characteristic ···················· 12
    12. Population on average in the population comparatively "large", linear or lanceolate ···················· 13
    12. Population with predominantly shorter, usually also narrower valves ···················· 15
13. Valves linear, width more than 4 μm ···················· 35. *N. intermedia*
13. Valves narrowly lanceolate, width less than 4 μm ···················· 14
    14. Ends rostrate, elongate ···················· 40. *N. subacicularis*
    14. Ends tapering, not elongate ···················· 41. *N. fruticosa*
15. Longer valves more or less lanceolate, at least not narrowly linear, striae difficult to resolve, because alveoli and transapical ribs form a very flat surface ···················· 42. *N. palea*
15. Longer valves in a population narrowly linear, striae 26—32/10 μm ···················· 43. *N. perminuta*
    16. Fibulae somewhat narrowed, cuneate and root-like leading into the transapical ribs, striae fewer than 20/10 μm ···················· 44. *N. amphibia*
    16. Fibulae broader, combined with more than one stria, striae more dense, difficult to visible in the light microscope ···················· 17

17. Striae more dense, more than 20/10 μm ·· 18
17. Striae comparatively widely spaced, less than 20/10 μm ·· 19
    18. Valves longer in comparison, lanceolate or linear-lanceolate with pointed to slightly capitate rounded ends ··· 45. *N. fonticola* in part
    18. Valves more or less laceolate, tribes in fresh water with moderate electrolyte levels ········ 46. *N. fossilis*
19. Valves always small, striae coarse, 16—19/10 μm, puncta not distinguishable because they lie in closely packed double rows ·· 47. *N. valdestriata*
19. Not with above combination characteristic ··· 20
    20. Valves always small, elliptical to linear-elliptical, ends bluntly rounded ············· 48. *N. inconspicua*
    20. Not with above combination characteristic ·· 21
21. Longer valves in the population linear, not lanceolate, valve ends short and cuneate ··· 49. *N. gessneri*
21. The longest valves in the population tending to lanceolate or linear-lanceolate shape, valve ends appear to be more elongate cuneate ··· 22
    22. Valves tend to linear-lanceolate with rostrate poles ··· 50. *N. tubicola*
    22. Valves distincly lanceolate ·· 23
23. Valves of the length usually much shorter than 40 μm, found in freshwater with moderate electrolyte concentrations; double rows of areolae in the raphe keel region ····································· 45. *N. fonticola*
23. Not with above combination of characteristics, particularly areolae rows are always simple ················ 25
    24. Valves lanceolate to linear-lanceolate, usually short, barely reaching 30 μm, and barely exceeding 3 μm wide, ends not rounded capitate ··· 51. *N. frustulum*
    24. Not with above combination of characteristics ·· 26
25. Valves narrow lanceolate, never concave on both sides, width 2.5—3 μm······················· 52. *N. radicula*
25. Valves broader, often with both sides somewhat concave in the center, width 3—6 μm ··· 53. *N. capitellata*

Key to the species of Sect. *Nitzschiellae*

1. Apical axis straight, with center raphe endings ··· 2
1. Apical axis sigmoid ··· 3
    2. The centre fibulae usually equidistant ··· 54. *N. acicularis*
    2. A space between the middle two fibulae·· 55. *N. draveillensis*
3. Striae coarse, usually fewer than 20/10 μm ·· 56. *N. lorenziana*
3. Striae more closely spaced, not able to be resolved by light microscope ····························· 57. *N. reversa*

Key to the species of Sect. *Epithemioideae*

Only 1 species——*Nitzschia epithemoides* in China.

Key to the species of Sect. unnamed

Only 1 species——*Nitzschia guadalupensis* in China.

Key to the species of *Hantzschia*

1. Raphe without central endings, middle fibulae equidistant or weak wider ··· 2
1. Raphe with central endings, quite a wide space between the two centre fibulae ·· 4

2. Frustules with a weak sigmoid, valves without the typical *Hantzschia*-like dorsiventral form ⋯⋯⋯ 3
2. Frustules not sigmoid, valves typically *Hantzschia*-like ⋯⋯⋯⋯⋯⋯⋯⋯⋯⋯⋯⋯⋯⋯ 1. *H. vivax*
3. Valves longer, ends elongated, rostrate ⋯⋯⋯⋯⋯⋯⋯⋯⋯⋯⋯⋯⋯⋯⋯⋯⋯⋯ 2. *H. spectabilis*
3. Valves shorter, ends obvious cuneate, capitate ⋯⋯⋯⋯⋯⋯⋯⋯⋯⋯⋯⋯⋯ 3. *H. nitzschioides*
    4. Fibulae narrow, usually combined with one transapical rib, predominantly marine ⋯⋯⋯⋯⋯ 5
    4. Fibulae in part broader and combined with more than one transapical rib ⋯⋯⋯⋯⋯⋯⋯⋯⋯ 6
5. Fibulae barely elongated ⋯⋯⋯⋯⋯⋯⋯⋯⋯⋯⋯⋯⋯⋯⋯⋯⋯⋯⋯ 4. *H. distinctepunctata*
5. Fibulae more or less elongated ⋯⋯⋯⋯⋯⋯⋯⋯⋯⋯⋯⋯⋯⋯⋯⋯⋯⋯⋯⋯ 5. *H. virgata*
    6. Coarsely punctate, number of the striae less than double number of the fibulae ⋯⋯⋯⋯⋯⋯⋯ 7
    6. Fine punctate, number of the striae more than double number of the fibulae ⋯⋯⋯⋯⋯⋯⋯ 13
7. Valves without the typical *Hantzschia*-like dorsiventral, both sides almost parallel ⋯⋯⋯⋯⋯ 6. *H. sinensis*
7. Valves with the typical *Hantzschia*-like dorsiventral ⋯⋯⋯⋯⋯⋯⋯⋯⋯⋯⋯⋯⋯⋯⋯⋯⋯⋯ 8
    8. Valves linear, usually more than 200 μm long ⋯⋯⋯⋯⋯⋯⋯⋯⋯⋯⋯⋯⋯⋯⋯⋯⋯⋯ 9
    8. Valves not obvious linear, length/width low ⋯⋯⋯⋯⋯⋯⋯⋯⋯⋯⋯⋯⋯⋯⋯⋯⋯⋯⋯ 10
9. Valve geniculate, length/width high ⋯⋯⋯⋯⋯⋯⋯⋯⋯⋯⋯⋯⋯⋯⋯⋯⋯⋯⋯ 7. *H. elongata*
9. Valve not geniculate, length/width low ⋯⋯⋯⋯⋯⋯⋯⋯⋯⋯⋯⋯⋯⋯⋯⋯⋯⋯⋯ 8. *H. longa*
    10. Valves with sharply narrowed ends, slightly elongate, taper-shape ⋯⋯⋯⋯⋯⋯⋯⋯⋯ 11
    10. Valves wider, more than 8 μm, with rostrate ends ⋯⋯⋯⋯⋯⋯⋯⋯⋯⋯⋯⋯⋯⋯⋯⋯⋯ 12
11. With obviously dorsiventral form, valves long, 100—118 μm ⋯⋯⋯⋯⋯⋯⋯⋯ 9. *H. barckhausenii*
11. Without obviously dorsiventral form, valves short, 54—89 μm ⋯⋯⋯⋯⋯⋯⋯⋯⋯⋯ 10. *H. yili*
    12. Ventral margin slightly convex, valve width 8—11 μm ⋯⋯⋯⋯⋯⋯⋯⋯ 11. *H. giessiana*
    12. Ventral margin obviously convex, valve width 14—16 μm ⋯⋯⋯⋯⋯⋯⋯⋯ 12. *H. compacta*
13. Valves almost without "shoulder" or not obviously ⋯⋯⋯⋯⋯⋯⋯⋯⋯⋯⋯⋯⋯⋯⋯⋯⋯ 14
13. Valves with "shoulder", with bluntly rounded or capitate ends ⋯⋯⋯⋯⋯⋯⋯⋯⋯⋯⋯⋯⋯ 15
    14. Valves large, with narrow rostrate ends ⋯⋯⋯⋯⋯⋯⋯⋯⋯⋯⋯⋯⋯ 13. *H. subrobusta*
    14. Valves short, with small capitate ends ⋯⋯⋯⋯⋯⋯⋯⋯⋯⋯⋯⋯⋯⋯ 14. *H. calcifuga*
15. Valves large, length/width high, gradually narrowed in the poles, with capitate ends ⋯⋯⋯⋯⋯ 16
15. Valves short, length/width low, sharply narrowed in the poles, with bluntly rounded ends ⋯⋯⋯⋯ 17
    16. Dorsal margin more or less convex, striae 13—15 /10 μm ⋯⋯⋯⋯⋯⋯⋯⋯⋯ 15. *H. vivacior*
    16. Dorsal margin not obviously convex in, striae 16—18 /10 μm ⋯⋯⋯⋯⋯⋯ 16. *H. pseudobardii*
17. Length/width high, valve width more than 10 μm ⋯⋯⋯⋯⋯⋯⋯⋯⋯⋯⋯⋯⋯⋯⋯⋯⋯⋯ 18
17. Length/width low, valve width less than 10 μm ⋯⋯⋯⋯⋯⋯⋯⋯⋯⋯⋯⋯ 17. *H. subrupestris*
    18. Valves large, with obvious dorsiventral ⋯⋯⋯⋯⋯⋯⋯⋯⋯⋯⋯⋯⋯⋯⋯ 18. *H. abundans*
    18. Valves short, without obvious dorsiventral ⋯⋯⋯⋯⋯⋯⋯⋯⋯⋯⋯⋯⋯⋯ 19. *H. amphioxys*

Key to the species of *Tryblionella*

1. Raphe not continuous, with central raphe endings, not easy to distinguish ⋯⋯⋯⋯⋯⋯⋯⋯⋯ 3
1. Raphe continuous, without central raphe endings, the number of fibulae always corresponds to the number of transapical ribs ⋯⋯⋯⋯⋯⋯⋯⋯⋯⋯⋯⋯⋯⋯⋯⋯⋯⋯⋯⋯⋯⋯⋯⋯⋯⋯⋯⋯⋯⋯⋯⋯ 2
    2. Valves comparatively long, linear, usually more than 7 μm wide ⋯⋯⋯⋯⋯⋯⋯⋯ 1. *T. angustata*
    2. Valves linear-lanceolate, lanceolate, 6—7 μm wide ⋯⋯⋯⋯⋯⋯⋯⋯⋯⋯⋯ 2. *T. angustatula*
3. Fibulae easily to differentiate ⋯⋯⋯⋯⋯⋯⋯⋯⋯⋯⋯⋯⋯⋯⋯⋯⋯⋯⋯⋯⋯⋯⋯⋯⋯⋯⋯ 5
3. Number of the fibulae is the same as the number of transapical ribs, difficult to differentiate ⋯⋯⋯⋯⋯ 4
    4. Width of the valves greater than 10 μm ⋯⋯⋯⋯⋯⋯⋯⋯⋯⋯⋯⋯⋯⋯⋯⋯ 3. *T. acuminata*

4. Width of the valves less than 9 μm ··· 4. *T. apiculata*
5. Striae composed of two rows areolae, interrupted by one sterna ··················· 5. *T. hungarica*
5. Striae formed from very fine areolae, occasionally broken. In some cases the striae are very difficult to resolve ··· 6
    6. Valves larger, transapical ribs usually visible at the edge on one side or distinguishable in outline ·········
    ·· 6. *T. littoralis*
    6. Valves smaller, transapical appear stepped, continuous or interrupted in the median area by longitudinal folds or displaced from each other ··· 7
7. Valves relatively narrow in relation to length, 8—11 μm wide, with relatively narrow and very dense transapical ribs ·· 7. *T. calida*
7. Valves relatively wider, usually more than 11 μm wide, with relatively wide and few transapical ribs ········
  ··· 8
    8. Valves larger, transapical ribs usually not visible ·· 8. *T. gracilis*
    8. Valves smaller, transapical ribs coarse, easily to visible ·· 9
9. Valves smaller, transapical ribs very dense ·· 9. *T. levidensis*
9. Valves larger, transapical ribs few dense ·· 10. *T. victoriae*

Key to the species of *Denticula*

1. Fibular partitions very low and seen only with careful focusing ·················· 1. *D. kuetzingii*
1. Fibular partitions across most of the valve width as high as the valve mantle ······················· 2
    2. Valves smaller, usually not more than 20 μm, the dense of transapical ribs more than 8/10 μm ···············
    ··· 2. *D. creticola*
    2. Valves larger, the dense of transapical ribs less than 6/10 μm ··································· 3
3. Striae more dense, 25—30/10 μm, usually difficult to differentiate ································· 3. *D. tenuis*
3. Striae usually less than 20/10 μm, usually easy to differentiate ·· 4
    4. Valves smaller, the dense of transapical ribs usually more than 4/10 μm ·················· 4. *D. elegans*
    4. Valves larger, the dense of transapical ribs usually less than 4/10 μm ······················ 5. *D. valida*

Key to the species of *Simonsenia*

1. Frustule smaller, Valves narrow lanceolate, transapical ribs 20—21/10 μm ·············· 1. *S. delognei*
1. Frustule larger, Valves lanceolate to elliptical-lanceolate, transapical ribs 13—17/10 μm ···· 2. *S. maolaniana*

Key to the species of *Cylindrotheca*

Only 1 species——*Cylindrotheca gracilis* in China.

Key to the genera of Rhopalodiaceae

1. The raphe is running on the ventral side of the valve, and proximally rises more or less into the valve face, apprear "V" ·· 1. *Epithemia*
1. The raphe usually on the keel near the dorsal side of the valve ···················· 2. *Rhopalodia*

Key to the species of *Epithemia*

1. Capitate septa on the fibular partitions distinctly seen in girdle view, central raphe ending usually rising towards the dorsal area over the half of the valve ··· 1. *E. argus*

1. Capitate septa on the fibular partitions not easily seen in girdle view ·· 2
    2. Raphe running along on the ventral valve edge, only central raphe ending distinguishable, transapical ribs parallel ··· 2. *E. frickei*
    2. Raphe curving towards the dorsal edge or lying in the valve face ·· 3
3. Striae 3 or fewer between each pair of fibulae ·· 5
3. Striae 3 or more between each pair of fibulae ··· 4
    4. The raphe branch running on the valve face over its entire length and curing towards the middle of the dorsal side ··· 3. *E. smithii*
    4. The raphe branch along the ventral edge of the valve for almost the entire length, only visible in the centre of the valve, but barely reaching the middle of the valve ·· 6
5. Valves with usually 5 or more fibulae/10 μm, striae more than 10/10 μm, dorsal side strongly convex, the raphe branch reaching the dorsal edge in the middle of the valve ·· 4. *E. sorex*
5. Valves with usually fewer than 5 fibulae/10 μm, striae less than 10/10 μm, dorsal side less convex, the raphe branch curving dorsally only in the middle of the valve ··· 5. *E. turgida*
6. Valves obviously dorsiventral ··· 6. *E. adnata*
6. Valves not obviously dorsiventral ··· 7. *E. arguiformis*

Key to the Sections *Rhopalodia*

1. Frustule in girdle valve linear or transapically inflated middle (predominantly fresh water forms) ··· Sect. *Gibba*
1. Frustule in girdle valve elliptical or more or less ovate (predominantly salt water forms) ··· Sect. *Gibberula*

Key to the species of Sect. *Gibba*

1. Frustule in girdle valve more or less swollen in the middle, valves bracket-shaped, bent ventrally at the ends and sharply rounded ·· 1. *R. gibba*
1. Frustule in girdle valve parallel, at most slightly swollen in the middle ·· 2
    2. Frustule larger, slightly narrow towards ends, with bluntly rounded ends ································· 3. *R. parallela*
    2. Frustule smaller, obviously narrow towards ends, with sharply rounded ····································· 2. *R. gracilis*

Key to the species of Sect. *Gibberula*

1. Puncta difficult to distinguish or indistinguishable by light microscope, always more than 30 puncta/ 10 μm on the striae ·· 4. *R. operculata*
1. Puncta distinguishable by light microscope with high magnification, usually fewer than 30 puncta/ 10 μm on the striae ··· 2
    2. Valve outline broad and sickle-shaped, fewer than 16 puncta/10 μm on the striae, valves with strongly convex dorsal edge and compact outline ··· 5. *R. musculus*
    2. Structure finer ··· 3
3. Ventral edge concave, outline of the valve sickle-shaped, keel weakly developed, raphe running along the valve edge, barely distinguishable in valve view ··· 6. *R. gibberula*
3. Ventral edge straight or slightly convex, valve outline a segment of a circle or lanceolate ·························· 4
    4. Smaller, length less than 40 μm, delicate forms with straight or concave ventral edge, single puncta on the striae ··· 7. *R. brebissonii*

4. Large forms with straight ventral edge, length more than 40 μm ··· 5
5. Double puncta irregularly arranged on the striae ·································· 8. *R. constricta*
5. Single puncta regularly arranged on the striae ····································· 9. *R. rupestris*

Key to the genera of Surirellaceae

1. Valves curved in a saddle shape, the apical axis of hypotheca and epitheca cross at right angles ·················
·· 4. *Campylodiscus*
1. Valves not curved in a saddle shape, the apical axis of hypotheca and epitheca run parallel ······················ 2
   2. Valves elongated, linear, apical axis S-shaped or straight, median area narrowly linear, but always distinctly bordered ··· 3. *Stenopterobia*
   2. Not with the above combination of characteristics ·· 3
3. Transapical undulations and striae interrupted, at least in the region of the median line ············ 2. *Surirella*
3. Transapical undulations and striae not interrupted in the region of the median line ············ 1. *Cymatopleura*

Key to the Species of *Cymatopleura*

1. Valves elliptica, isopolar, with broadly round end··· 1. *C. elliptica*
1. Valves panduriform, the fibulae continue transapically in narrow undulations, which reach to the axial area
··· 2
   2. Valves in the same plane, not twisted ·· 2. *C. solea*
   2. Valves slightly twisted in the apical axis ·· 3
3. valves constricted at the margin in the middle ·· 3. *C. aquastudia*
3. valves unconstricted at the margin in the middle ··· 4. *C. xinjiangiana*

Key to the Sections of *Surirella*

1. Valves with wings and alar canal, often with loops visible along the valve edge, outline often linear-oval, usually fresh water ·· Sect. *Robustae*
1. Usually small broadly elliptical species with psudoinfundibulae, without wings and alar canal, the fibulae are canal like structures ··· Sect. *Pinnatae*

Key to the species of Sect. *Pinnatae*

1. Apical axis isopolar··· 2
1. Apical axis heteropolar·· 5
   2. Valves large, robust·· 1. *S. gracilis*
   2. Valves delicate, finely-structured ··· 3
3. Valves dumbbell, broad in the end, narrow in the centre ····························· 2. *S. tientsinensis*
3. Valves narrow linear, gradually narrow towards ends ··· 4
   4. Shorter forms, length/breadth ratio of the girdle side around 5∶1, transapical undulations run to the median line·· 3. *S. angusta*
   4. Longer forms, narrow, length/breadth ratio of the girdle side up to 15∶1 ······················ 4. *S. lapponica*
5. Transapical undulations running paralle to the median line, length/breadth radio up to 5∶1, fibulae 60—80 /100 μm ··· 5. *S. minuta*
5. Not with the above combination of characteristics·· 6
   6. Valves broad, pear-shaped·· 7

6. Valves oval to linear-oval ··· 8
7. Valves with only weak concentric undulations or no concentric undulations at all ················ 6. *S. crumena*
7. Valves with strong concentric undulations, fibulae and psudoinfundibula form a narrow marginal ring ········
··· 7. *S. peisonis*
  8. Both valve poles wedge-shaped (noly distinct in large forms) ··· 9
  8. One or both valve poles broadly rounded in moderate to large forms ································ 11
9. Valves linear-elliptical ··· 8. *S. patella*
9. Valves broadly lanceolate ··· 10
  10. Psudoinfundibula with indistinct borders, valves with weak concentric undulations ············ 9. *S. ovalis*
  10. Psudoinfundibula with distinct borders, valves with strongly concentric undulations ····· 10. *S. brightwellii*
11. Transapical undulations or striae distinct ·· 12
11. Transapical undulations or striae indistinct ··· 13
  12. Valves usually less than 50 μm ··· 11. *S. subsalsa*
  12. Valves usually more than 70 μm, wave crests nearly as broad as the wave troughs ···························
  ··· 12. *S. striatula*
13. Valves linear-elliptical ·· 13. *S. visurgis*
13. Valves oval ·· 14. *S. brebissonii*

Key to the species of Sect. *Robustae*

1. Frustule strongly twisted around the apical axis ·· 15. *S. spiralis*
1. Frustule not or only slightly twisted around the apical axis ·· 2
  2. Apical axis isopolar ·· 3
  2. Apical axis heteropolar ·· 6
3. Valves lanceolate, median area narrowly lanceolate, edges of the wave crest with distinct spinules ············
  ·· 16. *S. bifrons*
3. Valves lanceolate to linear-lanceolate, transapical undulations visible in the median area ······················ 4
  4. Large robust forms with fewer than 20 alar canals/100 μm, wave crest usually with numerous very
    delicate spinutes ·· 17. *S. biseriata*
  4. Smaller robust forms with more than 20 alar canals/100 μm ·· 5
5. Valves linear to linear-lanceolate ··· 18. *S. linearis*
5. Valves elliptical-lanceolate, with numerous spinutes on the median area and transapical rib ····· 19. *S. helvetica*
  6. Valve with strong spines just in front of the broad pole and occasionally also just in front of the narrow
    pole ·· 20. *S. capronii*
  6. Typical spines not present, alar canals less than 35/100 μm, transapical undulations dictint in the median
    area ·· 7
7. Alar projection distinct ··· 8
7. Alar projection not as clear or indistinct, large forms ··· 21. *S. elegans*
  8. Alar canals more than 20/100 μm, large forms with distinct middle rib ································ 22. *S. tenera*
  8. Alar canals less than 20/100 μm ··· 9
9. Valves slightly twisted around apical axis ·· 23. *S. astridae*
9. Valves not twisted around apical axis ·· 10
  10. Alar canals less than 10/100 μm, transapical undulations very distinct ······················· 24. *S. robusta*
  10. Transapical undulations more delicate, Alar projections small and indistinct, alar canals usually more than
    10/100 μm ··· 25. *S. splendida*

Key to the species of *Stenopterobia*

1. Valves long and narrow, S shape ··· 2
1. Valves linear, straight ·· 1. *S. Delicatissima*
   2. Striae fewer than 18 /10 μm ·· 2. *S. anceps*
   2. Striae more than 18 /10 μm ··· 3. *S. intermedia*

Key to the species of *Campylodiscus*

1. Wave troughs very narrow, with the appearance of narrow, radial, sharply contoured ribs ····················· 2
1. Wave troughs broad, radial undulations irregular ··· 4
   2. Valves strongly saddle-shaped and twisted around the apical axis, with radial undulations ·····················
   ·· 1. *C. levanderi*
   2. Valves saddle-shaped, but not twist around the apical axis ·· 3
3. Radial undulations 20—30/100 μm, often reaching the centre of the valves ···························· 2. *C. noricus*
3. Radial undulations 10—20/100 μm, always with a large area free of undulations in the centre of the valve ··
   ·· 3. *C. hibernicus*
   4. Small forms, marginal radial undulations not interrupted, the centre of the valve with the same pattern as the marginal region ·· 4. *C. bicostatus*
   4. Large forms, marginal radial undulations interrupted by a hyaline concentric ring ············· 5. *C. clypeus*

附录 II 汉英术语对照表
（按笔画顺序排列）

上壳　epitheca，epivalve	远缘　distal margin
下壳　hypotheca，hypovalve	远缝端　distal raphe end
弓形的　arcuate	间质片层　frets
中央节　central nodule	附生（植物）的　epiphytic
中央壳面　median valve plane	附着（污泥）的　epipelic
中央横切面　median transapical plane	附着生物　epipelon
中缝端　central raphe ending	具隔室的　loculate
内共生的　endosymbiotic	披针形　lanceolate
分支孔板　vola (volae)	浅裂的　lobed
双列的　biseriate	环带　cingulum, girdle bands, copulae
双弧形　biarcuate	贯壳轴　pervalvar axis
头状的　capitate	顶轴　apical axis
对角线对称　nitzschioid symmetry	顶面　apical plane
末端裂缝　terminal fissure	带面　girdle view
龙骨　keel	浮游的　planktonic
龙骨壳缝系统　keeled raphe system	脊　ridge
龙骨突　fibulae	高电导率　high-conductivity
会聚的　convergent	偏转　deflected
光镜　light microscopy (LM)	等极的　isopolar
异极的　heteropolar	筛状的　cribrate
网状的　reticulate	筛板　cribrum
网眼孔　areola (areolae)	隔片　partition
肋纹　costae	新月形的　crescent-shaped
伸长的　elongate	腹侧的　ventral
壳体　frustule	腹板，胸骨　sternum
壳面　valvar plane	蜂窝状的　alveolate (alveoli)
壳套　mantle	管壳缝　canal raphe
壳缝骨（板）　raphe-sternum (raphe-sterna)	膜　hymen
拟孔　poroid	槽沟　groove
拟窗龙骨　fenestrated keel	横肋纹　transapical costae
极节　polar nodule	横轴　transapical axis
极缝端　polar raphe ending	镜面对称　hantzschioid symmetry
近缝端　proximal raphe end	螺旋舌　helictoglossa

参 考 文 献

Agardh C A. 1827. Aufzählung einiger in den östreichischen Ländern gefundenen neuen Gattungen und Arten von Algen, nebst ihrer Diagnostik und beigefügten Bemerkungen. Flora, 10(40): 625-640.

Aimas Kerim, Akbar Yimit, Wang K Q. 2000. Studies on diatoms in Urumqi areas(2). *Journal of Xinjiang Normal University*(Nat. Sci. Ed.), 19(3): 52-55 [阿力马斯·克里木, 艾克拜尔·依米提, 王克勤. 2000. 乌鲁木齐周围地区硅藻植物研究(2). 新疆师范大学学报(自然科学版), 19(3): 52-55].

Aleem A A, Hustedt F. 1951. Einige neue Diatomeen von der Südkuste Englands. *Botaniska Notiser*, 1951/1: 13-20.

Antoniades D, Hamilton P B, Douglas M S V., Smol J P. 2008. Diatoms of North America: The freshwater floras of Prince Patrick, Ellef Ringnes and northern Ellesmere Islands from the Canadian Arctic Archipelago. *Iconogr. Diatomol.*, 17: 1-649.

Archibald R E M. 1966. Some New and Rare Diatoms from South Africa 2. Diatoms from Lake Sibayi and Lake Nhlange in Tongaland(Natal). *Nova Hedwigia*, 7: 477-495.

Bao S K, Tan M C, Zhong Z X. 1986. A survey of the algal flora in the Jiuzhaigou Nature Reserve of Sichuan. *Journal of Xinan Normal University*, (3): 60-65 [包少康, 谭明初, 钟肇新. 1986. 四川九寨沟自然保护区藻类植物调查. 西南师范大学学报, (3): 60-65].

Bao W M, Reimer C W. 1992. New taxa of the diatoms from Changbaishan Mountain, China. *Bull. Bot. Res.*, 12(4): 357-361 [包文美, 瑞墨尔·查. 1992. 中国长白山硅藻的新分类单位. 植物研究, 12(4): 357-361].

Bao W M, Wang Q X, Reimer C W. 1992. Diatoms from the Changbaishan Mountain area. *Bull. Bot. Res.*, 12(2): 125-143 [包文美, 王全喜, 瑞墨尔·查. 1992. 长白山地区硅藻的研究. 植物研究, 12(2): 125-143].

Bao W M, Wang Q X, Shi X L. 1989. Studies on the phytoplankton in the Gaoleng-Yilan Section of the Songhua River. *Journal of Harbin Normal University*(Nat. Sci. Ed.), 5(1): 75-93 [包文美, 王全喜, 施心路. 1989. 松花江高楞——依兰江段浮游藻类调查. 哈尔滨师范大学(自然科学版), 5(1): 75-93].

Bi L J, Hu Z Y, Liu G X. 2001. The review on early literature of freshwater algae systematics in China and the chronology. *In*: Study of Phycology in China. Wuhan: Wuhan Press[毕列爵, 胡征宇, 刘国祥. 2001. 我国淡水藻类系统分类学的早期文献概况及编年表. 载于刘永定等《中国藻类学研究》. 武汉: 武汉出版社: p. 262-272].

Bi L J, Liang J J. 1994. *The History of Phycology in China*. Beijing: Science Press[毕列爵, 梁家骥. 1994. 中国藻类学史. 北京: 科学出版社].

Bleisch M. 1863. Über einige in den Jahren 1856-62 in der Gegend von Strehlen gefundene Diatomeen. Abhandlungen Schlesische Gesellschaft fur Vaterlandische Kulture Abth. fur Naturwissenschaften und Medicin, *Breslau*, 1862(2): 75-84.

Bourrelly P, Manguin É. 1952. Algues d'eau douce de la Guadeloupe et dépendances: recueillies par la Mission P. Allorge en 1936. Paris: Société d'Édition d'Enseignement Superiéur.: pp. 1-282.

Bramberger A J, Haffner G D, Hamilton P B, *et al.* 2006. An examination of species within the genus *Surirella* from the Malili Lakes, Sulawesi Island, Indonesia, with descriptions of 11 new taxa. Diat. Res., 21(1): 1-56.

Brébisson L A de, Godey L L. 1835 '1836'. Algues des environs de Falaise, décrites et dessinées par MM. de Brébisson et Godey. Mémoires de la Société Académique des Sciences, Artes et Belles-Lettres de Falaise, 1835: 1-62, 256-269 [corrections], pls. I-VIII.

Brébisson L A de. 1838. *Considerations sur les diatomées et essai d'une classification des genres et des espèces appartenant à celle famille, par A. de Brébisson, auteur de la Flore de Normandie, etc. pp. [i]*, [1]-20, [4, err.]. Falaise & Paris: Brée l'Ainée Imprimeur-Libraire; Meilhac.

Brébisson A de. 1869. Extrait d'un Essai Monographique sur les Vanheurckia, nouveau genre appartenant à la tribu des Diatomacées Naviculées. Annales de la Société Phytologique et Micrographique de Belgique 1: 201-206, 1 pl.

Brun J. 1880. *Diatomées des Alpes et du Jura et de la region suisse et française des environs de Genève.* pp. 1-146, 9 pls. Genève et

Paris.

Brun J. 1891. Diatomées espèces nouvelles marines, fossiles ou pélagiques. *Mémoires de la Société de Physique et d'Histoire Naturelle de Genève* 31 (part 2, no.1): 1-47, pl. 11-22.

Bukhtiyarova L N. 1995. Novye taksonomischeskie kombinatsii diatomovykh vodoroslei (Bacillariophyta). *Algologia*, 5(4): 417-424.

Cahoon L B, Laws R A. 1993. Benthic Diatoms From the North Carolina Continental Shelf: Inner and Mid Shelf. *J. Phycol.*, 28: 257-263.

Cai S X. 1985. The succession of diatom flora in one natural water body. *J. Shanxi Agric. Univ.*, 5(1): 17-30[蔡石勋. 1985. 一个天然水体硅藻区系的演替. 山西农业大学学报, 5(1): 17-30].

Cao X, Strojsová A, Znachor P, Zapomelová E. 2005. Detection of extracellular phosphatases in natural spring phytoplankton of a shallow eutrophic lake (Donghu, China). *European Journal of Phycology*, 40: 251-258.

Carruthers W. 1864. *The Diatomaceae. In*: Gray J E. Handbook of British Water weeds or Algae, R. Hardwick. London, 123 pp.

Chen C, Hu X H, Wang C L. 1996. A preliminary study on the algal flora in the area of Wuyang River. *Journal of Guizhou Normal University* (Nat. Sci. Ed.), 14(1): 22-30 [陈椽, 胡晓红, 王承录. 1996. 贵州施秉舞阳河藻类植物初步研究. 贵州师范大学学报 (自然科学版), 14(1): 22-30].

Chen G, Gao S Z. 1986. A survey of the diatom in the Liupanshan Mountain, Ningxia. *Journal og Ningxia Agricultural College*, (1-2): 60-79 [陈功, 高淑贞. 1986. 宁夏六盘山自然保护区硅藻调查. 宁夏农学院学报, (1-2): 60-79].

Chen Y W, Gao X Y, Qin B Q. 1998. The summer phytoplankton species composition in northern part of West Taihu Lake. *Journal of Lake Sciences*, 10(4): 35-40 [陈宇炜, 高锡云, 秦伯强. 1998. 西太湖北部夏季藻类种间关系的初步研究. 湖泊科学, 10(4): 35-40].

Chin T G. 1951. A list of Chinese diatoms from 1847-1946, *Amoy fisheries bulletin*, 1(5): 41-143 [金德祥. 1951. 中国矽藻目录. 厦门水产学报, 1(5): 41-143].

Chin T G. 1978. A discussion on the phylogenesis of Diatoms. *Journal of the Xiamen University*, 2: 31-50 [金德祥. 1978. 硅藻分类系统的探讨. 厦门大学学报, 2: 31-50].

Cholnoky B J. 1956. Neue und seltene Diatomeen aus Afrika. II. Diatomeen aus dem Tugela-Gebiete in Natal. *Österreichische Botanische Zeitschrift*, 103: 53-97.

Cholnoky B J. 1963. Ein Bertrag zur Kenntnis der Diatomeenflora von Holländisch- Neuguinea. *Nova Hedwigia*, 5: 157-198.

Cleve-Euler A. 1951. Die Diatomeen von Schweden und Finnland. Teil I. Centricae. *K Svenska Vet Akad*. Fjärde Serien, 2(1): 1-163.

Cleve-Euler A. 1953a. Die Diatomeen von Schweden und Finnland. Teil II. Arraphideae. Brachyraphideae. *K Svenska Vet Akad*. Fjärde Serien, 4(1): 1-158.

Cleve-Euler A. 1953b. Die Diatomeen von Schweden und Finnland. Teil III. Monoraphideae. Biraphideae 1. *K Svenska Vet Akad*. Fjärde Serien, 4(5): 1-255.

Cleve-Euler A. 1955. Die Diatomeen von Schweden und Finnland. Teil IV. Biraphideae 2. *K Svenska Vet Akad*. Fjärde Serien, 5(4): 1-232.

Cleve-Euler A. 1952. Die Diatomeen von Schweden und Finnland. Teil V. (Schluss.). *K Svenska Vet-Akad*. Fjärde Serien, 3(3): 1-153.

Cleve-Euler A. (Cleve). 1895. On recent freshwater Diatoms from Lule Lappmark in Sweden. Bihang till *K Svenska Vet-Akad*. Handlingar 21 (Afd. III, 2): 44 pp., 1 pl.

Cleve P T, Möller J D. 1878. *Diatoms*. Part III, No. 109-168. Upsala: Esatas Edquists Boktryckeri.

Cleve P T, Möller J D. 1879. *Diatoms*. Part IV, No.169-216. Upsala: Esatas Edquists Boktryckeri.

Cleve P T, Grunow A. 1880. Beiträge zur Kenntniss der arctischen Diatomeen. *K Svenska Vet-Akad. Handl.* Ser., 4, 17(2): 1-121, pls. I-VII.

Cocquyt C, Vyverman W. 1993. *Surirella sparsipunctata* Hustedt and *S. sparsipunctata* var. *laevis* Hustedt (Bacillariophyceae), a light and electron microscopical study. *Hydrobiologia*, 269/270: 97-101.

Cocquyt C, Jahn R. 2007a. *Surirella engleri* O. Müller- a study of its original infraspecific types, variability and distribution. *Diat. Res.*, 22(1): 1-16.

Cocquyt C, Jahn R. 2007b. *Surirella nyassae* O. Müller, *S. malombae* O. Müller and *S. chepurnovii* Cocquyt & R. Jahn sp. nov.(Bacillariophyta)- typification and variability of three closely related East African diatoms. *Nova Hedwigia*, 84(3-4): 529-548.

Compere P. 1984. *Nitzschia fragilariiformis* a new species from NW Sudan forming ribbon-like colonies. 8th Diatom-Symposium: 253-258.

Coste M, Ricard M. 1980. Observation en microscopie photonique de quelques *Nitzschia* nouvelles ou intéressantes dont la striation est à la limite du pouvoir de résolution. *Cryptogamie, Algologie*, 1(3): 187-212.

Cox E J. 1993. Diatom systematics – a review of past and present practice and a personal vision for future development. *Nova Hedwigia, Beiheft*, 106: 1-20.

Deng X Y, Shen Z G, Zhou M, et al. 1997. A preliminary study on the algae vegetation of Xishuangbanna in Yunnan province. *Southwest China Journal of Agricultural Sciences*, 10(1): 85-90 [邓新晏, 沈宗庚, 邹敏, 等. 1997. 云南西双版纳藻类植物初报. 西南农业学报, 10(1): 85-90].

Deng C K, Yang C L, Wu M, Li Z H. 1983. Preliminary studies on diatoms from Lingqu River, Guangxi. *Journal of Guangxi Normal University*(Nat. Sci. Ed.), (1): 86-91 [邓春匡, 杨存亮, 吴孟, 李振海. 1983. 广西灵渠硅藻初报. 广西师范大学学报(自然科学版), (1): 86-91].

Deng X Y, Xu J H. 1996. Studies on algae of Fuxian Lake in Chengjiang County, Yunnan. *Journal of Yunnan University*(Nat. Sci. Ed.), 18(2): 139-145 [邓新晏, 许继宏. 1996. 澄江抚仙湖藻类植物研究. 云南大学学报(自然科学版), 18(2): 139-145].

De Toni G B. 1892. *Sylloge algarum omnium hucusque cognitarum. Vol. II. Sylloge Bacillariearum. Sectio II. Pseudoraphideae.* pp. [i-v], 491-817. Padova [Padua]: Sumptibus auctoris.

Dreβler M, Hübener T. 2006. Morphology and ecology of *Cyclostephanos delicates*(Genkal) Casper & Scheffler (Bacillariophyceae) in comparison with *C. tholiformis* Stoermer, Håkansson & Theriot. *Nova Hedwigia*, 82: 409-434.

Du G S, Wang J T, Wu D W, et al. 2001. Structure and density of the phytoplankton community of Miyun Reservoir. *Acta Phytoecologica Sinica*, 25(4): 501-504 [杜桂森, 王建厅, 武殿伟, 等. 2001. 密云水库的浮游植物群落结构与密度. 植物生态学报, 25(4): 501-504].

Ehrenberg C G. 1830. Organisation, Systematik und geographisches Verhältniss der Infusionsthierchen. Zwei verträge. *Abhandlungen der Königlichen Akademie der Wissenschaften zu Berlin*. 108 pp., 8 kupfertafeln(copper pls.).

Ehrenberg C G. 1832. Über die Entwickelung und Lebensdauer der Infusionsthiere; nebst ferneren Beiträgen zu einer Vergleichung ihrer organischen Systeme. *Abhandlungen der Königlichen Akademie Wissenschaften zu Berlin, Physikalische Klasse*, 1831: 1-154, pls. I-IV.

Ehrenberg C G. 1834. Dritter Beitrag zur Erkenntniss grosser Organisation in der Richtung des kleinsten Raumes. *Abhandlungen der Königlichen Akademie der Wissenschaften zu Berlin*, 1833: 145-336, pls. I-XIII.

Ehrenberg C G. 1836a. Remarks on the real occurrence of fossil Infusoria and their extensive diffusion. *Poggendorff's Annalen der Physik und Chemie*, 38(5): 1-213, 1 pl.

Ehrenberg C G. 1836b. Vorlaufige Mitteilung uber das wirkliche Vorkommer fossiler Infusorien und ihre grosse. *Verbreitung Ann. Phys. u. Chem.*, 38: 213-227.

Ehrenberg C G. 1837a. Über das Massenverhältnifs der jetzt lebenden Kiesel-Infusorien und über ein neues Infusorien-Conglomerat als Polirschiefer von Jastraba in Ungarn. *Abhandlungen der Königlichen Akademie der Wissenchaften zu Berlin, Physikalische Klasse*, 1836: 109-135, pl.1-2.

Ehrenberg C G. 1837b. Über ein aus fossilen Infusorien bestehendes, 1832 zu Brod verbacknes Bergmehl von den Grenzen Lapplands in Schweden. *Bericht über die zur Bekanntmachung geeigneten Verhandlungen der Königl. Preuß. Akademie der Wissenschaften zu Berlin*, 1837: 43-45.

Ehrenberg C G. 1838. *Die Infusionsthierchen als vollkommene Organismen: Ein Blick in das tiefere organische Leben der Natur.* pp.

i-xviii, [1-4], 1-547, [1]. Leipzig: Verlag von Leopold Voss.

Ehrenberg C G. 1840. Characteristik von 274 neuen Arten von Infusorien. *Bericht über die zur Bekanntmachung geeigneten Verhandlungen der Königlich-Preussischen Akademie der Wissenschaften zu Berlin*, 1840: 197-219.

Ehrenberg C G. 1843. Verbreitung und Einfluss des mikroskopischen Lebens in Süd-und Nord-Amerika. *Abhandlungen der Königlichen Akademie der Wissenschaften zu Berlin*, 1841: 291-466, pls. 1-4.

Ehrenberg C G. 1845. Vorläufige zweite Mettheilung über die weitere Erkenntnifs der Beziehungen des kleinsten organischen Lebens zu den vulkanischen Massen der Erde. *Bericht über die zur Bekanntmachung geeigneten Verhandlungen der Königlich-Preussischen Akademie der Wissenschaften zu Berlin*, 1845: 133-157.

Ehrenberg C G. 1849. Passatstaub und Blutregen. Ein Grofses organisches unsichtbares Wirken und Leben in der Atmosphäre. *Abhanlungen der Königlichen Akademie der Wissenschaften zu Berlin*, 1847: 269-460.

Fan Y W, Bao W M, Wang Q X. 2001. Studied on Aulonoraphidinales from Wudalian-chi Lake. *Bull. Bot. Res.*, 21: 239-244 [范亚文, 包文美, 王全喜. 2001. 五大连池管壳缝目硅藻研究初报. 植物研究, 21(2): 239-244].

Fan Y W, et al. 2004. A New Combination and Two Varieties of Polystigmate *Gomphonema*(Gomphonemaceae Bacillariophyta)from Heilongjiang Province, China. *Chin J Ocean Limn*, 22(2): 198-203.

Fan Y W, Hu Z Y. 2004. Studied on Aulonoraphidinales from Xingkaihu Lake in Heilongjiang province. *Acta Hydrobiologica Sinca*, 28(4): 421-425 [范亚文, 胡征宇. 2004. 黑龙江省兴凯湖地区管壳缝目硅藻初步研究. 水生生物学报, 28(4): 421-425].

Fan Y W. 2004. Studied on Aulonoraphidinales(Surirellales)from Heilongjiang province. Harbin: Northeast Forestry University Press: 109 pp. [范亚文. 2004. 黑龙江省管壳缝目植物研究. 哈尔滨: 东北林业大学出版社: 109 pp].

Fang Y C, Tian C. 1992. Investigation on freshwater diatom in Shenyang Area. *Journal of Shenyang Universiry*(Nat. Sci. Ed.), (4): 42-46 [房英春, 田春. 1992. 沈阳地区淡水硅藻调查. 沈阳大学学报(自然科学版), (4): 42-46].

Fang Y C, Liu G C, Su B L, Tian C. 2001. Investigation on freshwater Cyanophyta, Bacillariophyta and Chlorophyta of alga in Shenyang Area. *Journal of Shenyang Agricultural University*, 32(6): 442-445[房英春, 刘广纯, 苏宝玲, 田春. 2001. 沈阳地区淡水蓝藻门、硅藻门和绿藻门藻类调查报告. 沈阳农业大学学报, 32(6): 442-445].

Felix E A, Rushforth S R. 1979. The Algal Flora of the Great Salt Lake. Utah. USA. *Nova Hedwigia*, 31: 163-195.

Ferguson Wood E J. 1961. Studies on Australian and New Zealand Diatoms IV: Descriptions of Further Sedentary Species. *Transactions of the Royal Society of New Zealand*, 88: 669-698.

Ferguson Wood E J. 1961. Studies on Australian and New Zealand Diatoms V: The Rawson Collection of Recent Diatoms. *Transactions of the Royal Society of New Zealand*, 88: 699-712.

Flower R J. 2005. A taxonomic and ecological study of diatoms from freshwater habitats in the Falkland Islands, South Atlantic. *Diat. Res.*, 20: 23-96.

Foged N. 1980. Diatoms in Egypt. *Nova Hedwigia*, 33: 629-707.

Fontell C W. 1917. Süsswasserdiatomeen Ober-Jämtland in Schweden. *Arkiv för Botanik*, 14(21): 68 pp., 2 pls.

Fourtanier E, Kociolek J P. 1999. Catalogue of the diatom genera. *Diat. Res.*, 14(1): 1-190.

Fott B. 1971. *Algenkunde*. Stuttgart: Erlangen-Nürnberg, 2. Aufl. 581 pp.

Frenguelli J. 1942. Diatomeas del Neuguén(Patagonia). XVII. Contribución al conocimiento de las diatomeas argentinas. *Rev. Mus. La Plata, Nueva Serie, Sección Botánica*, 5(20): 73-219, 12 pl.

Gandhi H P. 1964. Notes on the Diatomaceae of Ahmedabad and its Environs V The Diatomflora of Chandola and Kankaria Lakes. *Nova Hedwigia*, 8: 347-402.

Gandhi H P. 1966. The Fresh-Water Diatomflora of the Jog-Falls. Mysore State. *Nova Hedwigia*, 11: p. 89-149.

Gao S Z. 1987. The diatoms of Mt. Huashan. *Wuhan Bot. Res.*, 5(4): 329-338 [高淑贞. 1987. 华山的硅藻. 武汉植物学研究, 5(4): 329-338].

Gao S Z, Chen G. 1988. A research on diatom-flora in Helan Mountains, Ningxia Autonomous Region. *Wuhan Bot. Res.*, 6(2): 113-119 [高淑贞, 陈功. 1988. 宁夏回族自治区贺兰山地区硅藻分布. 武汉植物研究, 6(2): 113-119].

Gao X，Tian J Y. 2000. The list of freshwater phytoplankton in the Yellow River Delta. *Transactions of Oceanology and Limnology*，（3）：65-77 [高霞，田家怡. 黄河三角洲淡水浮游植物名录. 海洋湖沼通报，（3）：65-77].

Garcia-Baptista M. 1993. Observations on the genus *Hantzschia* Grunow at a sandy beach in Rio Grande do Sul，Brazil. *Diat. Res.*，8(1)：31-43.

Gasse F. 1986. East African diatoms：Taxonomy. ecological distribution. *Bibliotheca Diatomologia*，11：201 pp.

Germain H. 1981. *Flore des Diatomées Diatomophycées*. Paris：Société nouvelle des éditions boubée：411 pp.

Germain H. 1984. The Central Nodule of Nitzschiae obtusae Grunow. 8th Diatom-Symposium：227-235.

Gmelin J F. 1791. *Carolia Linne···Systema Naturae per regna tria naturae secundum classes，ordines，genera species cum characteribus，differentiis，synonymis，locis. Ed. 13，Tomus I. Pars VI.* pp. 3021-3910. Lipsiae [Lepizig]：Georg Emanuel Beer.

Great Lakes Diatoms. Center for Great Lakes and Aquatic Sciences. Available online at http：//www.umich.edu/~phytolab/Great LakesDiatomHomePage/top.html.Consulted on 2009-02-23.

Gregory W. 1856. Notice of some new species of British Fresh-water Diatomaceae. *Quarterly Journal of Microscopical Science*，new series，London 4：1-14，pl. I.

Gregory W. 1857. On the post-Tertiary diatomaceous sand of Glenshira. Part II. Containing an account of a number of additional undescribed species. *Transactions of the Microscopical Society of London*，5：67-88，pl. 1.

Grunow A. 1862. Die Österreichischen Diatomaceen nebst Anschluss einiger neuen Arten von andern Lokalitäten und einer kritischen Uebersicht der bisher bekannten Gattungen und Arten. Erste Folge. Epithemieae，Meridioneae，Diatomeae，Entopyleae，Surirelleae，Amphipleureae. Zweite Folge. Familie Nitzschieae. *Verhandlungen der Kaiserlich-Königlichen Zoologisch-Botanischen Gesellschaft in Wien*，12：315-472，545-588.

Grunow A. 1877. New diatoms from Honduras，with notes by F. Kitton. *Monthly Microscopical Journal*，London，18：165-186，pls. 193-196.

Grunow A. 1878. Algen und Diatomaceen aus dem Kaspischen Meere. *In*：Schneider O. Naturwissenschaftliche Beiträge zur Kenntnis der Kaukasusländer，auf Grund seiner Sammelbeute，Dresden：pp. 98-132.

Grunow A. 1880. Bemerkungen zu den Diatomeen von Finnmark，dem Karischen Meere und vom Jenissey nebst Vorarbeiten für Monographie der Gattungen *Nitzschia*，*Achnanthes*，*Pleurosigma*，*Amphiprora*，*Plagiotropis*，*Hyalodiscus*，*Podosira* und einiger *Navicula*-Gruppen. *Kongliga Svenska Vetenskaps-Akademiens* Handligar，ser.，4 17(2)：16-121；pl. I-VII.

Guiry M D，Guiry G M. 2015. *AlgaeBase*. World-wide electronic publication，National University of Ireland，Galway. http：//www.algaebase.org [2015-7-20].

Guo J，Liu S C，Lin J H. 1999. Five newly-recorded species in the genus *Nitzschia* from China. *Acta Phytotaxonomica Sinica*，37(5)：526-528 [郭健，刘师成，林加涵. 1999. 我国首次记录的菱形藻属植物. 植物分类学报，37(5)：526-528].

Guo Y Q，Xie S Q. 1994. A new species of Bacillariophyta from Taishan of Shandong. *Acta Phytotaxonomica Sinica*，32(3)：271-272 [郭玉清，谢淑琦. 1994. 山东泰山硅藻一新种. 植物分类学报，32(3)：271-272].

Guo Y Q，Xie S Q，Liu A W，Li Z H. 1996. Study on diatoms from Tai Mountain of Shandong province. *Journal of Shanxi University*(Nat. Sci. Ed.)，19(2)：215-220 [郭玉清，谢淑琦，刘安文，李志红. 1996. 山东泰山硅藻研究. 山西大学学报（自然科学版），19(2)：215-220].

Guttinger W. 1987-1998. *Collection of SEM Micrographs of Diatoms*，Series 1-9. Pura：Via Soriscio.

Gutwinski R. 1891. *Flora Glonów Okolic Lwowa*(*Flora algarum agri Leopoliensis*). Kraków：Drukarnia Uniwersytety Jagiellonskiego：pp. 76-111.

Håkansson H. 1979. Examination of diatom type material of C. A. Agardh. *In*：Simonsen R. Proceedings of the Fifth Symposium on Recent and Fossil Diatoms，Antwerp，September 3-8，1978. *Nova Hedwigia*，Beihefte，64：163-168.

Hantzsch C A. 1860. Neue Bacillarien：*Nitzschia vivax* var. *elongata*，*Cymatopleura nobilis*. *Hedwigia*，2(7)：1-40，pl. 6.

Hasle G R. 1994. *Pseudo-Nitzschia* as a Genus Distinct from *Nitzschia*(Bacillariophyceae). *J. Phycol.*，30：1036-1039.

Hasle G R. 1995. *Pseudo-Nitzschia Pugens* and *P. Multiseries*(Bacillariophyceae) Nomenclatural History. Morphology and Distribution. *J. Phycol.*，31：428-435.

Hartley B. 1996. *An Atlas of Brithsh Diatoms* (ed. by Sims P A). Dorchester: Dorset Press: 601pp.

Heiberg P A C. 1863. *Conspectus criticus* Diatomacearum Danicarum. Wilhelm Priors Forlag, Kjøbenhavn, 135pp.

Henderson M V, Reimer C W. 2003. Bibliography on the Fine Structure of diatom Frustules (Bacillariophyceae), II & Deletions, Addenda and Corrigenda for Bibliography I. *Diatom monographs*, 3: 1-372.

Hendey N I. 1954. A preliminary check-list of British marine diatoms. *Journal of the Marine Biological Association of the United Kingdom*, 33: 537-560.

Heurck C. van 1881. *Synopsis des Diatomées de Belgique Atlas.* pls. XXXI-LXXVII. Anvers: Ducaju et Cie.

Heurck H. van 1885. *Synopsis des Diatomées de Belgique.* Texte. pp. 1-235. Anvers: Martin Brouwers & Co.

Heurck H. van 1896. *A treatise on the Diatomaceae.* Translated by W.E. Baxter. London: William Wesley & Son. pp. 1-558, pls. 1-35.

Hoek C, Mann K G, Jahns H M. 1995. *Algae* (*An introduction to phycology*). Cambridge: Cambridge University: p. 132-159.

Hu B F, Li Z, Shi J H, Xie S L. 2004. Study on stream algae in Shuishentang Spring, Shanxi. *Journal of Shanxi University* (Nat. Sci. Ed.), 27(4): 402-405 [胡变芳, 李砧, 史兼华, 谢树莲. 2004. 山西水神堂泉藻类植物的研究. 山西大学学报(自然科学版), 27(4): 402-405].

Hu H J, Li Y Y, Wei Y X, Zhu H Z, Chen J Y, Shi Z X. 1980. *Freshwater Algae of China.* Shanghai: Science and Technology Press: 525pp [胡鸿钧, 李尧英, 魏印心, 朱蕙忠, 陈嘉佑, 施之新. 1980. 中国淡水藻类. 上海: 上海科学技术出版社: 525pp].

Hu H J, Wei Y X. 2006. *Freshwater Algae of China.* Beijing: Science Press: 1023pp [胡鸿钧, 魏印心. 2006. 中国淡水藻类. 北京: 科学出版社: 1023pp].

Huang C Y, Cai Z R. 1984. Diatom floras in the Miocene Shengxian formation of Shengxian, Zhejiang province. *Acta Palaeontologica Sinica*, 23(3): 358-372 [黄成彦, 蔡祖仁. 1984. 浙江中新世嵊县组的硅藻植物群. 古生物学报, 23(3): 358-372].

Huang C Y, Liu S C, Cheng Z D, et al. 1998. *Atlas of Limnetic Fossil Diatoms of China.* Beijing: China Ocean Press: 164pp [黄成彦, 刘师成, 程兆第, 等. 1998. 中国湖相化石硅藻图集. 北京: 海洋出版社: 164 pp].

Hustedt F. 1930. *Bacillariophyta* (*Diatomeae*). - *In:* Pascher A. Die Süsswasserflora Mitteleuropas, Heft 10. (2. Aufl.). Jena: Gustav Fischer: 466pp.

Hustedt F. 1911. Beiträge zur Algenflora von Bremen IV. Bacillariaceen aus der Wumme. *Abh. Naturw. Ver. Bremen*, 20: 257-315, 3 pls.

Hustedt F. 1927. Bacillariales aus dem Aokikosee in Japan. *Archiv für Hydrobiologie*, 18: 155-172.

Hustedt F. 1934. Die Diatomeenflora von Poggenpohls Moor bei Dötlingen in Oldenburg. *Abhandlungen und Vorträgen der Bremer Wissenschaftlichen Gessellschaft*, 8/9: 362-403.

Hustedt F. 1942. Süßwasser-Diatomeen des indomalayischen Archipels und der Hawaii-Inslen. *Internationale Revue der gesamten Hydrobiologie und Hydrographie*, 42(1/3): 1-252.

Hustedt F. 1943. Die Diatomeenflora einiger Hochgebirgsseen der Landschaft Davos in den schweizer Alpen. *Internationale Revue der gesamten Hydrobiologie und Hydrographie*, 43: 124-197, 225-280.

Hustedt F. 1953a. Diatomeen aus dem Naturschutzgebiet Seeon. *Archiv für Hydrobiologie*, 47(4): 625-635.

Hustedt F. 1953b. Diatomeen aus der Oase Gafsa in Südtunesien, ein Beitrag zur Kenntnis der Vegetation afrikanischer Oasen. *Archiv für Hydrobiologie*, 48(2): 145-153.

Hustedt F. 1959. Die Diatomeenflora des Salzlackengebietes im österreichischen Burgenland. *Österreichischen Akademie der Wissenschaften, Mathematische und Naturwissenschaftliche, Kl. Abt.* 1, 168(4/5): 387-452, 1 pl.

Hustedt F. 1922a. Bacillariales aus Innerasien. Gesammelt von Dr. Sven Hedin. *In:* Hedin S. Southern Tibet, discoveries in former times compared with my own researches in 1906-1908. *Lithographic Institute of the General Staff of the Swedish Army*, Stockholm, 6(3): 107-152, pls. 9-10.

Hustedt F. 1922b. Die Bacillariaceen-Vegetation des Lunzer Seengebietes (Nieder-Österreich). *Internationale Revue der gesamten Hydrobiologie und Hydrographie*, 10(1-2): 40-74, 233-270, pl. III.

Hustedt F. 1957. Die Diatomeenflora des Flußsystems der Weser im Gebiet der Hansestadt Bremen. *Abh. Naturw. Ver. Bremen*, 34(3): 181-440.

Hustedt F. 1959. Die Diatomeenflora des Salzlackengebietes im österreichischen Burgenland. *Österreichischen Akademie der Wissenschaften, Mathematische und Naturwissenschaftliche, Kl. Abt.* 1, 168(4/5): 387-452, 1 pl.

Huvane J K, Cooper S R. 2001. Diatoms as indicators of environmental change in sediment cores from Northeastern Florida Bay. *Bulletins of American Paleontology*, 361: 145-158.

Index Nominum Algarum, University Herbarium, University of California, Berkeley. *Compiled by Paul Silva.* Available online at http://ucjeps.herb.berkeley.edu/INA.html. [2015-09-08].

ITIS(The Integrated Taxonomic Information System). 2000. Catalogue of Life: *2007 Annual Checklist.* Available online at http://www.catalogueoflife.org/annual -checklist/ browse_taxa.php? selected_taxon=4207. [2014-12-08].

Jao C C. 1964. Some fresh-water algae from southern Tibet. *Oceanlogia limnol. Sin.*, 6(2): 169-190 [饶钦止. 1964. 西藏西南部地区的藻类. 海洋与湖沼, 6(2): 169-189].

Jao C C, Zhu H Z, Li Y Y. 1973. Notes on the freshwater algae of the Mt. Jolmolungma region in South Tibet. *China Sci. Bull.*, 18(1): 30-32 [饶钦止, 朱惠忠, 李尧英. 1973. 我国西藏南部珠穆朗玛峰地区藻类概要. 科学通报, 18(1): 30-32].

Järnefelt H. 1925. Zur Limnologie einiger Gewässer Finnlands. *Annales Botanici Societatis Zoologicae Botanicae Fennicae, Vanamo, Helsinki*, 2: 185-352.

Johansen R J, Rushforth R S. 1981. Diatoms of surface waters and soils of selected oil shale lease areas of Eastern Utah. *Nova Hedwigia*, 34: 333-390.

Jurilj A. 1949. Nove Dijatomeje-Surirellaceae-iz Ohridskog Jezera i njihovo filogenetsko znacenje.(New Diatoms-Surirellaceae- of Ochrida Lake in Yugoslavia and their phylogenetic significance). *Jugoslavenska Akademija Znanosti i Umjetnosti, Zagreb(Prirodoslovnih istrazivanja)*, 24: 171-259. [reprint pp. 3-94].

Kaczmarska I, Fryxell G A. 1984. The diatom genus *Nitzschia*: Morphologic variation of some small bicapitate species in two Gulf Stream Warm Core Rings. 8th Diatom-Symposium: 237-252.

Kapinga M R M, Gordon R. 1992a. Cell attachment in the motile colonial diatom *Bacillaria paxillifer*. *Diat. Res.*, 7(2): 215-220.

Kapinga M R M, Gordon R. 1992b. Cell motility rhythms in *Bacillaria paxillifer*. *Diat. Res.*, 7(2): 221-225.

Karsten G. 1899. Die Diatomeen der Kieler Bucht. *Wissenschaftliche Meeresuntersuchungen, Herausgegeben von der Kommission zur wissenschaftlichen Untersuchung der deutschen Meere in Keil und der biologischen Anstalt auf Helgoland*, 3: 1-205.

Karthick B, Hamilton P B, Kociolek J P. 2012. Taxonomy and biogeography of some *Surirella* Turpin(Bacillariophyceae)taxa from Peninsular India. *Nova Hedwigia, Beiheft*, 141: 81–116.

Kirchner O. 1878. Algen. *In*: Cohn, F. *Kryptogamen-Flora von Schlesien. Part 1.* Vol. 2, pp. i-iv [vii] 1-284. Breslau: J.U. Kern's Verlag.

Knattrup A, Yde M, Lundholm N, Ellegaard M. 2007. A detailed description of a Danish strain of *Nitzschia sigmoidea*, the type species of *Nitzschia*, providing a reference for future morphological and phylogenetic studies of the genus. *Diat. Res.*, 22(1): 105-116.

Kobayasi H, Nagumo T, Tanaka S. 1993. *Rhopalodia iriomotensis* sp. nov., a brackish diatom with shoehorn-shaped projectins on the canal raphe(Bacillariophyceae). *Nova Hedwigia, Beiheft*, 106: 133-141.

Kobayasi H, Kobayashi H. 1986. A study of *Epithemia amphicephala*(Östr.) comb. et stat. nov. and *E. reticulata* Kütz., with special reference to the areolar occlusion. 9th Diatom-Symposium: 459-480.

Kociolek J P, Stoermer E F. 1989a. Phylogenetic relationships and evolutionary history of the diatom genus *Gomphoneis*. *Phycologia*, 28(4): 438-454.

Kociolek J P, Stoermer E F. 1989b. Chromosome numbers in diatoms: a review. *Diat. Res.*, 4(1): 47-54.

Kociolek J P, Theriot E C, Williams D M. 1989. Inferring diatom phylogeny: a cladistic perspective. *Diat. Res.*, 4: 289-300.

Krammer K. 1988a. The Gibberula-group in the genus *Rhopalodia* O. Müller(Bacillariophyceae)I. Observations on the valve morphology. *Nova Hedwigia*, 46: 277-303.

Krammer K. 1988b. The Gibberula-group in the genus *Rhopalodia* O. Müller (Bacillariophyceae) II. Revision of the group and new taxa. *Nove Hedwigia*, 47: 159-205.

Krammer K. 1989a. Valve Morphology and Taxonomy in the Genus *Stenopterobia* (Bacillariophyceae). *Br. phycol. J.*, 24: 237-243.

Krammer K. 1989b. Functional Valve Morphology in Some *Surirella* Species (Bacillariophyceae) and a Comparisom with the Naviculaceae. *J. Phycol.*, 25: 159-167.

Krammer K, Lange-Bertalot H. 1986. *Bacillariophyceae*. 1. Teil: *Naviculaceae*. In: Ettl H, *et al*. Süßwasserflora von Mitteleuropa. Band 2/1. Heidelberg: Spektrum Akademischer Verlag: 876pp.

Krammer K, Lange-Bertalot H. 1987. Morphology and taxonomy of *Surirella ovalis* and related taxa. *Diat. Res.*, 2: 77-95.

Krammer K, Lange-Bertalot H. 1988, nachdr. 1997. *Bacillariophyceae*. 2. Teil: *Bacillariaceae, Epithemiaceae, Surirellaceae*. In: Ettl H. *et al*. Süßwasserflora von Mitteleuropa. Band 2/2. Heidelberg: Spektrum Akademischer Verlag: 611pp.

Krammer K, Lange-Bertalot H. 1991a. *Bacillariophyceae*. 3. Teil: *Centrales, Fragilariaceae, Eunotiaceae*. In: Ettl H. *et al*. Süßwasserflora von Mitteleuropa. Band 2/3. Heidelberg: Spektrum Akademischer Verlag: 599pp.

Krammer K, Lange-Bertalot H. 1991b. *Bacillariophyceae*. 4. Teil: *Achnanthaceae, Kritische Ergänzungen zu Achnanthes s. 1., Navicula s. str., Gomphonema. Gesamtliteraturverzeichnis*. Teil 1-4. In: Ettl H. *et al*. Süßwasserflora von Mitteleuropa. Band 2/4. Heidelberg: Spektrum Akademischer Verlag: 468pp.

Krammer K, Lange-Bertalot H. 2000. *Bacillariophyceae*. Part 5: *English and French translation of the keys*. In: Ettl H. *et al*. Süßwasserflora von Mitteleuropa. Band 2/5. Heidelberg: Spektrum Akademischer Verlag: 311pp.

Krammer K. 2000. *Diatoms of Europe*. Volume 1: *The genus Pinnularia*. Königstein: A. R. G. Gantner Verlag K. G.: 703pp.

Krammer K. 2002. *Diatoms of Europe*. Volume 3: *Cymbella*. Königstein: A. R. G. Gantner Verlag K. G.: 584pp.

Krammer K. 2003. *Diatoms of Europe*. Volume 4: *Cymbopleura, Delicata, Navicymbula, Gomphocymbellopsis, Afrocymbella*. Königstein: A. R. G. Gantner Verlag K. G.: 703pp.

Krasske G. 1929. Beiträge zur Kenntnis der Diatomeenflora Sachsens Botanisches Archiv, 27(3/4): 348-380, 1 pl.

Kützing F T. 1834' 1833'. Synopsis diatomearum oder Versuch einer systematischen Zusammenstellung der Diatomeen. *Linnaea*, 8: 529-620, pls. XIII-XIX [79 figs].

Kützing F T. 1844. *Die Kieselschaligen Bacillarien oder Diatomeen*. pp. [i-vii], [1]-152, pls. 1-30. Nordhausen: zu finden bei W. Köhne.

Kützing F T. 1849. *Species Algarum*. Lipsiae. F.A. Brockhaus. 922 pp.

Kwandrans J. 2007. Diversity and ecology of benthic diatom communities in relation to acidity, acidification and recovery of lakes and rivers. *Diatom monographs*, 9: 1-169.

Lan D Z, Chen C H, Chen F. 1999. Characteristics and geological significance of diatoms in core from Jiulongjiang Estuary. *Journal of Oceanography in Taiwan Strait*, 18(3): 283-290 [蓝东兆, 陈承惠, 陈峰. 1999. 九龙江口岩心中的硅藻特征及其地质意义. 台湾海峡, 18(3): 283-290].

Lange-Bertalot H. 1976. Eine revision zur taxonomie der *Nitzschiae Lanceolatae* Grunow. *Nova Hedwigia*, 28: 253-307.

Lange-Bertalot H. 1978. Zur Systematik, Taxonomie und Ökologie des abwasserspezifisch wichtigen Formenkreises um "*Nitzschia thermalis*. *Nova Hedwigia*, 30: 635-652.

Lange-Bertalot H, Simonsen R. 1978. A Taxonomic Revision of the *Nitzschiae lanceolatae* Grunow 2. European and Related Extra-European Fresh Water and Brackish Water Taxa. *Bacillaria*, 1: 11-111.

Lange-Bertalot H. 1979. *Simonsenia*, a New Genus with Morphology Intermediate between *Nitzschia* and *Surirella*. *Bacillaria*, 2: 127-136.

Lange-Bertalot H. 1980. New Species, Combinations and Synonyms in the Genus *Nitzschia*. *Bacillaria*, 3: 41-77.

Lange-Bertalot H. 1993a. 85 neue taxa und über 100 weitere neu definierte Taxa ergänzend zur Süsswasserflora von Mitteleuropa, Vol. 2/1-4. *Bibl. Diatomol.*, 27: 1-164.

Lange-Bertalot H. 1993b. Observations on *Simonsenia* and some small species of *Denticula* and *Nitzschia*. In: P.A. Sims. Progress in diatom studies, Contributions to taxonomy, ecology and nomenclature. Special volume in honour of Robert Ross on the occasion of his 80th Birthday. *Nova Hedwigia*, Beiheft 106: 121-131.

Lange-Bertalot H, Krammer K. 1987. Bacillariaceae, Epithemiaceae, Surirellaceae. Neue und wenig bekannte taxa, neue Kombinationen und Synonyme sowie Bemerkungen und Erganzungen zu den Naviculaceae. *Bibl. Diatomol.*, 15: 1-289.

Lange-Bertalot H, Metzeltin D. 1996a. Indicators of Oligotrophy. *Iconogr. Diatomol.*, 2: 1-390.

Lange-Bertalot H, Metzeltin D. 1996b. Ultrastructure of *Surirella desikacharyi* sp. nov. and *Campylodiscus indianorum* sp. nov. (Bacillariophyta) in comparison with some other taxa of the "robustoid" type construction. *Nova Hedwigia*, 112: 321-328.

Lange-Bertalot H, Genkal S I. 1999. Diatomeen aus Sibirien, I: Inselln im Arktischen Ozean (Yugorsky-Shar Strait). *Iconogr. Diatomol.*, 6: 1-304.

Lange-Bertalot H. 2001. *Diatoms of Europe*. Volume 2: *Navicula sensu stricto*, 10 Genera Separated from Navicula sensu lato, *Frustulia*. Königstein: A. R. G. Gantner Verlag K. G.: 525pp.

Lange-Bertalot H, Cavacini P, Tagliaventi N, Alfinito S. 2003. Diatoms of Sardinia. *Iconogr. Diatomol.*, 12: 1-438.

Lange-Bertalot H. 1978. Zur Systematik, Taxonomie und Ökologie des abwasserspezifisch wichtigen Formenkreises um *Nitzschia thermalis*. *Nova Hedwigia*, 30: 635-652.

Lange-Bertalot H, Krammer K. 1993. Observations on *Simonsenia* and some small species of *Denticula* and *Nitzschia*. *Beihefte zum Botanischen Centralblatt*, 106: 121-131, 52 figs.

Lee R E. 1999. *Phycology*. Cambridge: Cambridge University Press: 614pp.

Lemmermann E. 1900. Beiträge zur Kenntnis der Planktonalgen. III. Neue Schwebalgen aus der Umgegend von Berlin. *Berichte der deutsche botanischen Gesellschaft*, 18: 24-32.

Levkov Z, Krstic S, Metzeltin D, Nakov T. 2007. Datoms of Lakes Prespa and Ohrid (Macedonia). *Iconogr. Diatomol.*, 16: 1-613.

Lewis F W. 1864. On some new and singular intermediate forms of Diatomaceae. *Proceedings of the Academy of Natural Sciences of Philadelphia*, 15: 336-346.

Li H L. 2003. Preliminary studies on the Aulonoraphidinales (Bacillariophyta) from Xinjiang in China. Master Dissertation, Shanghai Normal University [李海玲. 2003. 新疆管壳缝目硅藻的初步研究. 上海师范大学硕士学位论文].

Li J D, Li X F, Zhang X C, Fu C Q. 1997. preliminary study on the patterns of distribution of population of sessile naviculoid diatom. *Transaction of oceanology and limnology*, (2): 43-47 [李进道, 李向峰, 张学超, 傅成秋. 1997. 附着舟形硅藻种群分布型的初步研究. 海洋湖沼通报, (2): 43-47].

Li J, Fan Y W, Wang Z B, et al. 2007. Preliminary studies on Diatoms from Qixing River Wetland. *Bull. Bot. Res.*, 27(1): 25-33 [李晶, 范亚文, 王泽斌, 杨立萍. 2007. 黑龙江省七星河湿地硅藻植物的初步研究. 植物研究, 27(1): 25-33].

Li Y. L, Shi Z X, Xie P, et al. 2003a. New varieties of *Gomphonema* and *Cymbella* (Bacillariophyta) from Qinghai Province. *Acta Hydrobiol. Sin.*, 27(2): 147-148.

Li Y L, Xie P, Gong Z J, Shi Z X. 2003b. Gymbellaceae and Gomphonemataceae (Bacillariophyta) from the Hengduan Mountains region (southwestern China). *Nova Hedwigia*, 76(3-4): 507-536.

Li Y L, Xie P, Gong Z J, Shi Z X. 2004. A Survey of the Gomphonemaceae and Cymbellaceae (Bacillariophyta) from the Jolmolungma Mountain (Everest) Rejion of China. *J. Freshw. Ecol.*, 19(2): 189-194.

Li Y L, Gong Z J, Xie P, Shen J. 2005. New species and new records of fossil diatoms from the late Pleistocene of the Jianghan Plain, Hubei province. *Acta Micropalaeontologica Sinica*, 22(3): 304-310 [李艳玲, 龚志军, 谢平, 沈吉. 2005. 江汉平原晚更新世化石硅藻新种和新记录属种. 微体古生物学报, 22(3): 304-310].

Li Y L, Gong Z J, Xie P, Shen J. 2006. Distribution and morphology of two endemic Gomphonemoid species, *Gomphonema kaznakowi* mereschkowsky and *G. yangtzensis* Li nov. sp. in China. *Diat. Res.*, 21(2): 313-324.

Li Y L, Gong Z J, Xie P, Shen J. 2007. New species and new records of fossil diatoms from China. *Acta Hydrobiologica Sinica*, 31(3): 319-324 [李艳玲, 龚志军, 谢平, 沈吉. 2007. 中国硅藻化石新种和新记录种. 水生生物学报, 31(3): 319-324].

Lin B Q, Wang Q H, Liu Y. 1998. Effects of micro-habitat on the composition of diatom communities. *Acta Botanica Sinica*, 40(3): 277-281 [林碧琴, 王起华, 刘岩. 1998. 小生境对硅藻群落组成的影响. 植物学报, 40(3): 277-281].

Liu Q. 2015. Taxononical and Ecological studies on diatoms from Zoige Wetland and its nearby waters in Sichuan Province. Doctor

Dissertation, Zhejiang University [刘琪. 2015. 四川若尔盖湿地及其附近水域硅藻的分类及生态研究. 浙江大学博士学位论文].

Liu W K, Fan Y W. 2005. The primary study on diatom samples collection and specimens distribution from Heilongjiang Province. *Journal of Harbin Normal University*(Nat. Sci. Ed.), 21(3): 84-89 [刘文凯, 范亚文. 2005. 黑龙江省硅藻标本采集及种类分布的初步研究. 哈尔滨师范大学(自然科学学报), 21(3): 84-89].

Liu Y, You Q M, Wang Q X. 2006. Studies on freshwater diatoms from Kinmen Island in Fujian. *Wuhan Bot. Res.*, 24: 38-46 [刘妍, 尤庆敏, 王全喜. 2006. 福建金门岛的淡水硅藻初报. 武汉植物学研究, 24(1): 38-46].

Liu Y, You Q M, Wang Q X. 2009. Newly Recorded Species of Nitzschiaceae (Bacillariophyta) from Da'erbin Lake in Great Xing'an Mountains, China, *Wuhan Bot. Res.*, 27(3): 274-276[刘妍, 尤庆敏, 王全喜. 2009. 大兴安岭达尔滨湖菱形藻科(硅藻门)中国新记录植物. 武汉植物学研究, 27(3): 274-276].

Liu Y, Wang Q X, Yang X Q, Fan Y W. 2014. Newly recorded species of Aulonoraphidinales (Bacillariophyta) from Great Xing'an Mountains, China. *Journal of Shanghai Normal University*(Nat. Sci.), 43(3): 269-272 [刘妍, 王全喜, 杨晓清, 范亚文. 2014. 大兴安岭管壳缝目硅藻中国新记录. 上海师范大学(自然科学学报), 43(3): 269-272].

Liu Y. 2007. Studies on Diatoms of Da'erbin Lake and Swamps around it in Daxing'anling Mountains. Master Dissertation, Shanghai Normal University [刘妍. 2007. 大兴安岭达尔滨湖及其周围沼泽硅藻的研究. 上海师范大学硕士学位论文].

Liu Y. 2010. Taxonomical and Ecological studies on diatoms from wetlands in Great Xing'an Mt., China. Doctor Dissertation, Zhejiang University [刘妍. 2010. 大兴安岭沼泽硅藻分类生态研究. 浙江大学博士学位论文].

Liu J, Wei G F, Hu R, Zhang C W, Han B P. 2013. *Atlas of benthic diatom from Zhujiang River Dongjiang Basin*. Beijing: China Environment Press[刘静, 韦桂峰, 胡韧, 张成武, 韩博平. 2013. 珠江水系东江流域底栖硅藻图集. 北京: 中国环境出版社].

Lopez M M. 1980. Un nuevo subgénero de Surirella en sedimentos del Salar de Carcote, Chile. *Noticiario mensual, Museo Nacional de Historia Natural*, 24(281/282): 3-7.

Lowe R L. 1974. *Environmental Requirements and Pollution Tolerance of Freshwater Diatoms*. Cincinnati: U.S.A. Environmental Protection Agency: 333pp.

Luan Z, Fan Y W. 2007. Preliminary studies on diatoms from Songhua River of Harbin. *Journal of Harbin Normal University*(Nat. Sci. Ed.), 23(1): 83-85 [栾卓, 范亚文. 2007. 松花江哈尔滨段硅藻植物初报. 哈尔滨师范大学(自然科学学报), 23(1): 83-85].

Lundholm N, Daugbjerg N, Moestrup Ø. 2002. Phylogeny of the Bacillariaceae with emphasis on the genus *Pseudo-nitzschia*(Bacillariophyceae) based on partial LSU rDNA. *Eur. J. Phycol.*, 37: 115-134.

Ma Y, Wang S M. 1992. The recent 400-year diatom history of Daihai Lake, Innermongolia with additioinal reference to its paleoenvironmental significance. *Journal of Lake Sciences*, 4(2): 19-24 [马燕, 王苏民. 1992. 内蒙岱海近400年来的硅藻植物群及其古环境意义. 湖泊科学, 4(2): 19-24].

Manhart J R, Mccourt R M. 1992. Molecular Data and Species Concepts in the Algae. *J. Phycol.*, 28: 730-737.

Mann D G. 1977. The diatom genus *Hantzschia* Grunow –an appraisal. *Nova Hedwigia, Beiheft*, 54: 323-354.

Mann D G. 1978. Studies in the family *Nitzschiaceae*(Bacillariophyta). Ph. D. Dissertation, University of Bristol.

Mann D G. 1980a. *Hantzschia fenestrate* Hust. (Bacillariophyta) – *Hantzschia* or *Nitzschia*. *Br. Phycol. J.*, 15: 249-260.

Mann D G. 1980b. Studies in the diatom genus *Hantzschia* II. *H. distinctepunctata*. *Nova Hedwigia*, 33: 341-352.

Mann D G. 1981. Studies in the diatom genus *Hantzschia* 3. Infraspecific variation in *H. virgata*. *Ann. Bot.*, 47: 377-395.

Mann D G. 1984. *Nitzschia* Subgenus *Nitzschia*(Notes for a Monograph of the Bacillariaceae, 2). *In*: Ricard M. *Proceedings of the 8th international Diatom Symposium*, 215-226. Koenigstein: Koeltz Scientific Books.

Mann D G. 1986. Methods of sexual reproduction in *Nitzschia*: Systematic and evolutionary implications. (Notes for a monograph of the Bacillariaceae 3). *Diat. Res.*, 1: 193-203.

Mann D G. 1987. Sexual reproduction in *Cymatopleura*. *Diat. Res.*, 2: 97-112.

Mann D G. 1999. The species concept in diatoms. *Phycologia*, 38(6): 437-495.

Mann D G. 2000. Auxospore formation and neoteny in *Surirella angusta* (Bacillariophyta) and a modified terminology for cells of Surirellaceae. *Nova Hedwigia*, 71: 165-183.

Mar BEF. 2004. European Marine Biodiversity Datasets. Available online at http://www.marbef.org/data/imis.php?module = dataset. [2013-01-08].

Mayer A. 1913. Die Bacillariaceen der Regensburger Gewässer. Systematicher Teil. *Berichte des naturwissenschaftlichen (früher zoologisch-mineralogischen) Vereins zu Regensburg*, 14: 1-364, 30 pls.

Mayer A. 1936. Die bayerischen Epithemien. *Denkschriften der Königlich-Baierischen Botanischen Gesellschaft in Regensburg*, 20: 87-108, 8 pl.

Medlin L K, Williams D M, Sim P A. 1993. The evolution of the diatoms (Bacillariophyta). I. Origin of the group and assessment of the monophyly og its major divisions. *Eur. J. Phycol.*, 28: 261-275.

Medlin L K, Kaczmarska. 2004. Evolution of the diatoms: V. Morphological and cytological support for the major clades and a taxonomic revision. *Phycologia*, 43(3): 245-270.

Meister F. 1912. Die Kieselalgen der Schweiz. *Beiträge zur Kryptogamenflora der Schweiz. K.J. Wyss*, Bern., 4(1): 254 pp., 48 pls.

Mereschkowsky C. 1906. Diatomees du Tibet. Imperial Russkoe geograficheskoe obshchestvo. St. Petersberg. 40 pp.

Metzeltin D, Witkowski A. 1996. Diatomeen der Baren-Insel: Süβwasser-und marine Arten. *Iconogr. Diatomol.*, 4: 1-287.

Metzeltin D, Lange-Bertalot H. 1998. Tropical Diatoms of South America I: About 700 predominantly rarely known or new taxa representative of the neotropical flora. *Iconogr. Diatomol.*, 5: 1-695.

Metzeltin D, Lange-Bertalot H. 2002. Diatoms from the "Island Continent" Madagascar. *Iconogr. Diatomol.*, 11: 1-286.

Metzeltin D, Lange-Bertalot H, Garcia-Rodriguez F. 2005. Diatoms of Uruguay. *Iconogr. Diatomol.*, 15: 1-736.

Metzeltin D, Lange-Bertalot H. 2007. Tropical Diatoms of South America II: Special remarks on biogeographic disjunction. *Iconogr. Diatomol.*, 18: 1-879.

Michael K H, Barbara M W, Michael J S. 2008. Bacillariophyta (Diatoms) of the Bahamas. *Iconogr. Diatomol.*, 19: 1-303.

Mongolian Diatom Home Page. 2009. International Partnership for Research and Trainning in Mongolia-The diatom (Bacillariophyta) flora of ancient Lake Hovsgol. Available online at http://www.umich.edu/~mongolia/index.html.

Min H M, Ma J H. 2007. Benthic diatoms from the tidal flat of Shanghai, China, during 2005 summer. *Journal of Tropical and Subtropical Botany*, 15(5): 390-398 [闵华明, 马家海. 2007. 上海市滩涂夏季底栖硅藻初步研究. 热带亚热带植物学报, 15(5): 390-398].

Müller O F. 1786. *Animalcula infusoria fluviatilia et marina* quæ detexit, systematice descripsit et ad vivum delineari curavit Otho Fridericus Müller; sistet opus hoc posthumum, quod cum tabulis Aeneis L. in lucem tradit vidua ejus nobilissima cura Othonis Fabricii. pp. i-lvi, 1-367. Hauniæ [Copenhagen]: Typis N. Mölleri.

Müller [G F.] O. 1895. *Rhopalodia* ein neues Genus der Bacillariaceen. *Botanische Jahrbucher fur Systematik, Pflanzengeschichte und Pflanzengeographie*, 22: 54-71, 2 pl.

Müller O. 1900. Bacillariaceen aus den Natronthälern von El Kab (Ober-Aegypten). *Hedwigia*, 38(5-6): 274-288, 289-321, pls. X-XII. [pp. 274-288 published in issue 38(5), 1899; pp. 289-321 published in issue 38(6), 1900].

Müller O. 1909. Bacillariaceen aus SüdPatagonien. *Engler's Bot. Jb. für Systematik Pflanzengesch. und Planzengeogr.*, 43: 1-40.

Muschler R. 1908. Enumération des Algues marines et d'eau douce observées jusqu'à ce jour en Egypte. *Memoires, Institut Egyptien* (Cairo), 5(3): 141-237.

Nitzsch C L. 1817. Beitrag zur Infusorienkunde oder Naturbeschreibung der Zerkarien und Bazillarien (Title). *Neue Schriften der Naturforschenden Gesellschaft zu Halle*, 3(1): 1-128, 6 pls.

Norman G. 1861. On some undescribed species of Diatomaceae. *Transactions of the Microscopical Society*, London, New Series, 9: 5-9, pl. II.

Niu Y L, Zheng Y X, Zhao J C. 2006. A preliminary study on the algae in Hengshui Lake National Nature Reserve. *Journal of Hebei Normal University* (Nat. Sci. Ed.), 30(3): 336-342 [牛玉璐, 郑云翔, 赵建成. 2006. 衡水湖自然保护区藻类植物资源初步研究. 河北师范大学学报(自然科学版), 30(3): 336-342].

Østrup E. 1910. *Danske Diatoméer*. pp. 1-323, 5 pls. Kjøbenhavn [Copenhagen]: C.A. Reitzels Boghandel.

Paddock T B B. 1978. Observations on the valve structure of diatoms of the genus *Plagiodiscus* and on some associated species of Surirella. *Botanical Journal of the Linnean Society*, 76: 1-25.

Paddock T B B. 1985. Observations and comments on the diatoms *Surirella fastuosa* and *Campylodiscus fastuosus* and on other species of similar appearance. *Nova Hedwigia*, 41: 417-444.

Paddock T B B, Sims P A. 1977. A preliminary survey of the raphe structure of some advanced groups of diatoms (Epithemiaceae-Surirellaceae). *In*: Simonsen. R. *proceedings of the Fourth Symposium on Recent and Fossil Diatoms*, Oslo, 1976. *Nova Hedwigia*, Beiheft, 54: 291-322.

Paddock T B B, Sims P A. 1990. Micromorphology and evolution of the keels of raphe-bearing diatoms. *In*: Claugher D. *Scanning Electron Microscopy in Taxonomy and Functional Morphology*. Systematics Association, 41: 171-191.

Pantocsek J. 1902. Kieselalgen oder Bacillarien des Balaton. Resultate der Wissenschaftlichen Erforschung des Balatonsees, herausgegeben von der Balatonsee-Commission der Ung. Geographischen Gesellschaft.: Commissionsverlag von Ed. Hölzel. Wien., 2(2): 112 pp., 17 pl.

Patrick R, Reimer C W. 1975. The Diatoms of the United States, exclusive of Alaska and Hawaii. *Acad. Nat. Sci. Philadephia Monogr.*, 13(2/1): 1-213.

Pedicino N A. 1867. Pochi studi sulle Diatomee viventi presso alcune terme dell'isola d'Ischia. *Atti Accademia delle scienze fisiche e matematiche di napoli*, 3(20): 2 pl.

Peragallo H, Peragallo M. 1897-1908. *Diatomées marines de France et des districts maritimes voisins*. pp. 1-491, 1-48. Grez-sur-Loing: J. Tempère, Micrographe-Editeur.

Peragallo H, Peragallo M. 1900. *Diatomées marines de France et des districts maritimes voisins*. Atlas. pp. pls. 73-80. Grez-sur Loing(S.-et-M.): J. Tempère, Micrographe-Editeur.

Peragallo M. 1903. *Le Catalogue Général des Diatomées* [issued in fascicles at various dates]. Vol. 2 pp. 472-973. Clermont-Ferrand.

Petersen J B. 1928. The aërial algae of Iceland. *In*: Rosenvinge L K, Warming E. *The botany of Iceland. Vol. II. Part II.* Copenhagen & London: Wheldon and Wesley: pp. 328-447.

Pfitzer E. 1871. Untersuchungen uber Bau und Entwickelung der Bacillariaceen(Diatomaceen). *Botanische Abhandlungen aus dem Gebiet der Morphologie und Physiologie*. Vol. 2, 6 pls. Bonn: Herausg. von J. Hanstein. 189 pp.

Pickett-Heaps J D, Cohn S, Schimid A-M M, Tippit D H. 1988. Valve morphogenesis in *Surirella*(Bacillariaphyceae). *J. Phycol.*, 24(1): 35-49.

Pickett-Heaps J D, Schimid A-M M, Edgar L A. 1990. The cell biology of diatom valve formation. *Progress in Phycological Research*, 7: 168 pp.

Pouličková A, Jahn R. 2007. *Campylodiscus clypeus*(Ehrenberg)Ehrenberg ex Kützing: typification, morphology and distribution. *Diat. Res.*, 22(1): 135-146.

Pritchard A. 1861. A history of infusoria, including the Desmidiaceae and Diatomaceae, British and foreign. Fourth edition enlarged and revised by J. T. Arlidge, M.B., B.A. Lond.; W. Archer, Esq.; J. Ralfs, M.R.C.S.L.; W. C. Williamson, Esq., F.R.S., and the author. London: Whittaker and Co., Ave Maria Lane. pp. i-xii, 1- 968, 40 pls.

Protic G. 1899. Wissenschaftliche Mittheilungen aus Bosnien und der Hercegovina, 6: 717.

Qi Y Z. 1995. *Flora algarum sinicarum aquae dulcis*(Tomus IV): *Bacillariophyta-Centricae*. Beijing: Science Press: 104pp [齐雨藻. 1995. 中国淡水藻志(第四卷): 硅藻门－中心纲. 北京: 科学出版社, 104pp].

Qi Y Z, Xie S Q. 1984. Studies on bog diatoms in Shennongjia primitive forests area(1). *Journal of Jinan University*, (3): 86-92 [齐雨藻, 谢淑琦. 1984. 湖北神农架苔藓沼泽硅藻(上). 暨南理医学报, (3): 86-92].

Qi Y Z, Xie S Q. 1985. Studies on bog diatoms in Shennongjia primitive forests area(2). *Journal of Jinan University*, (1): 98-108 [齐雨藻, 谢淑琦. 1985. 湖北神农架苔藓沼泽硅藻(下). 暨南理医学报, (1): 98-108].

Qi Y Z, Yang J R. 1985. New data on the early Pleistocene fossil diatoms from Miyi, Sichuan, China. *Acta Micropalaeontologica Sinica*, 2(3): 283-290 [齐雨藻, 杨景荣. 1985. 四川米易早更新世化石硅藻的新资料. 微体古生物学报, 2(3): 283-290].

Qi Y Z, Zhang Z A. 1977. Studies on the taxonomy of diatoms by Scanning Electron Microscopy. *Acta Phytotaxonomica Sinica*, 15(2): 113-120 [齐雨藻, 张子安. 1977. 扫描电子显微镜下的硅藻分类研究. 植物分类学报, 15(2): 113-120].

Qi Y Z, Li J Y. 2004. *Flora algarum sinicarum aquae dulcis*(Tomus X): *Bacillariophyta-Araphidiales*, *Raphidionales*. Beijing: Science Press: 161pp [齐雨藻, 李家英. 2004. 中国淡水藻志(第十卷): 硅藻门 - 无壳缝目, 拟壳缝目. 北京: 科学出版社: 161pp].

Rabenhorst L. 1853. Die Süsswasser-Diatomaceen(Bacillarien.): für Freunde der Mikroskopie. pp. i-xii, 1-72. Leipzig: Eduard Kummer.

Rabenhorst L. 1848-1860. Die Algen Sachsens. Resp. Mittel-Europa's Gesammelt und herausgegeben von Dr. L. Rabenhorst, *Dec. 1-100. No. 1-1000*. [*Exsiccata*, *issued at various dates*]. Dresden.

Rabenhorst L. 1860. Erklärung der Tafel VI. *Hedwigia*, 2: 40, 6/6.

Rabenhorst L. 1862. Die Algen Europa's. Decas, 129-130.

Rabenhorst L. 1861-1882. Die Algen Europas, Fortsetzung der Algen Sachsens, resp. Mittel-Europas. *Decades I-CIX*, numbers *1-1600(or 1001-2600)*. Dresden.

Radzimowsky D O. 1928. Bemerkung Uber das Phytoplankton im Gestrupp des Sudlichen Bugs. *Académie des Sciences de l'Ukraine*, *Mémoires de la Classe des Sciences Physiques et Mathématiques*, 10, book 2: 85-98.

Rao D V S, Partensky F, Wohlgeschaffen G, *et al.* 1991. Flow Cytometry and Microscopy of Gametogenesis in *Nitzschia* pungens. A Toxic. Bloom- Forming. Marine Diatom. *J. Phycol.*, 27: 21-26.

Reichardt E. 1996. Die Identität von *Campylodiscus levanderi* Hust. *Diat. Res.*, 11(1): 81-87.

Reichardt E. 2001. Zwei in Europa unbekannte Campylodiscus-Arten(Bacillariophyceae)im Iffigsee, Schweiz. *Berichte der Bayerischen Botanischem Gesellschaft*, 71: 21-27.

Roper F C S. 1854. Some observations on the Diatomaceae of the Thames. *Transactions of the Microscopical Society*, New Series, London 2: 67-80, pl.VI.

Roper F C S. 1858. Notes on some new species and varieties of British marine Diatomaceae. *Quarterly Journal of Microscopical Science*, London 6: 17-25, pl. III.

Rosa T P. 2007. Ecological analysis of periphytic diatoms in Mediterranean coastal wetlands(Empordà wetlands, NE Spain). *Diatom monographs*, 7: 1-210.

Ross R. 1950. Report on diatom flora from Hawks Tor, Cornwall. *Philos. Trans.*, Ser, B, 234: 461-469.

Ross R, Cox E J, Karayeva N I, Mann D G, Paddock T B B, Simonsen R, Sims P A. 1979. An amended terminology for the siliceous components of the diatom cell. *Nova Hedwigia*, Beiheft, 64: 513-533.

Round F E. 1970. The genus *Hantzschia* with particular reference to *H. virgata* v. *intermedia* (Grun.) comb. Nov. Annals of Botany, new series, 34(134): 75-91.

Round F E, Sims P A. 1981. *The distribution of diatom genera in marine and freshwater environments and some evolutionary considerations*. In: Ross R. Proceedings of the Sixth International Symposium on Recent and Fossil Diatoms, Budapest, Hungary, 1980. Otto Koeltz, Koenigstein & Biopress, Bristol: 301-320.

Round F E, Crawford R M, Mann D G. 1990. *The diatoms*. Cambridge: Cambridge University Press: 747pp.

Ruck E C, Kociolek J P. 2004. Preliminary Phylogeny of the Family Surirellaceae(Bacillariophyta). *Bibl. Diatom.*, 50: 1-236.

Ruck E C, Theriot E C. 2011. Origin and Evolution of the Canal Raphe System in Diatoms. *Protist*, 162(5): 723-737.

Rumrich U, Lange-Bertalot H, Rumrich M. 2000. Diatoms of the Andes(from Venezuela to Patagonia / Tierra del Fuego). *Iconogr. Diatomol.*, 9: 1-673.

Sala S E. 1990. Valve morphology of *Surirella rorata* Frenguelli(Bacillariophyceae). *Diat. Res.*, 5(2): 219-236.

Schmid A-M M. 1979a. The development of structure in the shells of diatoms. *Nova Hedwigia*, Beiheft., 64: 219-236.

Schmid A-M M. 1979b. Influence of environmental factors on the development of the valve in diatoms. *Protoplasma*, 99: 99-115.

Schmid A-M M. 1980. Valve morphogenesis in diatoms: a pattern-related filamentous system in pennates and the effects of AMP, colchicines and osmoticpressure. *Nova Hedwigia*, 33: 811-847.

Schmid A-M M. 2000. *Value of pyrenoids in the systematics of the diatoms: their morphology and ultrastructure. In*: Economou-Amilli A. Proceedings of the Sixteenth International Diatom Symposium, Athens, Greece, 2000. Otto Koeltz, Koenigstein & Biopress, Bristol: 1-31.

Schmidt A, Fricke F. 1904. Atlas der Diatomaceen-kunde. Leipzig. O.R. Reisland Series VI (Heft 62-63): pls. 245-252.

Schmidt A, Hustedt F. 1911. Atlas der Diatomaceen-kunde. Leipzig. O.R. Reisland Series VI (Heft 69): pls. 273-276.

Schmidt A, Schmitz M, Fricke F, Müller O, Heiden H, Hustedt F. 1874-1959. *Atlas der Diatomaceen-Kunde.* 4 Bands, pls. 1-480. Leipzig.

Schumann J. 1862. Preussische Diatomeen. *Schriften der koniglichen physikalisch-okonomischen Gesellschaft zu Konigsberg*, 3 (Abh.): 166-192, pls. 8-9.

Shi Z X. 1991. New taxa of fossil diatoms from borehole 47 in the Jianghan Plain, Hubei. *Acta Micropalaeontologica Sinica*, 8(4): 449-459 [施之新. 1991. 江汉平原47号钻孔化石硅藻的新种类. 微体古生物学报, 8(4): 449-459].

Sims P A. 1983. A taxonomic Study of the Genus *Epithemia* with Special Reference to the Type Species *E. turgida* (Ehrenb.) Kütz. *Bacillaria*, 6: 211-235.

Sims P A, Paddock T B B. 1982. The fenestral fibula: A new structure in the diatoms. *Bacillaria*, 5: 7-42.

Simonsen R. 1979. The diatom system: ideas on phylogeny. *Bacillaria*, 2: 9-71.

Simonsen R. 1987. *Atlas and Catalogue of the Diatom Types of Friedrich Hustedt*. Vol. 1-3. Berlin: Gebrüder Borntraeger Verlagsbuchhandlung.

Siver P A, Hamilton P B, Stachura-Suchoples K, Kociolek J P. 2005. Diatoms of North America: The Freshwater Flora of Cape Cod, Massachussetts, U.S.A. *Iconogr. Diatomol.*, 14: 1-463.

Skuja H. 1937. Algae. Symbolae Sinicae: botanische Ergebnisse der Expedition der Akademie der Wissenschaften in Wein nach Sudwest-China, 1914-1918: 105 pp.

Skvortzow B W. (Skvortzov B V.) 1927. *Diatoms from Tientsin North China. Journal of Botany*, 65 (772): 102-109, 28 figs.

Smith W. 1851. Notes on the Diatomaceae, with descriptions of British species included in the genera *Campylodiscus*, *Surirella* and *Cymatopleura. Annals and Magazine of Natural History*, series 2 7: 1-14, pls. I-III.

Smith W. 1853. *A synopsis of the British Diatomaceae; with remarks on their structure, function and distribution; and instructions for collecting and preserving specimens*. Vol. 1 pp. [v]-xxxiii, 1-89, 31 pls. London: John van Voorst.

Smith W. 1856. *A synopsis of the British Diatomaceae; with remarks on their structure, functions and distribution; and instructions for collecting and preserving specimens*. Vol. 2 pp. [i-vi] - xxix, 1-107, pl. 32-60, 61-62, A-E. London: John van Voorst.

Stehr C M, Connell L, Baugh K A, et al. 2002. Morphological. Toxicological. and Genetic Differences Among *Pseudo-Nitzschia* (Bacillariophyceae) Species in Inland Embayments and Outer Coastal Waters of Washington State. USA. *J. Phycol.*, 38: 55-65.

Stephens C F, Gibson R A. 1980. Ultrastructural Studies of Some *Mastogloia* (Bacillariophyceae) Species Belonging to the Group Sulcatae. *Nova Hedwigia*, 33: 219-248.

Sterrenburg F A S. 2001. Transfer of *Surirella patrimonii* Sterrenburg to the genus *Petrodictyon*. *Diat. Res.*, 16(1): 109-111.

Stoermer E F, Jr Kreis R G, Andresen N A. 1999. Checklist of diatoms from the Laurentian Great lakes. II. *J. Great Lakes Res.*, 25: 515-566.

Stoermer E F, Smol J P. 2000. *The Diatoms: Applications for the Environmental and Earth Sciences*. Cambridge: Cambridge University Press.

Taylor F J. 1966. Phytoplankton of the South Western Indian Ocean. *Nova Hedwigia*, 12: 433-476.

Tempère J, Peragallo H. 1908. *Diatomées du Monde Entier*, Edition 2, 30 fascicules. Fascicule 2-7. Chez J. Tempère, Arcachon, Gironde. pp. 17-112.

Tempère J, Peragallo H. 1909. *Diatomées du Monde Entier*, Edition 2, 30 fascicules. Fascicule 8-12. pp. 113-208. Arcachon, Gironde: J. Tempère.

Tippit D H, Pickett-Heaps J D. 1977. Cell division in the pinnate diatom *Surirella ovalis. Journal of Cell Biology*, 73: 705-727.

Tian J Y. 1995. Planktons in the Xiaoqing River in Shandong province. *Transactions of Oceanology and Limnology*，（1）：39-46 [田家怡. 1995. 山东小清河的浮游植物. 海洋湖沼通报，（1）：39-46].

Tian Y M. 1994. A check-list of algae in the Lake Baiyangdian. *Ecologic Science*，（2）：26-39 [田玉梅. 1994. 白洋淀藻类植物名录. 生态科学，（2）：26-39].

Tudesque L，Rimet F，Ector L. 2008. A new taxon of the section Nitzschiae Lanceolatae Grunow：*Nitzschia costei* sp. nov. compared to *N. fonticola* Grunow，*N. macedonica* Hustedt，*N. tropica* Hustedt and related species. *Diat. Res.*，23(2)：483-501.

Turpin P J F. 1828. Observations sur le nouveau genre *Surirella*. *Mémoires du Musée d'Histoire Naturelle*，16：361-368.

Uppsala T K. 1964. Two Surirellaceae from South America. *Nova Hedwigia*，8：135-137.

Ussing A P，Gordon R，Ector L，Buczko K，Desnitskiy A G，Vanlandingham S L. 2005. The Colonial Diatom"Bacillaria paradoxa"：Chaotic Gliding Motility，Lindenmeyer Model of Colonial Morphogenesis，and Bibliography，with Translation of O. F. Muller (1783). "About a peculiar being in the beach water". *Diatom monographs*，5：1-139.

VanLandingham S L. 1969. Catalogue of the fossil and recent genera and species of diatoms and their synonyms. Part III. *Coscinophaena* through *Fibula*. 3301 Lehre，Verlag von J. Cramer 3：1087-1756.

VanLandingham S L，Kentucky L. 1966. Diatoms from Dry Lakes in Nye and Esmeralda Counties. Nevada. U.S.A. Mit deutscher Zusammenfassung. *Nova Hedwigia*，11：222-241.

VanLandingham S L. 1967-1979. Catalogue of the fossil and recent genera and species of diatoms and their synonyms. Part 2，1968：Bacteriastrum through Coscinodiscus. Part 8，1978：Rhoicosphenia through Zygoceros，Verlag Von J. Crammer，Lehre.

Van Goor A C J. 1925. Uber *Nitzschia actinastroides* (Lemm.). *Recueil des travaux botaniques néerlandais*，22：320-323.

Van Heurck. 1880. *Synopsis des Diatomées de Belgique Atlas*. pls. I-XXX. Anvers：Ducaju et Cie.

Wachnicka A H，Gaiser E E. 2007. Characterization of Amphora and Seminavis from South Florida，U.S.A. *Diat. Res.*，22(2)：387-455.

Wang C H，Xin X Y，Xie S Q. 1994. Study on attached diatoms of Mianshan Mountain Qingshui River. *Journal of Shanxi Universit* (Nat. Sci. Ed.)，17(3)：345-349 [王翠红, 辛晓云, 谢淑琦. 1994. 绵山清水河着生硅藻之研究. 山西大学学报（自然科学版），17(3)：345-349].

Wang G R. 1998. Holocene diatoms from the Delta of Pearl-River，South China. *Acta Palaeontologica Sinica*，37(3)：305-324 [汪桂荣. 1998. 珠江三角洲全新世硅藻. 古生物学报，37(3)：305-324].

Wang K Q. 1997. Studies on Fresh Algae in Xinjiang. *Arid Zone Res.*，14(2)：25-30 [王克勤. 1997. 新疆淡水藻类研究：乌鲁木齐南山八一林场硅藻初报. 干旱区研究，14(2)：25-30].

Wang M S，Tan M C. 1992. Studies on the diatoms from part area of Yunnan，China. *Journal of Xinan Normal University* (Nat. Sci. Ed.)，17(1)：127-138 [王明书, 谭明初. 1992. 云南部分地区硅藻调查研究. 西南师范大学学报（自然科学版），17(1)：127-138].

Wang K F，Guo X M. (translated). 1984. *The diatoms*. The Geological Publishing House，100pp.[王开发，郭蓄民译，（小泉 格编），1984. 硅藻. 北京：地质出版社：100pp.].

Wang T L，Lin J M. 1998. Some advances in molecular biology of diatoms. *Plant Physiology Communications*，34(2)：140-148 [王团老，林均民. 1998. 硅藻分子生物学研究的一些进展. 植物生理学通讯，34(2)：140-148].

Wang W. 1997. The diatoms. *Journal of biology*，14(1)：14，20 [王伟. 1997. 硅藻. 生物学杂志，14(1)：14，20].

Wang Z M. 1989. Investigation on diatoms collected around West Lake in Hangzhou (II). *Journal of Hangzhou Nornal College* (Nat. Sci. Ed.)，(6)：72-79 [王智敏. 1989. 杭州西湖风景区硅藻调查(II). 杭州师范学院（自然科学版），(6)：72-79].

Wang Z M. 1991. Investigation of diatoms in the Mountain Western Tianmu in Zhejiang. *Journal of Hangzhou Nornal College* (Nat. Sci. Ed.)，(3)：65-73 [王智敏. 1991. 浙江西天目山自然保护区硅藻调查. 杭州师范学院（自然科学版），(3)：65-73].

Wendker S，Geissler U. 1986. Investigations on Valve Morphology of Two Nitzschiae Lanceolatae. 9th Diatom-Symposium：469-480.

Werum M，Lange-Bertalot H. 2004. Diatoms in Springs from Central Europe and elsewhele under the influence of hydrogeology and anthropogenic impacts. *Iconogr. Diatomol.*，13：1-417.

Williams D M. 1993. Diatom nomenclature and future of taxonomic database studies. *Nova Hedwigia*，*Beiheft*，106：21-32.

Williams D M，Reid G. 2006. *Amphorotia* nov. gen.，a new genus in the family Eunotiaceae（Bacillariophyceae），based on Eunotia clevei Grunow in Cleve et Grunow. *Diatom monographs*，6：1-153.

Witkowski A，Lange-Bertalot H，Metzeltin D. 2000. Diatom Flora of Marine Coasts 1，*Iconogr. Diatomol.*，7：1-925.

Wo RMS. 2007. The World Register of Marine Species. Available online at http：//www.marinespecies.org. [2009-01-09].

Wu F S，Ge L，Cai Z P，Feng H Y. 2008. Phytoplankton Diversity of Algae in Suganhu Lake in Gansu Province. *Acta Bot. Boreal.-Occident. Sin.*，28（12）：2521-2526 [武发思，葛亮，蔡泽平，冯虎元. 2008. 甘肃省苏干湖浮游植物多样性研究. 西北植物学报，28（12）：2521-2526].

Xiang S D，Wu W W. 1999. Epipelic Algal Community on the Bottom of West Lake，Hangzhou. *Journal of Lake Sciences*，11（2）：177-183 [项思端，吴文卫. 1999. 杭州西湖湖底附泥藻群落. 湖泊科学，11（2）：177-183].

Xie S Q，Guo Y Q，Wang C H. 1991. Study on diatoms of Jinyang Lake. *Journal of Shanxi University*（Nat. Sci. Ed.），14（4）：412-418 [谢淑琦，郭玉清，王翠红. 1991. 晋阳湖硅藻之研究. 山西大学学报（自然科学版），14（4）：412-418].

Xie S Q，Xin X Y，Li T. 1993. Studies on plankton diatoms from Salt Lake of Yun Chen. *Journal of Shanxi University*（Nat. Sci. Ed.），6（3）：332-339 [谢淑琦，辛小云，李婷. 1993. 运城盐池浮游硅藻的研究. 山西大学学报（自然科学版），6（3）：332-339].

Xie S Q，Li T. 1994. A new diatom species of Salt Lake from Shanxi province. *Acta Phytotaxonomica Sinica*，32（3）：273-274 [谢淑琦，李婷. 1994. 山西省盐池硅藻一新种. 植物分类学报，32（3）：273-274].

Xin X Y，Xu J H，Xie S Q. 1997. Study on diatoms of Jinci. *Journal of Shanxi University*（Nat. Sci. Ed.），20（1）：101-106 [辛晓云，徐建红，谢淑琦. 1997. 晋祠硅藻之研究. 山西大学学报（自然科学版），20（1）：101-106].

Xiong Y X，Luo Y C，Lin Y G. 1992a. A preliminary study on algae in the Caohai nature reserve. *Collection of Guizhou Agriculture*，20（2）：85-98 [熊源新，罗应春，林跃光. 1992a. 草海自然保护区藻植物研究. 贵州农学院丛刊，20（2）：85-98].

Xiong Y X，Lin Y G，Luo Y C. 1992b. A preliminary list of common algae flora in Guizhou. *Collection of Guizhou Agriculture*，20（2）：99-116 [熊源新，林跃光，罗应春. 1992b. 贵州常见藻类植物初步名录. 贵州农学院丛刊，20（2）：99-116].

Yang J R. 1988. Lake Miocene fossil diatom flora from Yiliang，Yunnan，China. *Acta Micropalaeontologica Sinica*，5（2）：153-170 [杨景荣. 1988. 云南宜良晚中新世硅藻植物群. 微体古生物学报，5（2）：153-170].

Yang J G. 1989. New records of plants-Several new records of the diatoms from China. *Journal of Anhui Normal University*，（2）：89-90 [杨积高. 1989. 植物新记录-我国几种硅藻植物新记录. 安徽师大学报，（2）：89-90].

Yang J G. 1990a. New records of the diatoms from China. *Bull. Bot. Res.*，10（1）：81-85 [杨积高. 1990a. 我国硅藻植物新记录. 植物研究，10（1）：81-85].

Yang J G. 1990b. New materials of the diatoms from China. *Bull. Bot. Res.*，10（4）：11-12 [杨积高. 1990b. 我国硅藻植物新资料. 植物研究，10（4）：11-12].

Yang J G. 1995. Some new records of Bacillariophyta in China. *Bull. Bot. Res.*，15（3）：335-337 [杨积高. 1995. 我国几种新记录的硅藻. 植物研究，15（3）：335-337].

Yang J G. 1999. Two new records of genus *Surirella* in China. *Journal of Anhui Normal University*（Nat. Sci. Ed.），22（4）：316 [杨积高. 1999. 中国双菱藻属植物二新记录. 安徽师范大学学报（自然科学版），22（4）：316].

Yoe H R D，Chan A M，Suttle C A. 1995. Phylogeny of Aureococcus Anophagefferens and a Morphologically Similar Bloom-Forming Alga from Texas as Determined by 18s Ribosomal RNA Sequence Analysis. *J. Phycol.*，31：413-418.

Yoe H R D，Rex L，Lowe H R，et al. 1992. Effects of Nitrogen and Phosphorus on the Endosymbiont Load of *Rhopalodia gibba* and *Epithemia turgida*（Bacillariophyceae）. *J. Phycol.*，28：773-777.

You Q M. 2006. Preliminary Studies on Flora of Diatoms from Xinjiang in China. Master Dissertation，Shanghai Normal University [尤庆敏. 2006. 中国新疆硅藻区系分类初步研究. 上海师范大学硕士学位论文].

You Q M. 2009. Studies on Aulonoraphidinales（Bacillariophyta）in China. Doctor Dissertation，East China Normal University [尤庆敏. 2009. 中国管壳缝目硅藻的分类学研究. 华东师范大学博士学位论文].

You Q M，Li H L，Wang Q X. 2005. Preliminary studies on diatoms from Kanasi in Xinjiang Uighur Autonomous. *Wuhan Bot. Res.*，

23(3): 247-256 [尤庆敏, 李海玲, 王全喜. 2005. 新疆喀纳斯地区硅藻初报. 武汉植物学研究, 23(3): 247-256].

You Q M, Wang Q X, Shi Z X. 2008. Newly recorded species of Cymbellaceae (Bacillariophyta) in China. *Acta Hydrobiologica Sinica*, 32(5): 735-740 [尤庆敏, 王全喜, 施之新. 2008. 中国桥弯藻科(硅藻门)的新记录植物. 水生生物学报, 32(5): 735-740].

You Q M, Liu Y, Wang Y F, Wang Q X. 2009. Taxonomy and distribution of diatoms in the genera *Epithemia* and *Rhopalodia* from the Xinjiang, China. *Nova Hedwigia*, 89(3-4): 397-430.

You Q M, Wang Q X. 2011a. Newly recorded species of *Nitzschia* and *Denticula* (Bacillariophyta) from Xinjiang, China. *Acta Botanica Boreali-Occidentalia Sinica*, 31(2): 0417-0422 [尤庆敏, 王全喜. 2011a. 新疆菱形藻属和细齿藻属(硅藻门)中国新纪录植物. 西北植物学报, 31(2): 0417-0422].

You Q M, Wang Q X. 2011b. Four Newly Recorded Species of Surirellaceae (Bacillariophyta) in China, *Plant Science Journal*, 29(2): 260-264 [尤庆敏, 王全喜. 2011b. 双菱藻科(硅藻门)——四个中国新纪录种. 植物科学学报, 29(2): 260-264].

You Q M, Liu Y, Wang Q X. 2011. Newly Recorded Species of *Hantzschia* (Bacillariophyta) in China. *Bulletin of Botanical Research*, 31(2): 129-133 [尤庆敏, 刘妍, 王全喜. 2011. 菱板藻属(硅藻门)中国新纪录种. 植物研究, 31(2): 129-133].

You Q M, Kociolek J P, Wang Q X. 2015. The diatom genus *Hantzschia* (Bacillariophyta) in Xinjiang province, China. *Phytotaxa*, 197(1): 1-14.

You Q M, Kociolek J P, Yu P, Cai M J, Lowe R L, Wang Q X. 2016. A new species of *Simonsenia* from a karst landform, Maolan Nature Reserve, Guizhou Province, China. *Diatom Research*, 31(3): 269-275.

You Q M, Kociolek J P, Cai M J, Yu P, Wang Q X. 2017. Two new *Cymatopleura* taxa (Bacillariophyta: Surirellales) with torsion of the valve, from Xinjiang, China. *Fottea*, 17(2): 293-302.

Zhang C, Zou J Z. 1994. Preliminary study on the subspecific taxonomy of *Nitzschia pungens* Grunow in Chinese coastal waters. *Oceanologic et Limnologic Sinica*, 25(2): 216-219 [张诚, 邹景忠. 1994. 中国近海尖刺菱形藻种下分类的初步研究. 海洋与湖沼, 25(2): 216-219].

Zhang R C, Niu Y L, Zhao J C, et al. 2006. Algae composition and temporal and spatial distributions in Huaishahe River-Huaijiuhe River Nature Reserve. *Acta Bot. Boreal. –Occident. Sin.*, 26(8): 1663-1670 [张茹春, 牛玉璐, 赵建成, 等. 2006. 北京怀沙河、怀九自然保护区藻类组成及时空分布动态研究. 西北植物学报, 26(8): 1663-1670].

Zhang Y K, Tian Y M. 1995. Studies on phytoplankton in Lake Baiyangdian. *Acta Hydrobiologica Sinica*, 19(4): 317-326 [张义科, 田玉梅. 1995. 白洋淀浮游植物现状. 水生生物学报, 19(4): 317-326].

Zhang Z A. 1986. Taxonomic studies on diatoms of Hainan Island, China. *Journal of Jinan University*, 3: 88-94 [张子安. 1986. 海南岛硅藻分类研究. (一) 菱形藻属的研究. 暨南医理学报, 3: 88-94].

Zhang N Q, Wang Z D, Du M H, et al. 2006. Phytoplankton and Water Quality of Water Source Area of the Middle Line Project of South to North Water Diversion. *Chin. J. Appl. Environ. Biol.*, 12(4): 506-510 [张乃群, 王正德, 杜敏华, 等. 2006. 南水北调中线水源区浮游植物与水质研究. 应用与环境生物学报, 12(4): 506-510].

Zhao A P, Liu F Y, Wu B, et al. 2005. Phytoplankton from the Dianshan Lake in Shanghai. *Journal of Shanghai Normal University*, 34(4): 70-76 [赵爱萍, 刘福影, 吴波等. 2005. 上海淀山湖的浮游植物. 上海师范大学学报, 34(4): 70-76].

Zhao J. 1998. *China Nature Geography*. Beijing: Higher Education Press: 342pp [赵济. 1998. 中国自然地理(第三版). 北京: 高等教育出版社: 342pp].

Zhong Z X, Bao S K, Tan M C, et al. 1986. Preliminary investigation on the diatoms in Dai Lake area of Junyun Mountain, Beibei, Chongqing (Studies on the Chongqing algae flora III). *Journal of Southwest Teachers University*, (2): 103-121 [钟肇新, 包少康, 谭明初, 等. 1986. 北碚缙云山黛湖水域硅藻植物研究初报(重庆藻类植物区系研究三). 西南师范大学学报, (2): 103-121].

Zhou G T. 1987. Distribution and name list of main algae in Xining area. *Journal of Qinghai Normal University*, (1): 47-53 [周广泰. 1987. 西宁地区主要藻类的分布规律及其名录. 青海师范大学学报, (1): 47-53].

Zhu H Z, Chen J Y. 1989a. New Species and Varieties of the Diatoms from Suoxiyu, Hunan, China. *In*: Li S H, et al. The algal flora and aquatic fauna of the Wulingyuan Nature Reserve Area, Hunan, China. Beijing: Science Press: p. 33-37 [朱惠忠, 陈嘉

佑. 1989a. 索溪峪硅藻的新种和新变种. 见：黎尚豪，等"湖南武陵源自然保护区水生生物". 北京：科学出版社：p. 33-37]

Zhu H Z，Chen J Y. 1989b. The Diatoms of the Suoxiyu Nature Reserve Area，Hunan，China. *In*：Li S H，*et al*. The algal flora and aquatic fauna of the Wulingyuan Nature Reserve Area，Hunan，China. Beijing: Science Press: p. 38-60 [朱惠忠，陈嘉佑. 1989b. 索溪峪的硅藻研究. 见：黎尚豪，等. "湖南武陵源自然保护区水生生物". 北京：科学出版社：p. 38-60].

Zhu H Z，Chen J Y. 1994. Study on the Diatoms of the Wuling Mountain Region. *In*：Shi Z X，*et al*. Compilation of Reports on the Survey of Algal Resources in South-Western China. Beijing：Science Press：p. 79-130 [朱惠忠，陈嘉佑. 1994. 武陵山区硅藻的研究. 见：施之新，等"西南地区藻类资源考察专集". 北京：科学出版社：p. 79-130].

Zhu H Z，Chen J Y. 1995. New taxa of diatom (Bacillariophyta) from Xizang (Tibet) (Ⅰ). *Acta Phytotaxonomica Sinica*，33(5)：516-519 [朱惠忠，陈嘉佑. 1995. 西藏硅藻的新种类(Ⅰ). 植物分类学报，33(5)：516-519].

Zhu H Z，Chen J Y. 1996. New taxa of diatom (Bacillariophyta) from Xizang (Tibet) (Ⅱ). *Acta Phytotaxonomica Sinica*，34(1)：102-104 [朱惠忠，陈嘉佑. 1996. 西藏硅藻的新种类(Ⅱ). 植物分类学报，34(1)：102-104].

Zhu H Z，Chen J Y. 2000. *Bacillariophyta of the Xizang Plateau*. Beijing：Science Press：353pp [朱惠忠，陈嘉佑. 2000. 中国西藏硅藻. 北京：科学出版社：353pp].

Zidarova R，Van de Vijver B，Quesada A，de Haan M. 2010. Revision of the genus *Hantzschia* (Bacillariophyceae) on Livingston Island (South Shetland Islands，Southern Atlantic Ocean). Plant Ecology and Evolution，143：318-333.

中 名 索 引

二画

二列双菱藻　117
二列双菱藻缩小变种　118
二肋纹马鞍藻　129
二额双菱藻　116
刀形菱形藻　20

三画

土栖菱形藻　23
小片菱形藻　45
小头菱形藻　38
小头端菱形藻　46
小型细齿藻　71
小型细齿藻粗变种　71
马鞍藻属　127

四画

中型菱形藻　37
中型长羽藻　126
中型长羽藻头端变种　126
中华菱板藻　53
丰富菱板藻　58
化石菱形藻　43
双菱藻科　97
双菱藻属　104
巴克豪森菱板藻　54
爪维兰斯菱形藻　47
木那菱形藻　32
反曲菱形藻　48
长羽藻属　124
长命菱板藻　51
长菱板藻　53

五画

丝状菱形藻　20
冬生马鞍藻　128
卡普龙双菱藻　120
可疑菱形藻　28
布列双菱藻　114
布赖韦尔双菱藻　112

五画（续）

平行棒杆藻　91
平行棒杆藻巨大变种　92
平行棒杆藻扭曲变种　92
平庸菱形藻　44
弗里克窗纹藻　79
瓜德罗普菱形藻　49
石生棒杆藻　97

六画

光亮型窗纹藻　88
光亮窗纹藻　76
光亮窗纹藻长角变种　77
光亮窗纹藻伸长变种　78
光亮窗纹藻龟形变种　79
光亮窗纹藻高山变种　77
匈牙利盘杆藻　65
华丽菱形藻　35
华美细齿藻　72
华彩双菱藻　123
吉斯纳菱形藻　44
吉塞拉菱形藻　29
多变菱形藻　29
多样菱形藻　37
尖锥盘杆藻　63
尖端菱形藻　14
杂种菱形藻　27
毕列松棒杆藻　96
纤细棒杆藻　93
肌状棒杆藻　94
伊犁菱板藻　55
优美长羽藻　125

七画

两尖菱板藻　59
两尖菱板藻头端变型　60
两尖菱板藻相等变种　60
两栖菱形藻　42
折曲菱形藻　15
克劳斯菱形藻　21
卵圆双菱藻　111
库津细齿藻　69

· 161 ·

库津细齿藻汝牧变种　70
杆状菱形藻　36
杆状藻科　10
杆状藻属　10
谷皮菱形藻　40
谷皮菱形藻微小变种　41
近针形菱形藻　39
近线形菱形藻　33
近盐生双菱藻　113
近粘连菱形藻斯科舍变种　22
近强壮菱板藻　56
近石生菱板藻　58
针形菱形藻　46
阿斯特里双菱藻　122
拟巴德菱板藻　57

八画

侧生窗纹藻　84
侧生窗纹藻顶生变种　86
侧生窗纹藻顶生变种二齿变型　86
侧生窗纹藻萨克森变种　87
侧生窗纹藻蛆形变种　87
具条纹双菱藻　113
具盖棒杆藻　93
凯特菱形藻　27
奇异杆状藻　11
奇异杆状藻肿大变种　12
岸边盘杆藻　65
拉普兰双菱藻　108
波缘藻属　98
直菱形藻　18
直菱板藻　52
线形菱形藻　30
线形菱形藻细变种　31
线性双菱藻　118
线性双菱藻缢缩变种　119
细长双菱藻　105
细长盘杆藻　66
细长菱形藻　37
细尖盘杆藻　64
细齿藻属　68
细微菱形藻　41
细端菱形藻　17
细端菱形藻中等变种　17
茂兰西蒙森藻　73

细筒柱藻　74
驼峰棒杆藻　94

九画

弯曲菱形藻　24
弯曲菱形藻平片变种　25
弯曲菱形藻缢缩变种　26
弯曲菱形藻德洛变种　25
弯菱形藻　23
弯棒杆藻　89
弯棒杆藻偏肿变种　91
显点菱板藻　52
柔软双菱藻　121
柔软双菱藻具脉变种　122
泉生菱形藻　42
洛伦菱形藻　47
洛伦菱形藻细弱变种　48
派松双菱藻　110
狭窄盘杆藻　63
玻璃质菱形藻　32
盾状马鞍藻　130
科瑞提细齿藻　71
类S状菱形藻　15
类附生菱形藻　49
美丽双菱藻　120
美丽菱板藻　51
草鞋形波缘藻　100
草鞋形波缘藻细长变种　101
草鞋形波缘藻细尖变种　102
草鞋形波缘藻整齐变种　102
钝端菱形藻　19
毡帽菱形藻　27
剑形长羽藻　125
施密斯窗纹藻　79

十画

泰特尼斯双菱藻　106
窄双菱藻　106
脐形菱形藻　28
莱维迪盘杆藻　67
莱温德马鞍藻　127
诺克里马鞍藻　128
较长菱板藻　54
高山菱形藻　36

十一画

密集菱板藻 55
常见菱形藻 35
淡黄双菱藻 119
渐窄盘杆藻 62
渐窄盘杆藻尖变种 63
盘杆藻属 61
盘状双菱藻 111
粗壮双菱藻 122
粗壮双菱藻宽大变型 123
粗肋菱形藻 35
粗条菱形藻 43
维多利亚盘杆藻 68
维苏双菱藻 114
盖斯纳菱板藻 55
菱形藻属 12
菱形菱板藻 51
菱板藻属 49

十二画

强壮细齿藻 72
普通菱形藻 38
棒杆藻科 75
棒杆藻属 88
椭圆波缘藻 98
椭圆波缘藻冬生变种 99
椭圆波缘藻缢缩变种 99
短形菱形藻 22
窗纹藻属 75

十三画

微小双菱藻 108
微型菱形藻 21

暖温盘杆藻 66
缢缩棒杆藻 96
辐射菱形藻 45
鼠形窗纹藻 80
鼠形窗纹藻细长变种 81
嫌钙菱板藻 56
新疆波缘藻 103

十四画

管毛菱形藻 44
管壳缝目 10

十五画

额雷菱形藻 16
德洛西蒙森藻 73

十六画

膨大窗纹藻 81
膨大窗纹藻韦斯特曼变种 84
膨大窗纹藻头端变种 83
膨大窗纹藻具褶变种 83
膨大窗纹藻颗粒变种 83

十七画

螺旋双菱藻 116
簇生菱形藻 24

二十画

蠕虫状菱形藻 14
灌木菱形藻 39

二十二画

囊形双菱藻 109

学 名 索 引

A

Aulonoraphidinales 10

B

Bacillaria J. F. Gmelin 10
—*paradoxa* Gmelin 11
——var. *turmidula* Grunow 12
Bacillariaceae 10

C

Campylodiscus C. G. Ehrenberg ex F. T. Kützing 127
—*bicostatus* W. Smith 129
—*clypeus* Ehrenberg 130
—*hibernicus* Ehrenberg 128
—*levanderi* Hustedt 127
—*noricus* Ehrenberg 128
Cymatopleura W. Smith 98
—*aquastudia* Kociolek & You 103
—*elliptica* (Brébisson) W. Smith 98
——var. *constricta* Grunow 99
——var. *hibernica* (W. Smith) Van Heurck 99
—*solea* (Brébisson) W. Smith 100
——var. *apiculata* (W. Smith) Ralfs 102
——var. *gracilis* Grunow 101
——var. *regula* (Ehrenberg) Grunow 102
—*xinjiangiana* You & Kociolek 103
Cylindrotheca Rabenhorst 74
—*gracilis* (Brébisson ex Kützing) Grunow 74

D

Denticula F. T. Kützing 68
—*certicola* Lange-Bertalot & Krammer 71
—*elegans* Kützing 72
—*kuetzingii* Grunow 69
——var. *rumrichae* Krammer 70
—*tenuis* Kützing 71
——var. *crassula* (Nägeli) Hustedt 71
—*valida* (Pedicino) Grunow 72

E

Epithemia F. T. Kützing 75
—*adnata* (Kützing) Brébisson 84
——var. *porcellus* (Kützing) Patrick 87
——var. *proboscidea* Patrick 86
————f. *bidens* nov. comb. 86
——var. *saxonica* (Kützing) Patrick 87
—*argus* (Ehrenberg) Kützing 76
——var. *alpestris* (W. Smith) Grunow 77
——var. *longicornis* (Ehrenberg) Grunow 77
——var. *protracta* Mayer 78
——var. *testudo* Fricke 79
—*arguiformis* You & Wang 88
—*frickei* Krammer 79
—*smithii* Carruthers 79
—*sorex* Kützing 80
——var. *gracilis* Hustedt 81
—*turgida* (Ehrenberg) Kützing 81
——var. *capitata* Fricke 83
——var. *granulata* (Ehrenberg) Grunow 83
——var. *plicata* Meister 83
——var. *westermannii* (Ehrenberg) Grunow 84

H

Hantzschia Grunow 49
—*abundans* Lange-Bertalot 58
—*amphioxys* (Ehrenberg) Grunow 59
——var. *aequalis* Cleve-Euler 60
————f. *capitata* O. Müller 60
—*barckhausenii* Lange-Bertalot & Metzeltin 54
—*calcifuga* Reichardt & Lange-Bertalot 56
—*compacta* (Hustedt) Lange-Bertalot 55
—*distinctepunctata* Hustedt 52
—*elongata* (Hantzsch) Grunow 53
—*giessiana* Lange-Bertalot & Rumrich 55
—*longa* Lange-Bertalot 54

—*nitzschioides* Lange-Bertalot 51
—*pseudobardii* Q-M You & J. P. Kociolek 57
—*spectabilis* (Ehrenberg) Hustedt 51
—*sinensis* Q-M You & J. P. Kociolek 53
—*subrupestris* Lange-Bertalot 58
—*vivacior* Lange-Bertalot 57
—*subrobusta* Q-M You & J. P. Kociolek 56
—*virgata* (Roper) Grunow 52
—*vivax* (W. Smith) M. Peragallo 51
—*yili* Q-M You & J. P. Kociolek 55

N

Nitzschia A. H. Hassall 12
—*acicularis* (Kützing) W. Smith 46
—*acula* (Kützing) Hantzsch 14
—*alpina* Hustedt 36
—*amphibia* Grunow 42
—*bacilliformis* Hustedt 36
—*brevissima* Grunow 22
—*capitellata* Hustedt 46
—*clausii* Hantzsch 21
—*communis* Rabenhorst 38
—*commutata* Grunow 29
—*dissipata* (Kützing) Rabenhorst 17
——var. *media* (Hantzsch) Grunow 17
—*diversa* Hustedt 37
—*draveillensis* Coste & Ricard 47
—*dubia* W. Smith 28
—*eglei* Lange-Bertalot 16
—*elegantula* Grunow 35
—*epithemoides* Grunow 49
—*fasciculata* (Grunow) Grunow 24
—*filiformis* (W. Smith) Van Heurck 20
—*flexa* Schumann 15
—*fonticola* (Grunow) Grunow 42
—*fossilis* (Grunow) Grunow 43
—*frustulum* (Kützing) Grunow 45
—*fruticosa* Hustedt 39
—*gessneri* Hustedt 44
—*gisela* Lange-Bertalot 29
—*gracilis* Hantzsch 37
—*guadalupensis* Manguin 49
—*homburgiensis* Lange-Bertalot 27

—*hybrida* Grunow 27
—*inconspicua* Grunow 44
—*intermedia* Hantzsch ex Cleve & Grunow 37
—*kittlii* Grunow 27
—*linearis* W. Smith 30
——var. *Tenuis* 31
—*lorenziana* Grunow 47
——var. *subtilis* Grunow 48
—*microcephala* Grunow 38
—*monachorum* Lange-Bertalot 32
—*nana* Grunow 21
—*obtusa* W. Smith 19
—*palea* (Kützing) W. Smith 40
——var. *minuta* (Bleisch) Grunow 41
—*perminuta* (Grunow) M. Peragallo 41
—*radicula* Hustedt 45
—*recta* Hantzsch ex Rabenhorst 18
—*reversa* W. Smith 48
—*scalpelliformis* Grunow 20
—*sigma* (Kützing) W. Smith 23
—*sigmoidea* (Nitzsch) W. Smith 15
—*sinuata* (Thwaites) Grunow 24
——var. *delognei* (Grunow) Lange-Bertalot 25
——var. *tabellaria* (Grunow) Grunow 25
——var. *constricta* Chen & Zhu 26
—*solita* Hustedt 35
—*subacicularis* Hustedt 39
—*subcohaerens* var. *scotica* (Grunow) Van Heurck 22
—*sublinearis* Hustedt 33
—*terrestris* (Petersen) Hustedt 23
—*tubicola* Grunow 44
—*umbonata* (Ehrenberg) Lange-Bertalot 28
—*vermicularis* (Kützing) Hantzsch 14
—*valdecostata* Lange-Bertalot & Simonsen 35
—*valdestriata* Aleem & Hustedt 43
—*vitrea* Norman 32

R

Rhopalodia O. Müller 88
—*brebissonii* Krammer 96
—*constricta* (W. Smith) Krammer 96
—*gibba* (Ehrenberg) O. Müller 89
——var. *ventricosa* (Kützing) H. et M. Paragallo 91

—*gibberula* (Ehrenberg) O. Müller 94
—*gracilis* O. Müller 93
—*musculus* (Kützing) O. Müller 94
—*operculata* (Agardh) Håkansson 93
—*parallela* (Grunow) O. Müller 91
——var. *distorta* Fricke 92
——var. *ingens* Fricke 92
—rupestris (W. Smith) Krammer 97

S

Stenopterobia A. de Brébisson ex H. Van Heurck 124
—*anceps* (Lewis) Brébisson ex Van Heurck 125
—*delicatissima* (Lewis) Brébisson ex Van Heurck 125
—*intermedia* (Lewis) Brébisson ex Van Heurck 126
——var. *capitata* Fontell 126
Surirella P. J. F. Turpin 104
—*angusta* Kützing 106
—*astridae* Hustedt 122
—*bifrons* Ehrenberg 116
—*biseriata* Brébisson 117
——var. *diminuta* Cleve-Euler 118
—*brebissonii* Krammer & Lange-Bertalot 114
—*brightwellii* W. Smith 112
—*capronii* Brébisson 120
—*crumena* Brébisson ex Kützing 109
—*elegans* Ehrenburg 120
—*gracilis* Grunow 105
—*helvetica* Brun 119
—*lapponica* A. Cleve 108
—*linearis* W. Smith 118
——var. *constricta* Grunow 119
—*ovalis* Brébisson 111

—*minuta* Brébisson 108
—*patella* Kützing 111
—*peisonis* Pantocsek 110
—*robusta* Ehrenberg 122
——f. *lata* Hustedt 123
—*spiralis* Kützing 116
—*splendida* (Ehrenberg) Kützing 123
—*striatula* Turpin 113
—*subsalsa* W. Smith 113
—*tenera* Greyory 121
——var. *nervosa* A. Schmidt 122
—*tientsinensis* Skvortzow 106
—*visurgis* Hustedt 114
Simonsenia H. Lange-Bertalot 73
—*delognei* (Grunow) Lange-Bertalot 73
—*maolaniana* You & Kociolek 73
Surirellaceae 97

T

Tryblionella W. Smith 61
—*acuminata* W. Smith 63
—*angustata* W. Smith 62
——var. *acuta* (Grunow) nov. comb. 63
—*angustatula* (Lange-Bertalot) nov. comb. 63
—*apiculata* Gregory 64
—*calida* (Grunow) D. G. Mann 66
—*gracilis* W. Smith 66
—*hungarica* (Grunow) D. G. Mann 65
—*levidensis* W. Smith 67
—*littoralis* (Grunow) D. G. Mann 65
—*victoriae* Grunow 68

图版 I

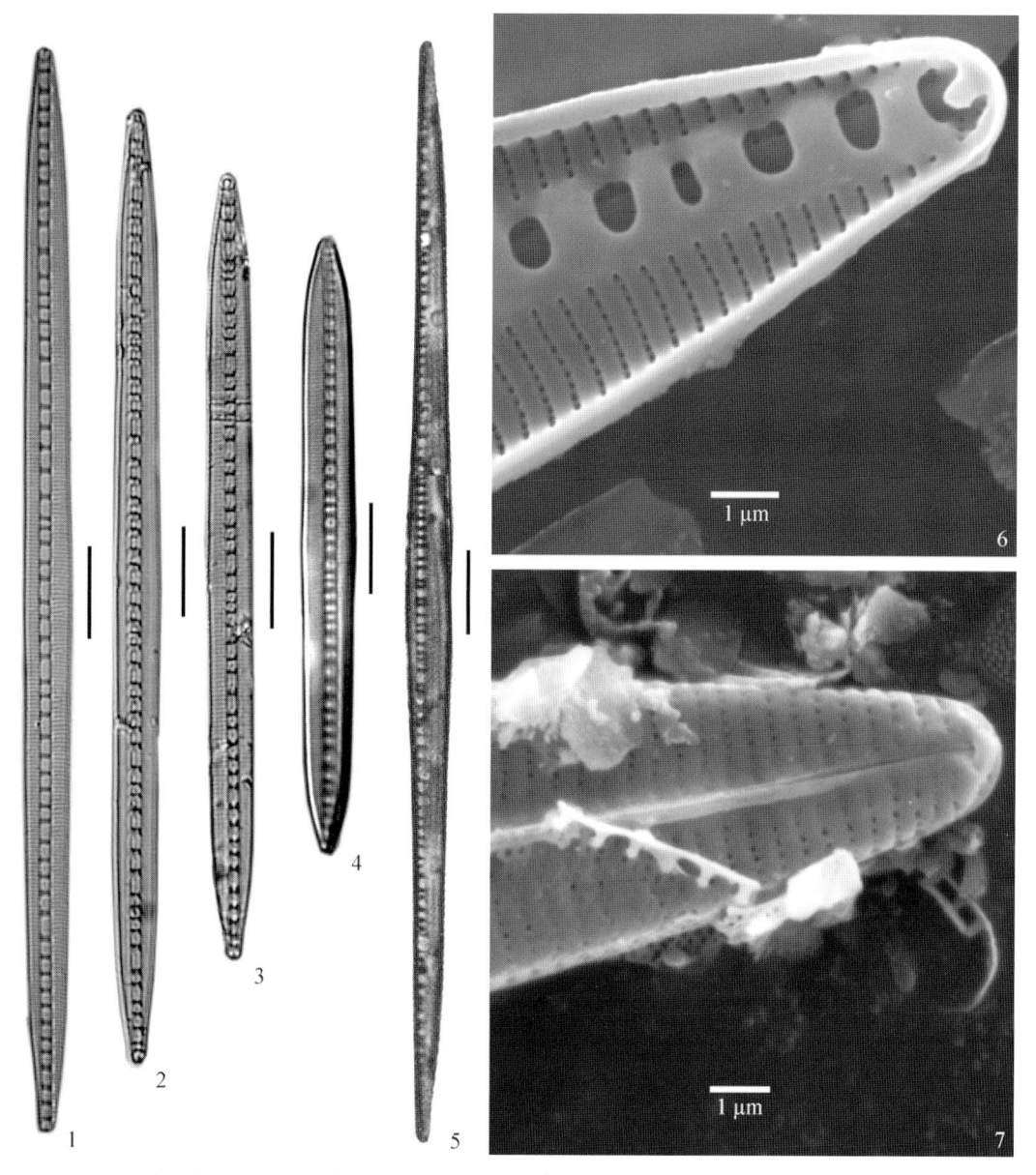

1—4，6—7. 奇异杆状藻（原变种）*Bacillaria paradoxa* var. *paradoxa* Gmelin，6. 示内壳面顶端结构（SEM），龙骨突和螺旋舌形态，7. 示外壳面顶端结构（SEM），孔纹和壳缝形态；5. 奇异杆状藻肿大变种 *Bacillaria paradoxa* var. *turmidula*(Grunow) De Toni(引自 Fan, 2004)。光镜(LM)照片，比例尺(Scale bars)=10 μm，电镜(SEM)照片比例尺见图。

图版 II

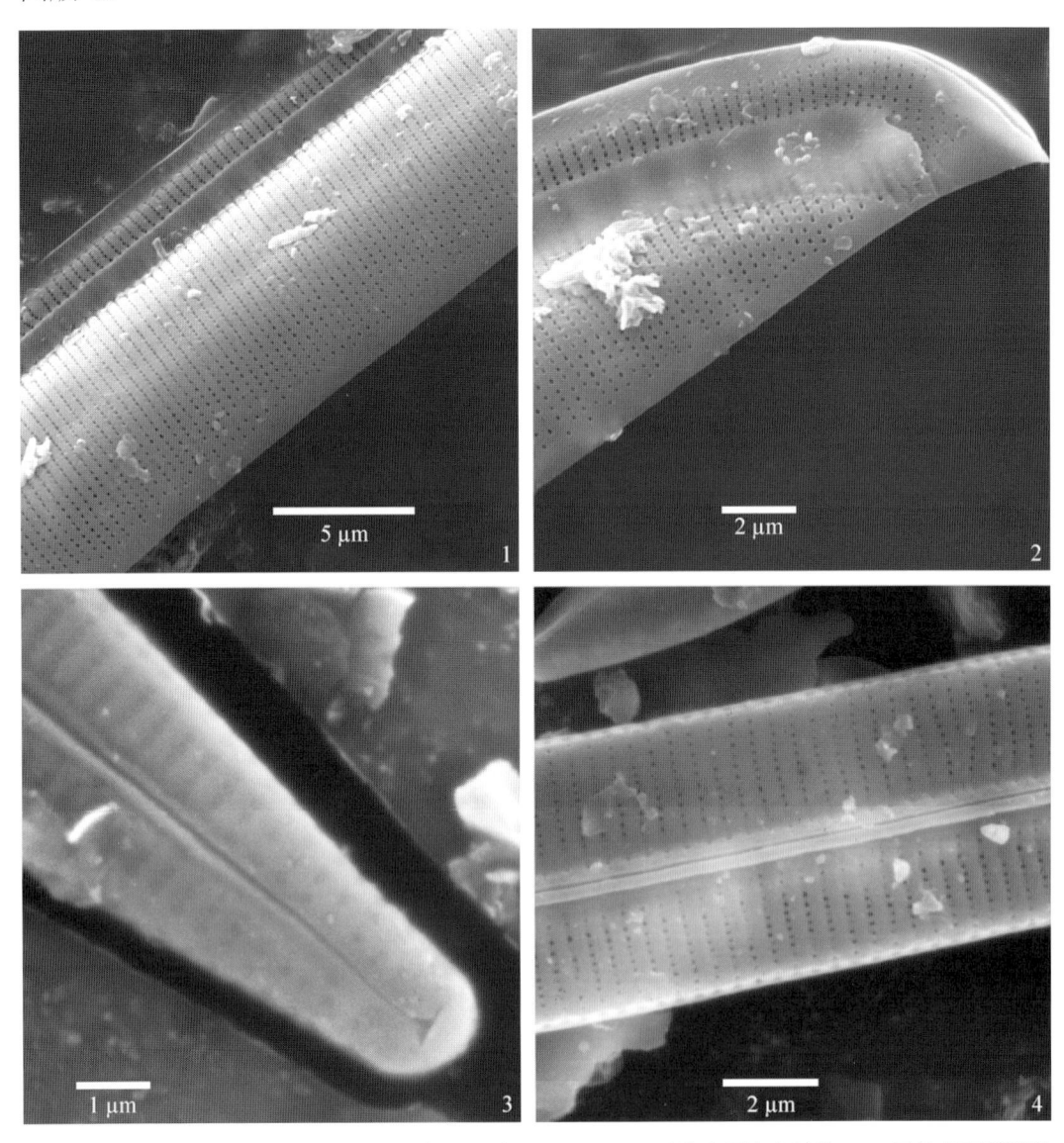

1—2. 类 S 状菱形藻 *Nitzschia sigmoidea* (Nitzsch) W. Smith, 1. 示外壳面中部结构, 2. 示外壳面顶端结构; 3—4. 奇异杆状藻（原变种）*Bacillaria paradoxa* var. *paradoxa* Gmelin, 3. 示外壳面顶端结构, 壳缝末端形状, 4. 示外壳面中部结构。所有图为电镜（SEM）照片, 比例尺见图。

图版 III

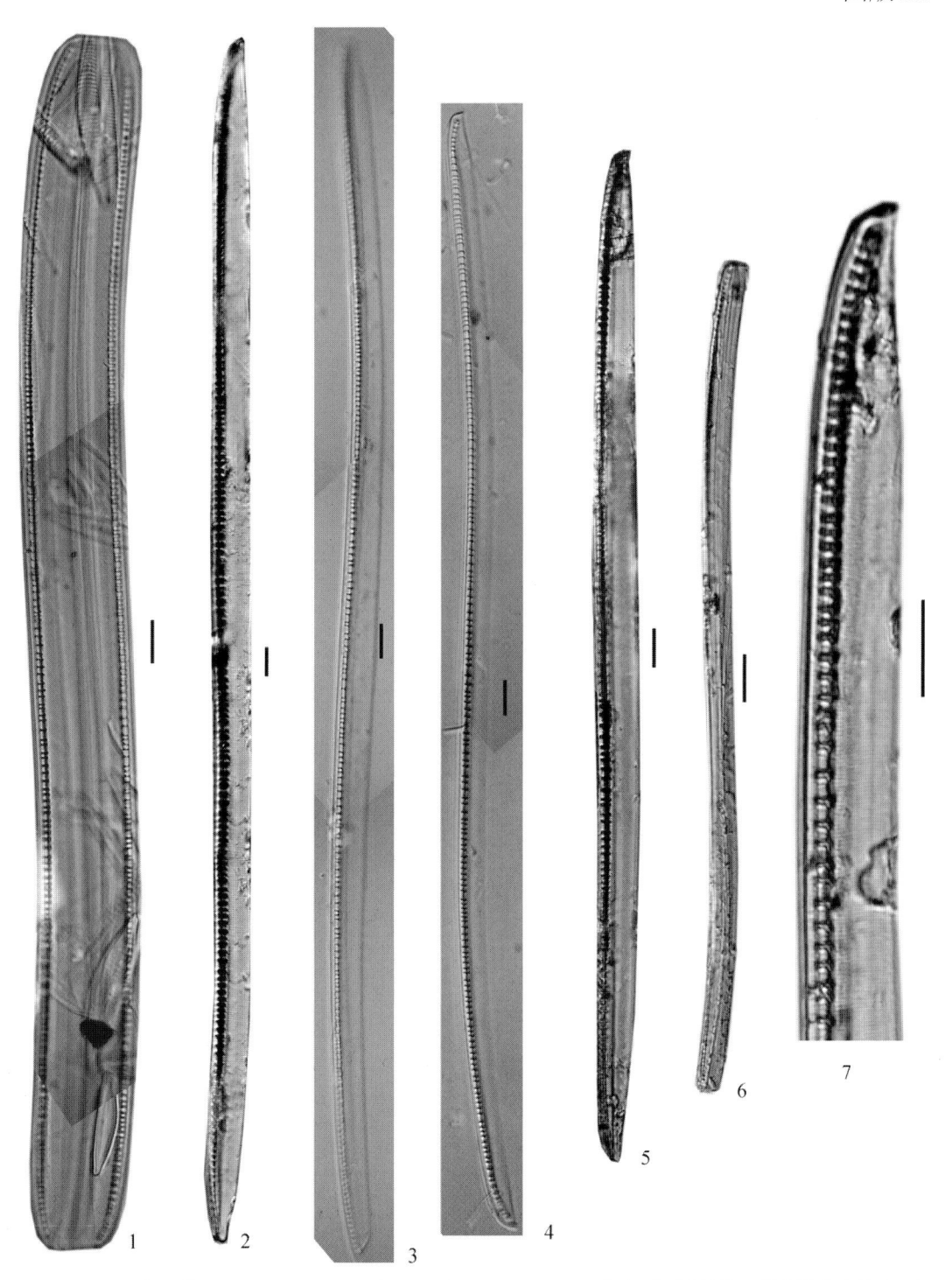

1—7. 类 S 状菱形藻 *Nitzschia sigmoidea* (Nitzsch) W. Smith，7. 示外壳面顶端结构。所有图为光镜(LM)照片，比例尺(Scale bars)=10 μm。

图版 IV

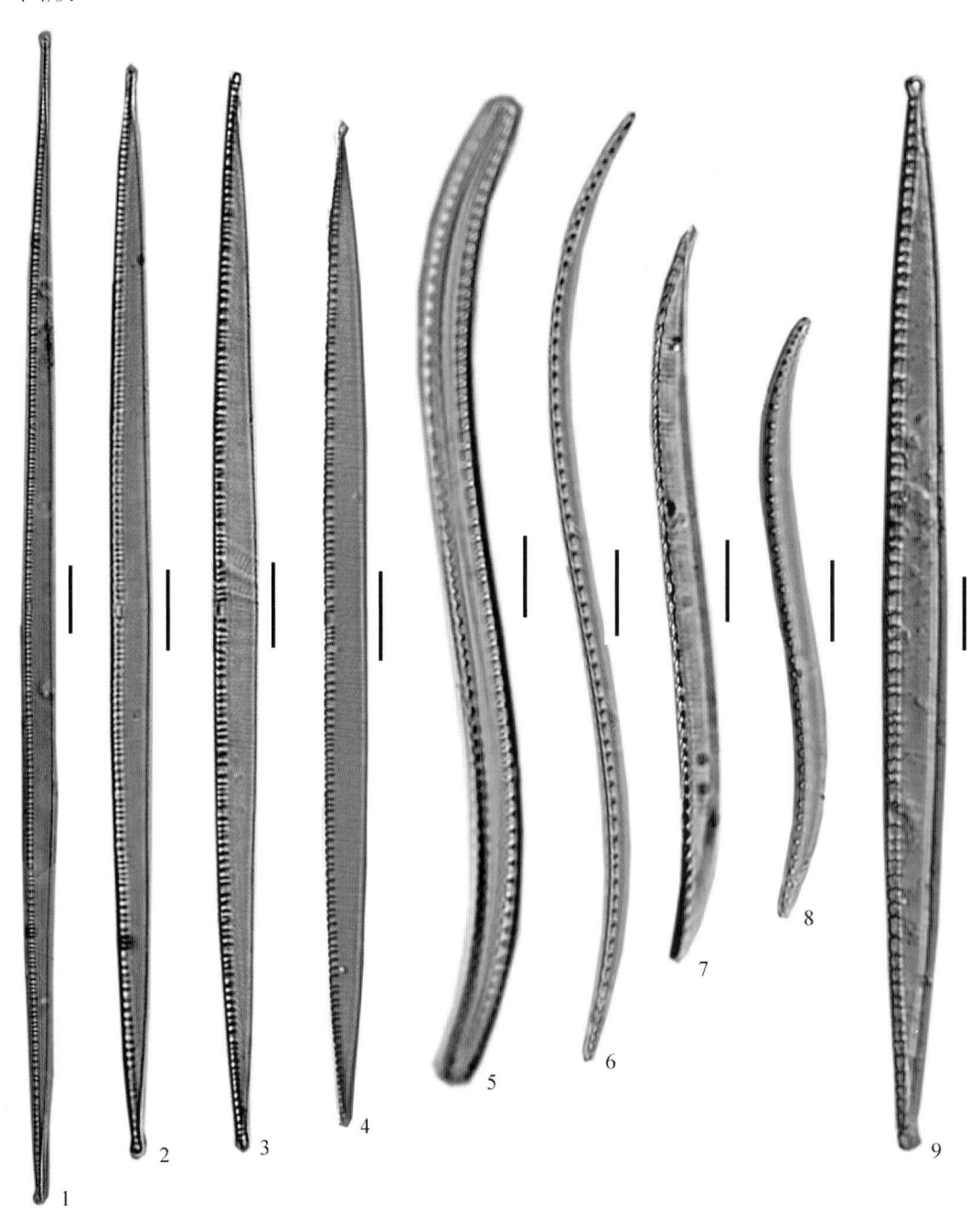

1—4. 蠕虫状菱形藻 *Nitzschia vermicularis* (Kützing) Hantzsch；5—8. 折曲菱形藻 *Nitzschia flexa* Schumann；9. 尖端菱形藻 *Nitzschia acula* Kützing Hantzsch。所有图为光镜(LM)照片，比例尺(Scale bars)=10 μm。

图版 V

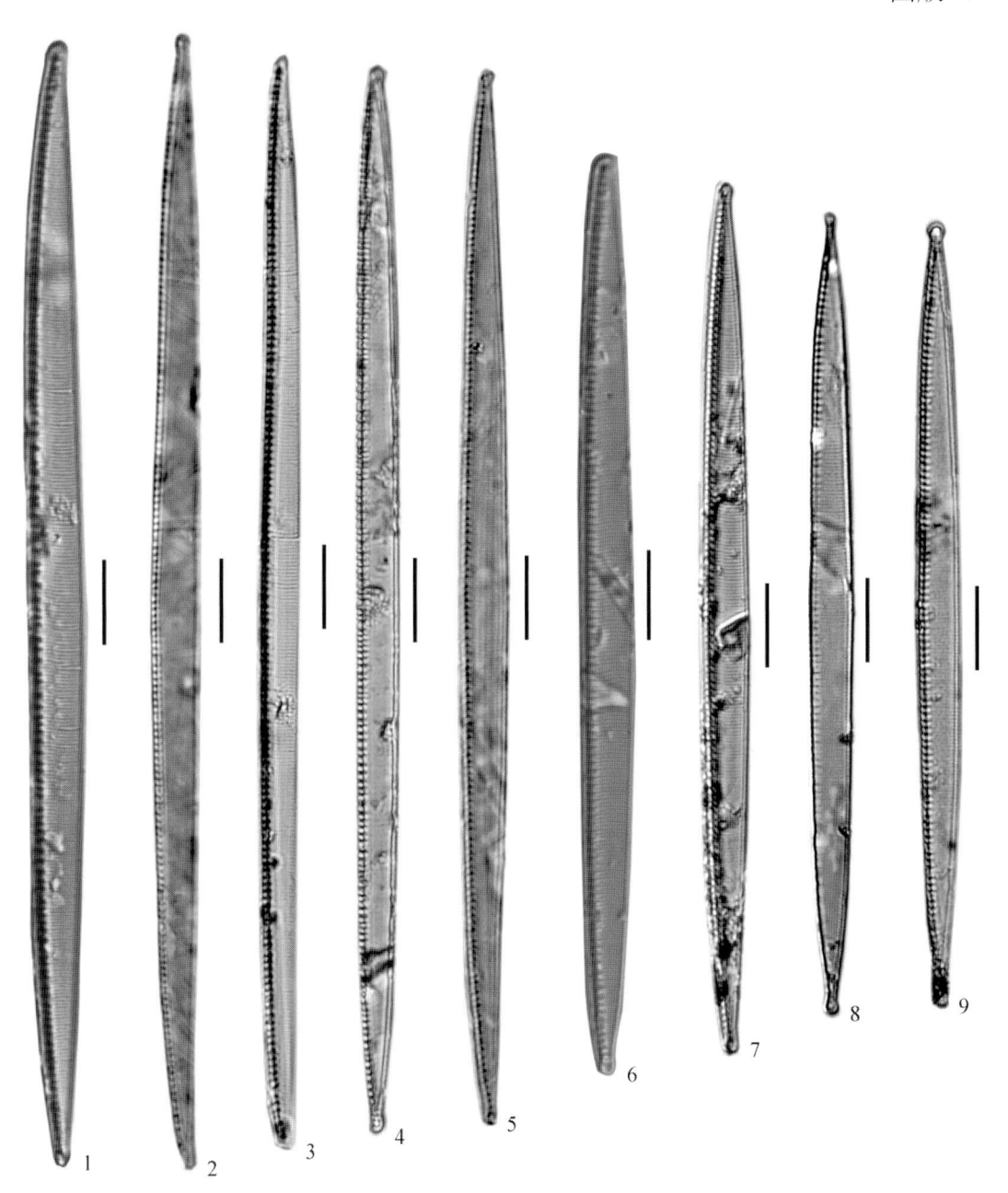

1—9. 额雷菱形藻 *Nitzschia eglei* Lange-Bertalot。所有图为光镜(LM)照片，比例尺(Scale bars)=10 μm。

图版 VI

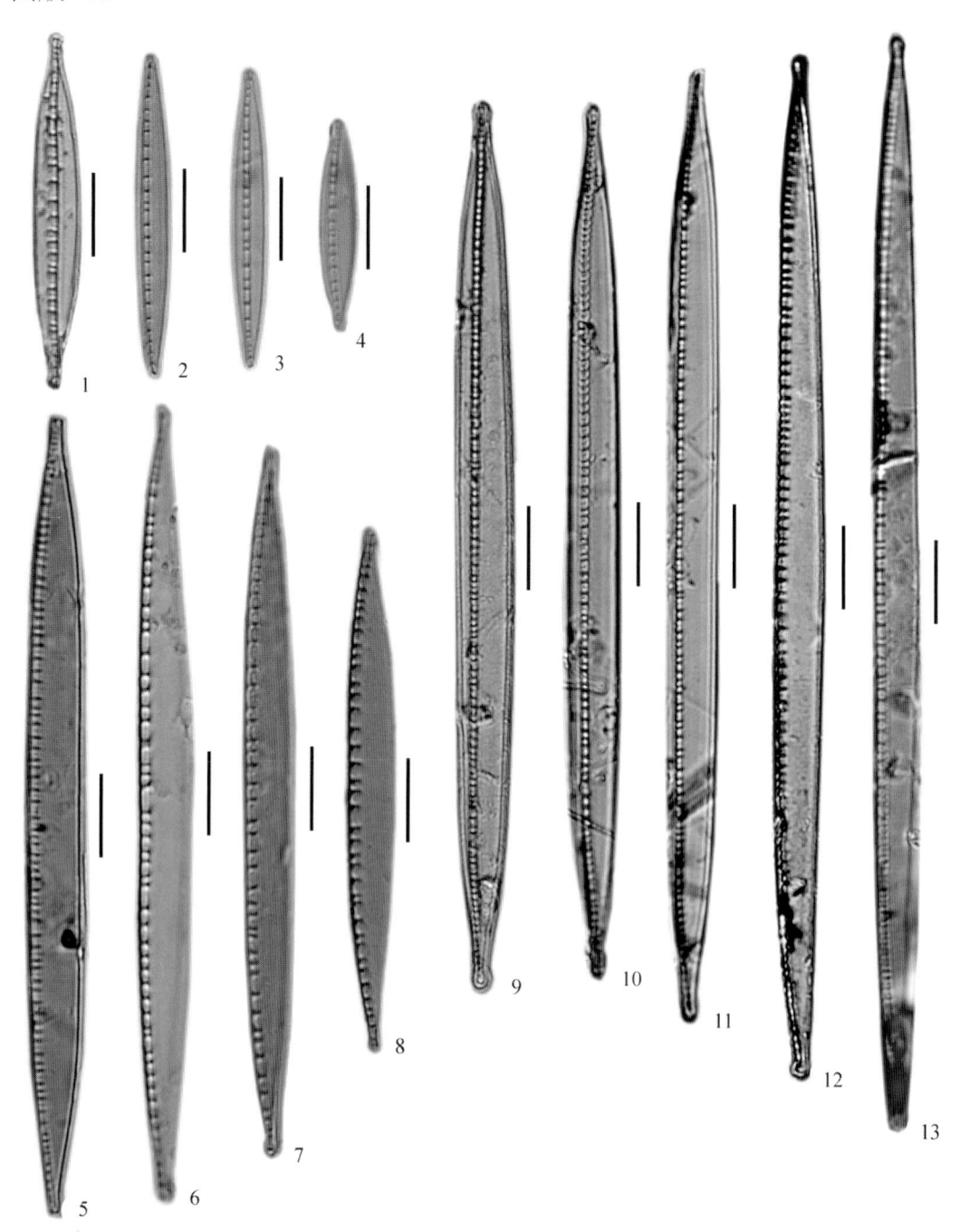

1—4. 细端菱形藻（原变种）*Nitzschia dissipata* var. *dissipata* (Kützing) Grunow；5—8. 直菱形藻 *Nitzschia recta* Hantzsch；9—13. 细端菱形藻中等变种 *Nitzschia dissipata* var. *media* (Hantzsch) Grunow。所有图为光镜(LM)照片，比例尺(Scale bars)=10 μm。

图版 VII

1—4. 细端菱形藻（原变种）*Nitzschia dissipata* var. *dissipata* (Kützing) Grunow；1—2. 示内壳面；3—4. 示内壳面中部结构，可见组成线纹的孔纹和龙骨突形态。所有图为电镜(SEM)照片，比例尺见图。

图版 VIII

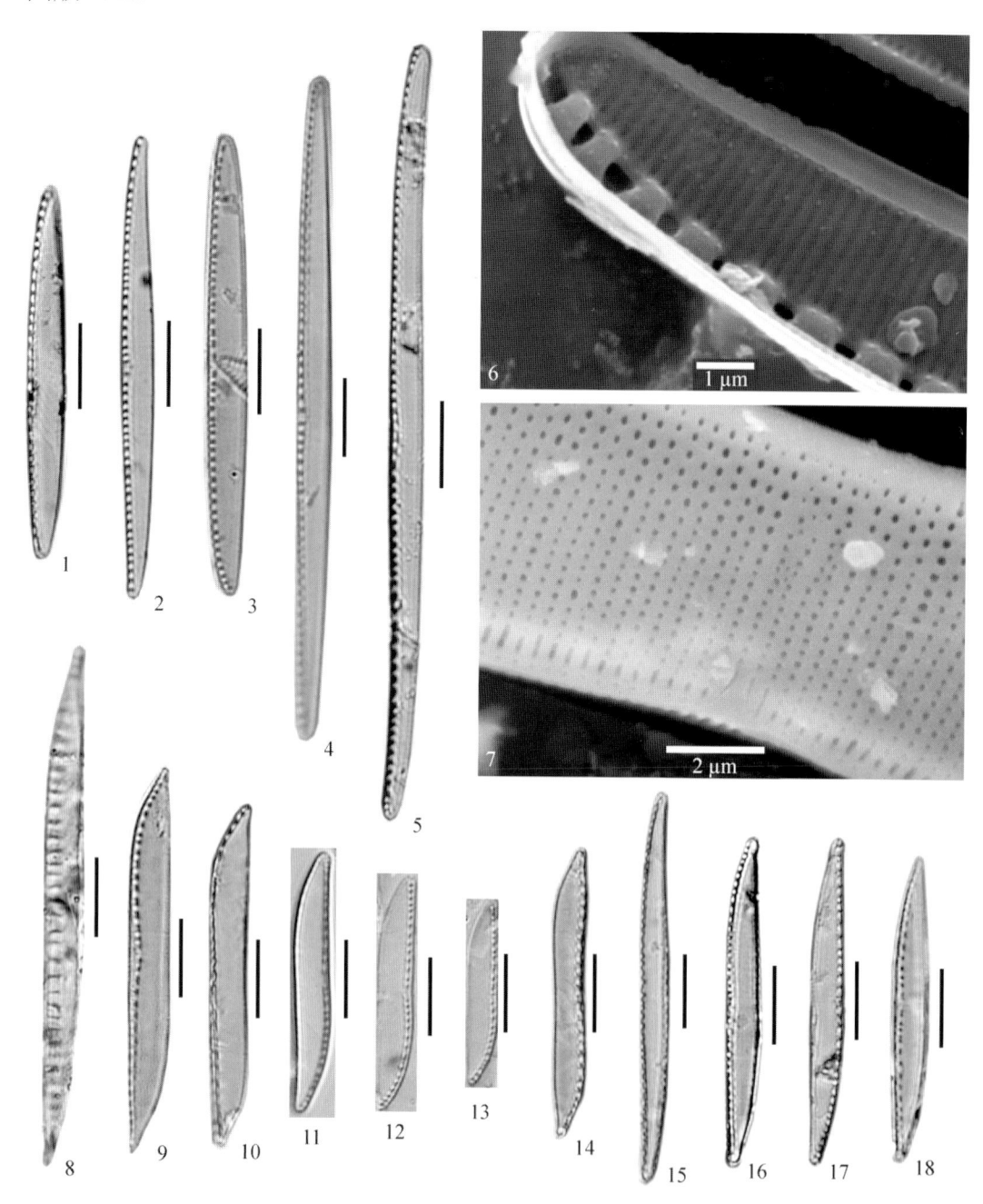

1—7. 丝状菱形藻 *Nitzschia filiformis* (W. Smith) Van Heurck，6. 示内壳面顶端结构(SEM)，7.示外壳面中部结构(SEM)；8. 簇生菱形藻 *Nitzschia fasciculata* (Grunow) Grunow；9—13. 克劳斯菱形藻 *Nitzschia clausii* Hantzsch；14. 土栖菱形藻 *Nitzschia terrestris* (Petersen) Hustedt；15—18. 近粘连菱形藻斯科舍变种 *Nitzschia subcohaerens* var. *scotica* (Grunow) Van Heurck。光镜(LM)照片，比例尺(Scale bars)=10 μm，电镜(SEM)照片比例尺见图。

图版 IX

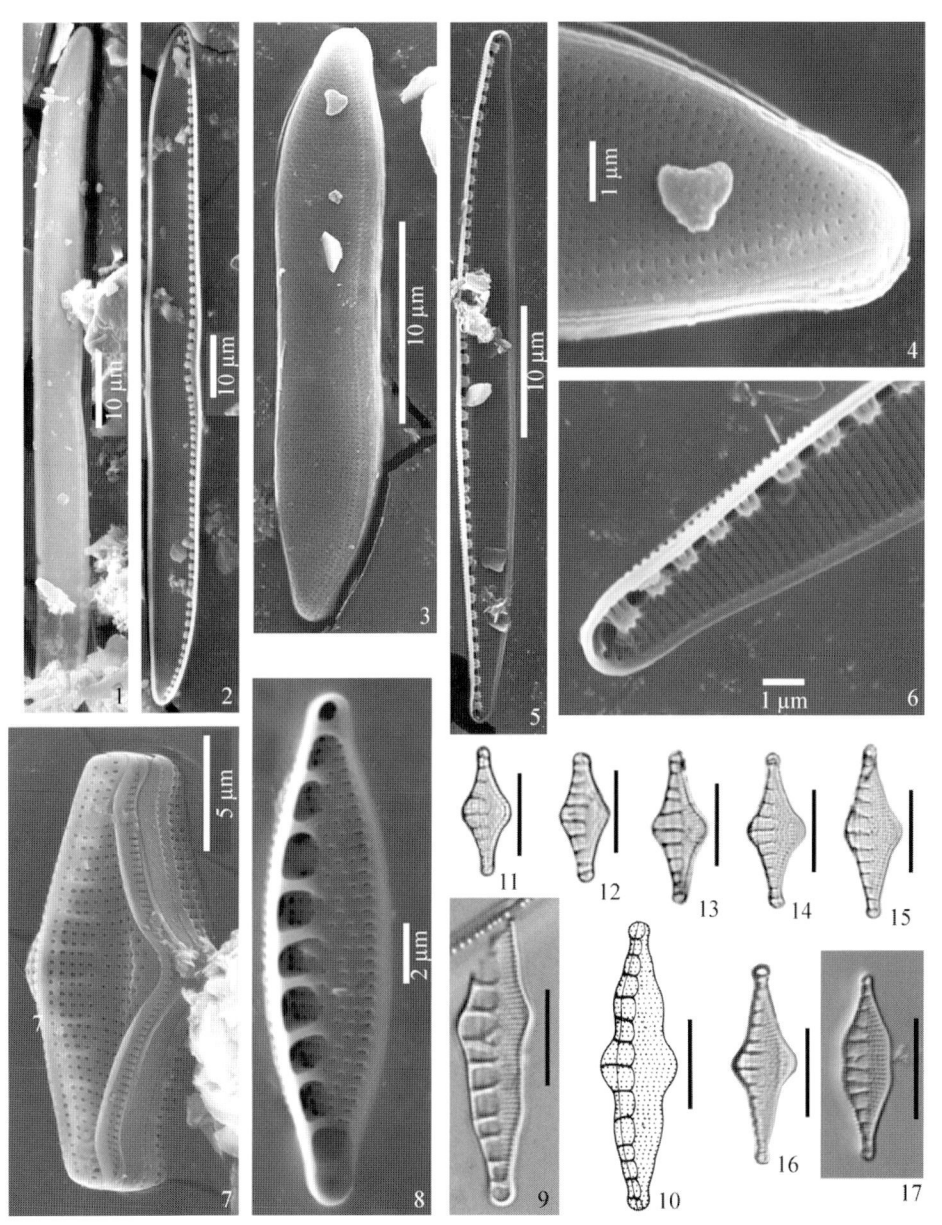

1—2. 微型菱形藻 *Nitzschia nana* Grunow，1. 示外壳面(SEM)，2. 示内壳面(SEM)；3—4. 短形菱形藻 *Nitzschia brevissima* Grunow，3. 示外壳面(SEM)，4. 示外壳面顶端(SEM)；5—6. 近粘连菱形藻斯科舍变种 *Nitzschia subcohaerens* var. *scotica* (Grunow) Van Heurck, 5. 示内壳面(SEM)，6. 示内壳面顶端(SEM)；7, 11—16. 弯曲菱形藻平片变种 *Nitzschia sinuata* var. *tabellaria* (Grunow) Grunow, 7. 示外壳面(SEM)；8. 弯曲菱形藻德洛变种 *Nitzschia sinuata* var. *delognei* (Grunow) Lange-Bertalot, 示内壳面(SEM)；9—10. 弯曲菱形藻(原变种) *Nitzschia sinuata* var. *sinuata* (Thwaites?) Grunow, 其中 10 引自 Zhu 和 Chen (2000)；17. 弯曲菱形藻缢缩变种 *Nitzschia sinuata* var. *constricta* Chen & Zhu。光镜(LM)照片，比例尺(Scale bars)=10 μm，电镜(SEM)照片比例尺见图。

图版 X

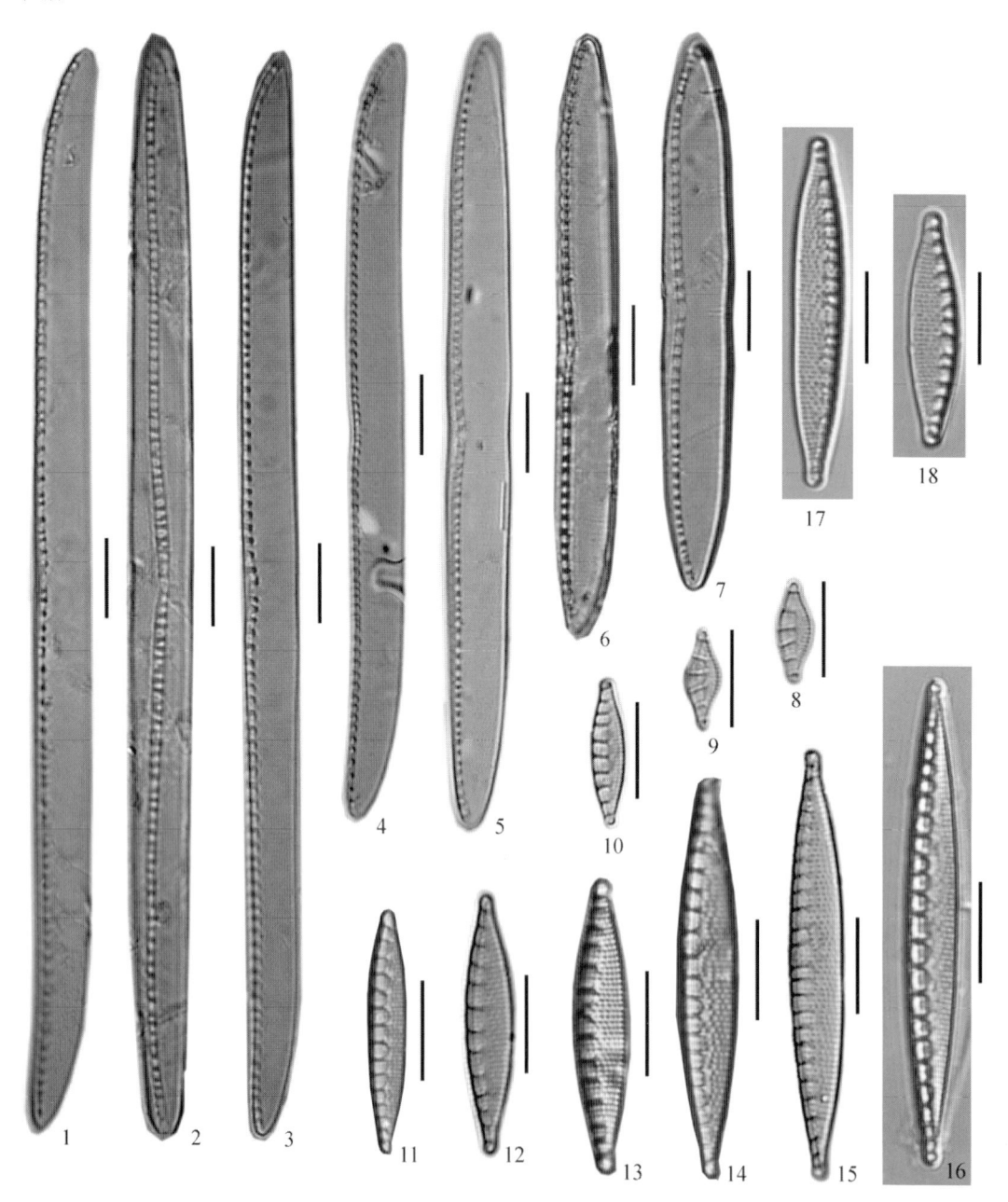

1—3. 钝端菱形藻 *Nitzschia obtusa* W. Smith；4—7. 刀形菱形藻 *Nitzschia scalpelliformis* Grunow；8—18. 弯曲菱形藻德洛变种 *Nitzschia sinuata* var. *delognei* (Grunow) Lange-Bertalot。所有图为光镜(LM)照片，比例尺(Scale bars)=10 μm。

图版 XI

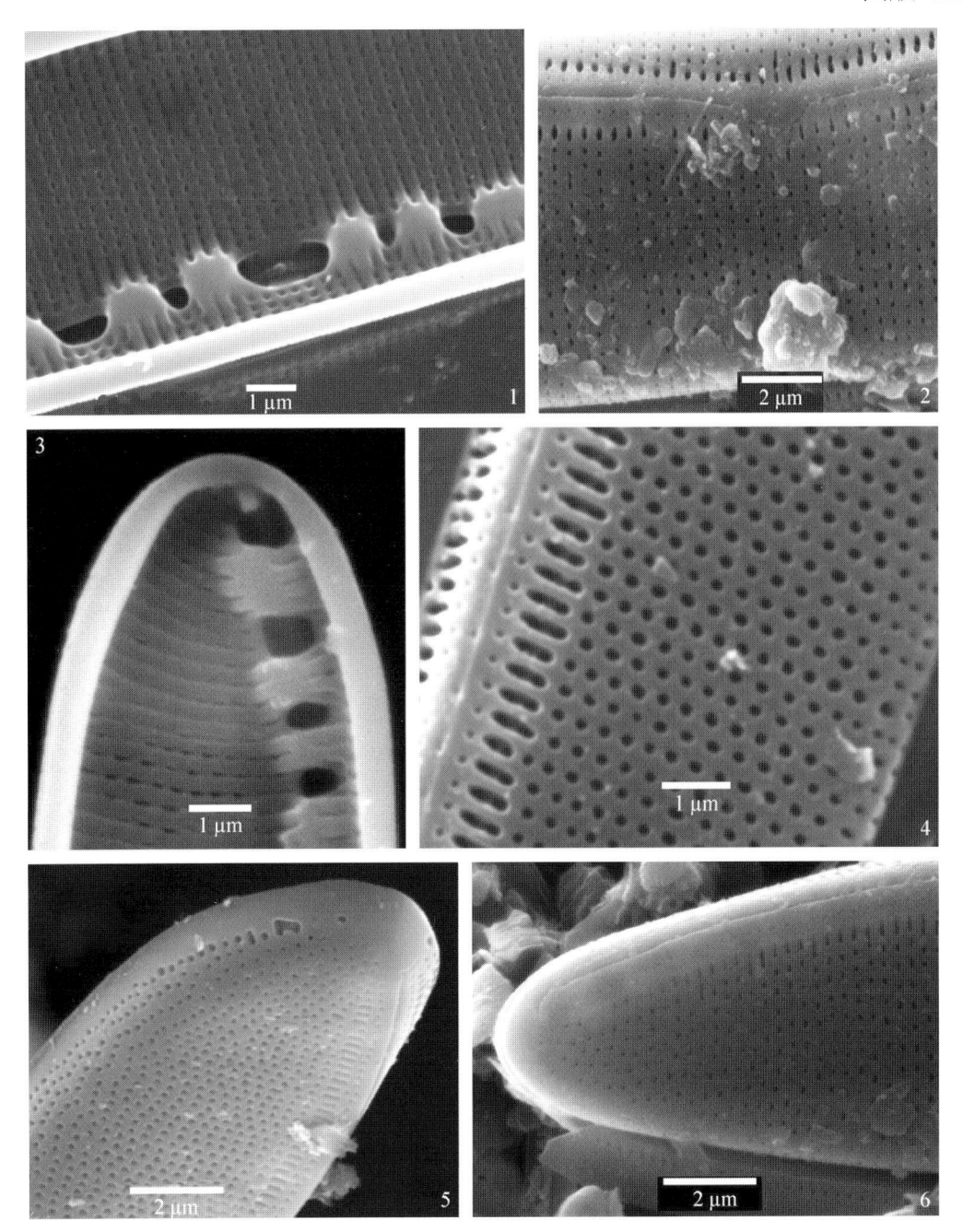

1—6. 钝端菱形藻 *Nitzschia obtusa* W. Smith，1. 示内壳面中部，近缝端和龙骨突形态，2. 示外壳面中部，近缝端形态，3. 示内壳面顶端，螺旋舌形态，4. 示外壳面线纹排列，5—6. 示外壳面顶端，远缝端形态。所有图为电镜(SEM)照片，比例尺见图。

图版 XII

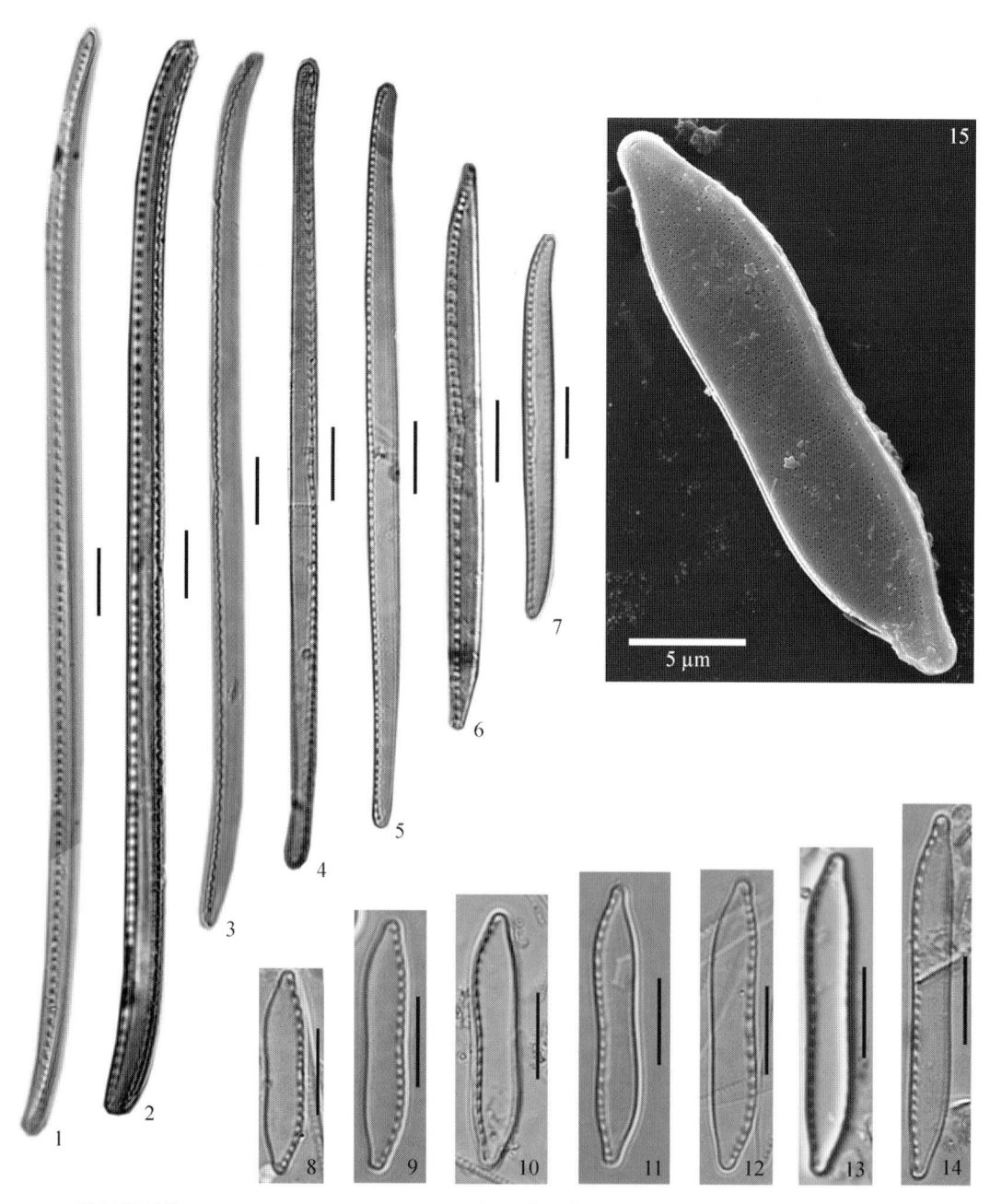

1—7. 微型菱形藻 *Nitzschia nana* Grunow；8—15. 短形菱形藻 *Nitzschia brevissima* Grunow，15. 示外壳面(SEM)。光镜(LM)照片，比例尺(Scale bars)=10 μm，电镜(SEM)照片比例尺见图。

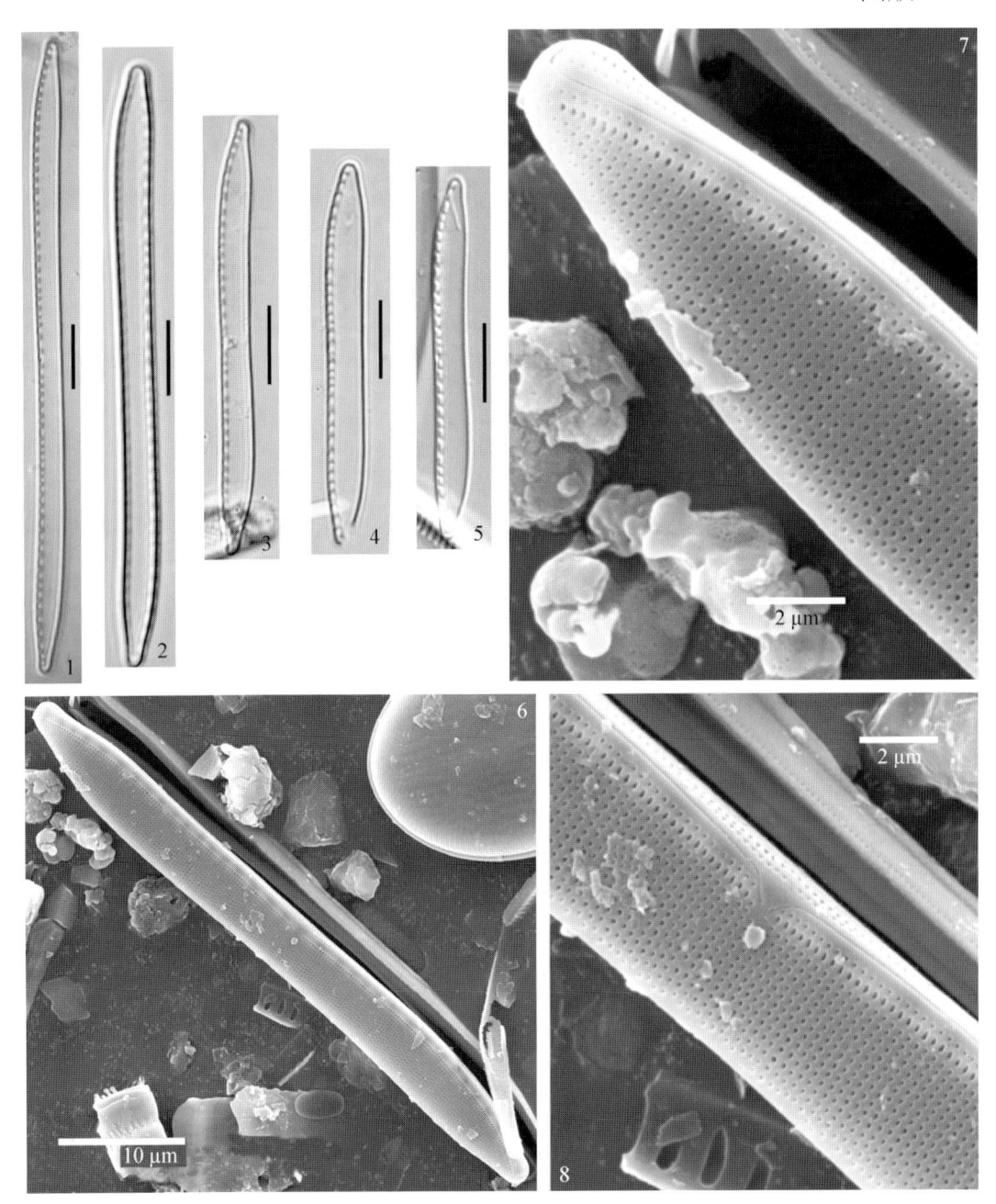

图版 XIII

1—8. 土栖菱形藻 *Nitzschia terrestris* (Petersen) Hustedt，6. 示外壳面(SEM)，7. 示外壳面顶端(SEM)，远缝端形态，8. 示外壳面中部(SEM)，近缝端形态。光镜(LM)照片，比例尺(Scale bars)=10 μm，电镜(SEM)照片比例尺见图。

图版 XIV

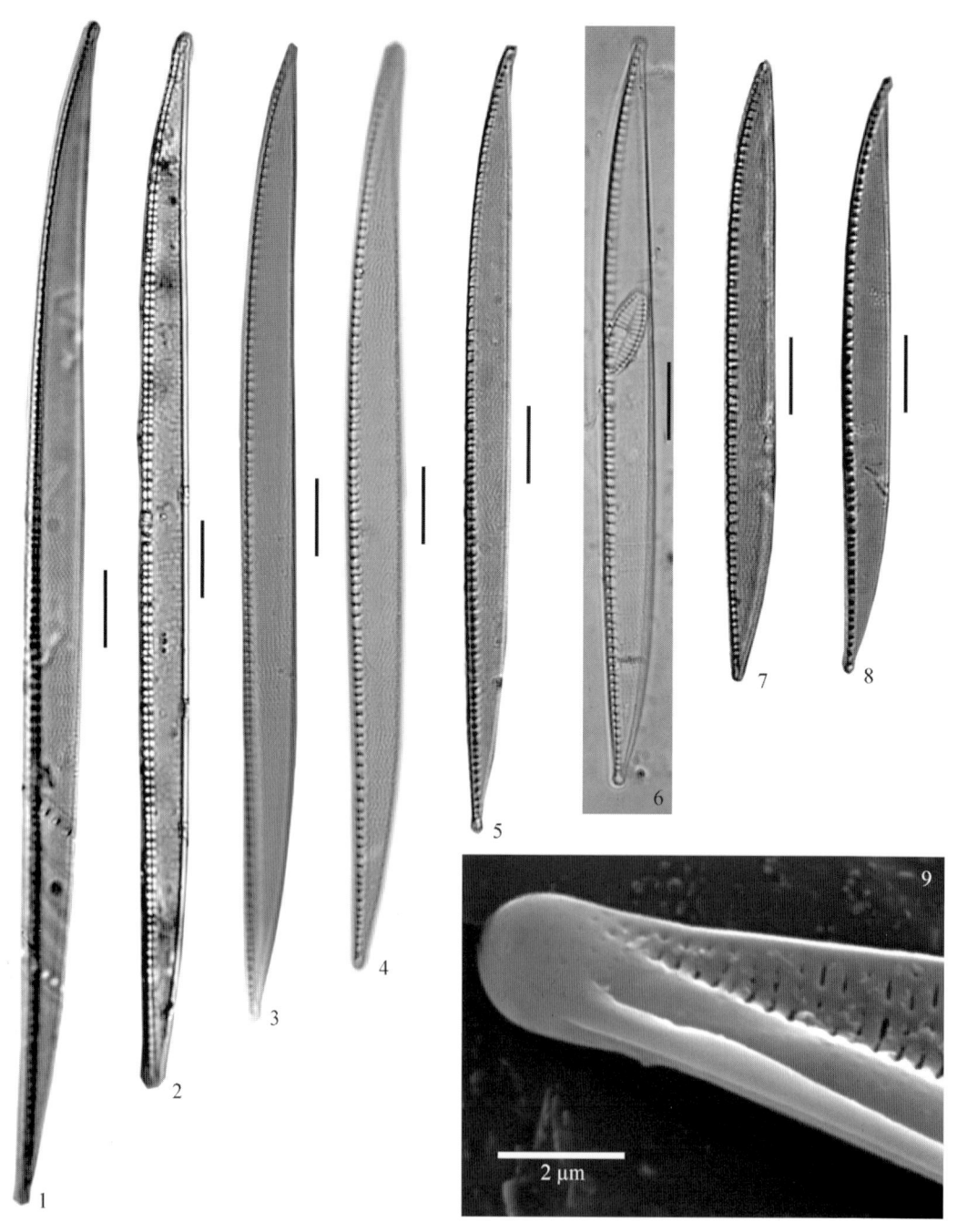

1—9. 弯菱形藻 *Nitzschia sigma* (Kützing) W. Smith，9. 示外壳面顶端(SEM)。光镜(LM)照片，比例尺 (Scale bars)=10 μm，电镜(SEM)照片比例尺见图。

图版 XV

1—2. 弯菱形藻 *Nitzschia sigma* (Kützing) W. Smith，1. 示外壳面，孔纹形状及排列方式，2. 示内壳面，龙骨突形状及孔纹排列；3—4. 多变菱形藻 *Nitzschia commutata* Grunow，3. 示外壳面，孔纹形态及排列，4. 示内壳面，龙骨突及孔纹形态及排列。所有图为电镜（SEM）照片，比例尺见图。

图版 XVI

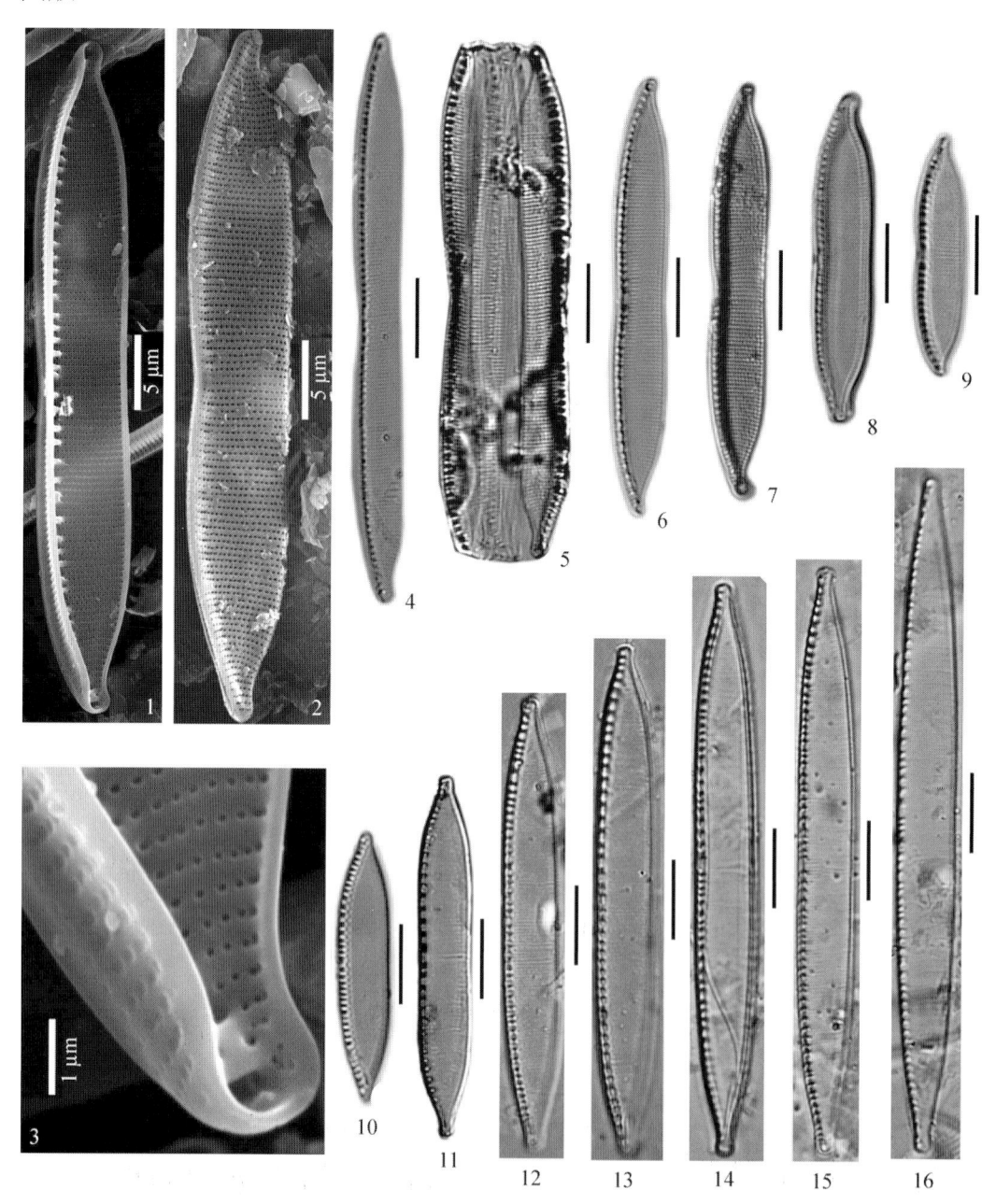

1—9. 多变菱形藻 *Nitzschia commutata* Grunow，1. 示内壳面（SEM），2. 示外壳面（SEM），3. 示内壳面顶端（SEM），螺旋舌结构；10—16. 脐形菱形藻 *Nitzschia umbonata*（Ehrenberg）Lange-Bertalot。光镜（LM）照片，比例尺（Scale bars）=10 μm，电镜（SEM）照片比例尺见图。

图版 XVII

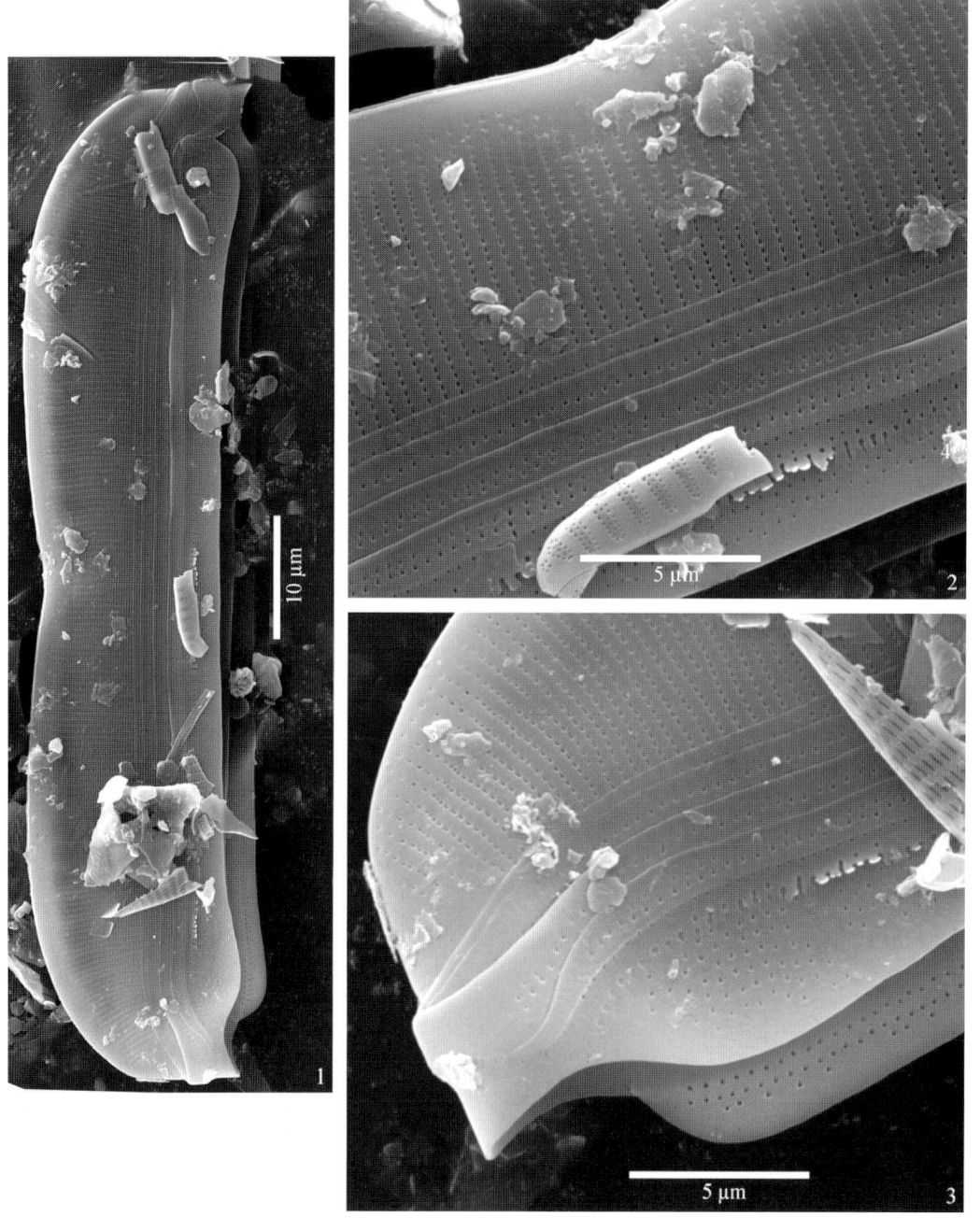

1—3. 凯特菱形藻 *Nitzschia kittlii* Grunow，1. 示外壳面，2. 示外壳面中部，孔纹形态及排列，3. 示外壳面顶端，孔纹及环带结构。所有图为电镜(SEM)照片，比例尺见图。

图版 XVIII

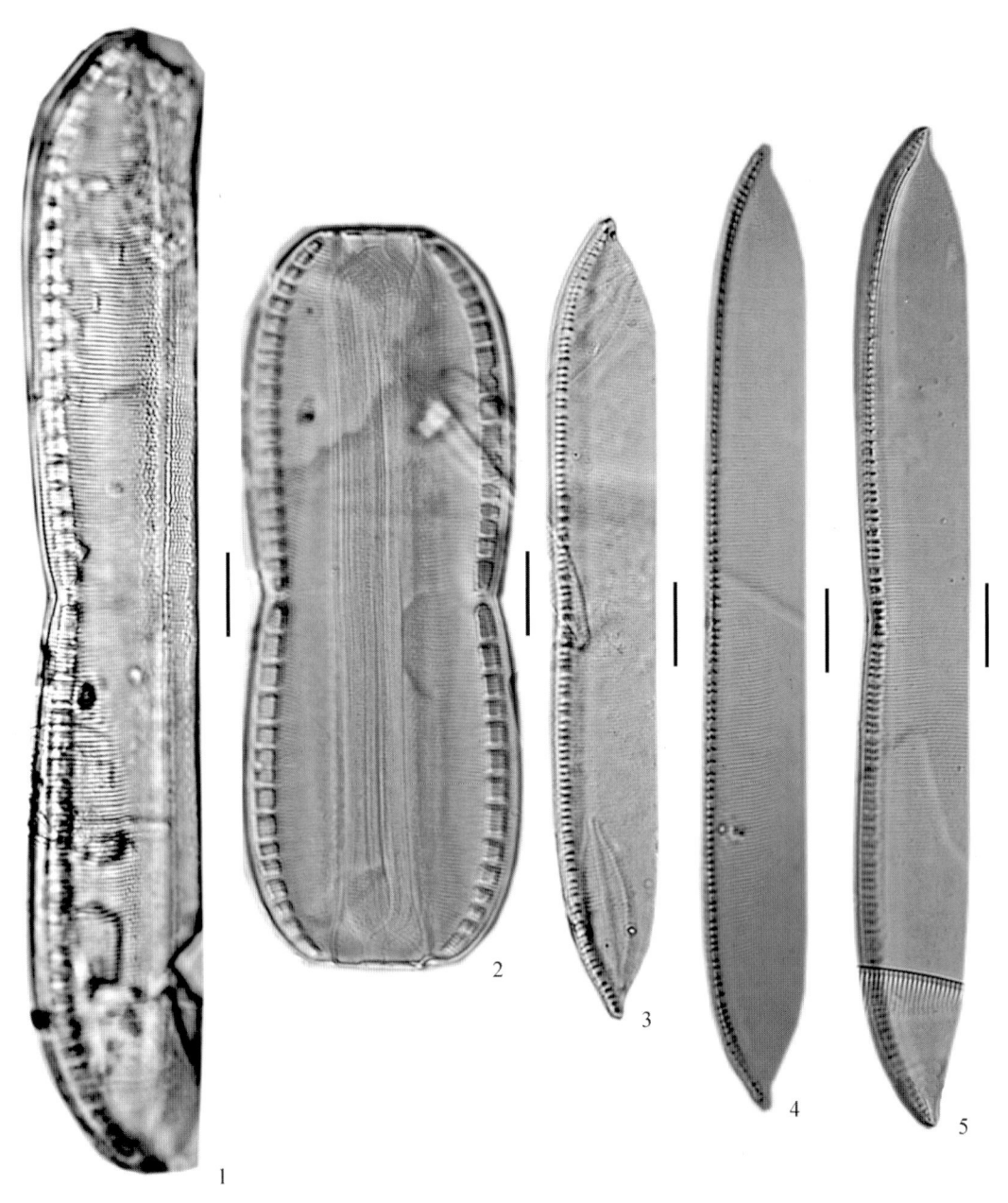

1—2. 凯特菱形藻 *Nitzschia kittlii* Grunow；3—5. 可疑菱形藻 *Nitzschia dubia* W. Smith。所有图为光镜（LM）照片，比例尺（Scale bars）=10 μm。

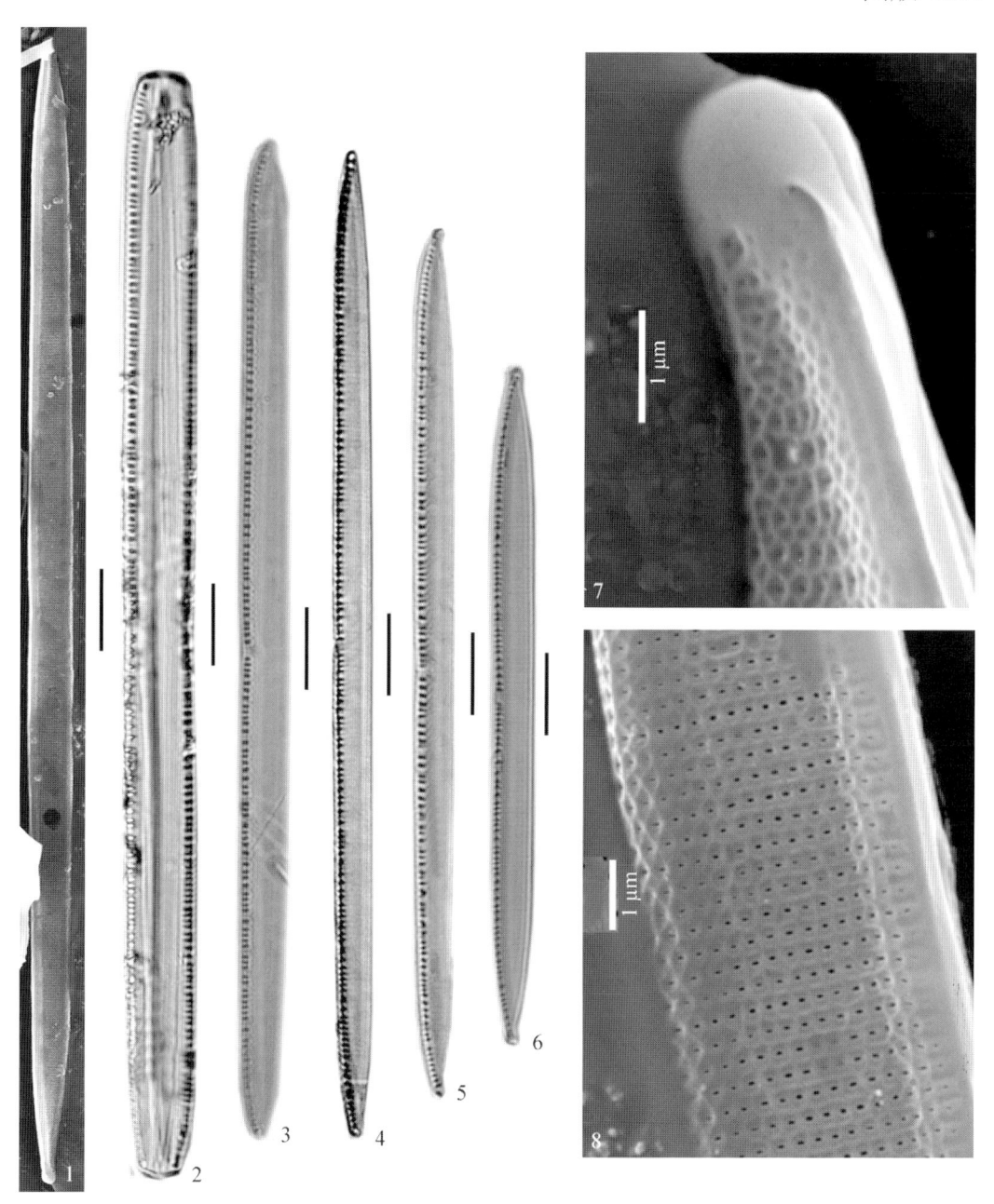

图版 XIX

1—8. 线形菱形藻 *Nitzschia linearis* W. Smith，1. 示外壳面(SEM)，7. 示外壳面顶端(SEM)，8. 示外壳面中部(SEM)，孔纹形态及排列。光镜(LM)照片，比例尺(Scale bars)=10 μm，电镜(SEM)照片比例尺见图。

图版 XX

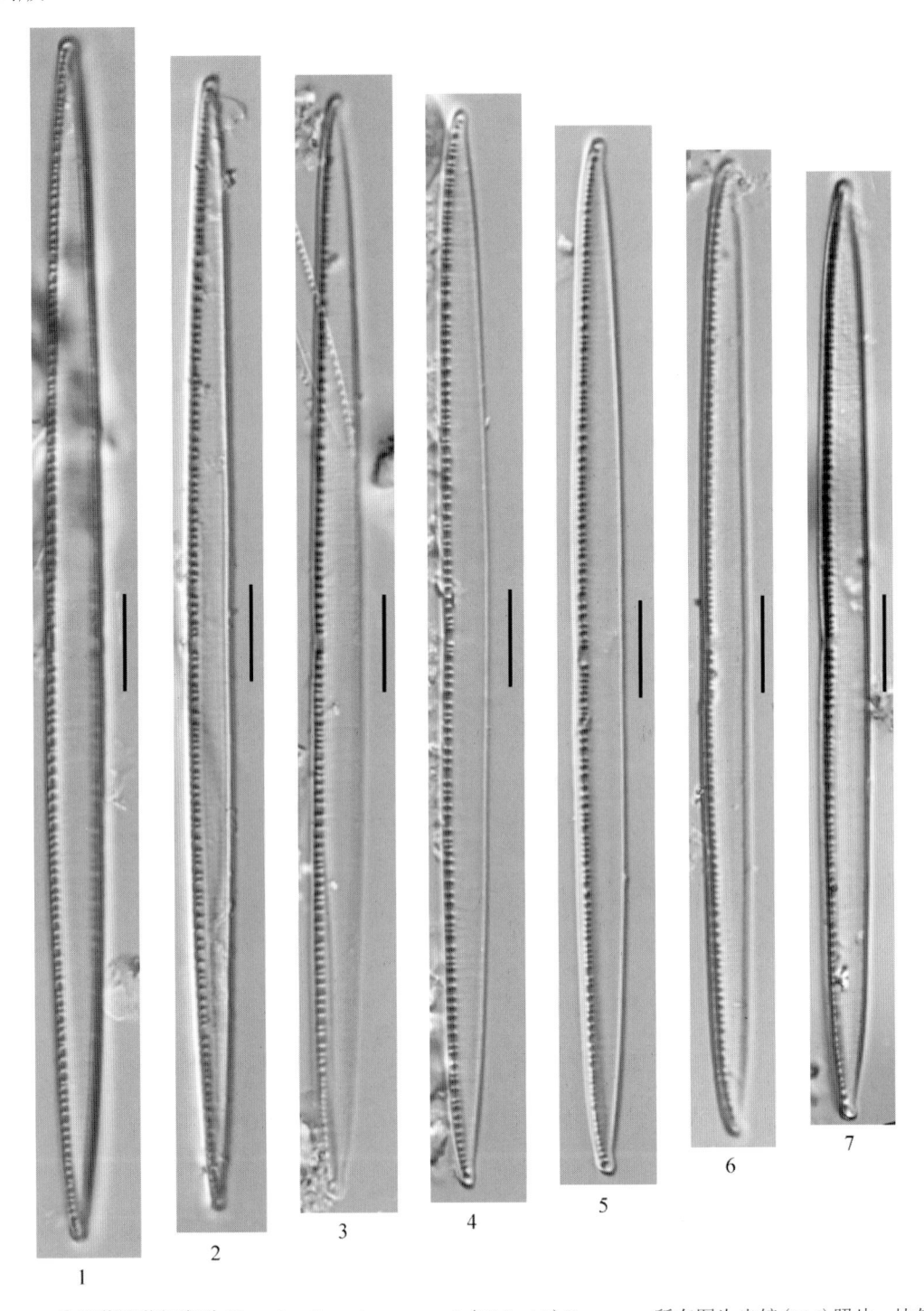

1—7. 线形菱形藻细变种 *Nitzschia linearis* var. *tenuis* (W. Smith) Grunow。所有图为光镜(LM)照片，比例尺(Scale bars)=10 μm。

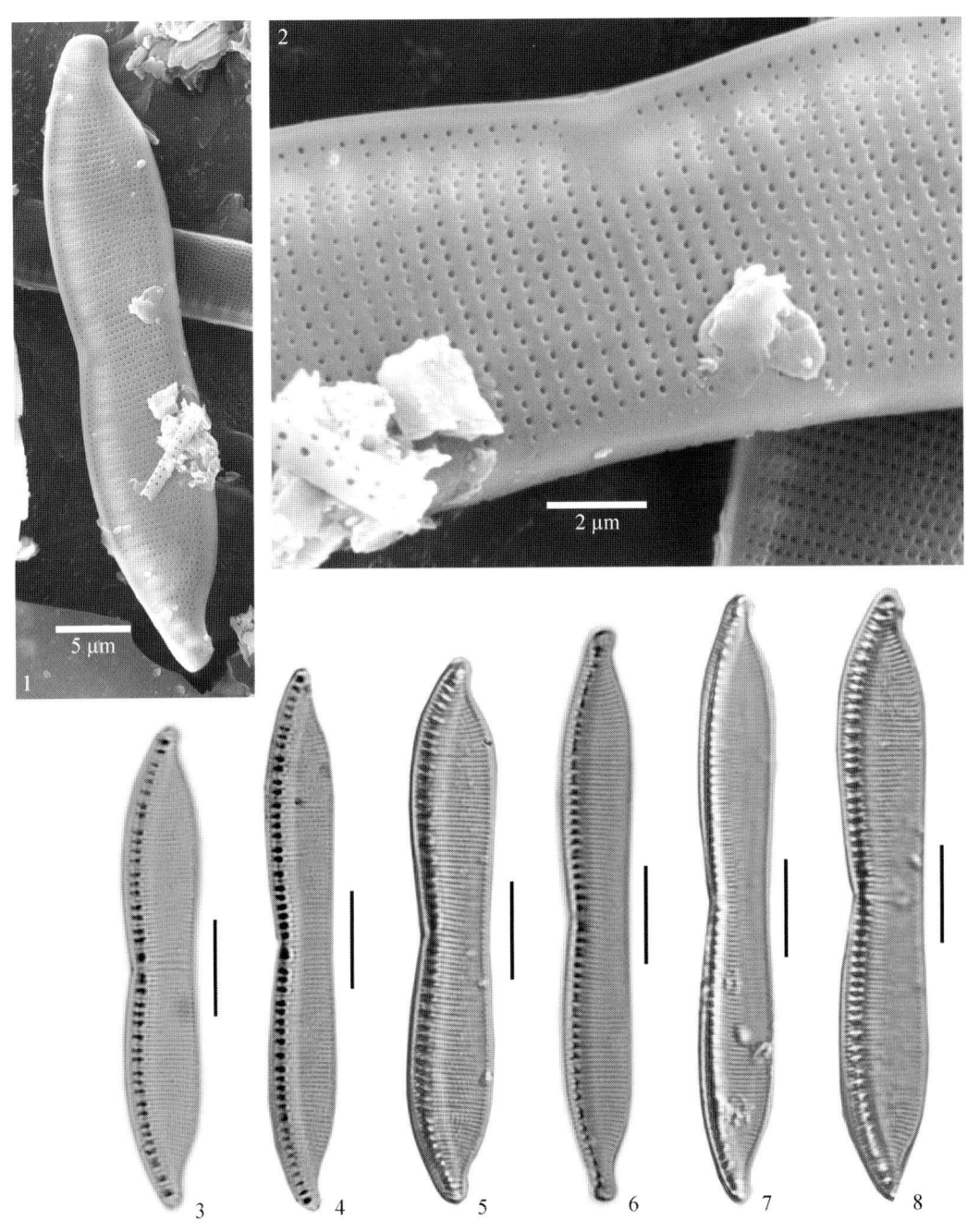

图版 XXI

1—8. 杂种菱形藻 *Nitzschia hybrida* Grunow，1. 示外壳面(SEM)，2. 示外壳面中部(SEM)，近缝端及孔纹形态。光镜(LM)照片，比例尺(Scale bars)=10 μm，电镜(SEM)照片比例尺见图。

图版 XXII

1—3. 玻璃质菱形藻 *Nitzschia vitrea* Norman，1. 示外壳面中部(SEM)，孔纹形态及排列，2. 示外壳面顶端(SEM)，孔纹及远缝端形态，3. 示外壳面(SEM)。所有图为电镜(SEM)照片，比例尺见图。

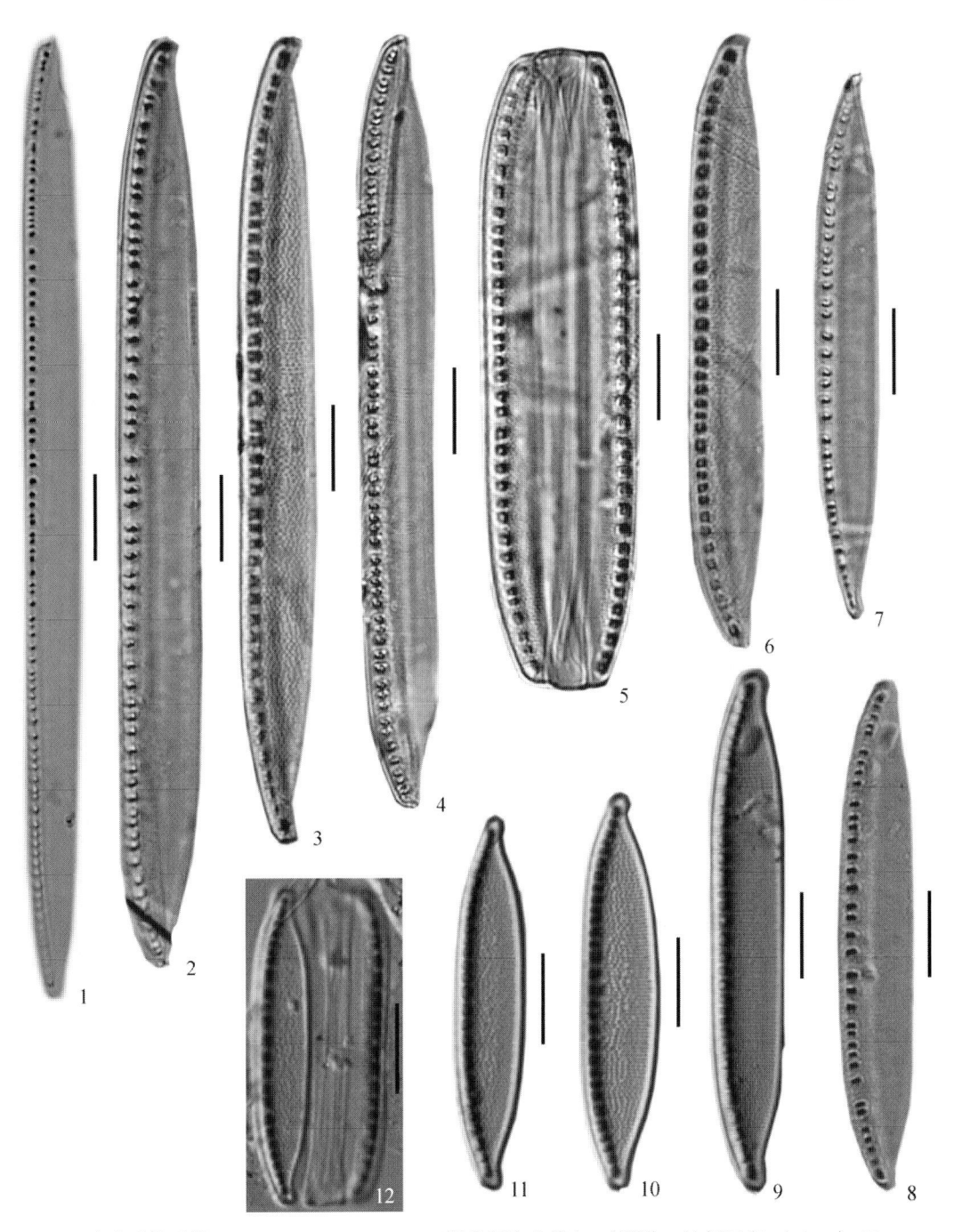

图版 XXIII

1—12. 玻璃质菱形藻 *Nitzschia vitrea* Norman。所有图为光镜(LM)照片，比例尺(Scale bars)=10 μm。

图版 XXIV

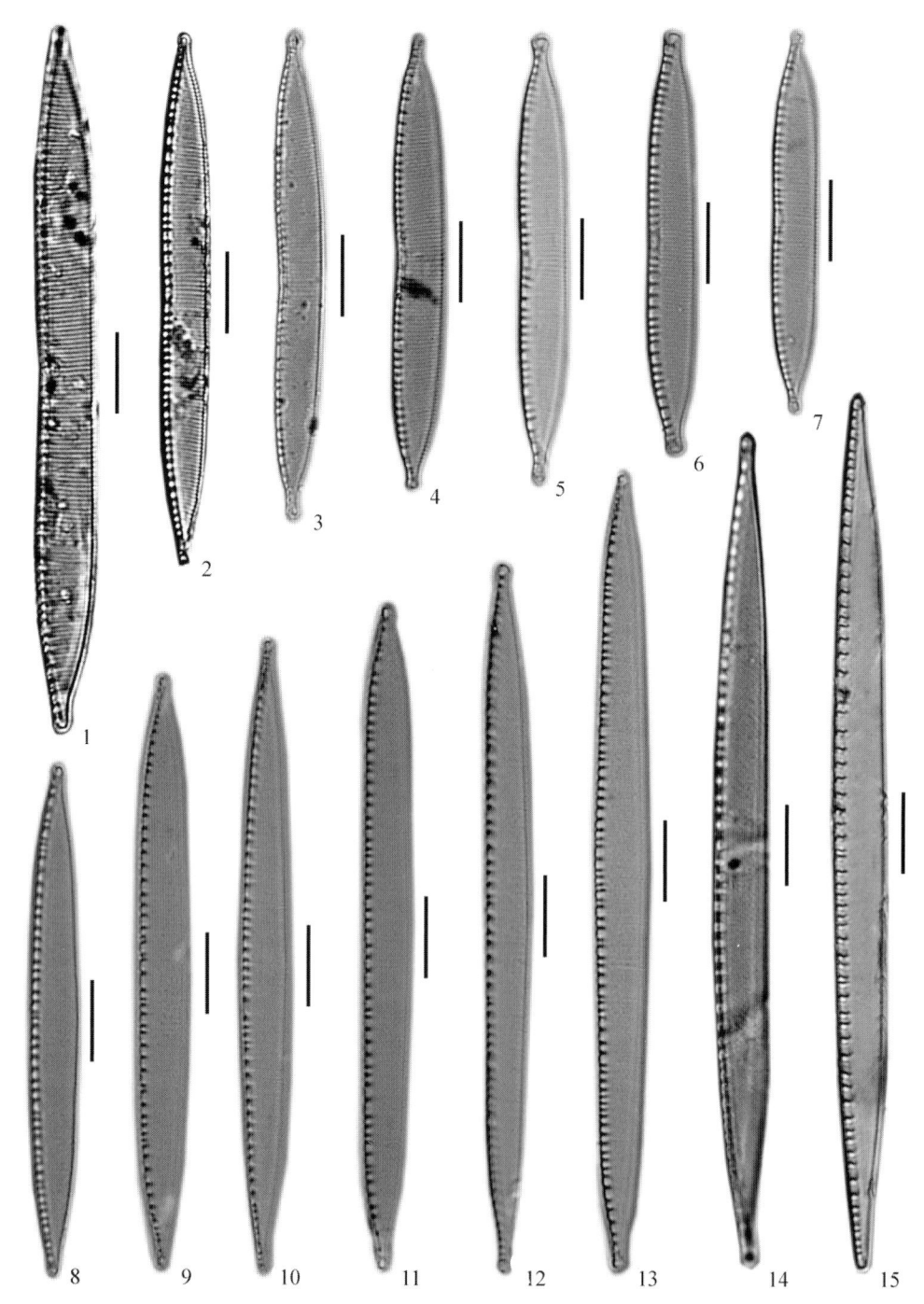

1—7. 吉塞拉菱形藻 *Nitzschia gisela* Lange-Bertalot；8—13. 近线形菱形藻 *Nitzschia sublinearis* Hustedt；14—15. 木那菱形藻 *Nitzschia monachorum* Lange-Bertalot。所有图为光镜（LM）照片，比例尺（Scale bars）=10 μm。

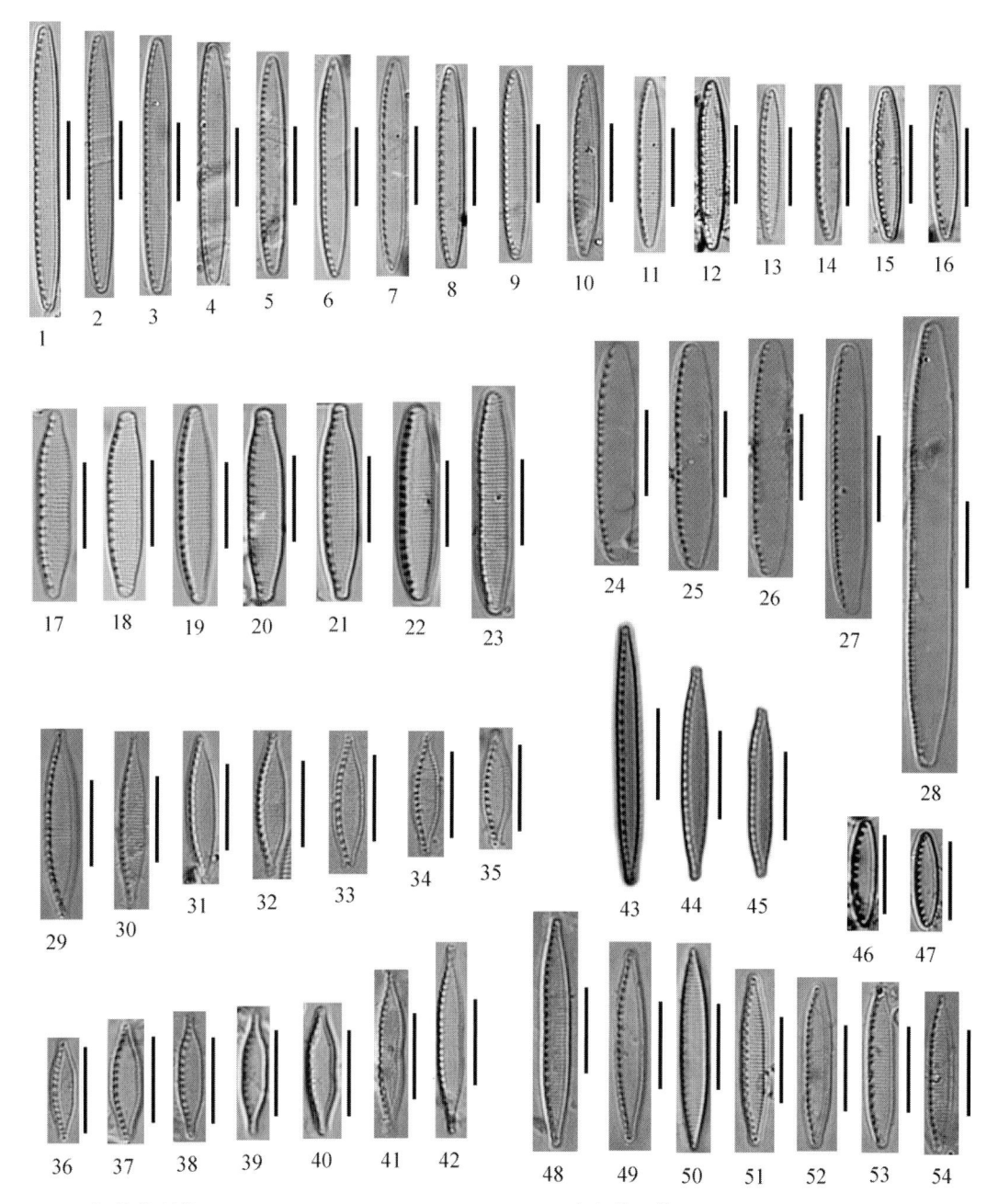

图版 XXV

1—16. 杆状菱形藻 *Nitzschia bacilliformis* Hustedt；17—23. 高山菱形藻 *Nitzschia alpina* Hustedt；24—28. 普通菱形藻 *Nitzschia communis* Rabenhorst；29—42. 华丽菱形藻 *Nitzschia elegantula* Grunow；43—45. 细微菱形藻 *Nitzschia perminuta* (Grunow) M. Peragallo；46—47. 平庸菱形藻 *Nitzschia inconspicua* Grunow；48—54. 小片菱形藻 *Nitzschia frustulum* (Kützing) Grunow。所有图为光镜(LM)照片，比例尺(Scale bars)=10 μm。

图版 XXVI

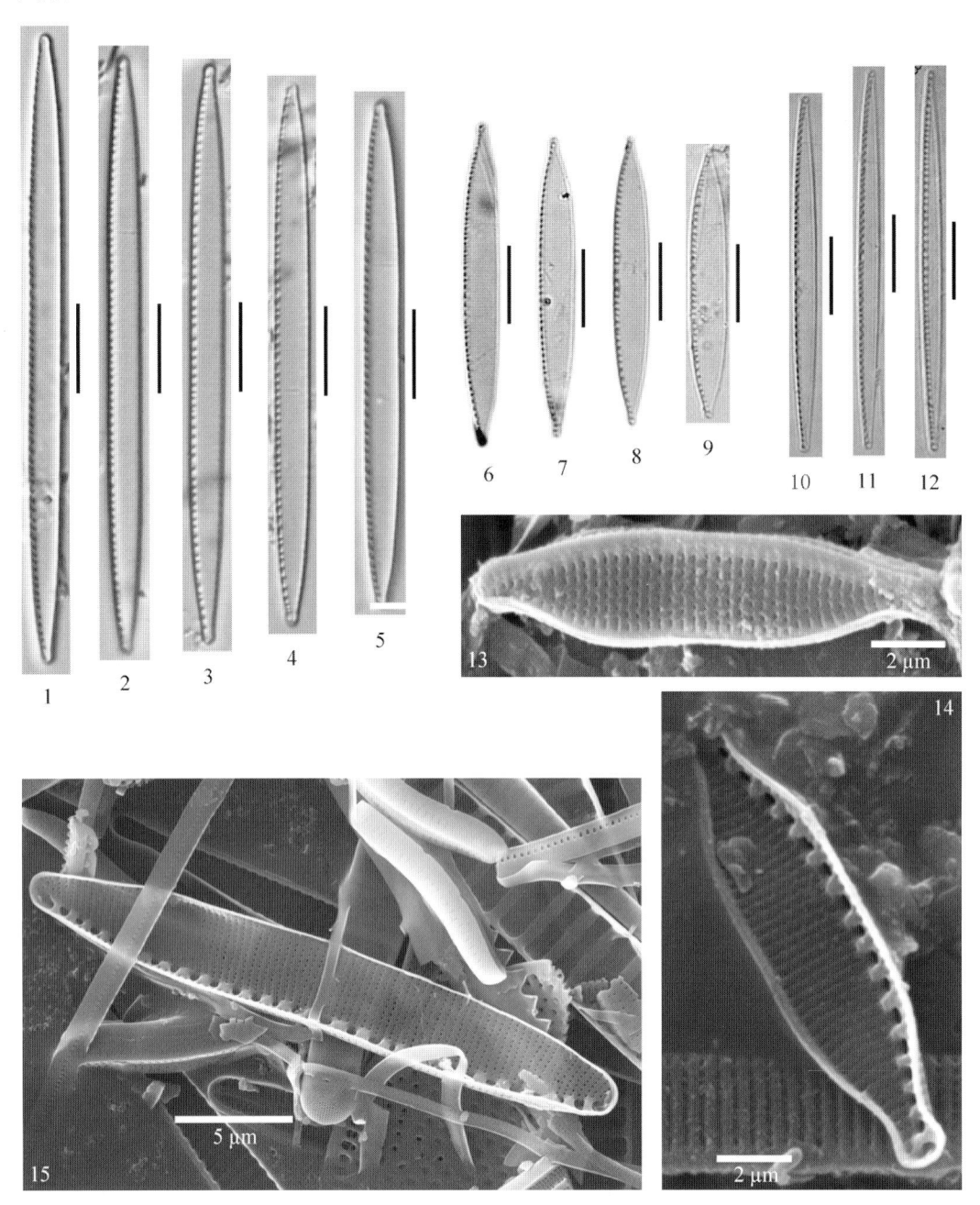

1—5. 灌木菱形藻 *Nitzschia fruticosa* Hustedt；6—9. 谷皮菱形藻微小变种 *Nitzschia palea* var. *minuta* (Bleisch) Grunow；10—12. 辐射菱形藻 *Nitzschia radicula* Hustedt；13—14. 小头菱形藻 *Nitzschia microcephala* Grunow，13. 示外壳面(SEM)，孔纹形态及排列，14. 示内壳面(SEM)，孔纹及龙骨突形态；15. 普通菱形藻 *Nitzschia communis* Rabenhorst，示内壳面(SEM)，孔纹和龙骨突形态。光镜(LM)照片，比例尺(Scale bars)=10 μm，电镜(SEM)照片比例尺见图。

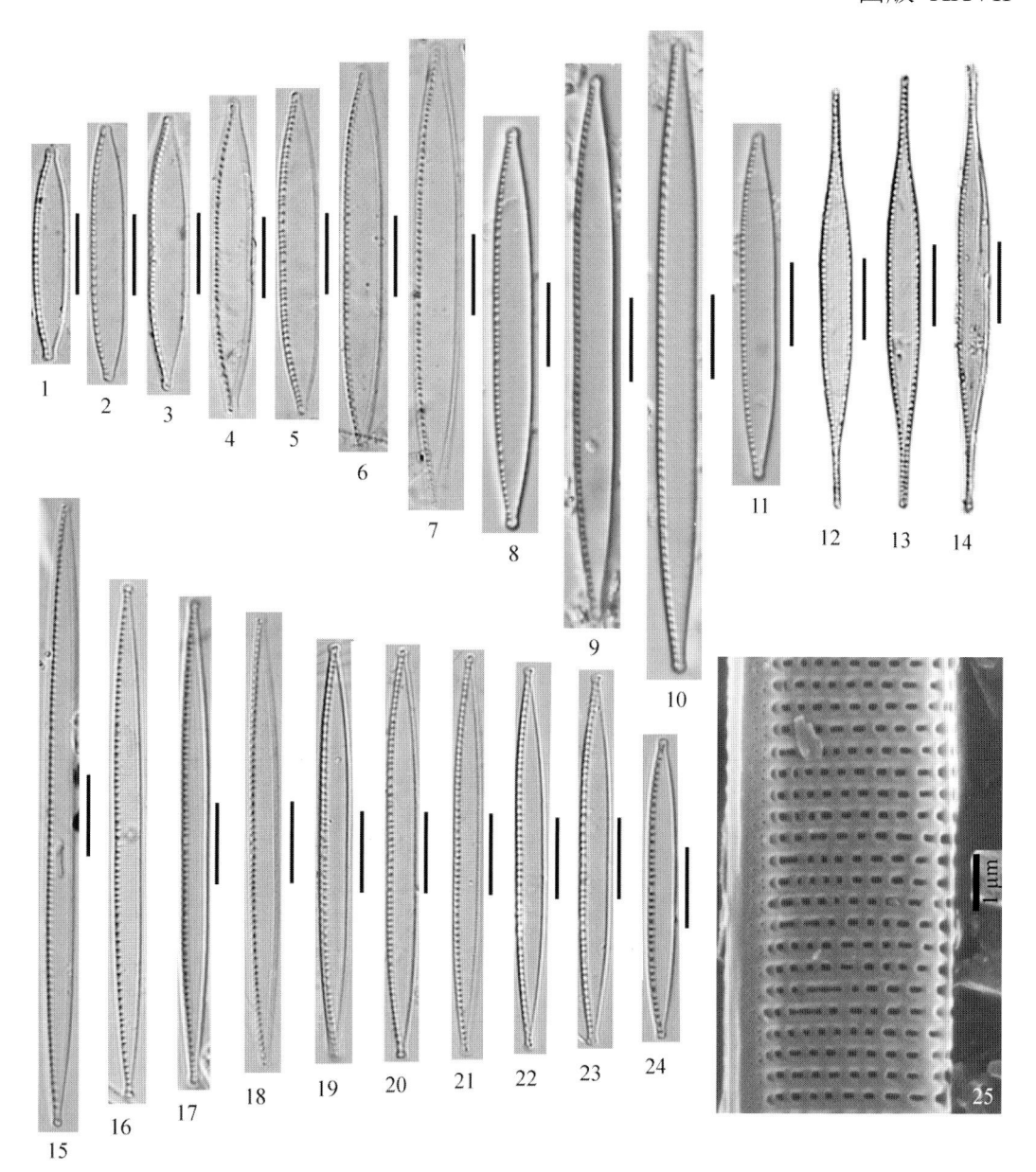

图版 XXVII

1—11. 谷皮菱形藻 *Nitzschia palea* (Kützing) W. Smith；12—14. 近针形菱形藻 *Nitzschia subacicularis* Hustedt；15—25. 多样菱形藻 *Nitzschia diversa* Hustedt，25. 示外壳面中部(SEM)，孔纹形态及排列。光镜(LM)照片，比例尺(Scale bars)=10 μm，电镜(SEM)照片比例尺见图。

图版 XXVIII

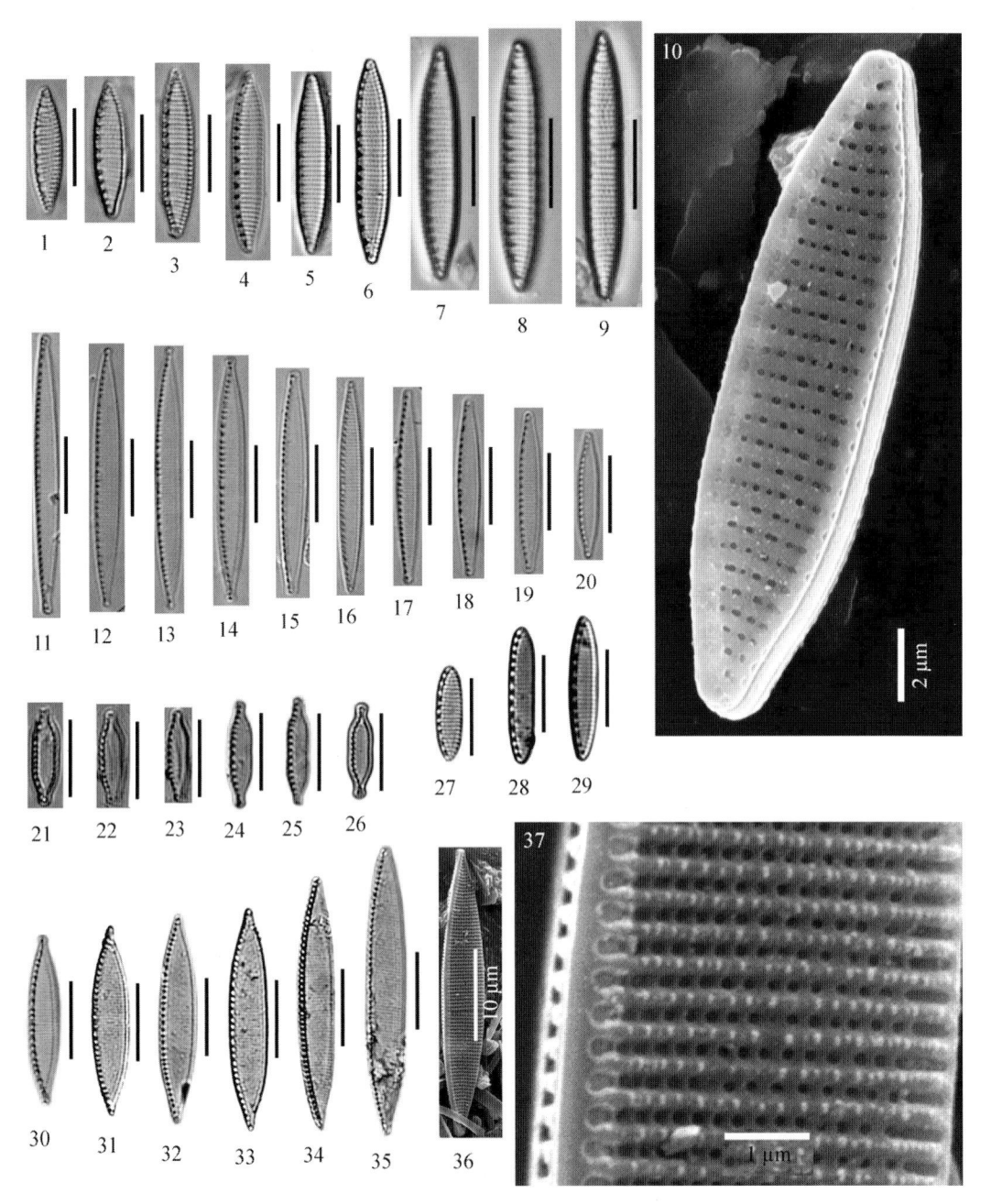

1—10. 两栖菱形藻 *Nitzschia amphibia* Grunow，10. 示外壳面（SEM）；11—20. 泉生菱形藻 *Nitzschia fonticola* (Grunow) Grunow；21—26. 小头菱形藻 *Nitzschia microcephala* Grunow；27. 粗条菱形藻 *Nitzschia valdestriata* Aleem & Hustedt；28—29. 粗肋菱形藻 *Nitzschia valdecostata* Lange-Bertalot & Simonsen；30—37. 常见菱形藻 *Nitzschia solita* Hustedt，36. 示外壳面（SEM），37. 示外壳面中部（SEM），孔纹形态及排列。光镜（LM）照片，比例尺（Scale bars）=10 μm，电镜（SEM）照片比例尺见图。

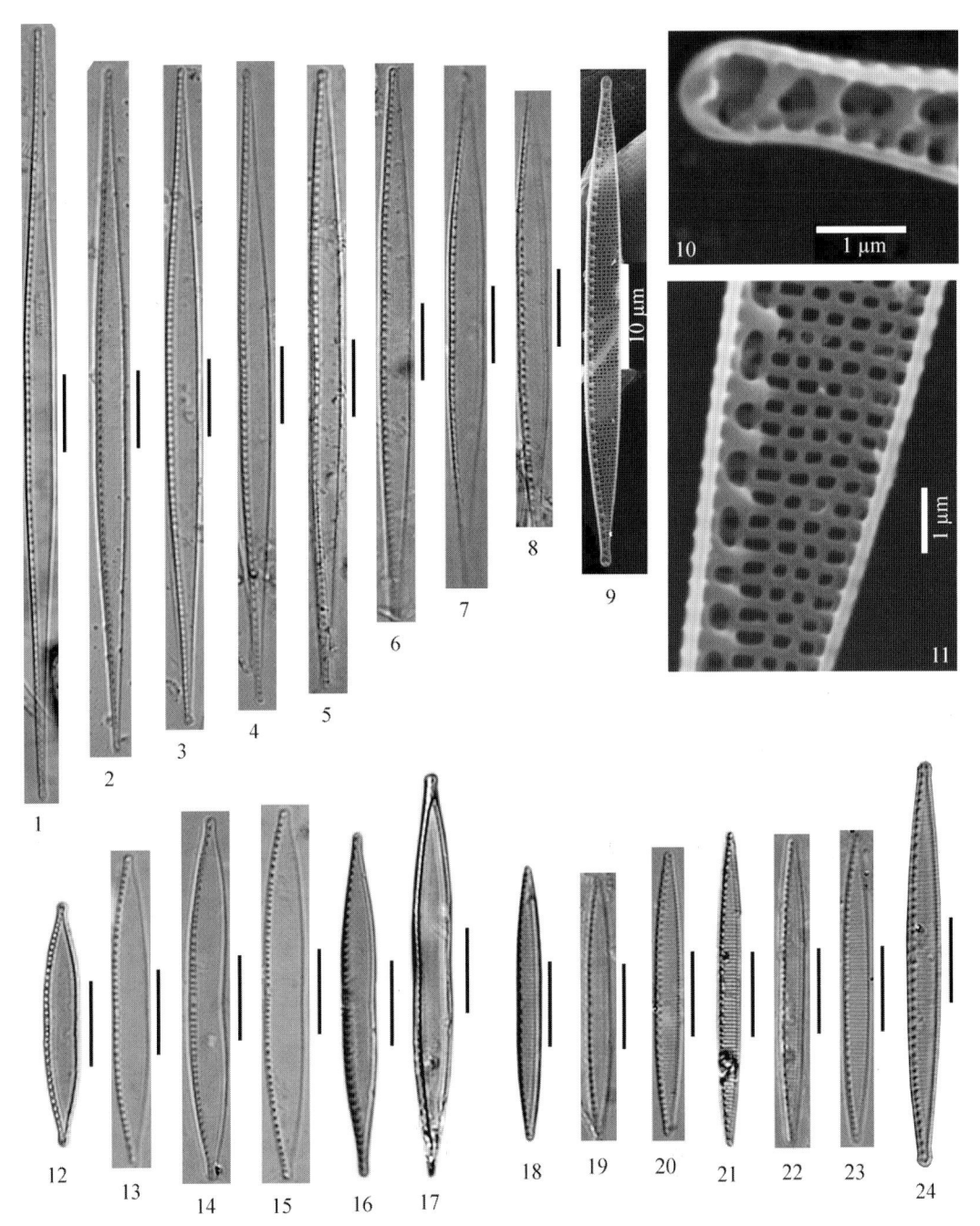

图版 XXIX

1—11. 细长菱形藻 *Nitzschia gracilis* Hantzsch，9. 示内壳面(SEM)，10. 示内壳面顶端(SEM)，螺旋舌结构，11. 示内壳面(SEM)，孔纹及龙骨突形态；12—17. 小头端菱形藻 *Nitzschia capitellata* Hustedt；18—24. 化石菱形藻 *Nitzschia fossilis* (Grunow) Grunow。光镜(LM)照片，比例尺(Scale bars)=10 μm，电镜(SEM)照片比例尺见图。

图版 XXX

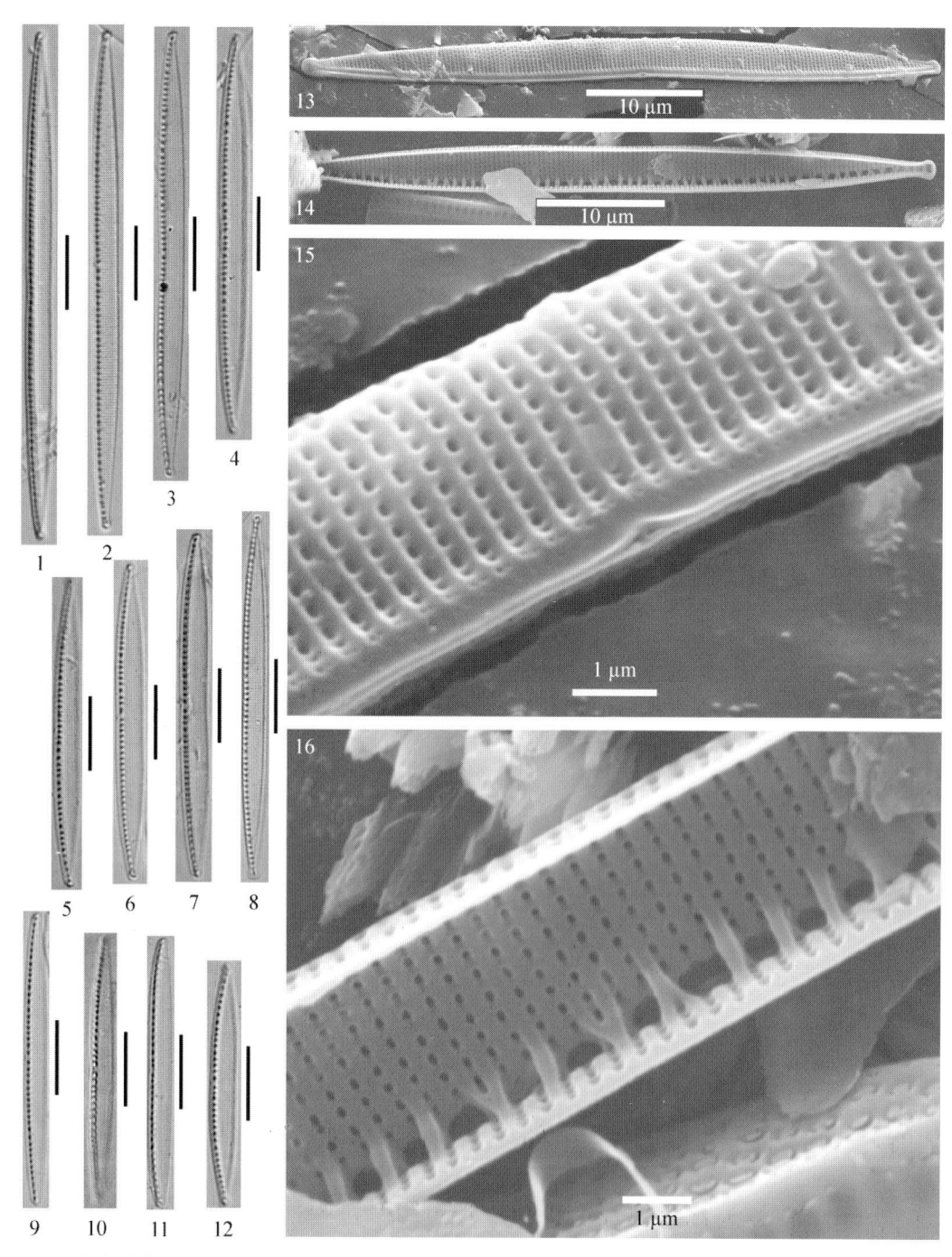

1—16. 吉斯纳菱形藻 *Nitzschia gessneri* Hustedt，13. 示外壳面(SEM)，14. 示内壳面(SEM)，15. 示外壳面中部(SEM)，孔纹及近缝端形态，16. 示内壳面(SEM)，孔纹及龙骨突形态。光镜(LM)照片，比例尺(Scale bars)=10 μm，电镜(SEM)照片比例尺见图。

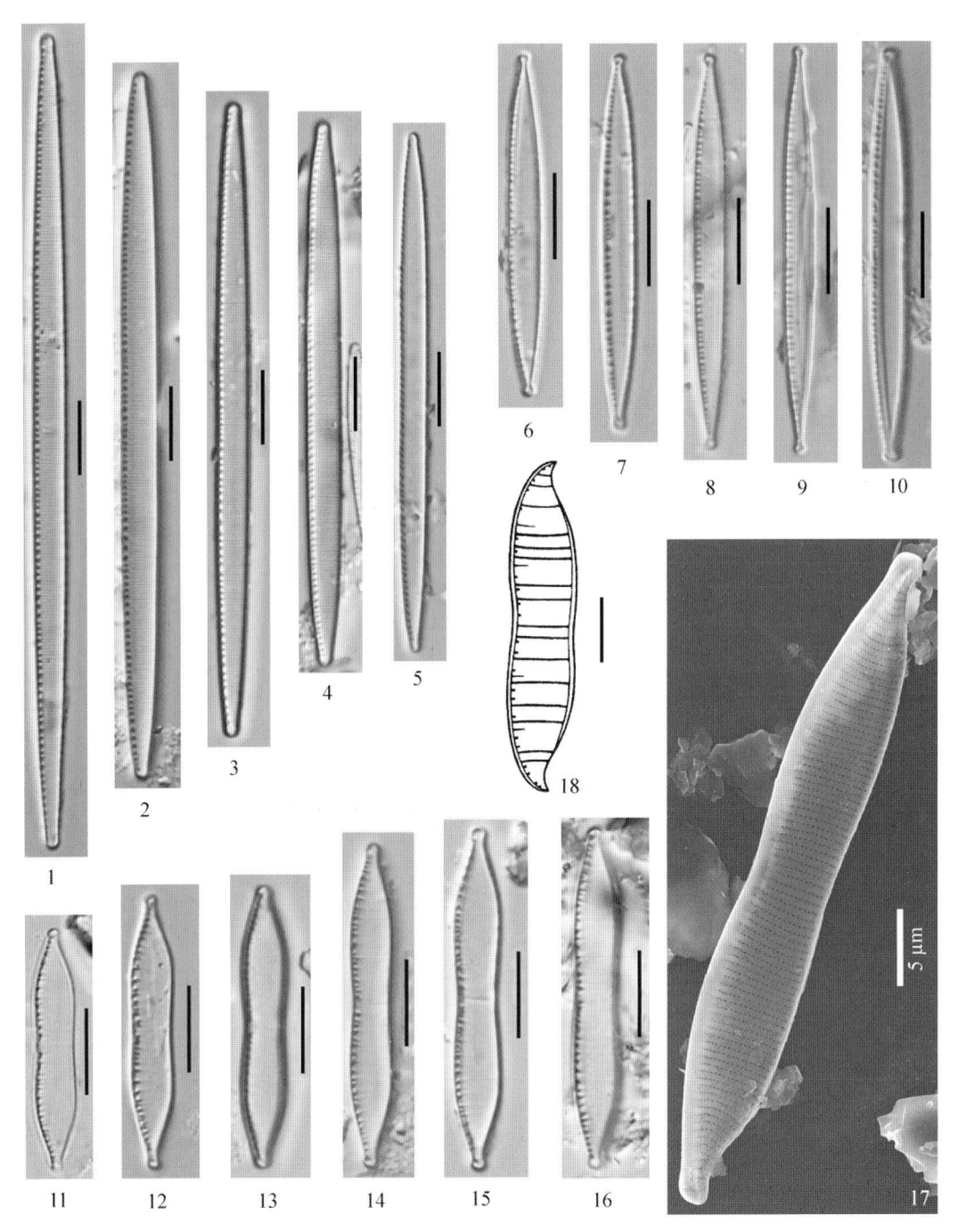

图版 XXXI

1—5. 中型菱形藻 *Nitzschia intermedia* Hantzsch ex Cleve & Grunow；6—10. 管毛菱形藻 *Nitzschia tubicola* Grunow；11—17. 毡帽菱形藻 *Nitzschia homburgiensis* Lange-Bertalot，17. 示外壳面（SEM）；18. 类附生菱形藻 *Nitzschia epithemoides* Grunow，手绘图，引自 Zhu 和 Chen（2000）。光镜（LM）照片，比例尺（Scale bars）=10 μm，电镜（SEM）照片比例尺见图。

图版 XXXII

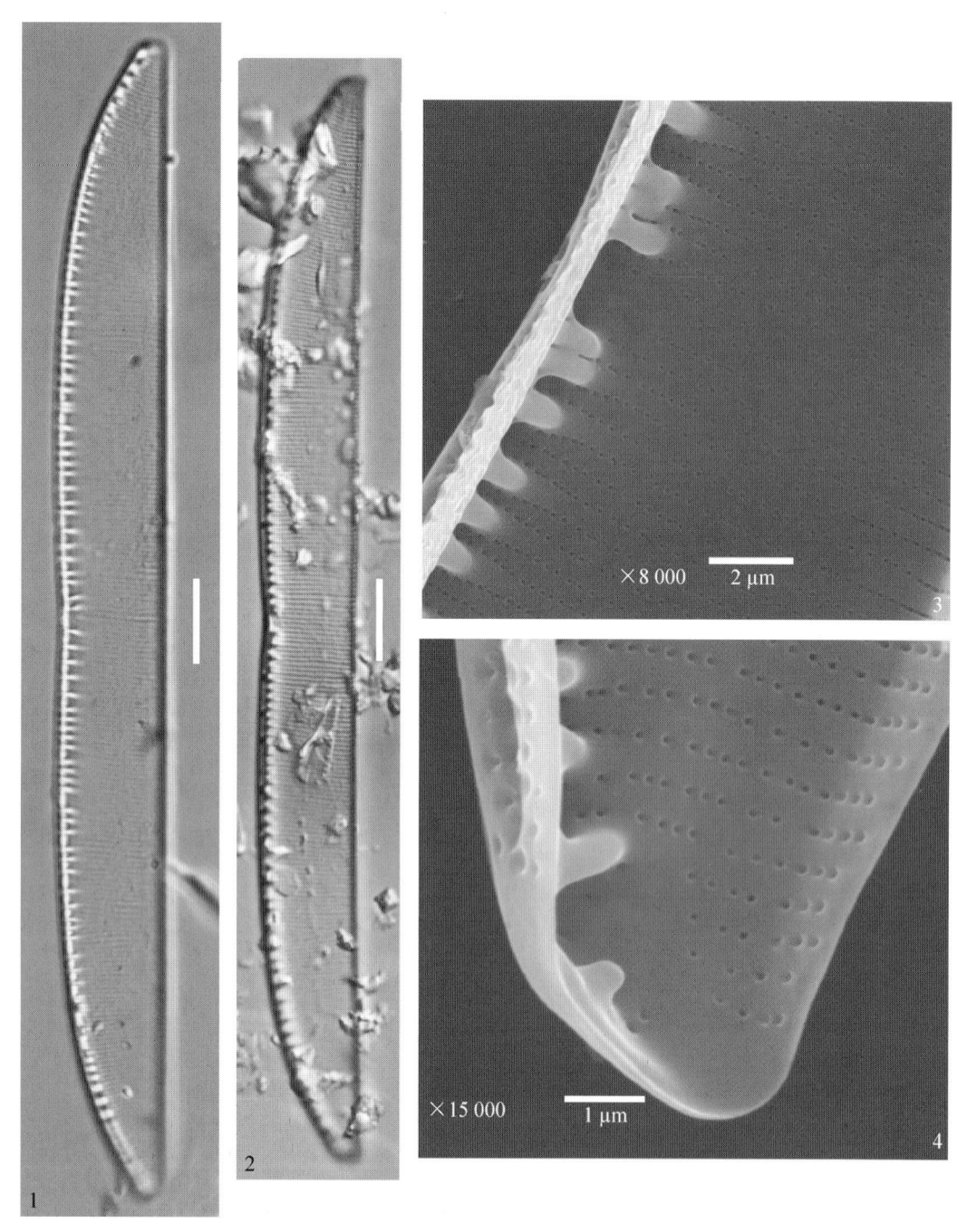

1—4. 瓜德罗普菱形藻 *Nitzschia guadalupensis* Manguin，3. 示内壳面(SEM)，线纹及龙骨突形态，4. 示内壳面顶端(SEM)，螺旋舌结构。光镜(LM)照片，比例尺(Scale bars)=10 μm，电镜(SEM)照片比例尺见图。

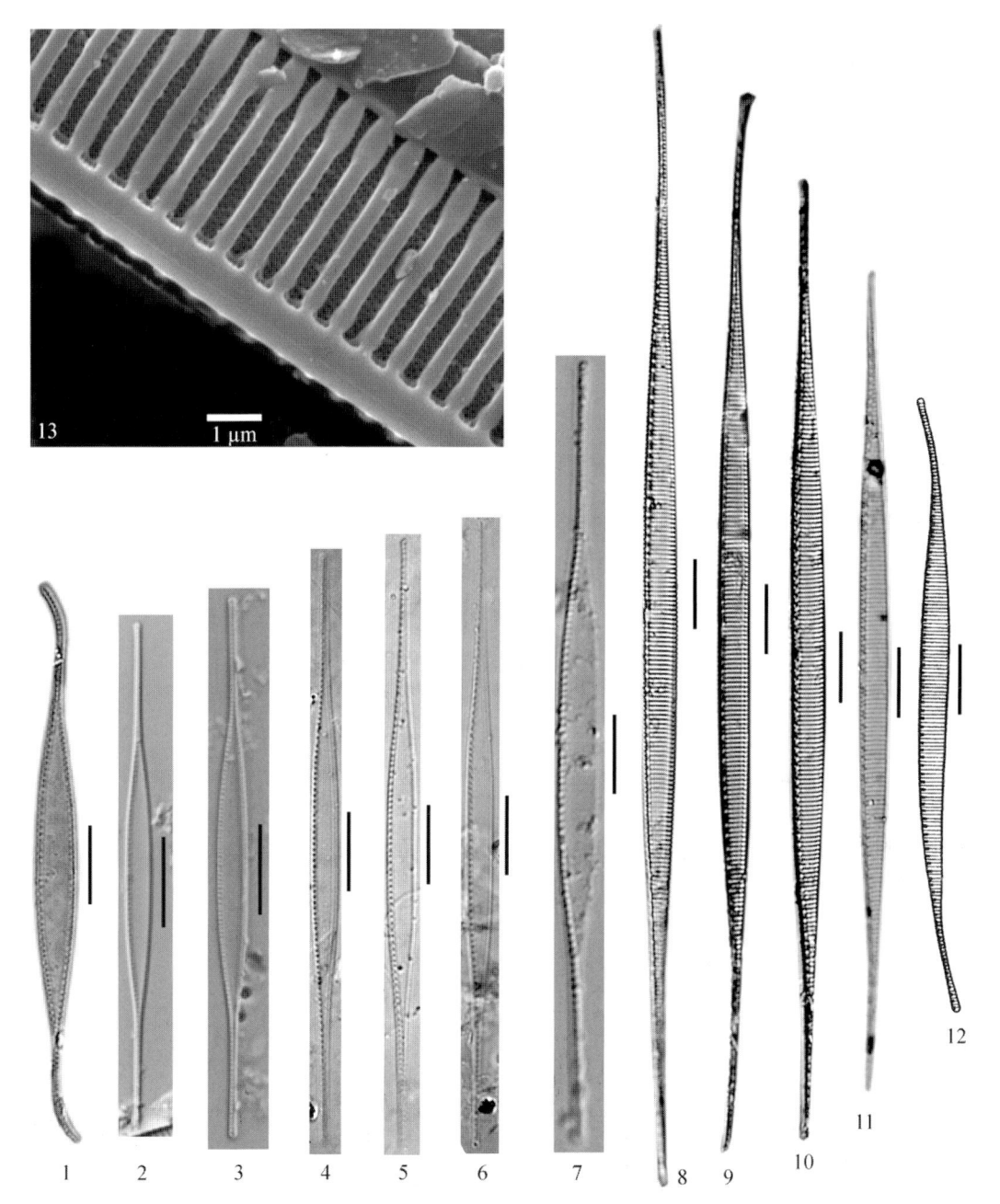

图版 XXXIII

1. 反曲菱形藻 *Nitzschia reversa* W. Smith；2—3.(?) 爪维兰斯菱形藻 *Nitzschia draveillensis* Coste & Ricard；4—6. 针形菱形藻 *Nitzschia acicularis* (Kützing) W. Smith；7. 爪维兰斯菱形藻 *Nitzschia draveillensis* Coste & Ricard；8—11, 13. 洛伦菱形藻 *Nitzschia lorenziana* Grunow, 13. 示外壳面(SEM)，孔纹及肋纹形态, 12. 洛伦菱形藻细弱变种 *Nitzschia lorenziana* var. *subtilis* Grunow, 手绘图，引自 Zhu 和 Chen (2000)。光镜(LM)照片，比例尺(Scale bars)=10 μm，电镜(SEM)照片比例尺见图。

图版 XXXIV

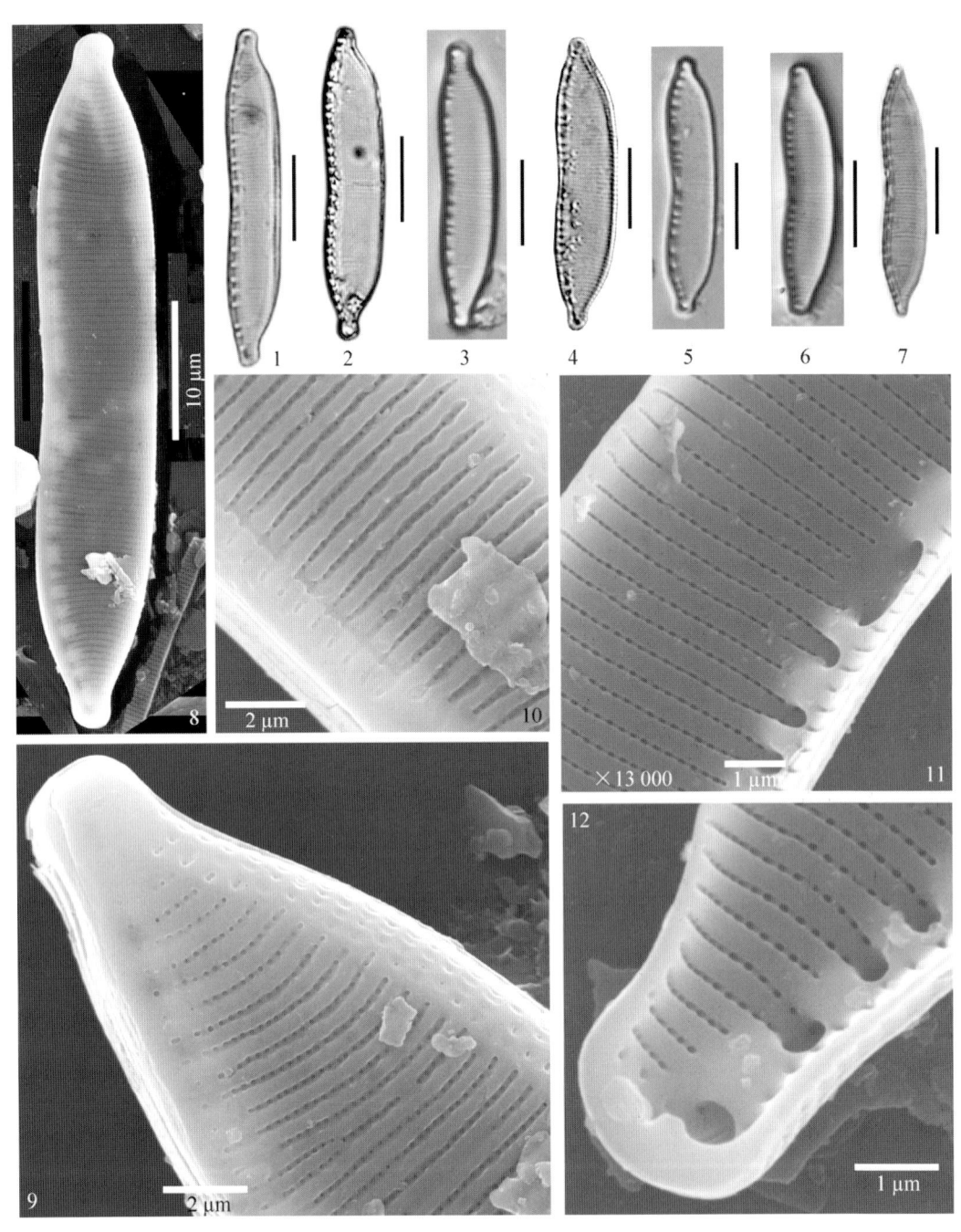

1—12. 两尖菱板藻（原变种）*Hantzschia amphioxys* var. *amphioxys* (Ehrenberg) Grunow，8. 示外壳面 (SEM)，9. 示外壳面顶端(SEM)，孔纹及顶端壳缝，10. 示外壳面中部(SEM)，孔纹形态，11. 示内壳面中部(SEM)，孔纹和近缝端形态，12. 示内壳面顶端(SEM)，龙骨突和螺旋舌形态。光镜(LM)照片，比例尺(Scale bars)=10 μm，电镜(SEM)照片比例尺见图。

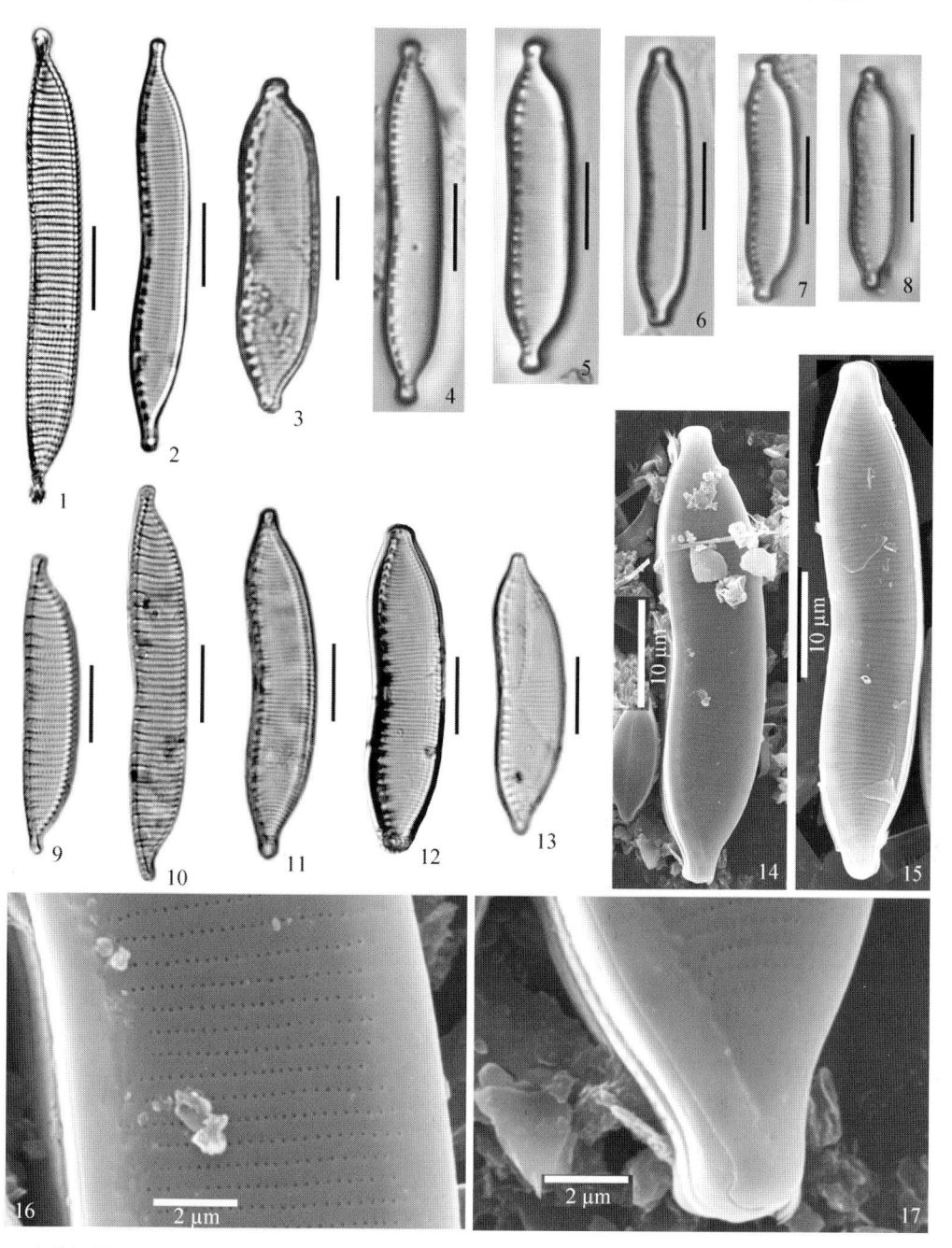

图版 XXXV

1. 显点菱板藻 *Hantzschia distinctepunctata* Hustedt；2. 两尖菱板藻头端变型 *Hantzschia amphioxys* f. *capitata* O. Müller；3—8. 两尖菱板藻相等变种 *Hantzschia amphioxys* var. *aequalis* Cleve-Euler；9—10. 直菱板藻 *Hantzschia virgata* (Roper) Grunow；11—17. 仿密集菱板藻 *Hantzschia paracompacta* Lange-Bertalot，14—15. 示外壳面，16. 示外壳面中部(SEM)，孔纹形态，17. 示外壳面顶端(SEM)，孔纹及顶端壳缝。光镜(LM)照片，比例尺(Scale bars)=10 μm，电镜(SEM)照片比例尺见图。

图版 XXXVI

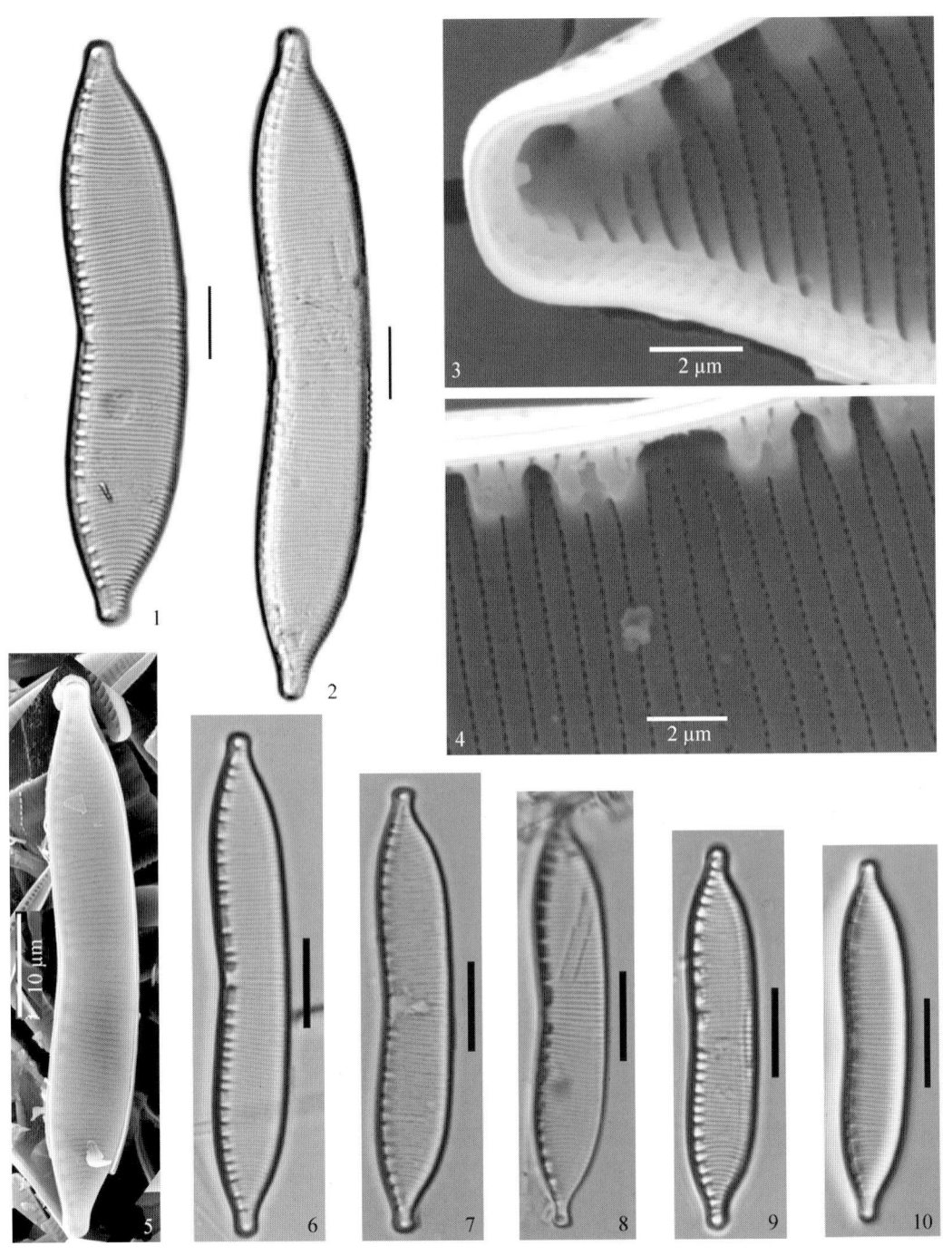

1—4. 密集菱板藻 *Hantzschia compacta* (Hustedt) Lange-Bertalot，3. 示内壳面顶端（SEM），孔纹和螺旋舌形态，4. 示内壳面中部（SEM），龙骨突和孔纹形态；5—10. 丰富菱板藻 *Hantzschia abundans* Lange-Bertalot，5. 示外壳面（SEM）。光镜（LM）照片，比例尺（Scale bars）=10 μm，电镜（SEM）照片比例尺见图。

图版 XXXVII

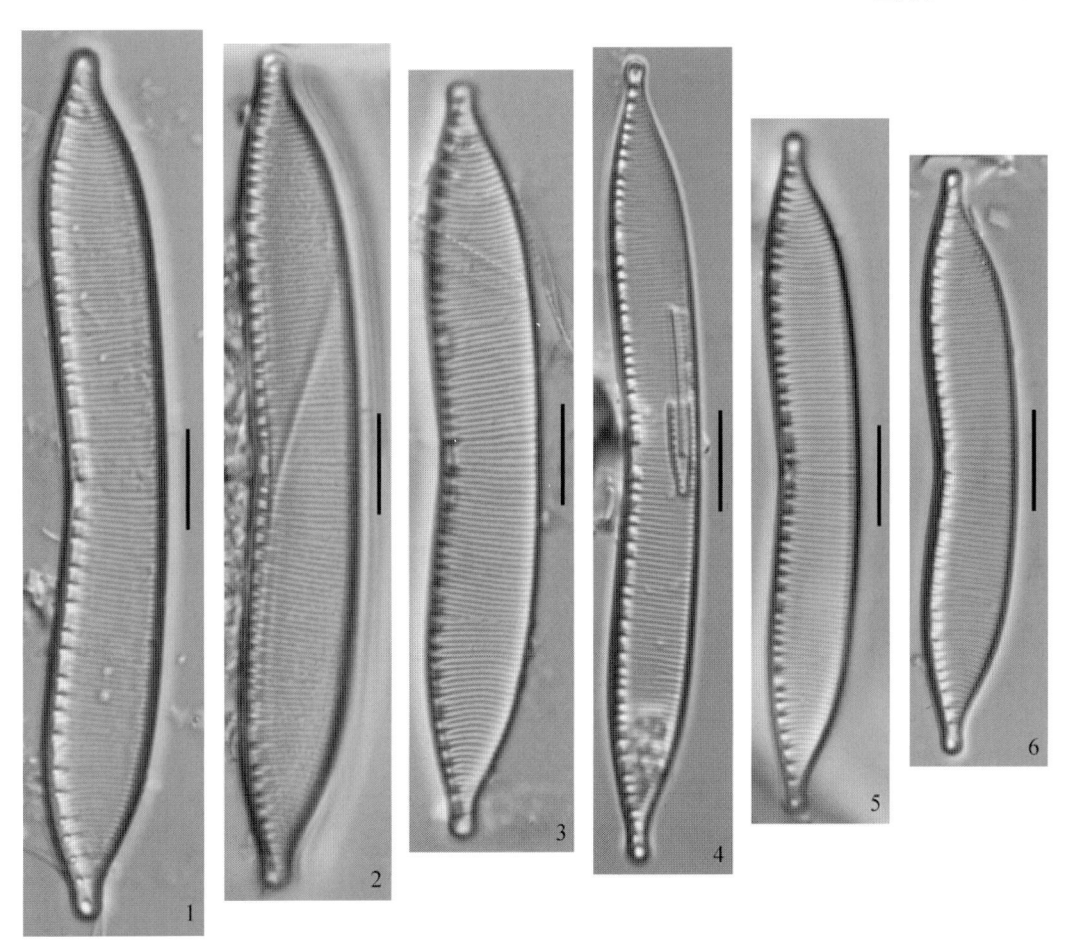

1—3. 近石生菱板藻 *Hantzschia subrupestris* Lange-Bertalot；4—6. 丰富菱板藻 *Hantzschia abundans* Lange-Bertalot。所有图为光镜(LM)照片，比例尺(Scale bars)=10 μm。

图版 XXXVIII

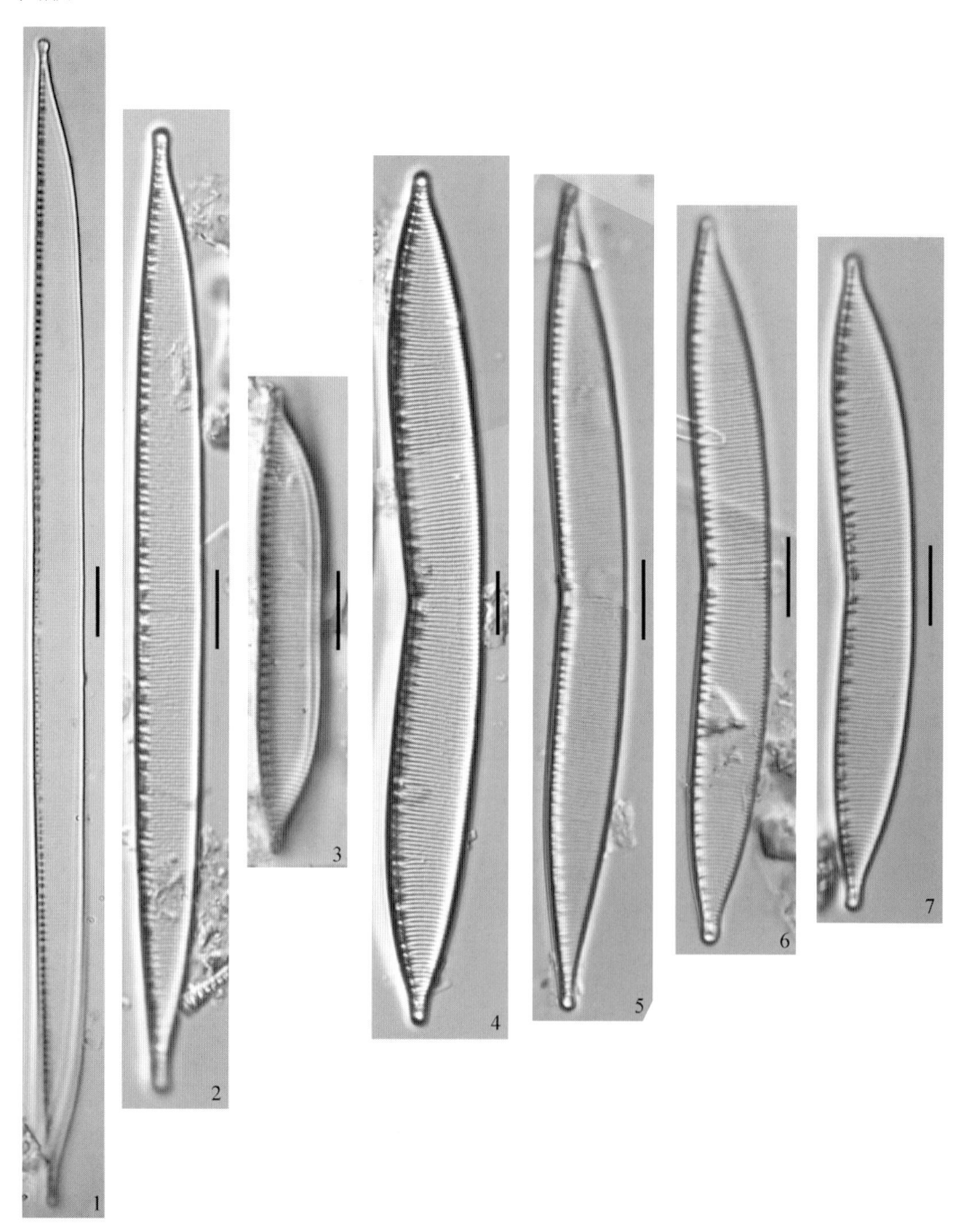

1—3. 长命菱板藻 Hantzschia vivax (W. Smith) M. Peragallo；4—7. 活跃菱板藻 Hantzschia vivacior Lange-Bertalot。所有图为光镜（LM）照片，比例尺（Scale bars）=10 μm。

图版 XXXIX

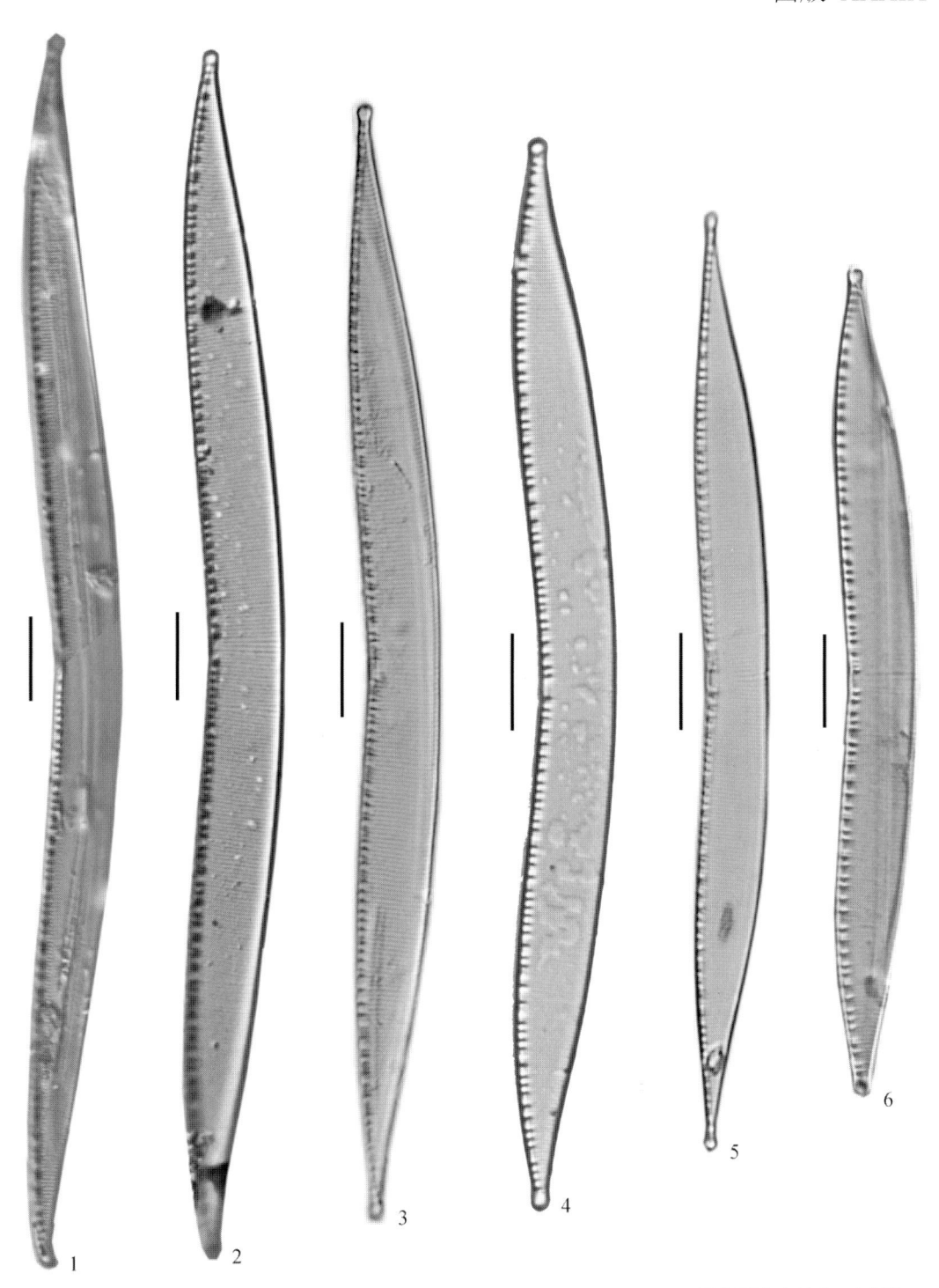

1—6. 巴克豪森菱板藻 *Hantzschia barckhausenii* Lange-Bertalot & Metzeltin。所有图为光镜(LM)照片，比例尺(Scale bars)=10 μm。

图版 XL

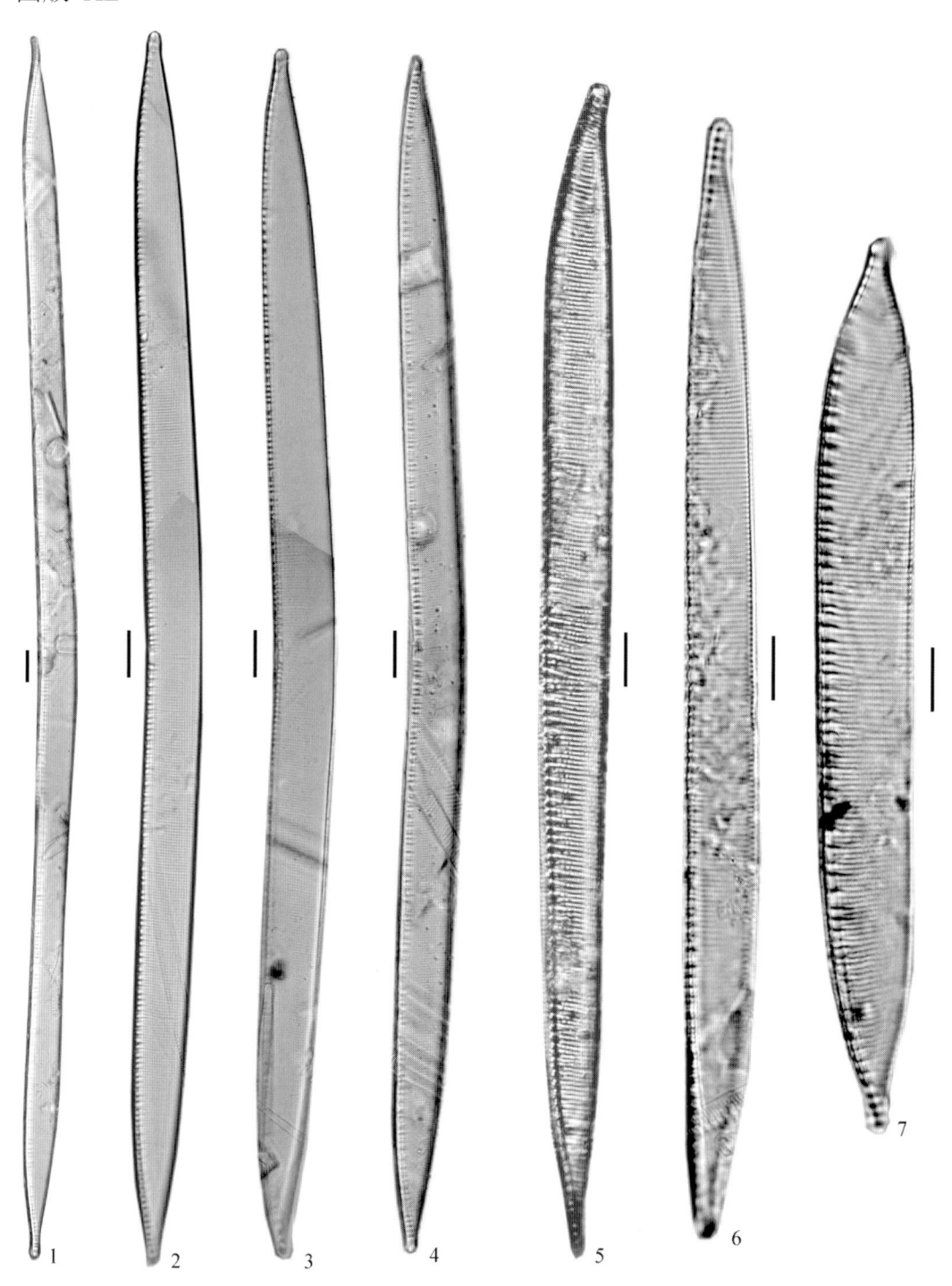

1. 长菱板藻 *Hantzschia elongata* (Hantzsch) Grunow；2—4：较长菱板藻 *Hantzschia longa* Lange-Bertalot；5—7：美丽菱板藻 *Hantzschia spectabilis* (Ehrenberg) Hustedt。所有图为光镜（LM）照片，比例尺（Scale bars）=10 μm。

图版 XLI

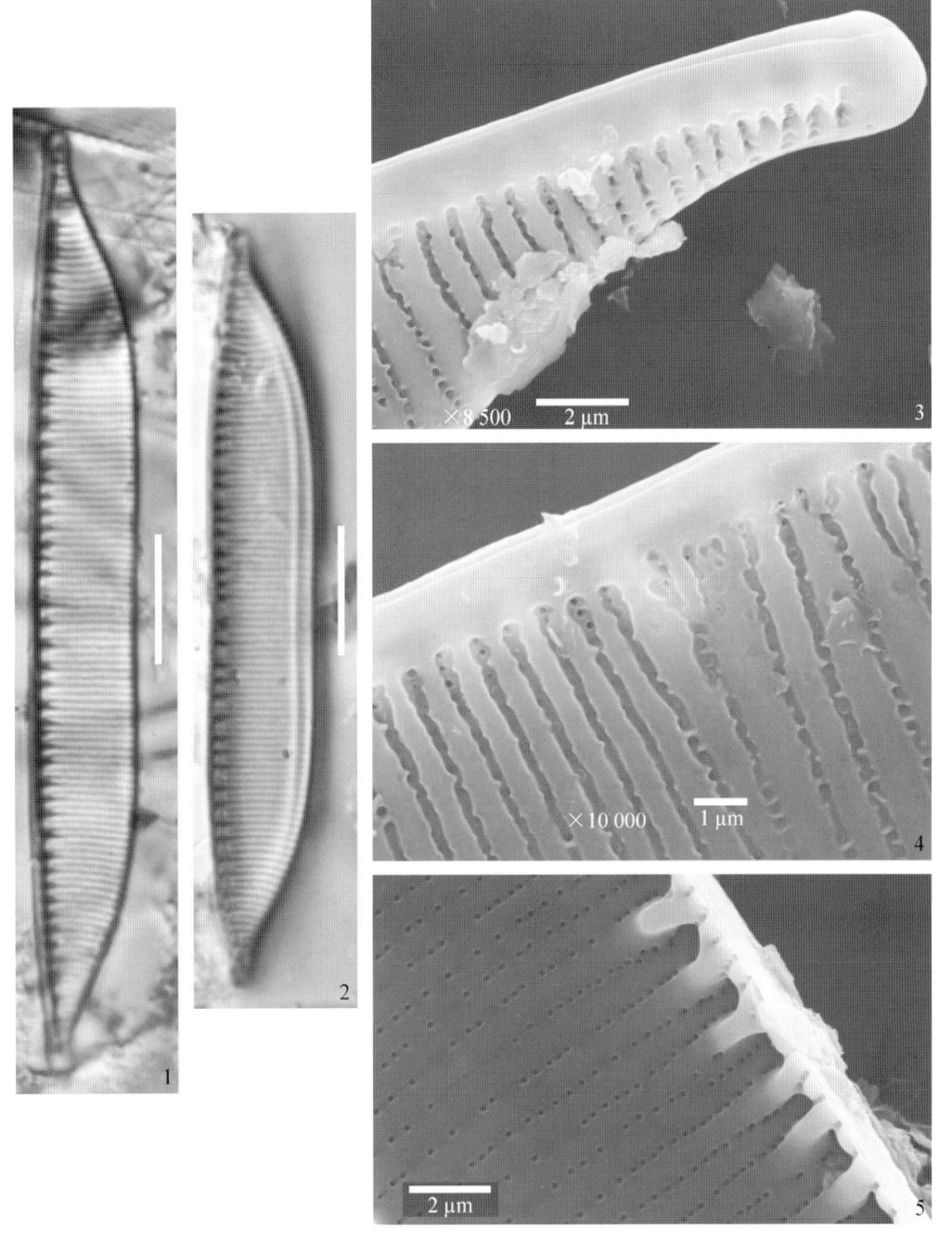

1—5. 盖斯纳菱板藻 *Hantzschia giessiana* Lange-Bertalot & Rumrich，3. 示外壳面顶端，孔纹和壳缝形态（SEM），4. 示外壳面中部，孔纹和近缝端形态（SEM），5. 示内壳面中部，龙骨突和孔纹形态（SEM）。光镜（LM）照片，比例尺（Scale bars）=10 μm，电镜（SEM）照片比例尺见图。

图版 XLII

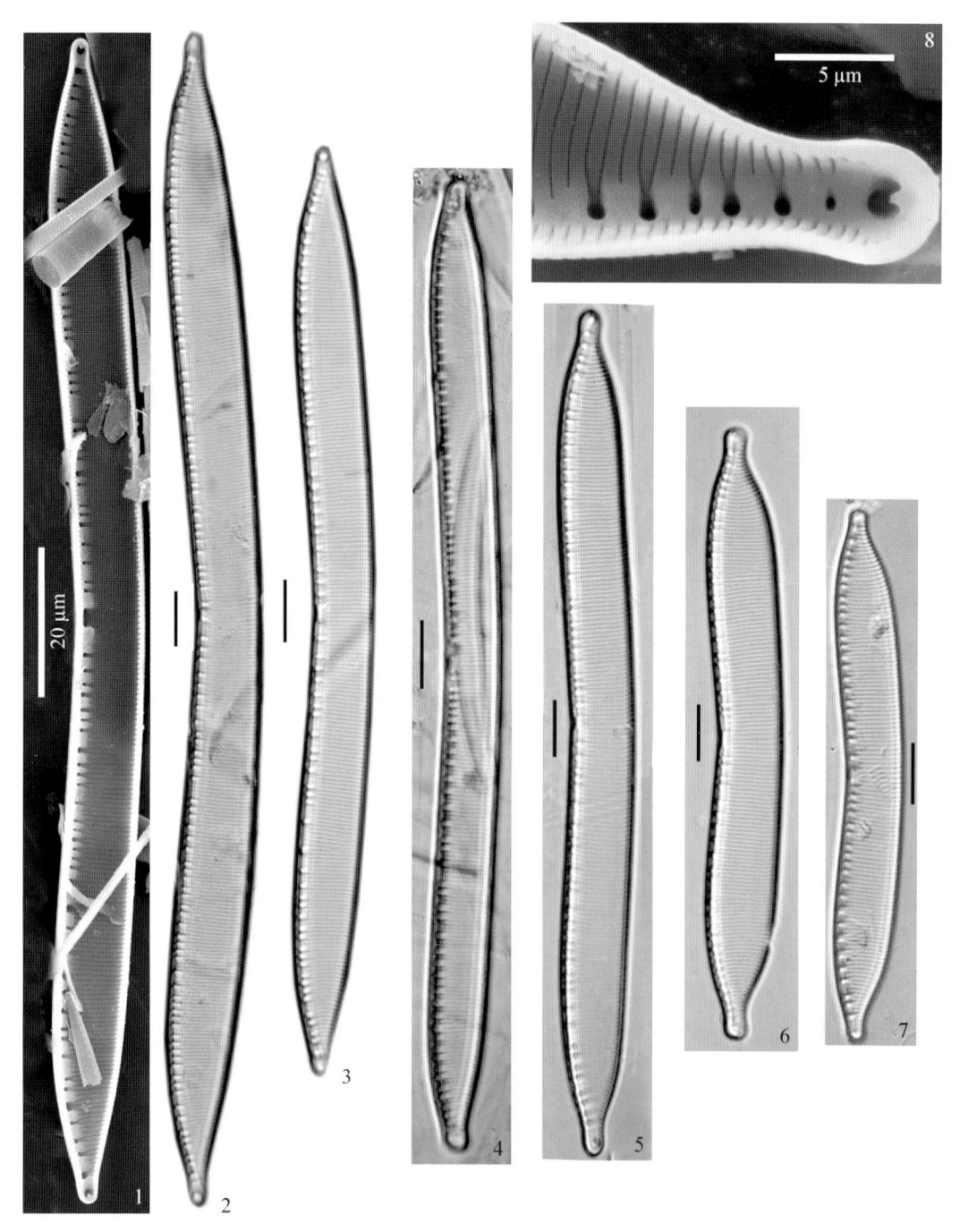

1—3. 近强壮菱板藻 *Hantzschia subrobusta* Q-M You & J. P. Kociolek，1. 示内壳面（SEM），龙骨突形态；
4—8. 嫌钙菱板藻 *Hantzschia calcifuga* Reichardt & Lange-Bertalot，8. 示内壳面顶端（SEM），龙骨突和螺旋舌形态。光镜（LM）照片，比例尺（Scale bars）=10 μm，电镜（SEM）照片比例尺见图。

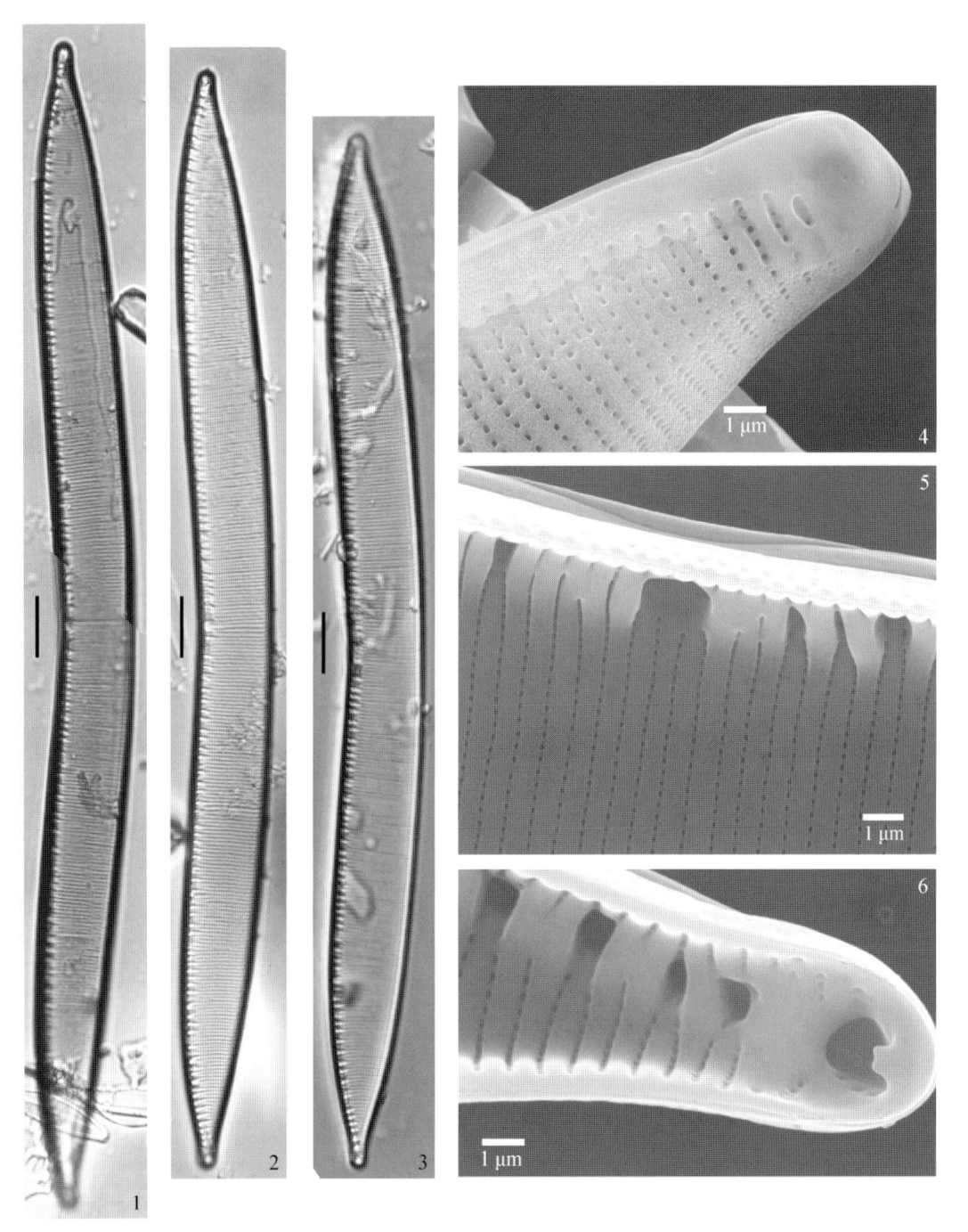

图版 XLIII

1—6. 近强壮菱板藻 *Hantzschia subrobusta* Q-M You & J. P. Kociolek，4. 示外壳面顶端(SEM)，孔纹和壳缝形态，5. 示内壳面中部(SEM)，龙骨突和孔纹形态，6. 示内壳面顶端(SEM)，螺旋舌形态。光镜(LM)照片，比例尺(Scale bars)=10 μm，电镜(SEM)照片比例尺见图。

图版 XLIV

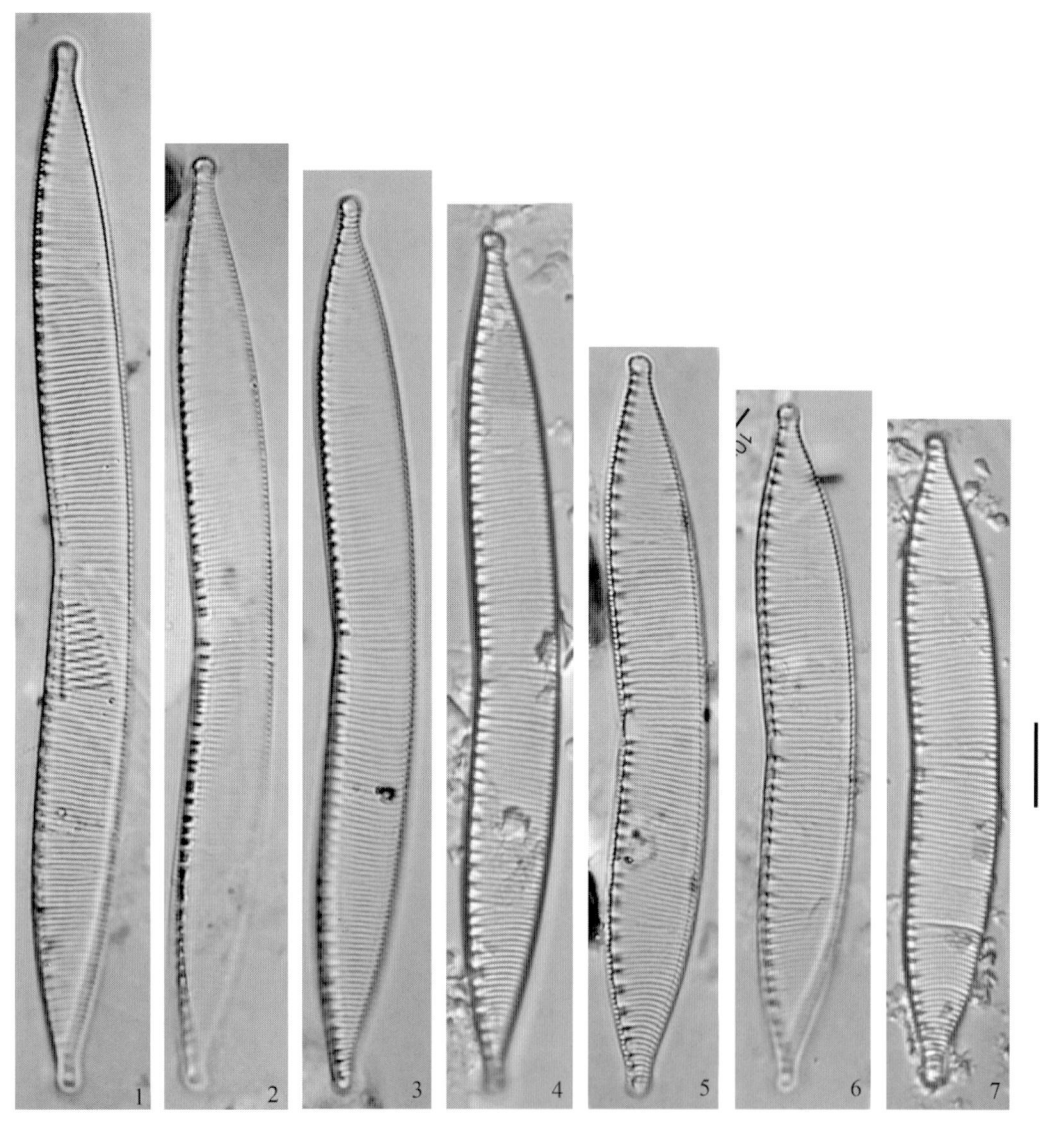

1—7. 活跃菱板藻 *Hantzschia vivacior* Lange-Bertalot。所有图为光镜(LM)照片，比例尺(Scale bars)=10 μm。

图版 XLV

1—4. 活跃菱板藻 *Hantzschia vivacior* Lange-Bertalot。1. 示外壳面中部(SEM)，孔纹和近缝端形态，2. 示外壳面顶端(SEM)，壳缝形态，3. 示内壳面中部(SEM)，孔纹和龙骨突形态，4. 示内壳面顶端(SEM)，孔纹和龙骨突形态。所有图为电镜(SEM)照片，比例尺见图。

图版 XLVI

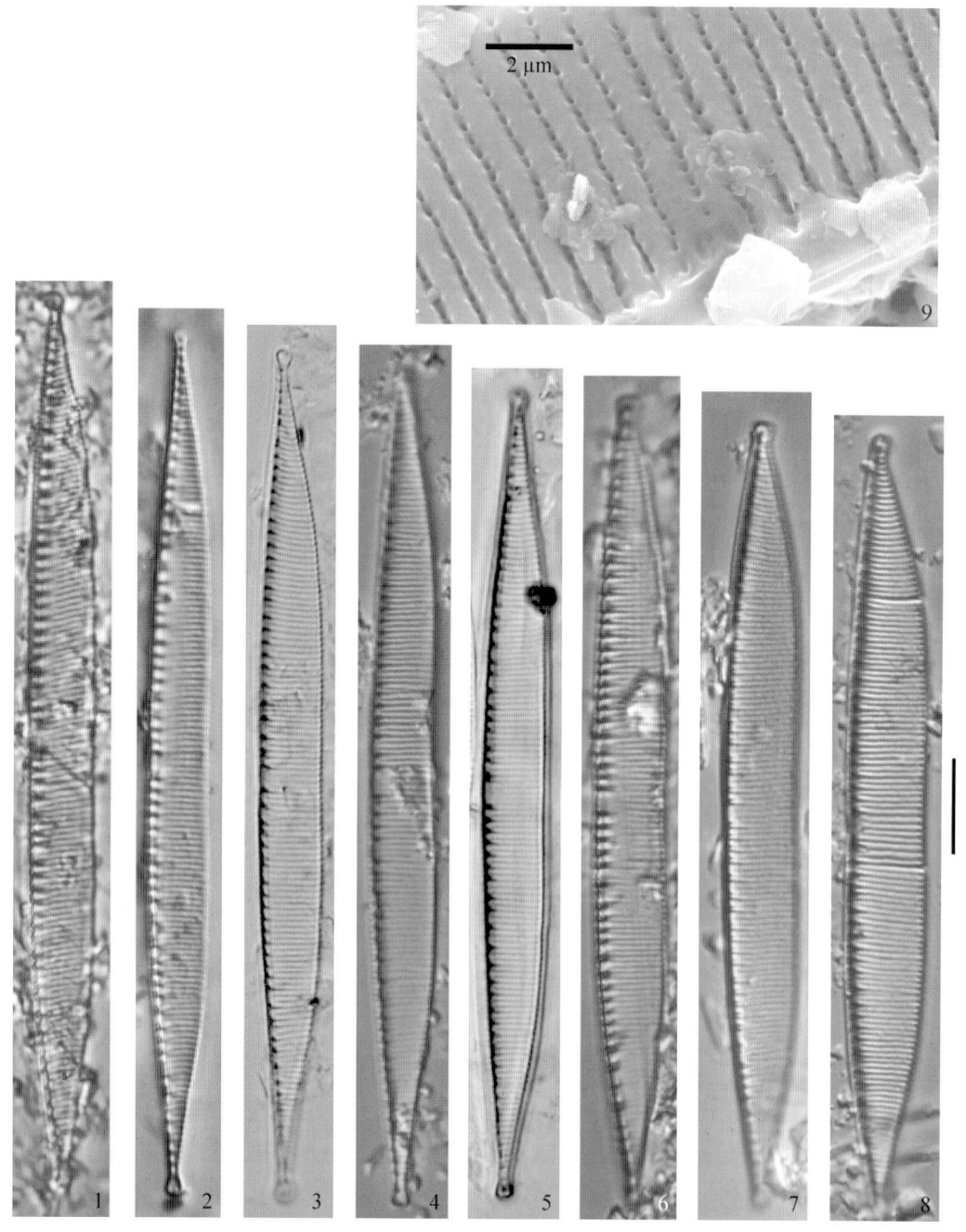

1—9. 菱形菱板藻 *Hantzschia nitzschioides* Lange-Bertalot，9. 示外壳面中部(SEM)，孔纹形态。光镜(LM)照片，比例尺(Scale bars)=10 μm，电镜(SEM)照片比例尺见图。

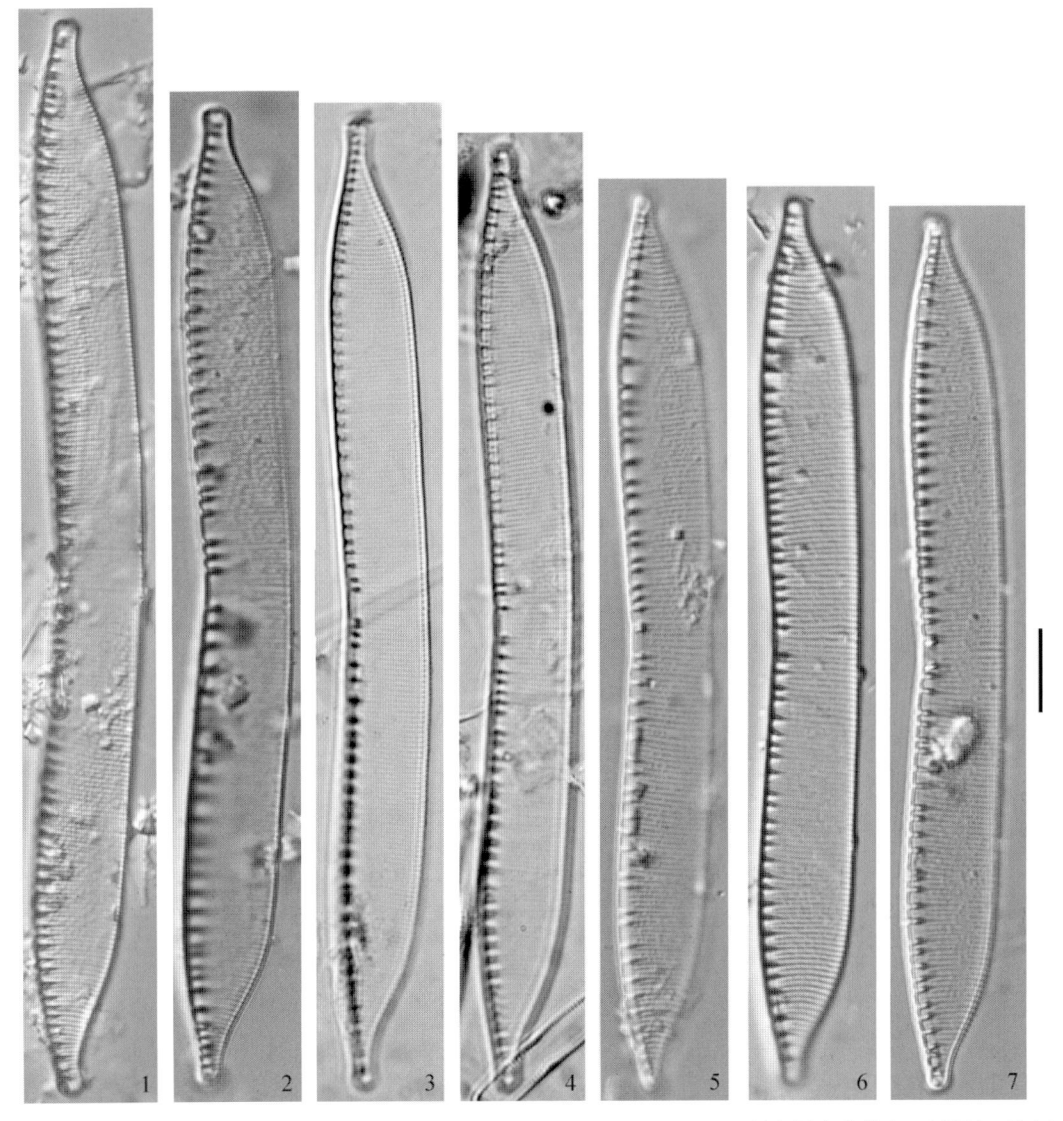

图版 XLVII

1—7. 拟巴德菱板藻 *Hantzschia pseudobardii* Q-M You & J. P. Kociolek。所有图为光镜(LM)照片，比例尺(Scale bars)=10 μm。

图版 XLVIII

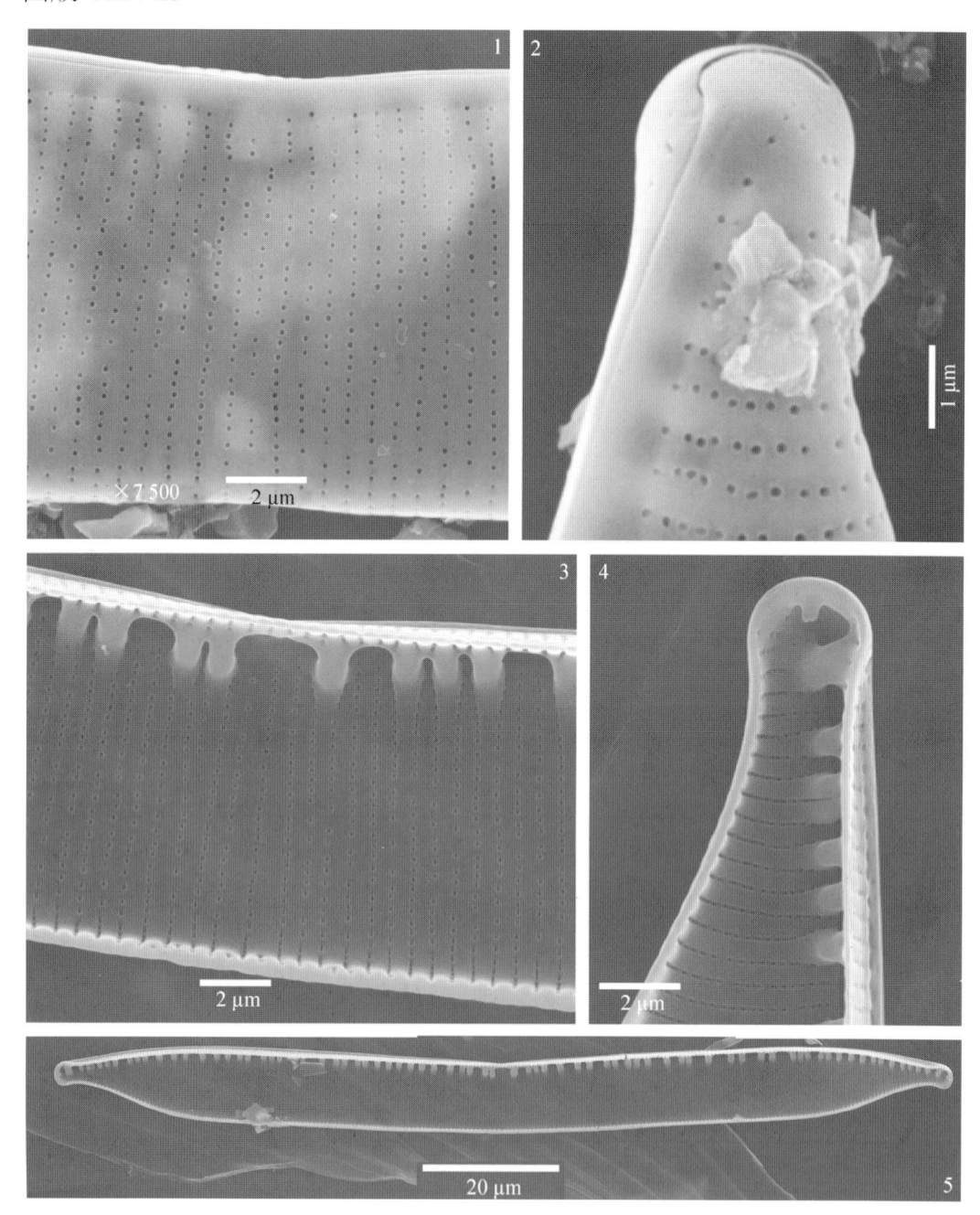

1—5. 拟巴德菱板藻 *Hantzschia pseudobardii* Q-M You & J. P. Kociolek，1. 示外壳面中部(SEM)，孔纹形态，2. 示外壳面顶端(SEM)，孔纹和壳缝形态，3. 示内壳面中部(SEM)，孔纹和龙骨突形态，4. 示内壳面顶端(SEM)，龙骨突和螺旋舌形态，5. 示内壳面(SEM)。所有图为电镜(SEM)照片，比例尺见图。

图版 XLIX

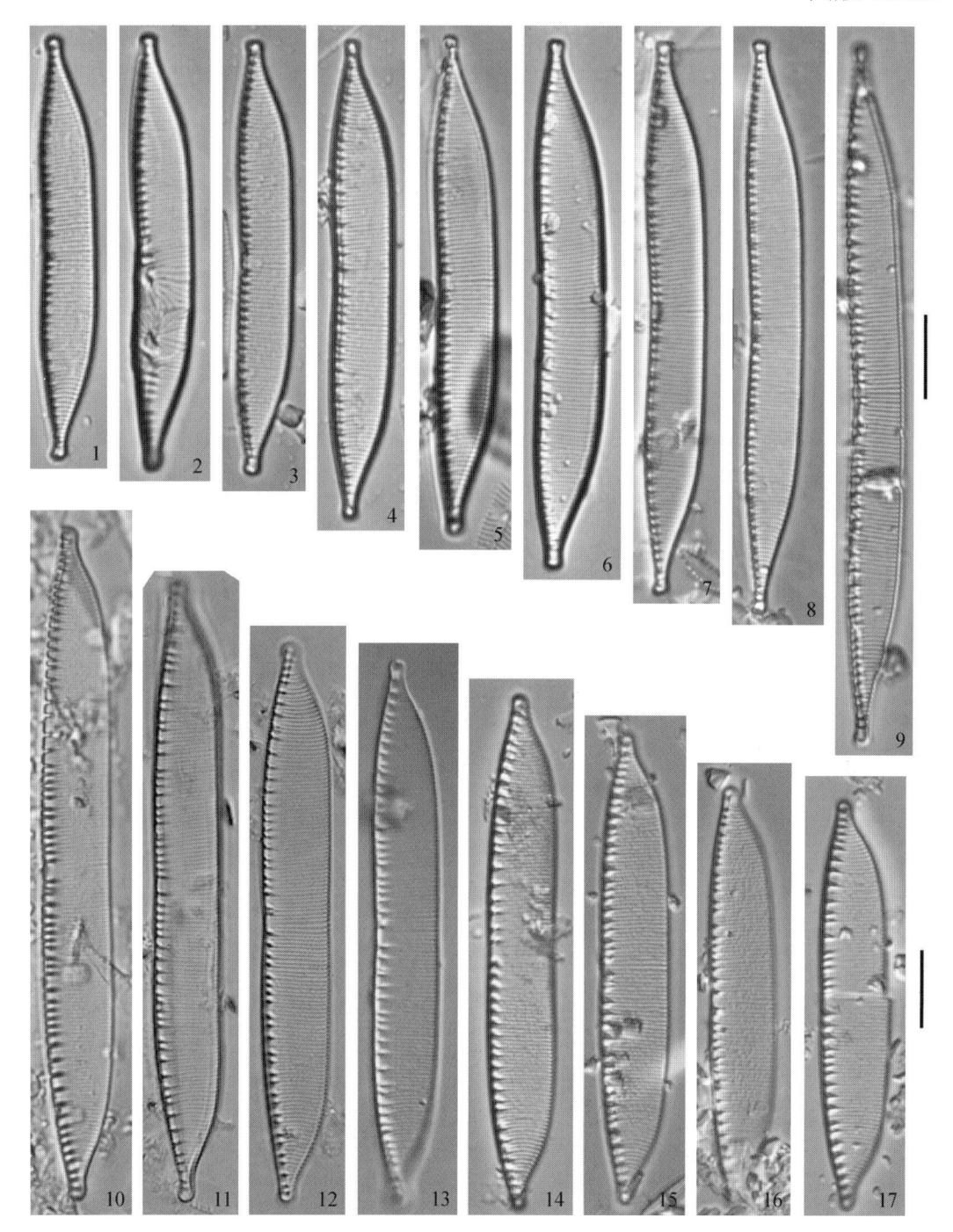

1—9. 伊犁菱板藻 *Hantzschia yili* Q-M You & J. P. Kociolek；10—17. 中华菱板藻 *Hantzschia sinensis* Q-M You & J. P. Kociolek。所有图为光镜(LM)照片，比例尺(Scale bars)=10 μm。

图版 L

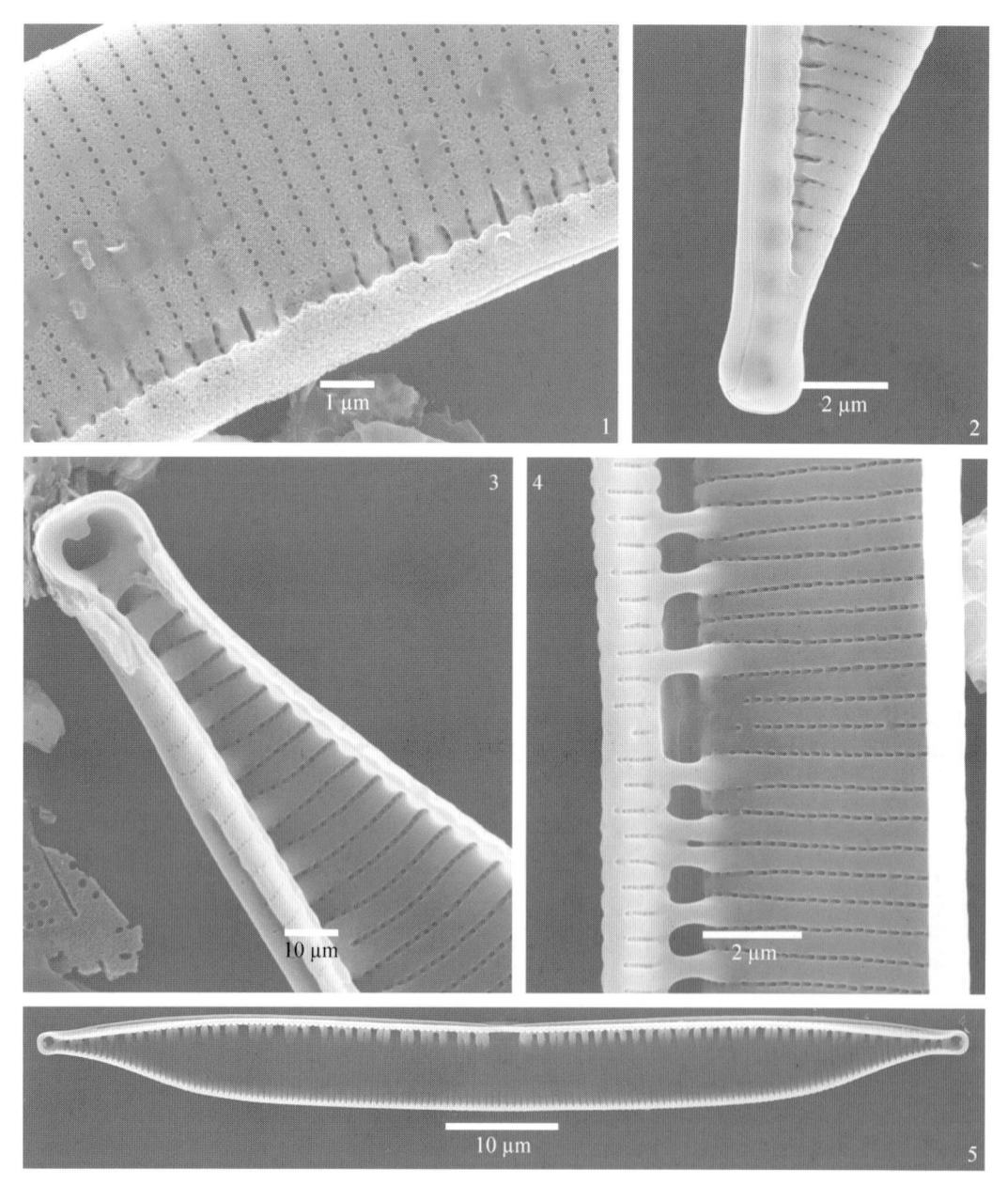

1—5. 伊犁菱板藻 *Hantzschia yili* Q-M You & J. P. Kociolek，1. 示外壳面中部(SEM)，示孔纹形态，2. 示外壳面顶端(SEM)，孔纹和壳缝末端形态，3. 示内壳面顶端(SEM)，孔纹和螺旋舌形态，4. 示内壳面中部(SEM)，近缝端和龙骨突形态，5. 示内壳面(SEM)。所有图为电镜(SEM)照片，比例尺见图。

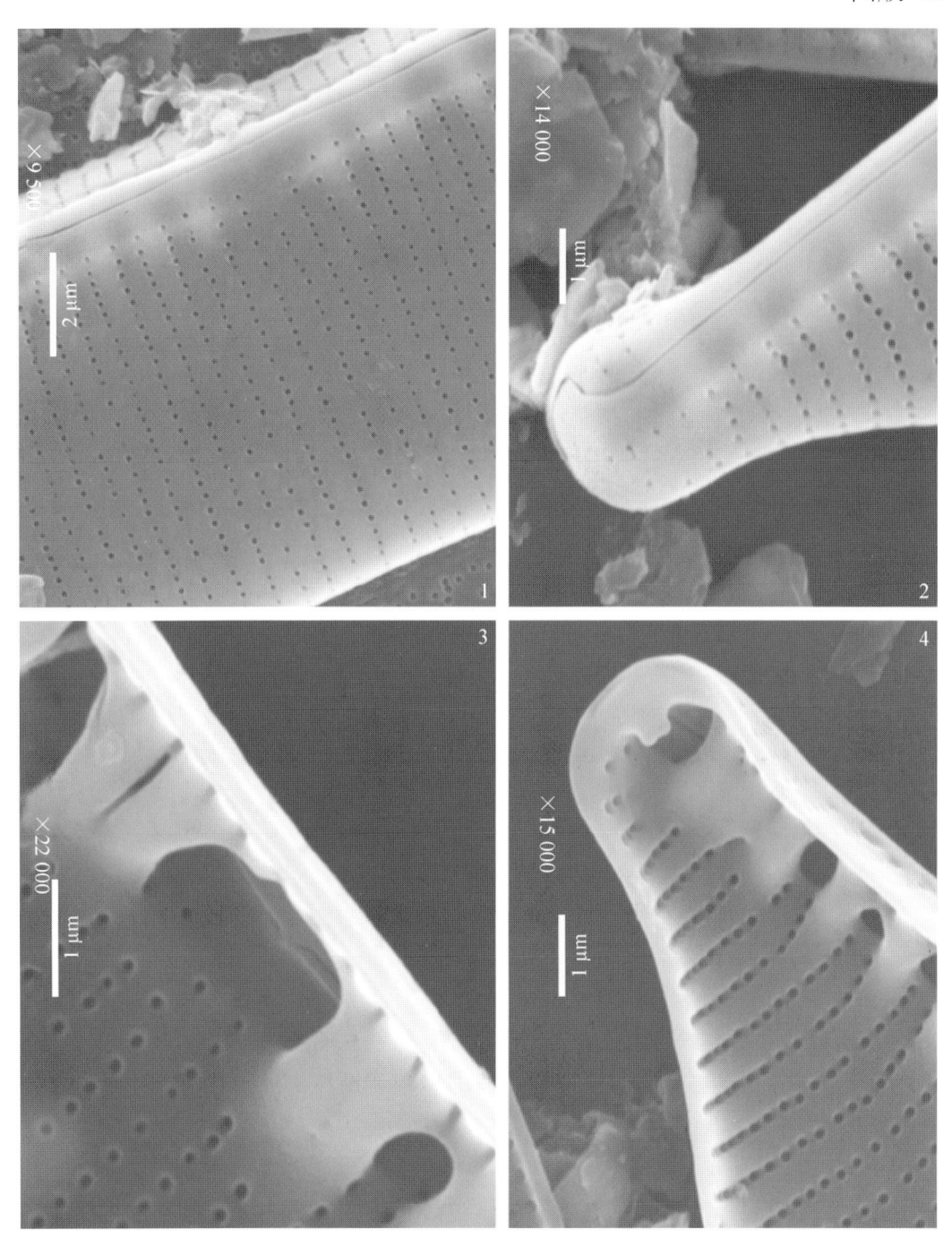

图版 LI

1—4. 中华菱板藻 *Hantzschia sinensis* Q-M You & J. P. Kociolek，1. 示外壳面中部(SEM)，孔纹和壳缝形态，2. 示外壳面顶端(SEM)，孔纹和壳缝形态，3. 示内壳面中部(SEM)，龙骨突和近缝端形态，4. 示内壳面顶端(SEM)，龙骨突和螺旋舌形态。所有图为电镜(SEM)照片，比例尺见图。

图版 LII

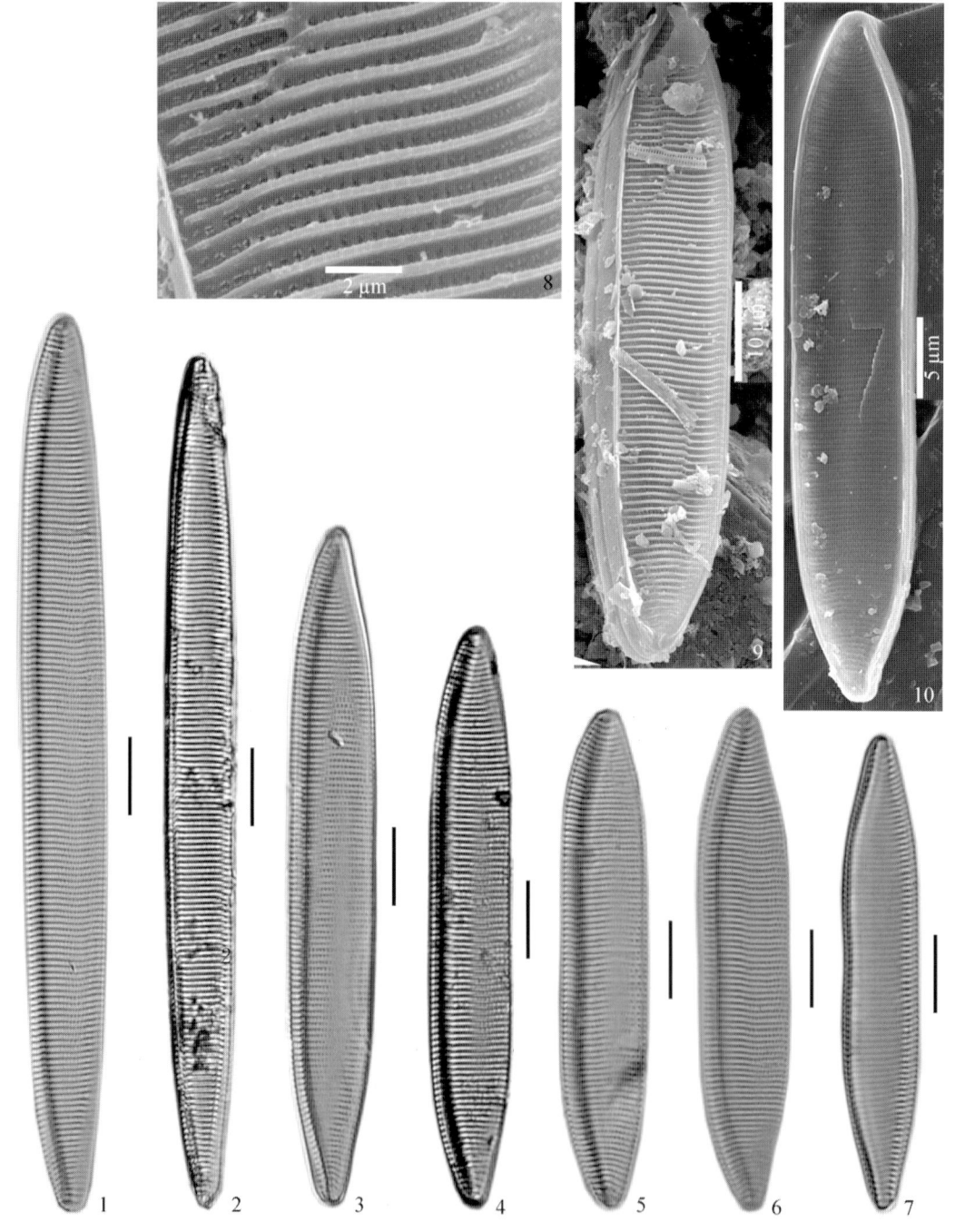

1—10. 渐窄盘杆藻(原变种)*Tryblionella angustata* var. *angustata* W. Smith, 8. 示外壳面中部(SEM), 孔纹和肋纹形态, 9—10. 示外壳面(SEM)。光镜(LM)照片, 比例尺(Scale bars)=10 μm, 电镜(SEM)照片比例尺见图。

图版 LIII

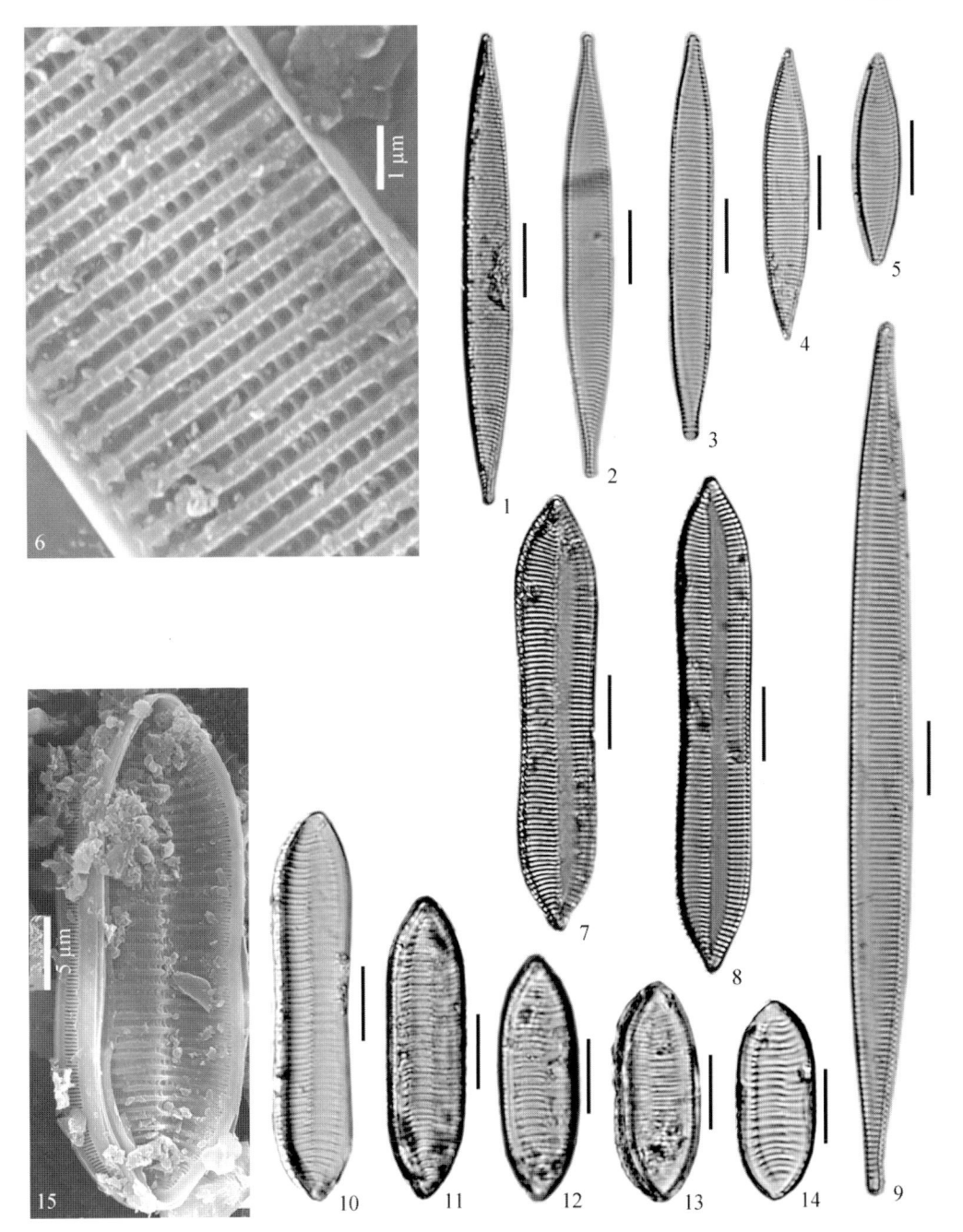

1—6. 狭窄盘杆藻 *Tryblionella angustatula* (Lange-Bertalot) Q-M You & Q-X Wang nov. comb.，6. 示外壳面中部，孔纹和肋纹形态；7—8. 尖锥盘杆藻 *Tryblionella acuminata* W. Smith；9. 渐窄盘杆藻尖变种 *Tryblionella angustata* var. *acuta* (Grunow) Bukhtiyarova；10—15. 莱维迪盘杆藻 *Tryblionella levidensis* W. Smith，15. 示外壳面(SEM)，波曲壳面。光镜(LM)照片，比例尺(Scale bars)=10 μm，电镜(SEM)照片比例尺见图。

图版 LIV

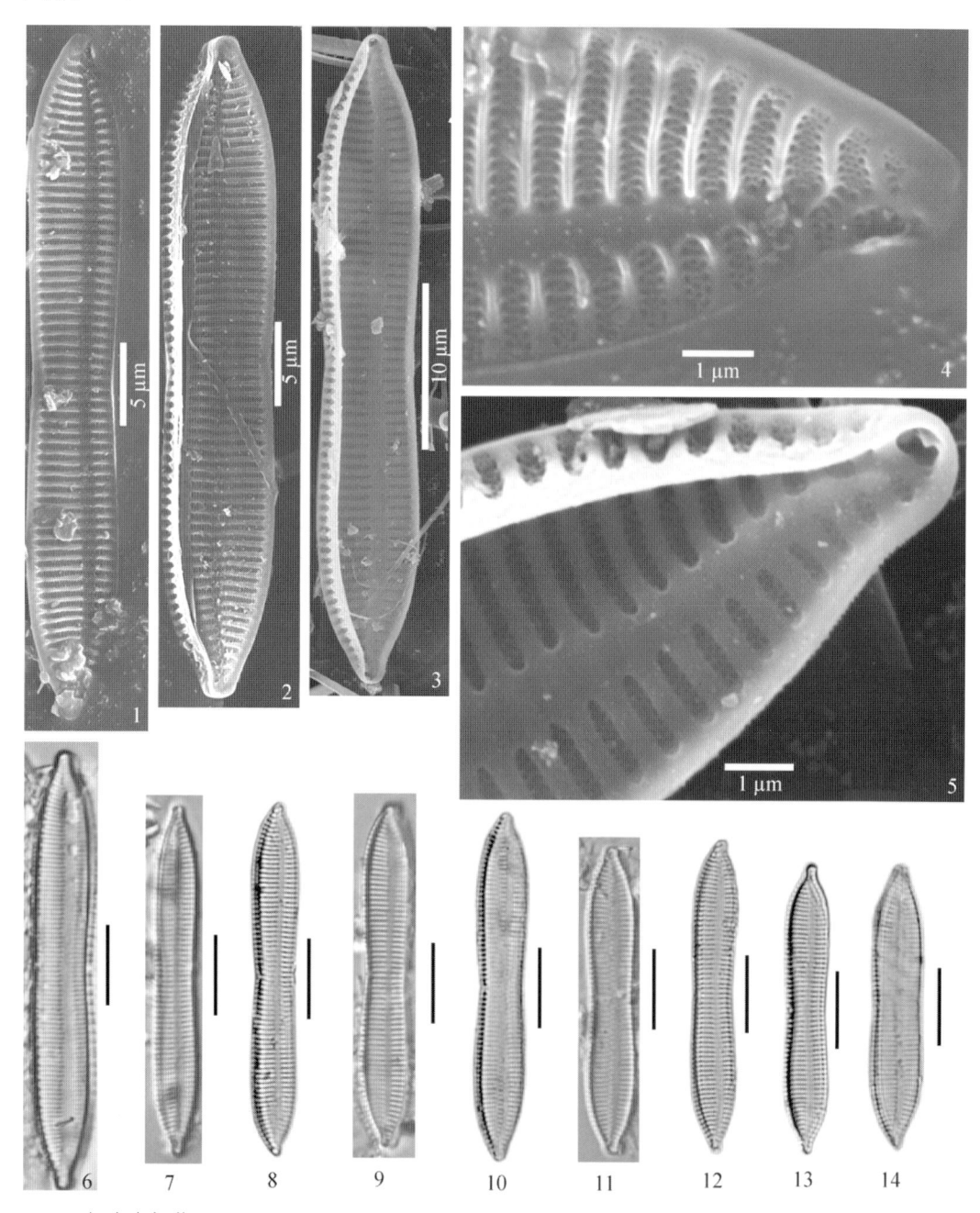

1—14. 细尖盘杆藻 *Tryblionella apiculata* Gregory，1—3. 示外壳面(SEM)，4. 示外壳面顶端(SEM)，孔纹和肋纹形态，5. 示内壳面顶端(SEM)，孔纹和肋纹形态。光镜(LM)照片，比例尺(Scale bars)=10 μm，电镜(SEM)照片比例尺见图。

图版 LV

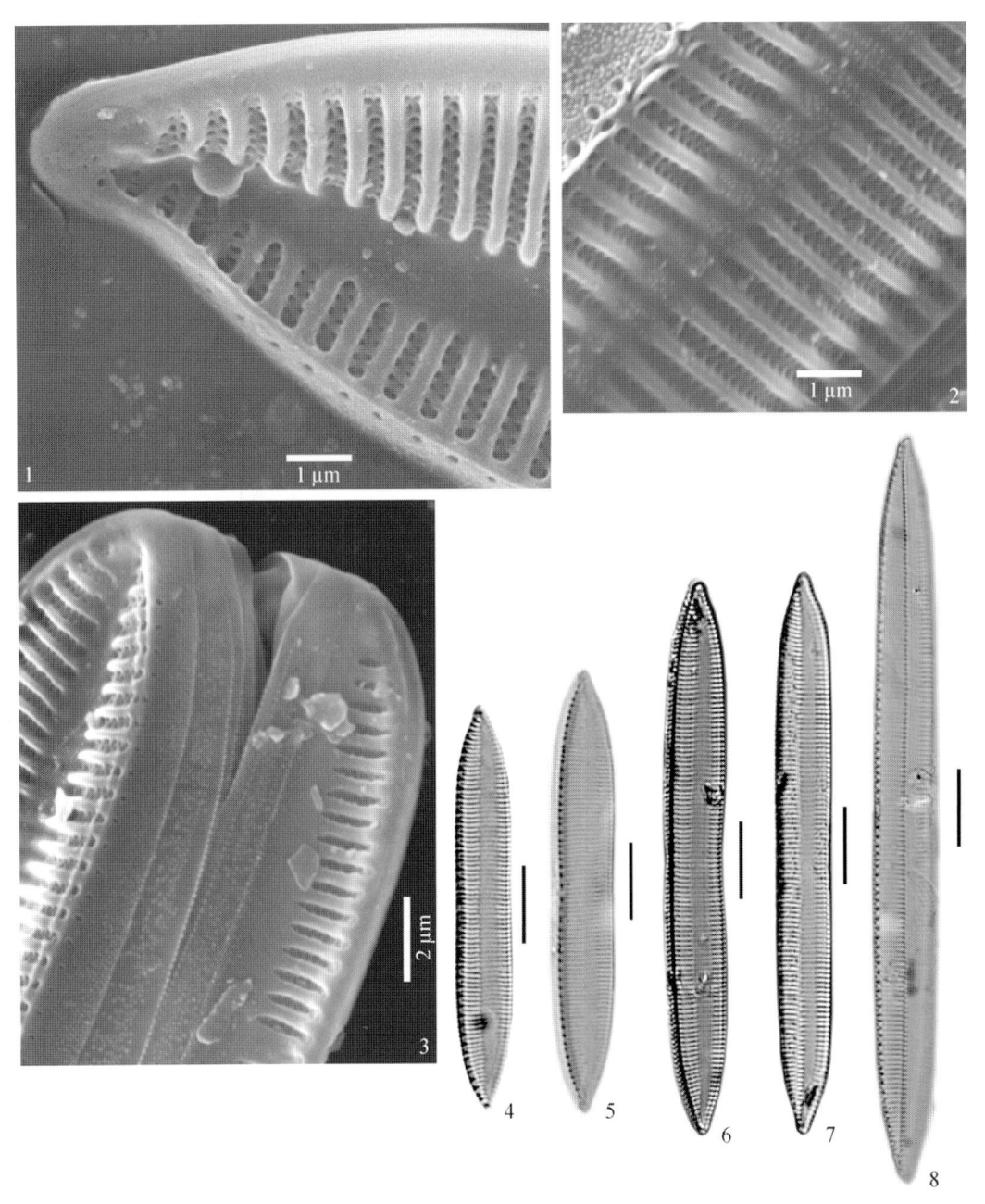

1—8. 匈牙利盘杆藻 *Tryblionella hungarica* (Grunow) Frenguelli，1. 示外壳面顶端，孔纹和肋纹形态 (SEM)，2. 示外壳面中部(SEM)，孔纹和肋纹形态，3. 示外壳面壳套面(SEM)。光镜(LM)照片，比例尺(Scale bars)=10 μm，电镜(SEM)照片比例尺见图。

图版 LVI

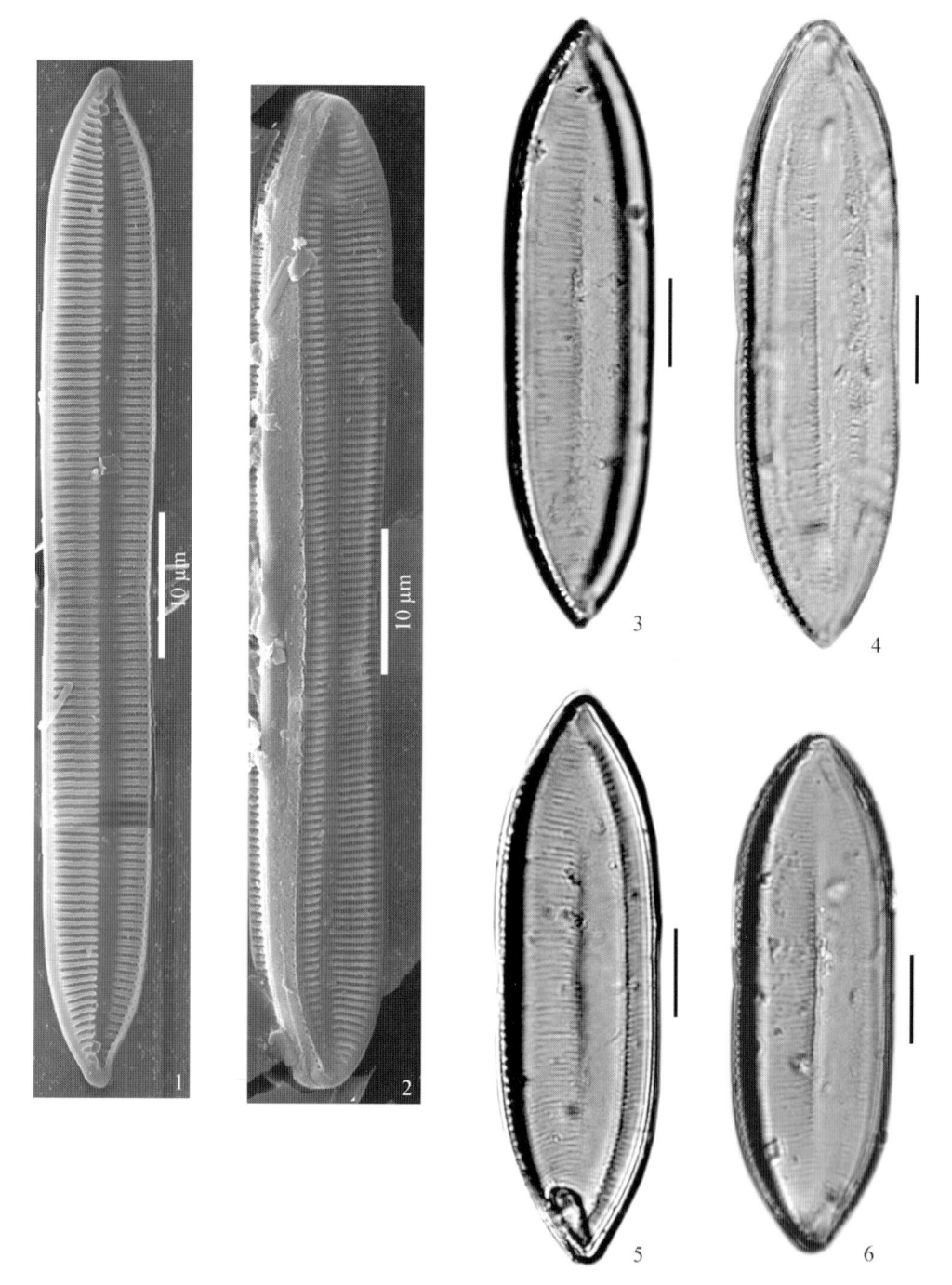

1—2. 匈牙利盘杆藻 *Tryblionella hungarica* (Grunow) D. G. Mann，1—2. 示外壳面（SEM），波曲壳面；
3—6. 维多利亚盘杆藻 *Tryblionella victoriae* Grunow。光镜（LM）照片，比例尺（Scale bars）=10 μm，电镜（SEM）照片比例尺见图。

图版 LVII

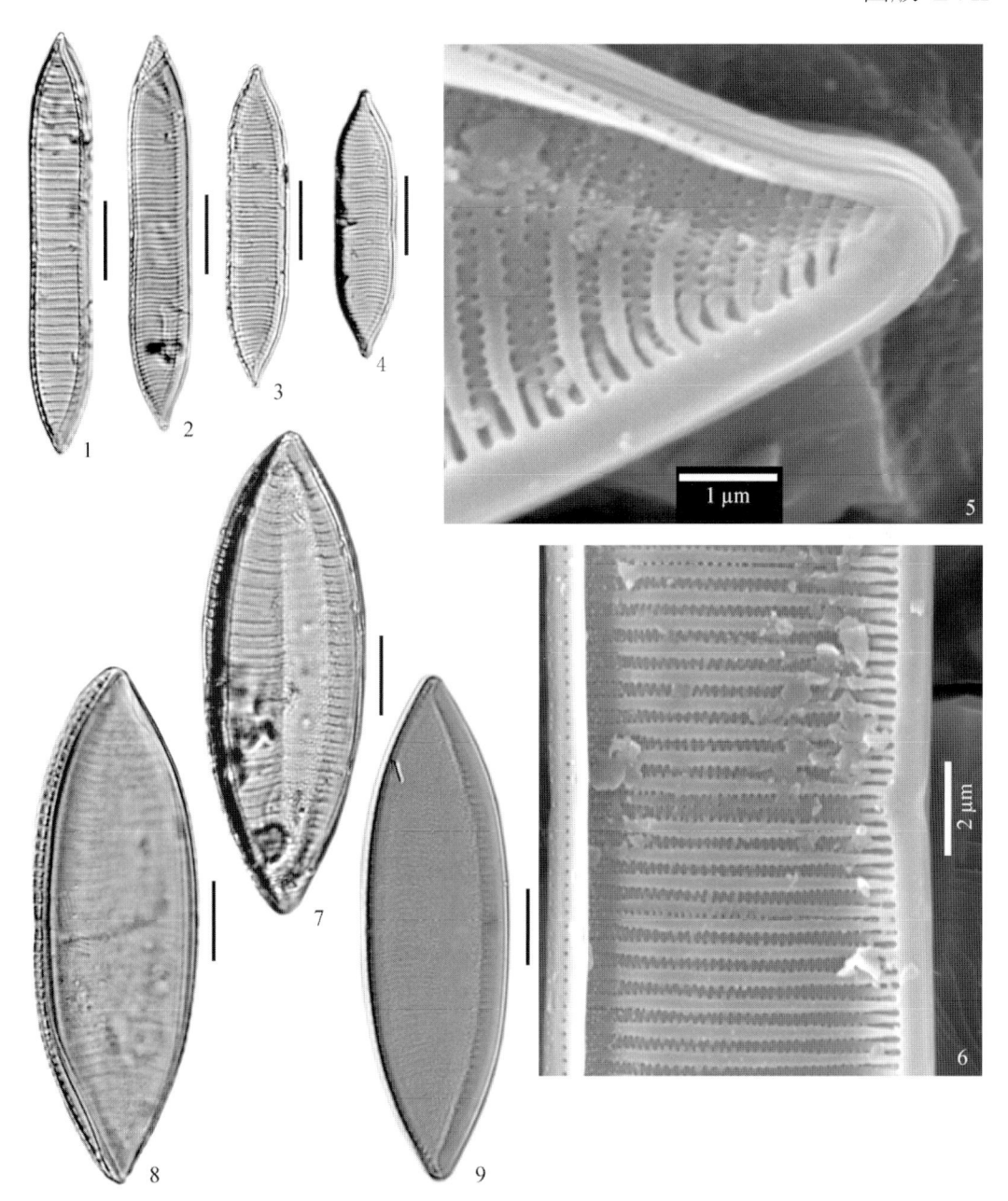

1—6. 暖温盘杆藻 *Tryblionella calida* (Grunow) D. G. Mann，5. 示外壳面顶端(SEM)，孔纹和肋纹形态，6. 示外壳面中部(SEM)，孔纹和肋纹形态；7—9：岸边盘杆藻 *Tryblionella littoralis* (Grunow) D. G. Mann。光镜(LM)照片，比例尺(Scale bars)=10 μm，电镜(SEM)照片比例尺见图。

图版 LVIII

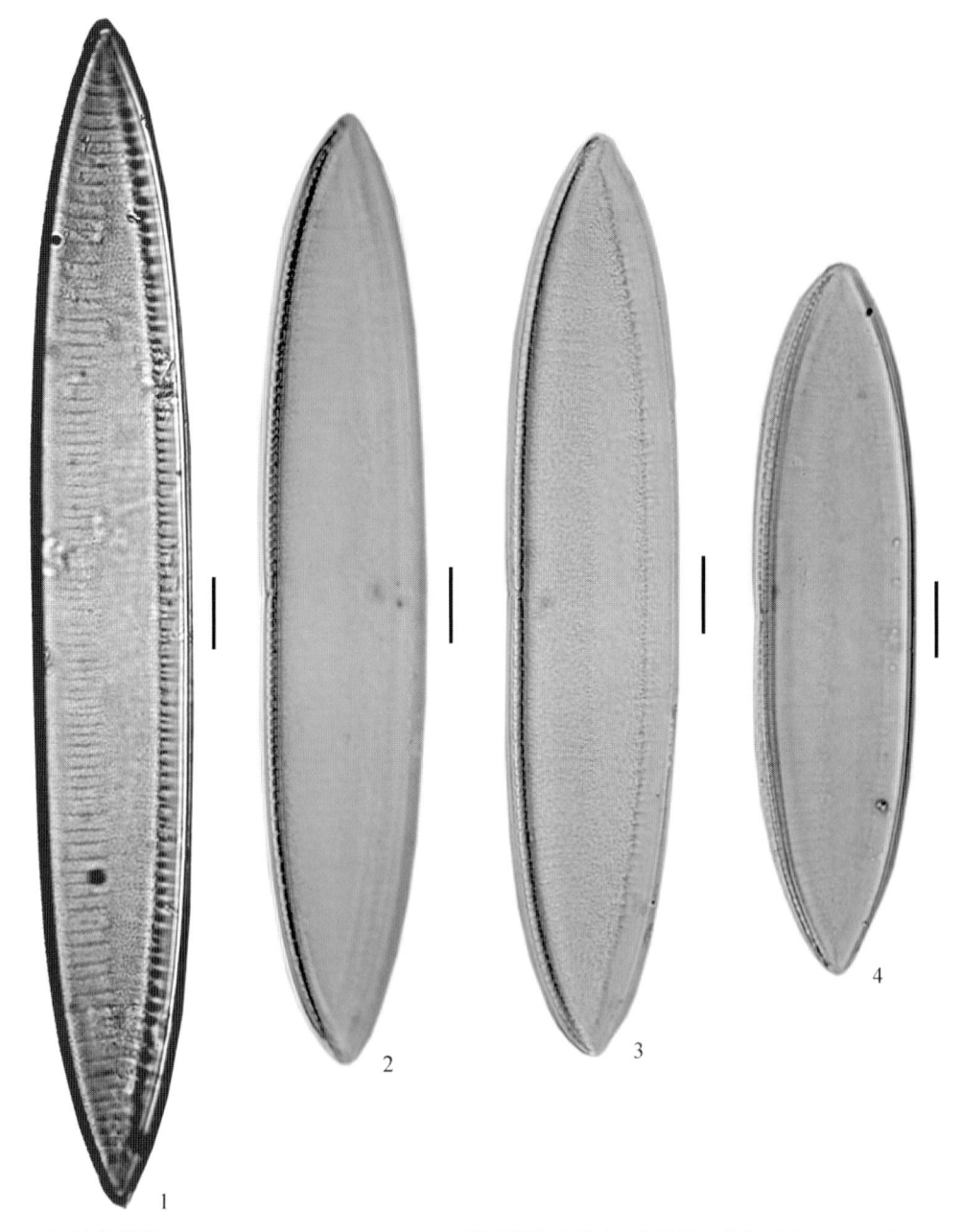

1—4. 细长盘杆藻 *Tryblionella gracilis* W. Smith。所有图为光镜(LM)照片，比例尺(Scale bars)=10 μm。

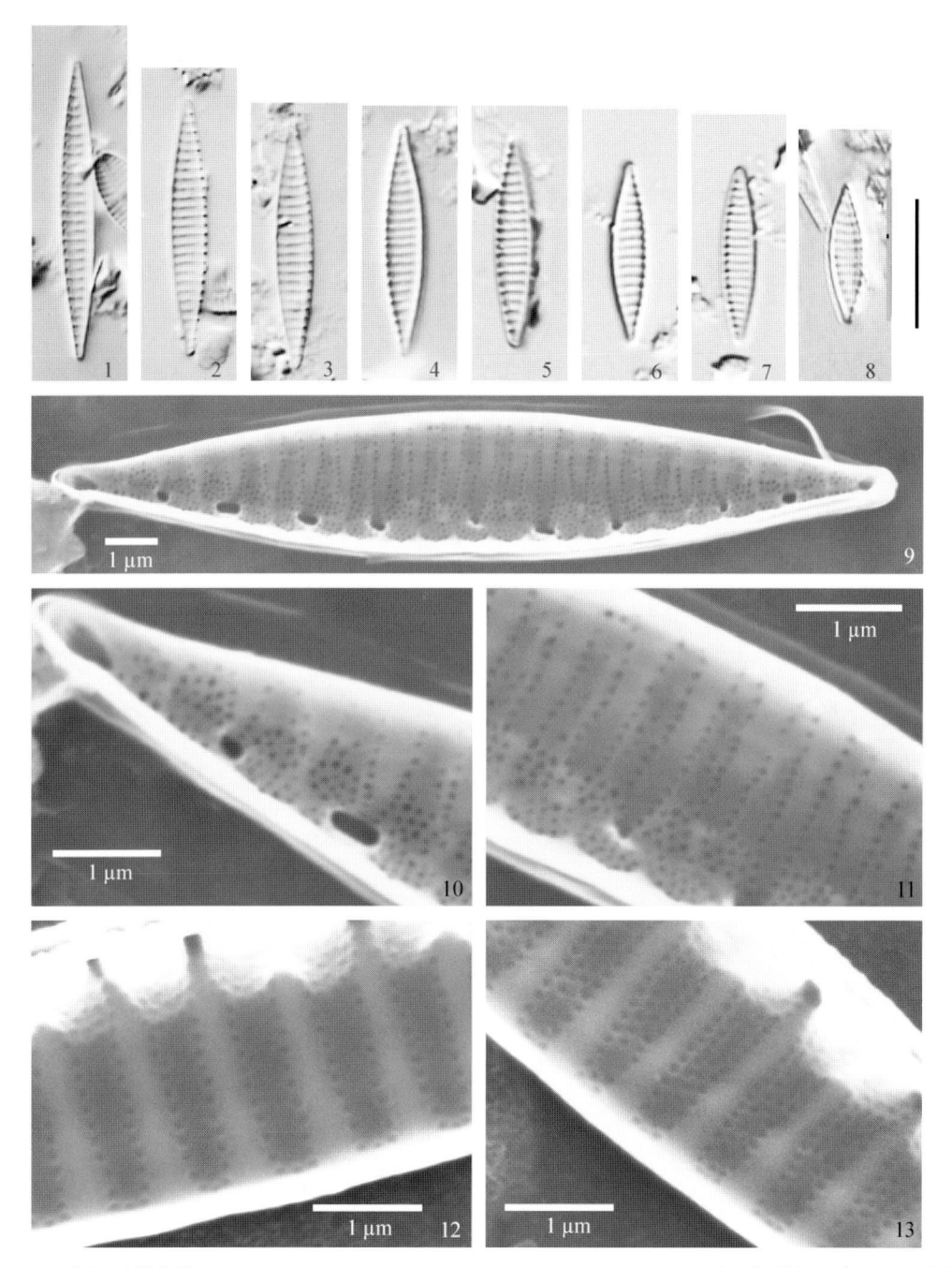

图版 LIX

1—13. 茂兰西蒙森藻 *Simonsenia maolaniana* Q-M You & J.P. Kociolek，9. 示内壳面(SEM)，10. 示内壳面末端(SEM)，11—13. 示内壳面中部(SEM)。光镜(LM)照片，比例尺(Scale bars)=10 μm，电镜(SEM)照片比例尺见图。

图版 LX

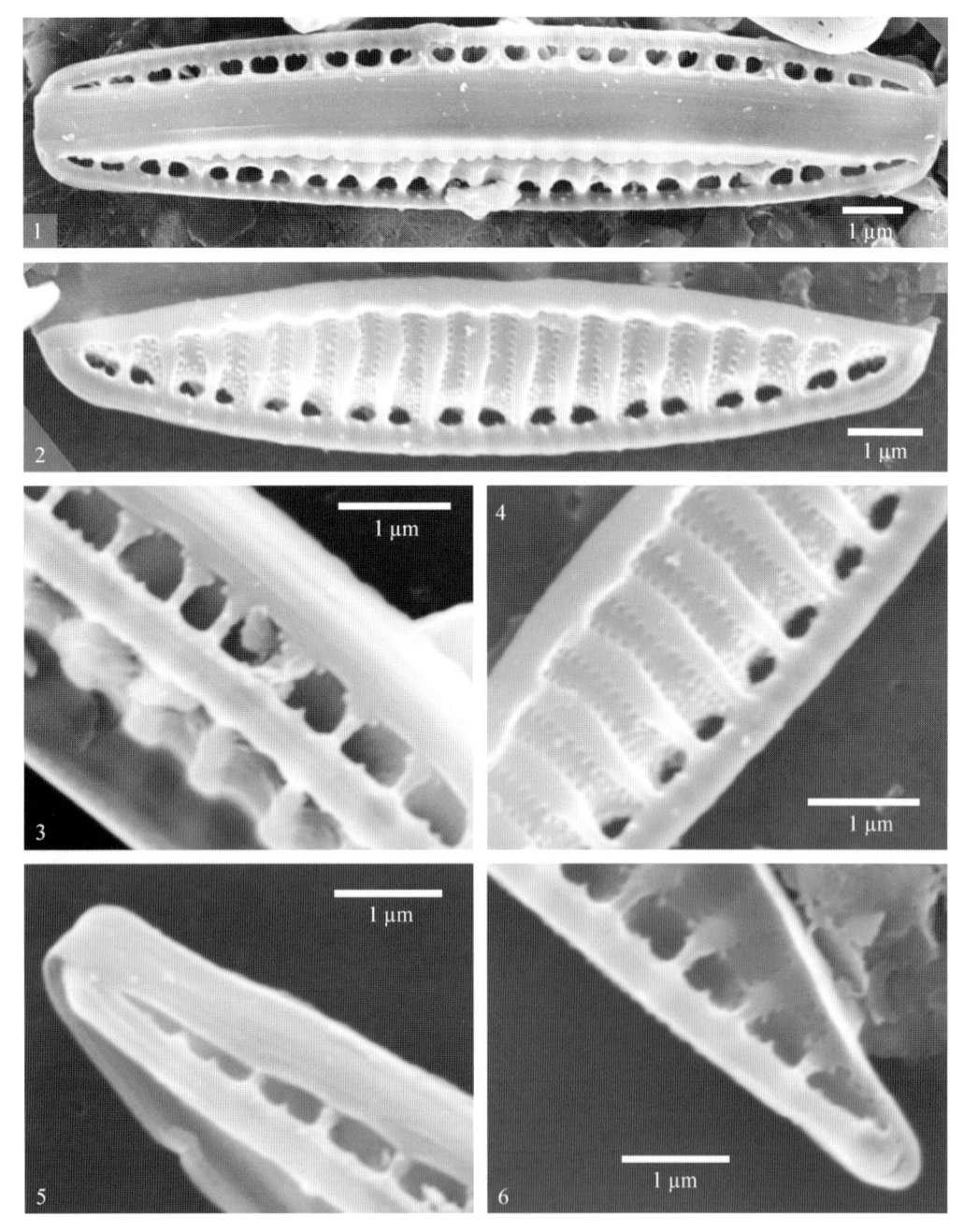

1—6. 茂兰西蒙森藻 *Simonsenia maolaniana* Q-M You & J.P. Kociolek，1. 示壳体（SEM），2. 示外壳面（SEM），3—4. 示外壳面中部（SEM），5—6. 示外壳面末端（SEM）。所有图为电镜（SEM）照片，比例尺见图。

图版 LXI

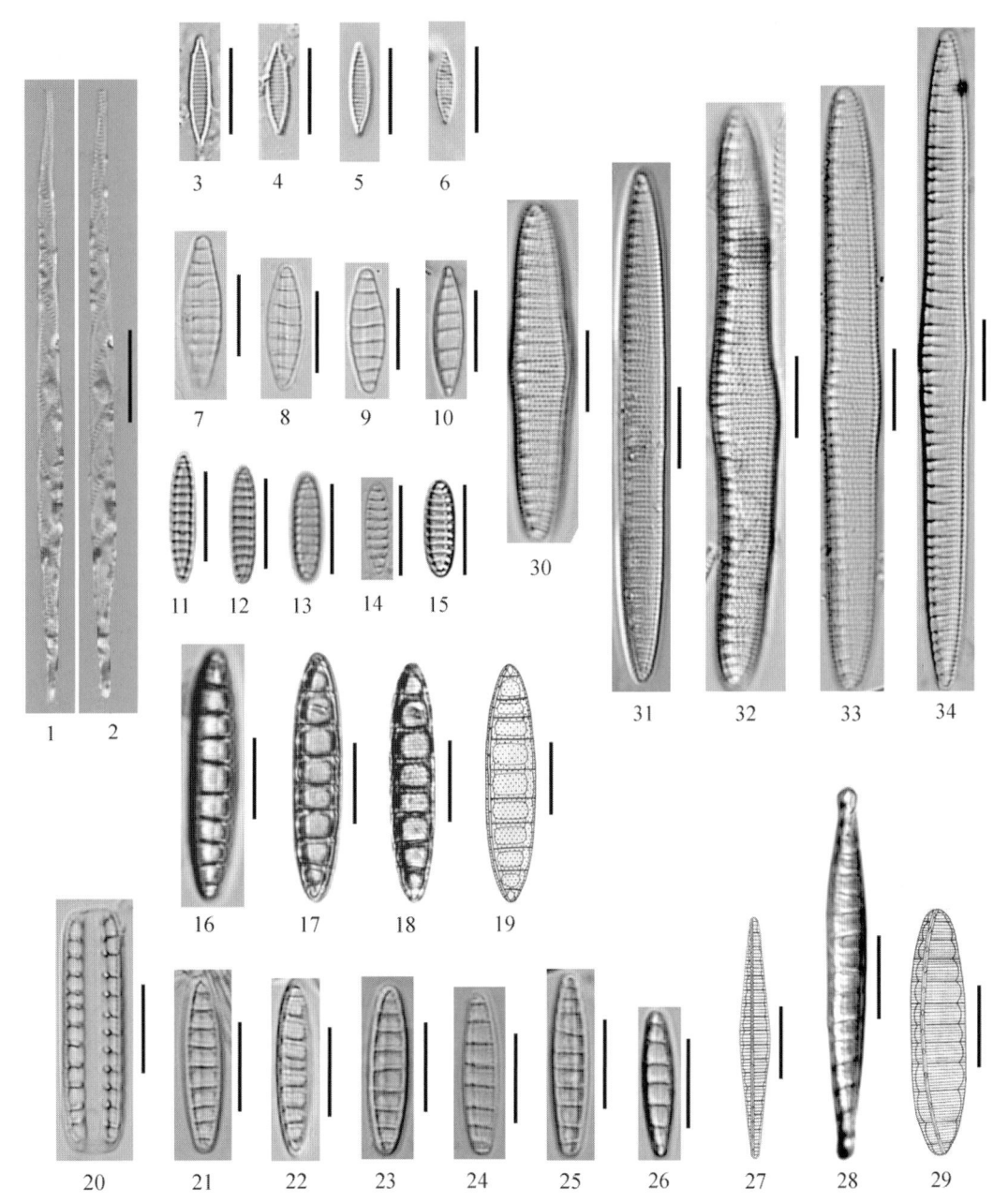

1—2. 细筒柱藻 *Cylindrotheca gracilis* (Brébisson ex Kützing) Grunow；3—6. 德洛西蒙森藻 *Simonsenia delognei* (Grunow) Lange-Bertalot；7—10. 小型细齿藻（原变种）*Denticula tenuis* var. *tenuis* Kützing；11—15. 科瑞提细齿藻 *Denticula creticola* (Østrup) Lange-Bertalot & Krammer；16—19. 强壮细齿藻 *Denticula valida* (Pedicino) Grunow；20—26. 华美细齿藻 *Denticula elegans* Kützing；27—28. 小型细齿藻（原变种）*Denticula tenuis* var. *tenuis* Kützing；29. 小型细齿藻粗变种 *Denticula tenuis* var. *crassula* (Nägeli) Hustedt；30—34. 库津细齿藻汝牧变种 *Denticula kuetzingii* var. *rumrichae* Krammer。其中 19, 27, 29 为手绘图，引自 Zhu 和 Chen (2000)，所有图为光镜 (LM) 照片，比例尺 (Scale bars) =10 μm。

图版 LXII

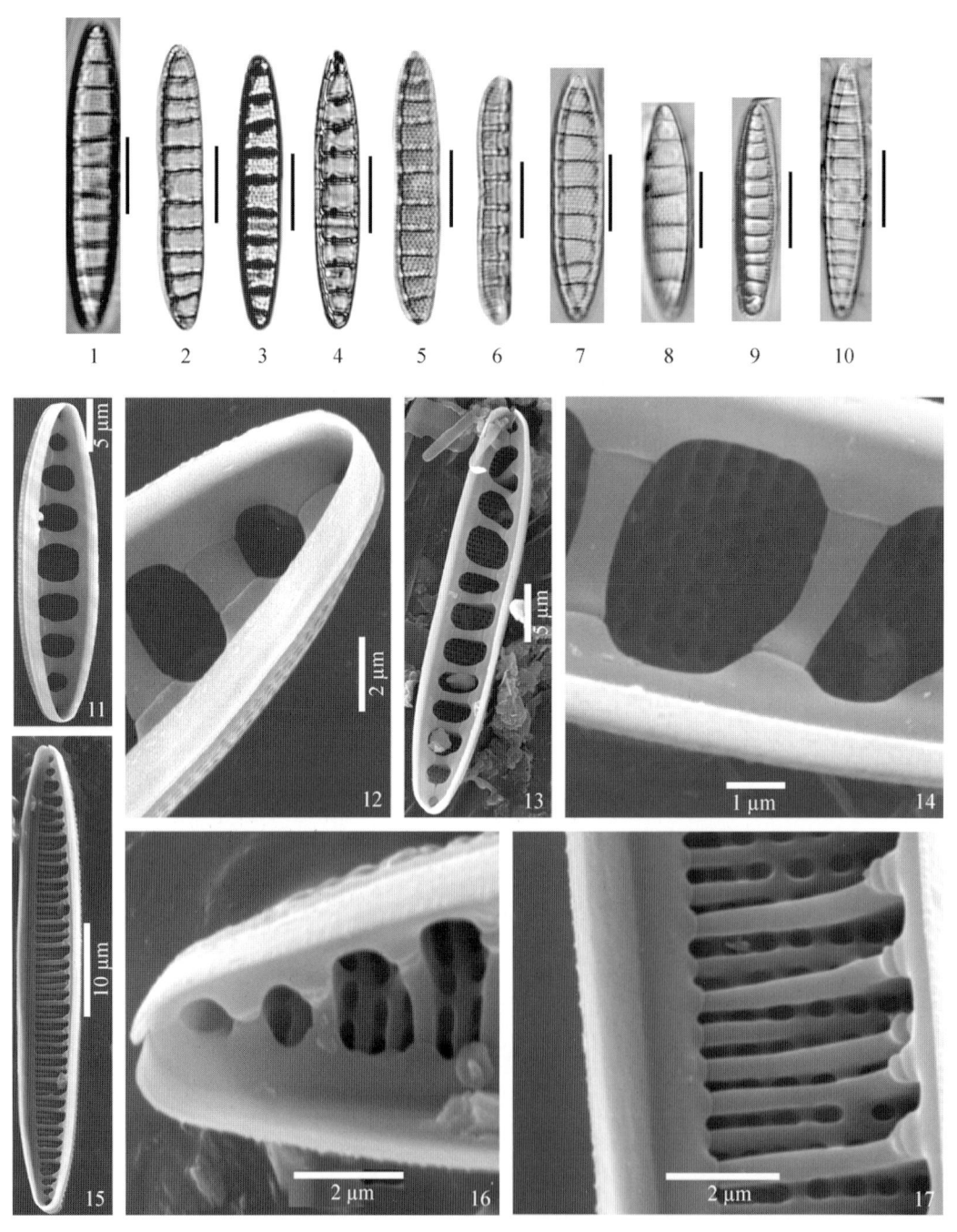

1—14. 强壮细齿藻 *Denticula valida* (Pedicino) Grunow, 11, 13. 示内壳面 (SEM), 12. 示内壳面顶端, 龙骨突形态, 14. 示内壳面中部, 龙骨突和孔纹形态; 15—17. 库津细齿藻（原变种）*Denticula kuetzingii* var. *kuetzingii* Grunow, 15. 示内壳面 (SEM), 16. 示内壳面顶端 (SEM), 龙骨突和孔纹形态, 17. 示内壳面中部 (SEM), 龙骨突和孔纹形态。光镜 (LM) 照片, 比例尺 (Scale bars) =10 μm, 电镜 (SEM) 照片比例尺见图。

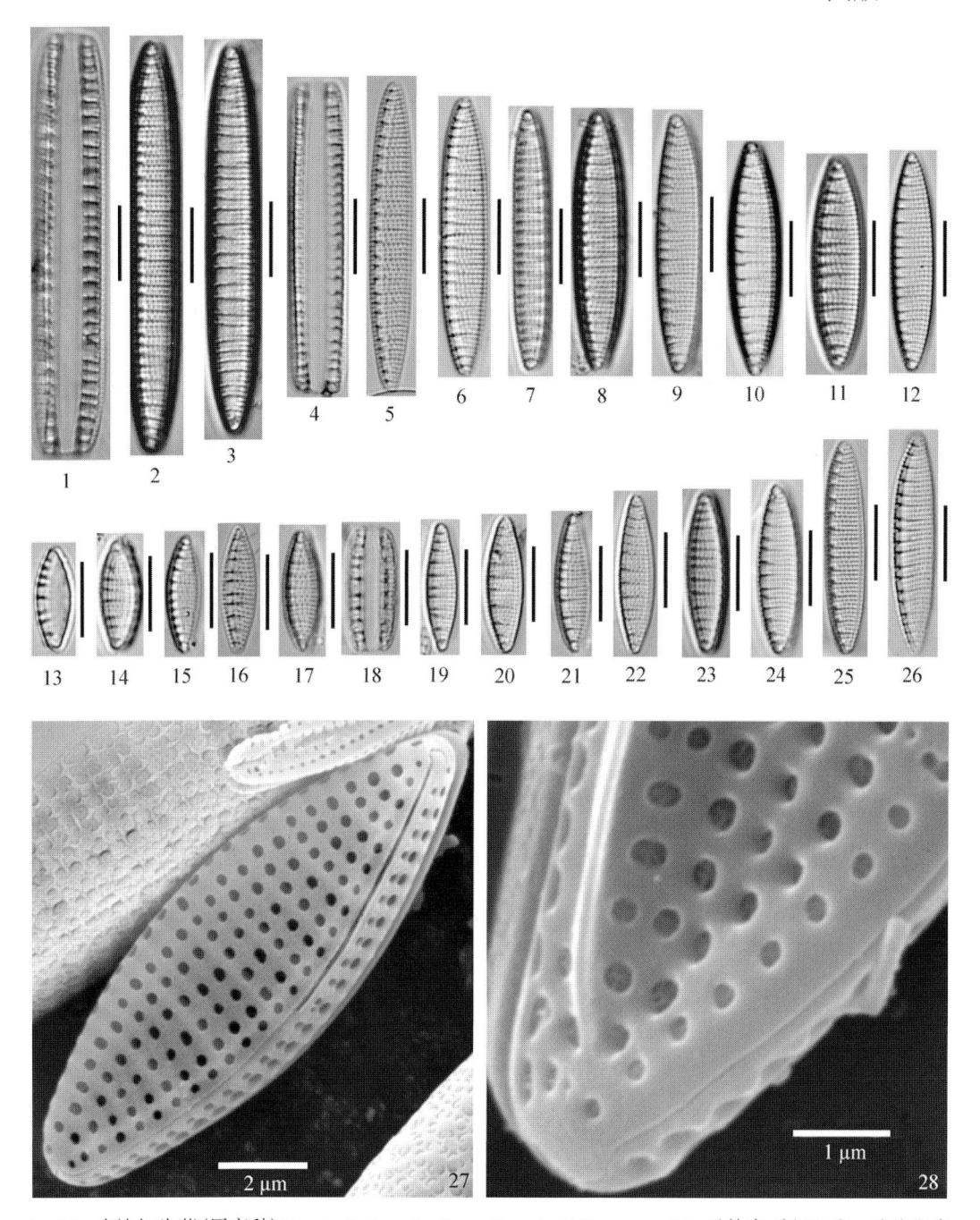

图版 LXIII

1—28. 库津细齿藻（原变种）*Denticula kuetzingii* var. *kuetzingii* Grunow，27. 示外壳面(SEM)，孔纹和壳缝形态；28. 示外壳面顶端(SEM)，孔纹形态。光镜(LM)照片，比例尺(Scale bars)=10 μm，电镜(SEM)照片比例尺见图。

图版 LXIV

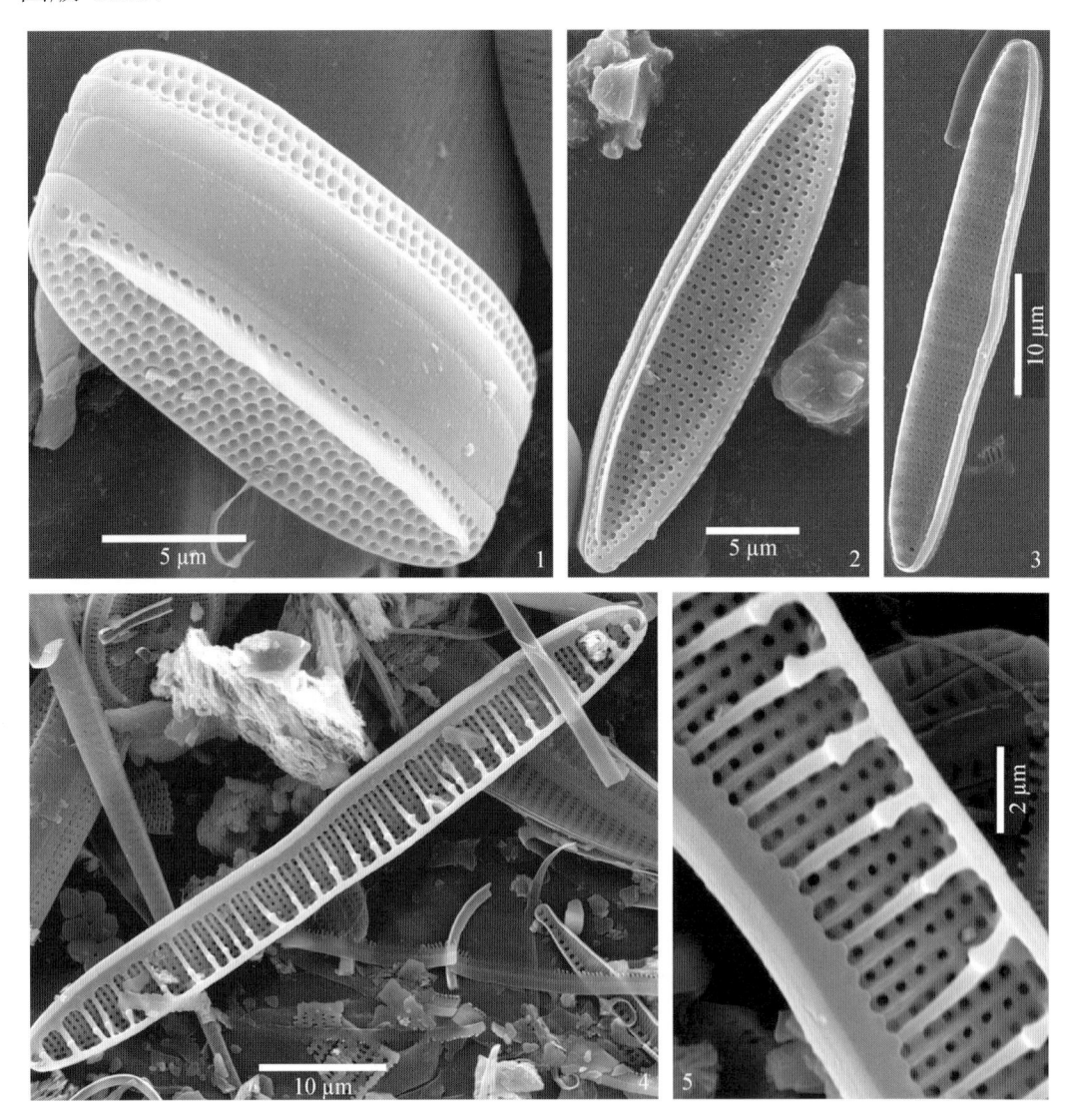

1—5. 库津细齿藻（原变种）*Denticula kuetzingii* var. *kuetzingii* Grunow，1—3. 示外壳面（SEM），4. 示内壳面（SEM），5. 示内壳面中部（SEM），孔纹和龙骨突形态。所有图为电镜（SEM）照片，比例尺见图。

图版 LXV

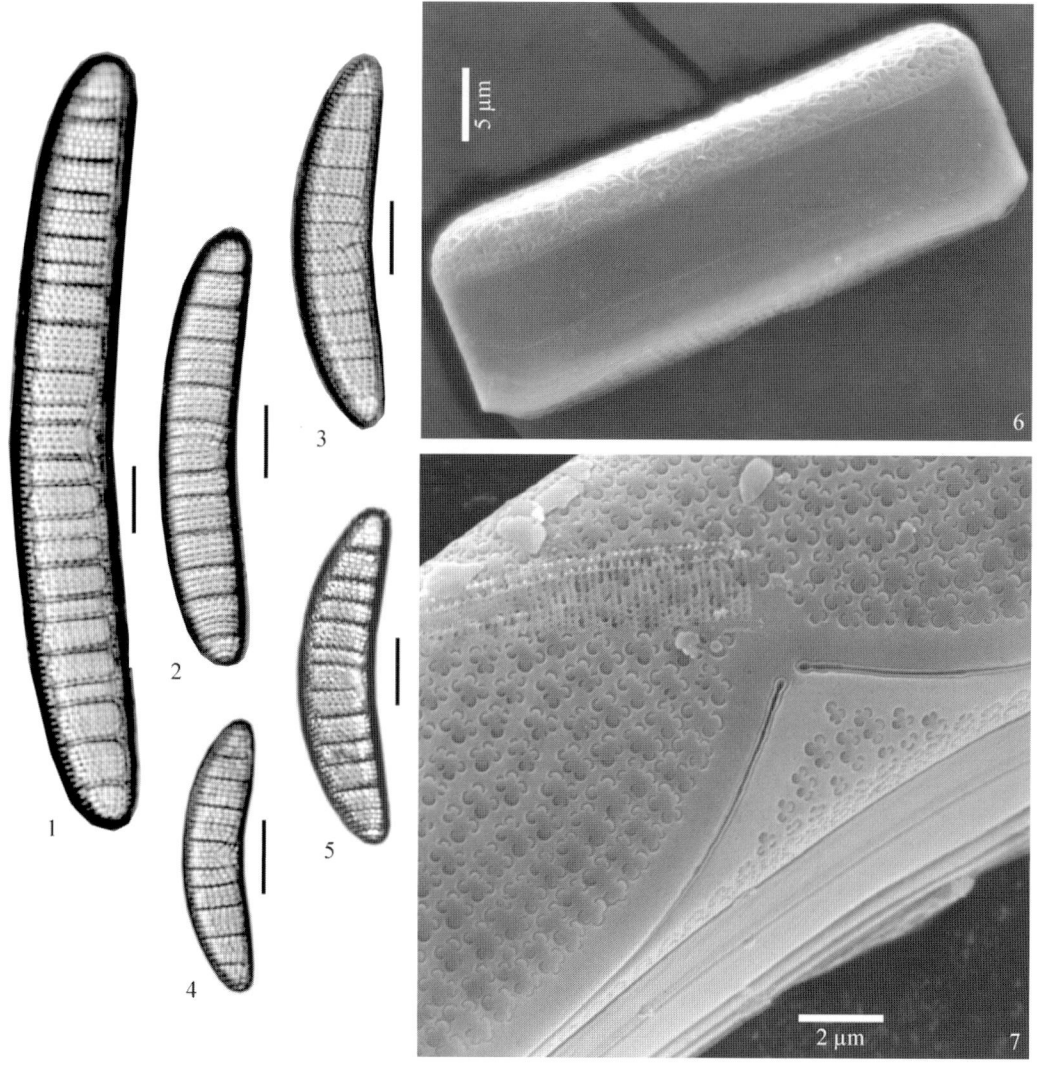

1—2. 侧生窗纹藻（原变种）*Epithemia adnata* var. *adnata* (Kützing) Brébisson；3—7. 侧生窗纹藻萨克森变种 *Epithemia adnata* var. *saxonica* (Kützing) Patrick，6. 示带面观(SEM)，7. 示外壳面中部(SEM)，孔纹和近缝端形态。光镜(LM)照片，比例尺(Scale bars)=10 μm，电镜(SEM)照片比例尺见图。

图版 LXVI

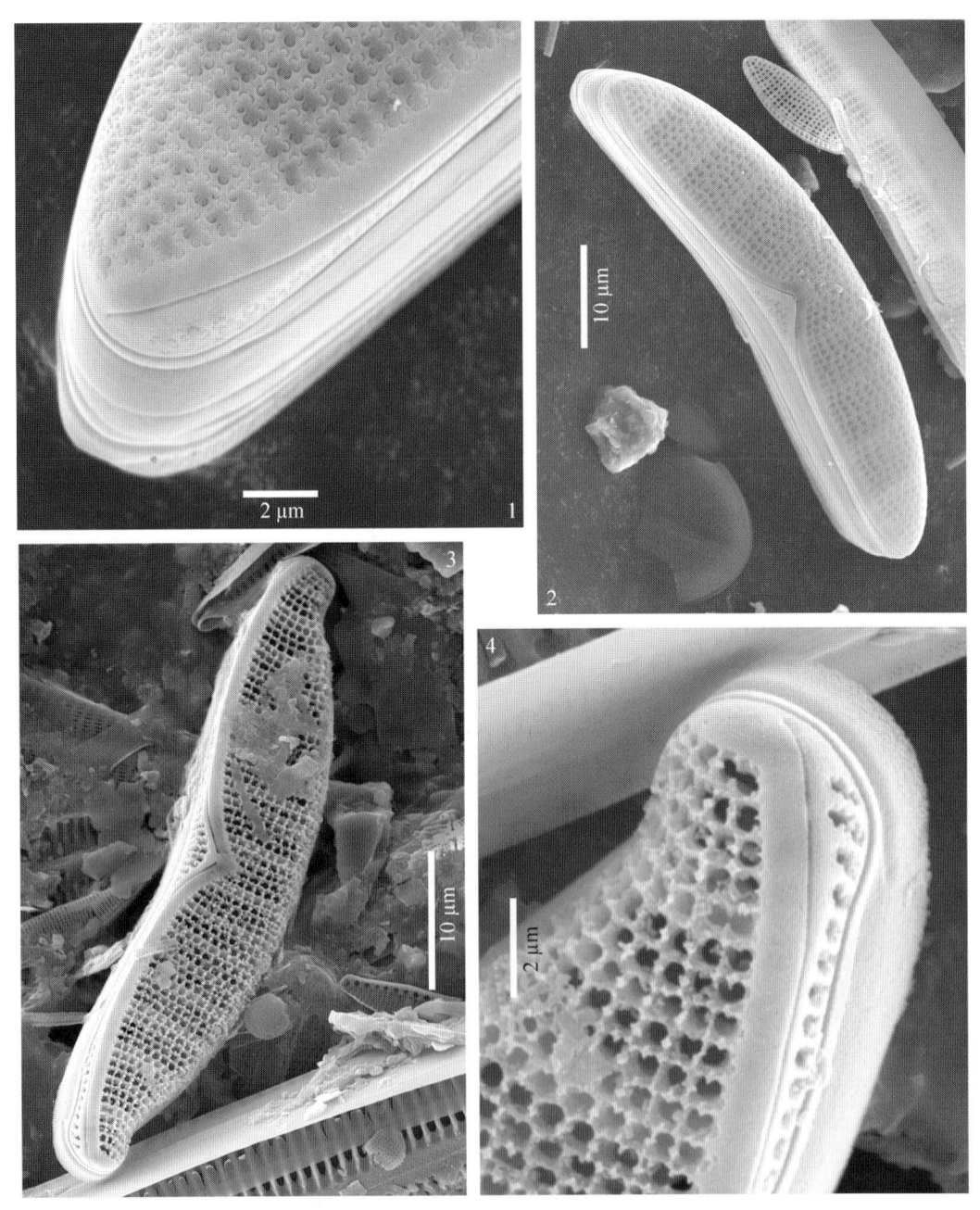

1—2. 侧生窗纹藻萨克森变种 *Epithemia adnata* var. *saxonica* (Kützing) Patrick，1. 示外壳面顶端(SEM)，孔纹和远缝端形态，2. 示外壳面(SEM)。3—4. 侧生窗纹藻蛆形变种 *Epithemia adnata* var. *porcellus* (Kützing) Patrick，3. 示外壳面(SEM)，4. 示外壳面顶端(SEM)，孔纹和远缝端形态。所有图为电镜(SEM)照片，比例尺见图。

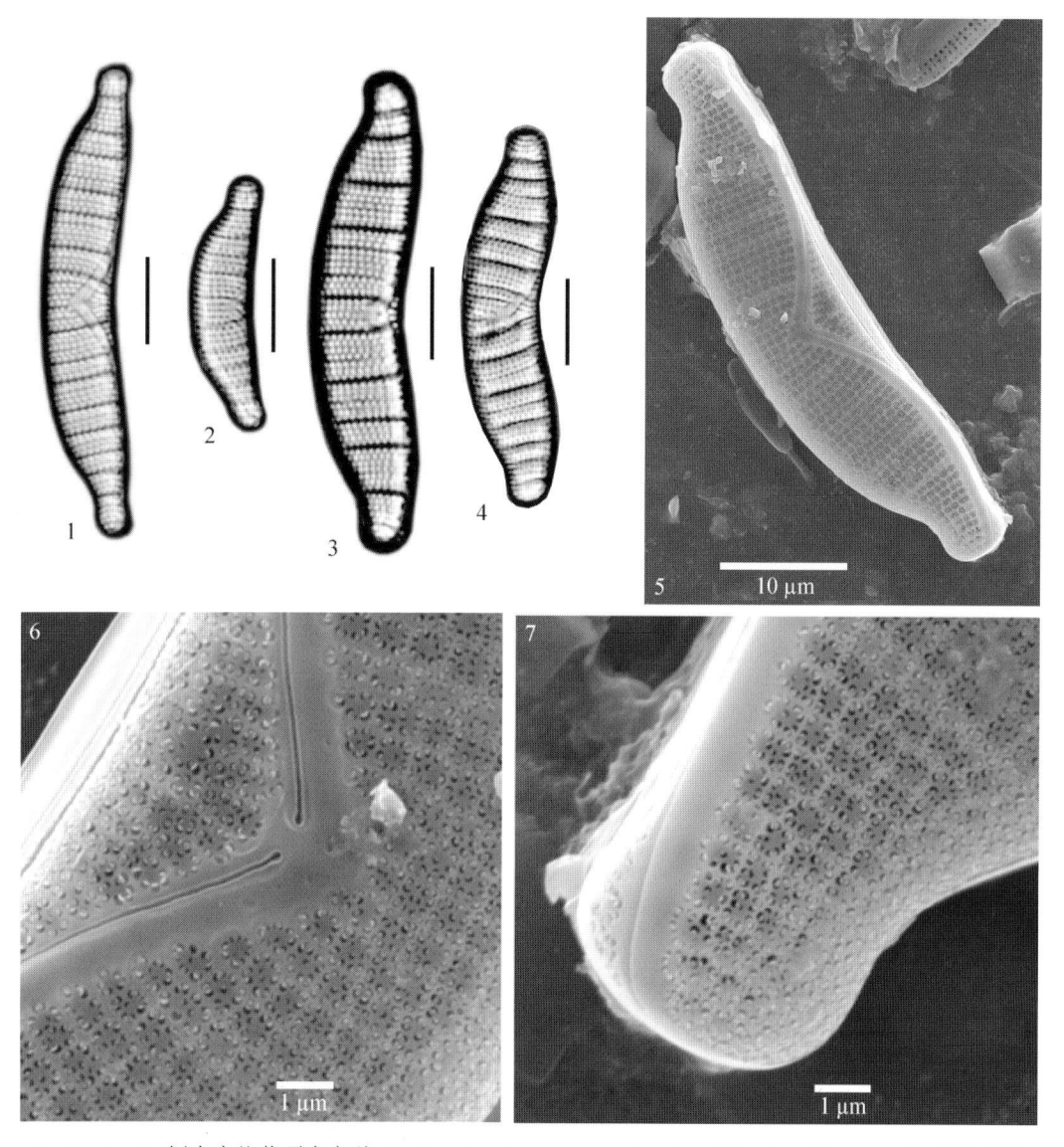

图版 LXVII

1—2，5—7. 侧生窗纹藻顶生变种 *Epithemia adnata* var. *proboscidea* (Kützing) Hendey，5. 示外壳面 (SEM)，6. 示外壳面中部(SEM)，孔纹和近缝端形态，7. 示外壳面顶端(SEM)，孔纹和远缝端形态；3—4. 侧生窗纹藻蛆形变种 *Epithemia adnata* var. *porcellus* (Kützing) Patrick。光镜(LM)照片，比例尺(Scale bars)=10 μm，电镜(SEM)照片比例尺见图。

图版 LXVIII

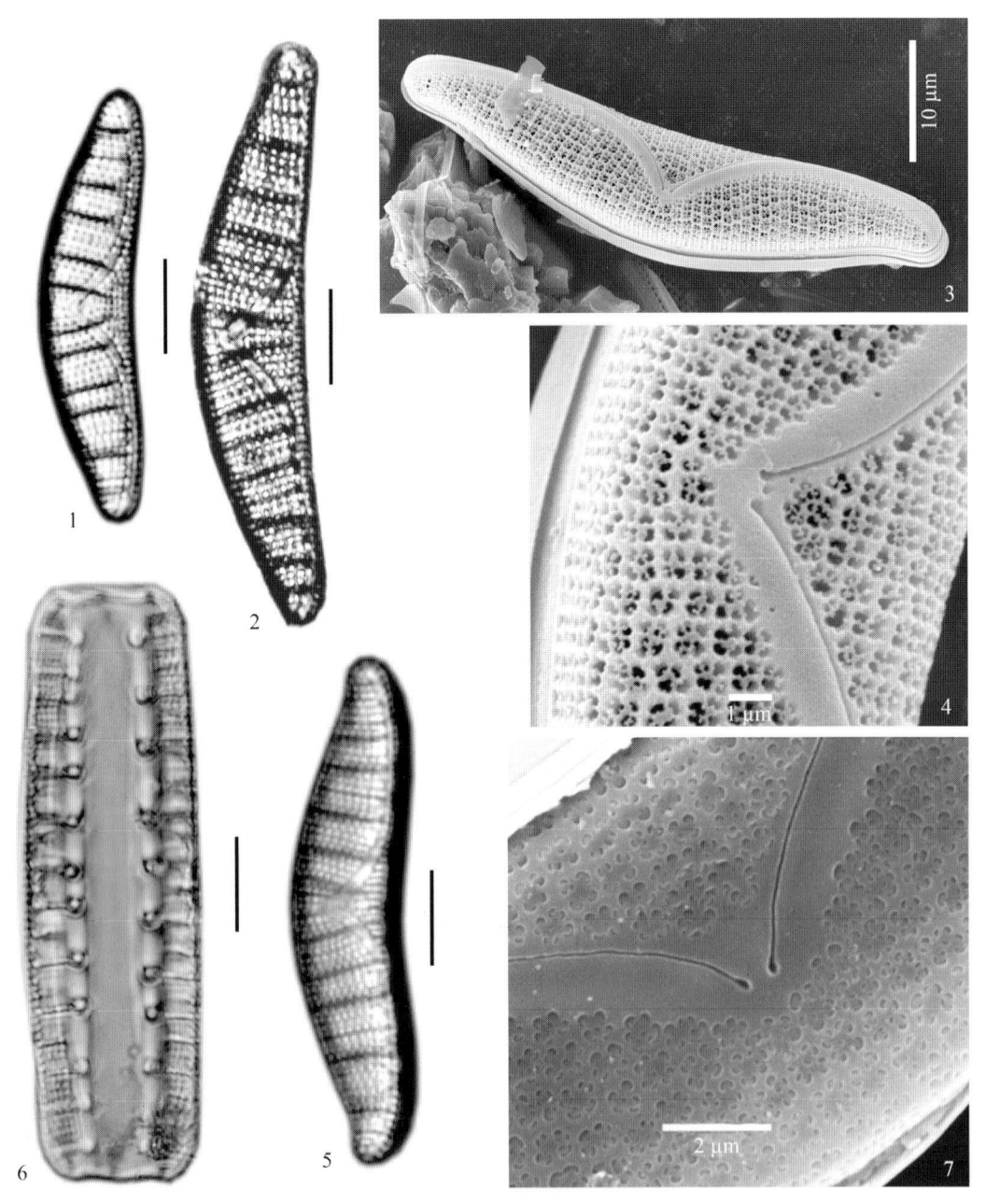

1—4. 光亮窗纹藻（原变种）*Epithemia argus* var. *argus* (Ehrenberg) Kützing，3. 示外壳面(SEM)，4. 示外壳面中部(SEM)，孔纹和近缝端形态；5—7. 光亮窗纹藻高山变种 *Epithemia argus* var. *alpestris* (W. Smith) Grunow，7. 示外壳面中部(SEM)，孔纹和近缝端形态。光镜(LM)照片，比例尺(Scale bars)=10 μm，电镜(SEM)照片比例尺见图。

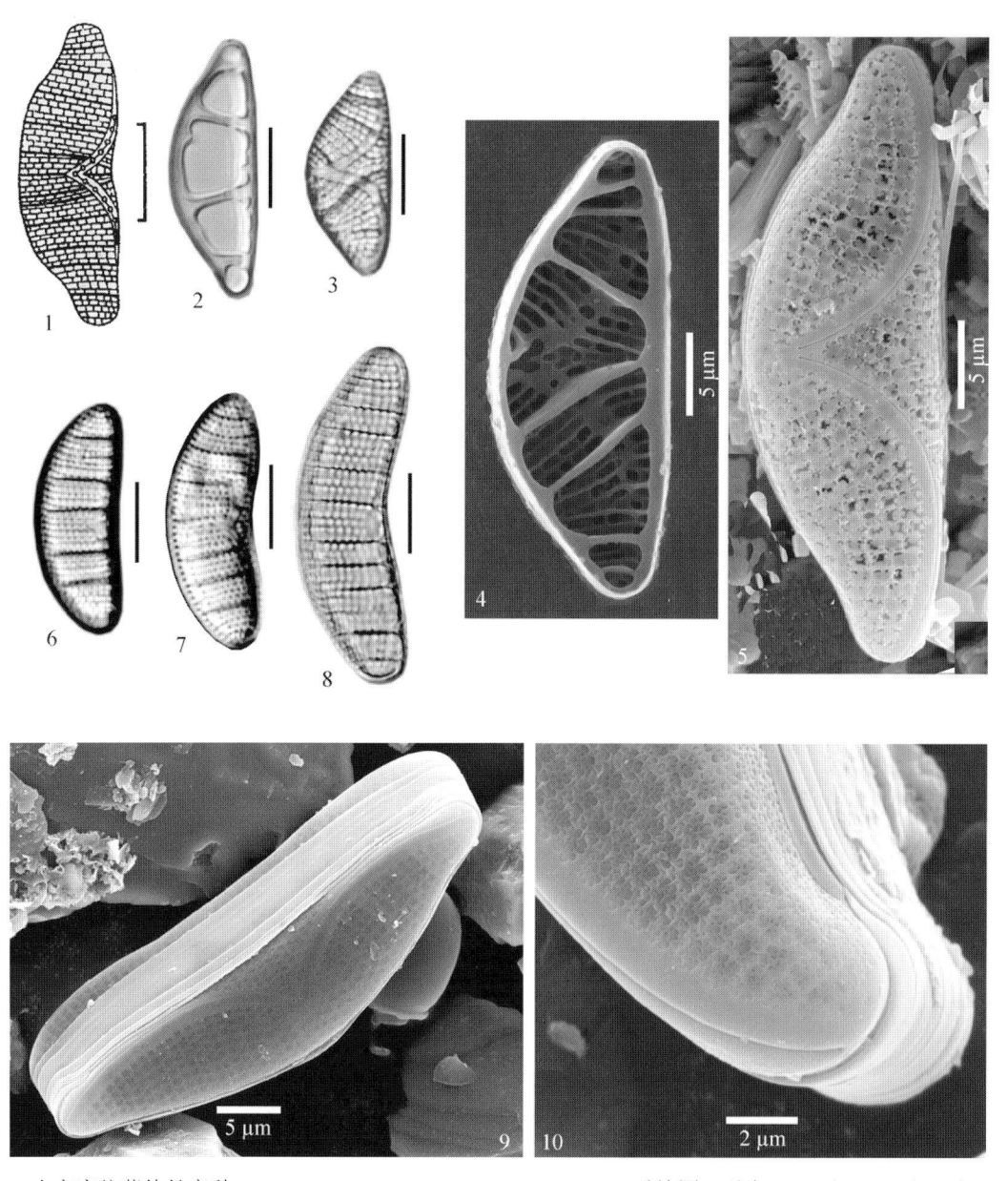

图版 LXIX

1. 光亮窗纹藻伸长变种 *Epithemia argus* var. *protracta* Mayer，手绘图，引自 Zhu 和 Chen (2000)；2—5. 光亮窗纹藻龟形变种 *Epithemia argus* var. *testudo* Fricke，4. 示内壳面 (SEM)，肋纹形态，5. 示外壳面 (SEM)，孔纹和壳缝形态；6—8. 弗里克窗纹藻 *Epithemia frickei* Krammer；9—10. 光亮窗纹藻高山变种 *Epithemia argus* var. *alpestris* (W. Smith) Grunow，9. 示外壳面 (SEM)，10. 示外壳面顶端 (SEM)，孔纹和远缝端形态。光镜 (LM) 照片，比例尺 (Scale bars) =10 μm，电镜 (SEM) 照片比例尺见图。

图版 LXX

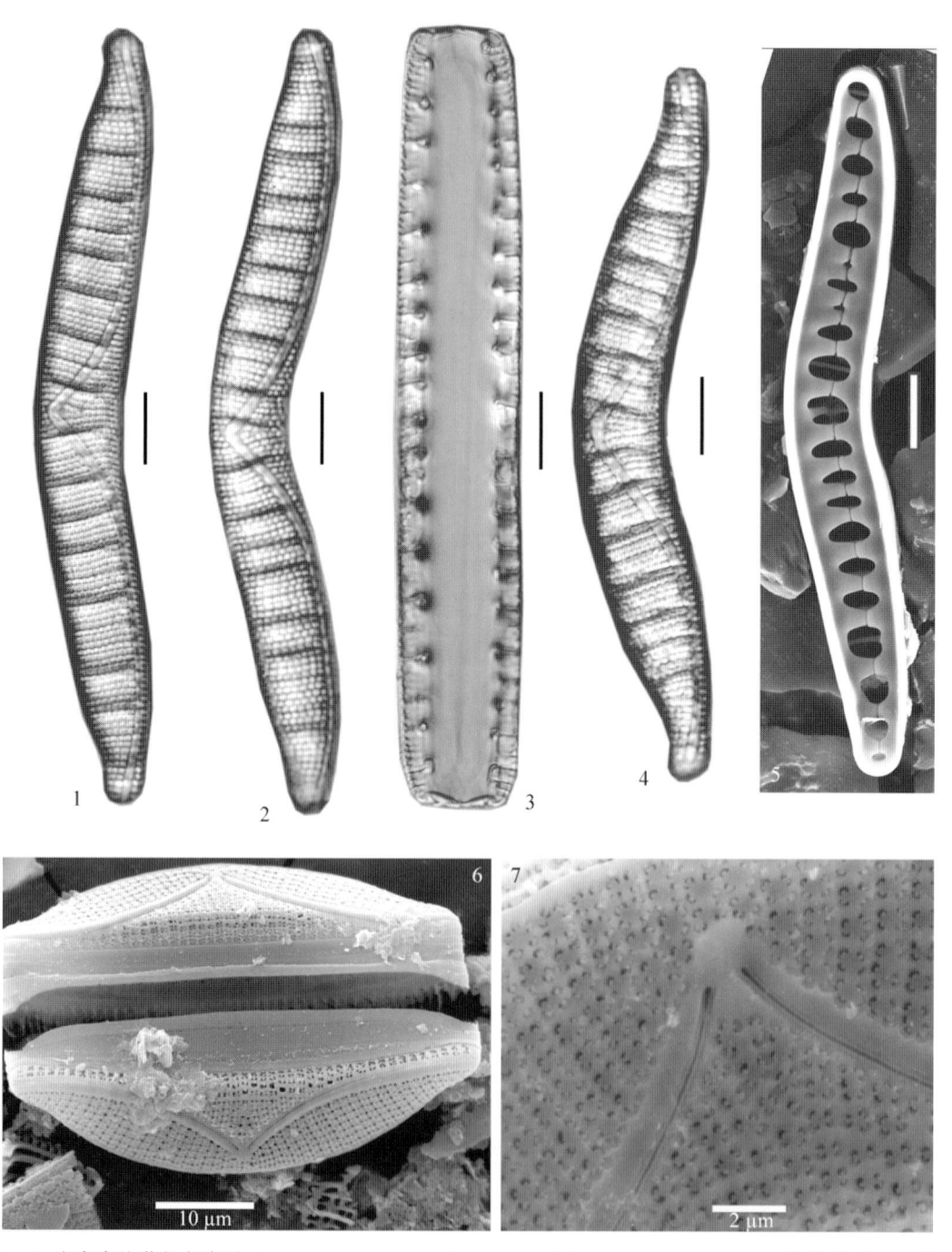

1—5. 光亮窗纹藻长角变种 Epithemia argus var. longicornis (Ehrenberg) Grunow，3. 示带面观，5. 示内壳面(SEM)，肋纹形态；6—7. 鼠形窗纹藻(原变种) Epithemia sorex var. sorex Kützing，6. 示外壳面(SEM)，壳体形态，7. 示外壳面中部(SEM)，孔纹和近缝端形态。光镜(LM)照片，比例尺(Scale bars)=10 μm，电镜(SEM)照片比例尺见图。

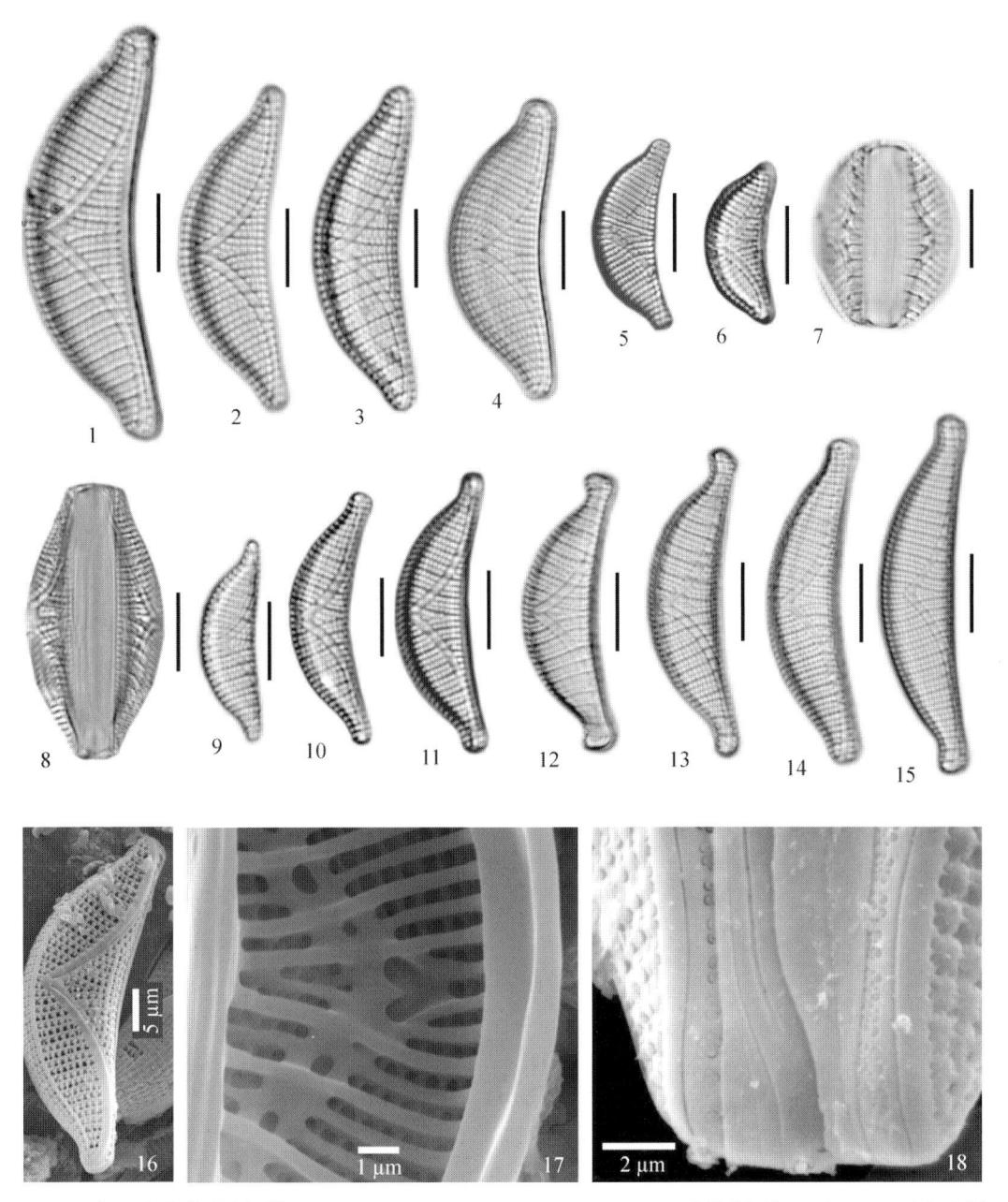

图版 LXXI

1—18. 鼠形窗纹藻（原变种）*Epithemia sorex* var. *sorex* Kützing，16. 示外壳面（SEM），17. 示内壳面中部（SEM），肋纹和管状壳缝形态，18. 示壳套面顶端（SEM）。光镜（LM）照片，比例尺（Scale bars）=10 μm，电镜（SEM）照片比例尺见图。

图版 LXXII

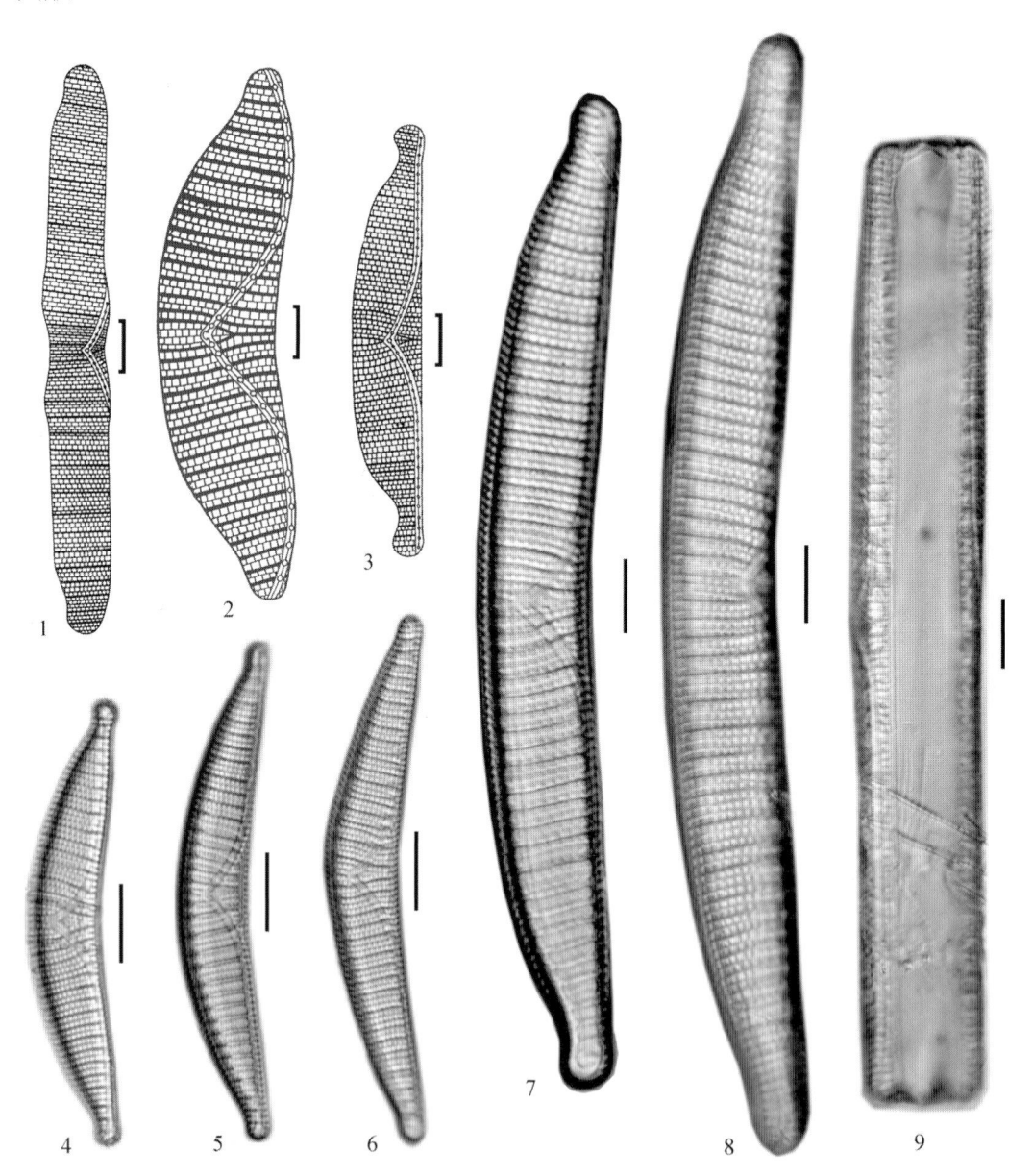

1. 侧生窗纹藻顶生变种二齿变型 Epithemia adnata var. proboscidea f. bidens (A. Cleve) Q-M You & Q-X Wang nov. comb.; 2. 膨大窗纹藻具褶变种 Epithemia turgida var. plicata Meister; 3—6. 膨大窗纹藻头端变种 Epithemia turgida var. capitata Fricke; 7—9. 膨大窗纹藻颗粒变种 Epithemia turgida var. granulata (Ehrenberg) Brun。其中，1，2，3 为手绘图，引自 Zhu 和 Chen (2000)，其他图为光镜 (LM) 照片，所有图比例尺 (Scale bars) =10 μm。

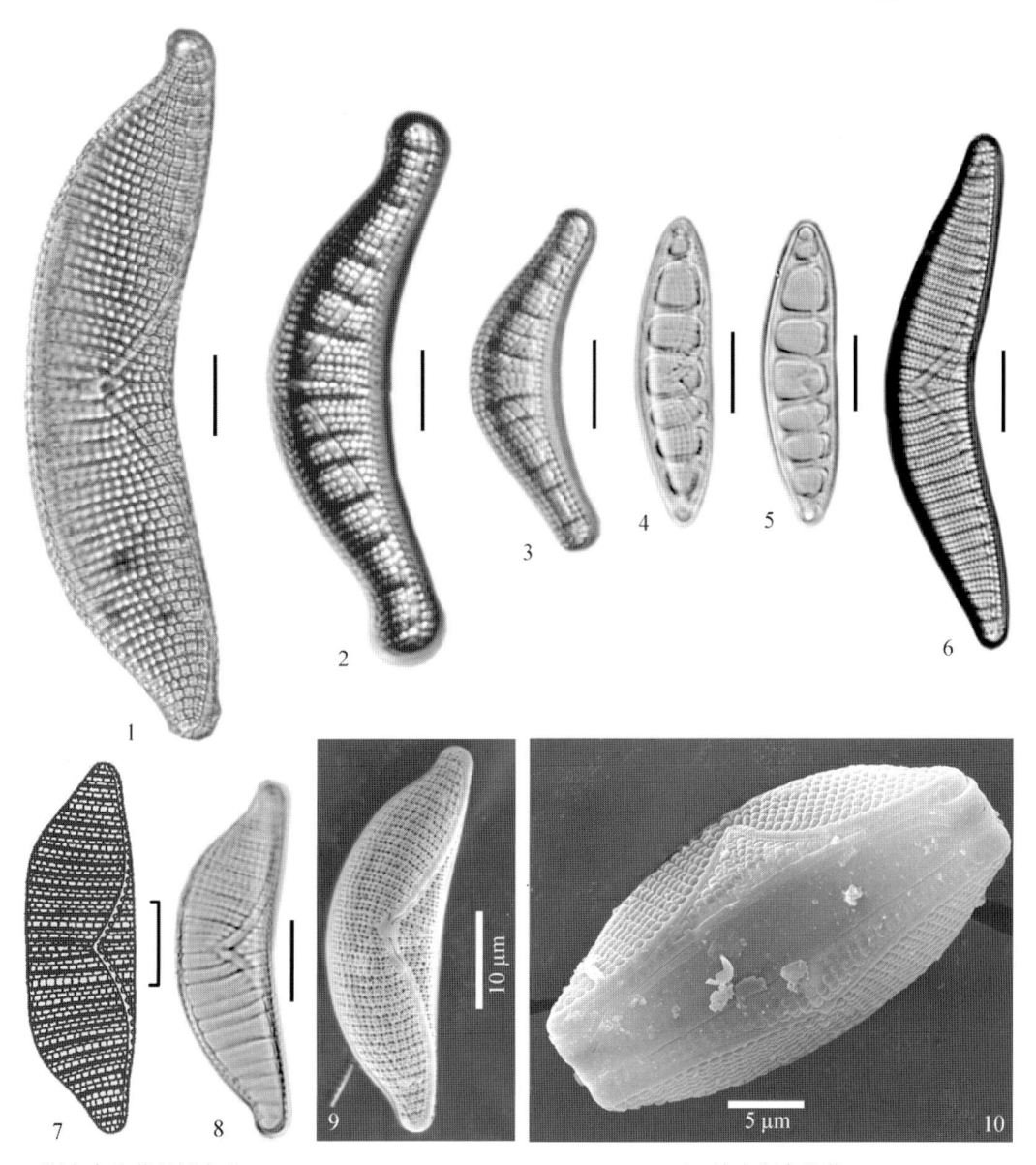

图版 LXXIII

1. 膨大窗纹藻具褶变种 *Epithemia turgida* var. *plicata* Meister；2—3. 施密斯窗纹藻 *Epithemia smithii* Carruthers；4—5. 光亮型窗纹藻 *Epithemia arguiformis* Q-X You & Q-X Wang；6. 鼠形窗纹藻细长变种 *Epithemia sorex* var. *gracilia* Hustedt；7—10. 膨大窗纹藻韦斯特曼变种 *Epithemia turgida* var. *westermannii* (Ehrenberg) Grunow，9. 示外壳面(SEM)，10. 示壳体(SEM)。其中，7 为手绘图，引自 Zhu 和 Chen (2000)，其他除注明外，均为光镜(LM)照片，手绘图和光镜照片的比例尺(Scale bars)=10 μm，电镜(SEM)照片比例尺见图。

图版 LXXIV

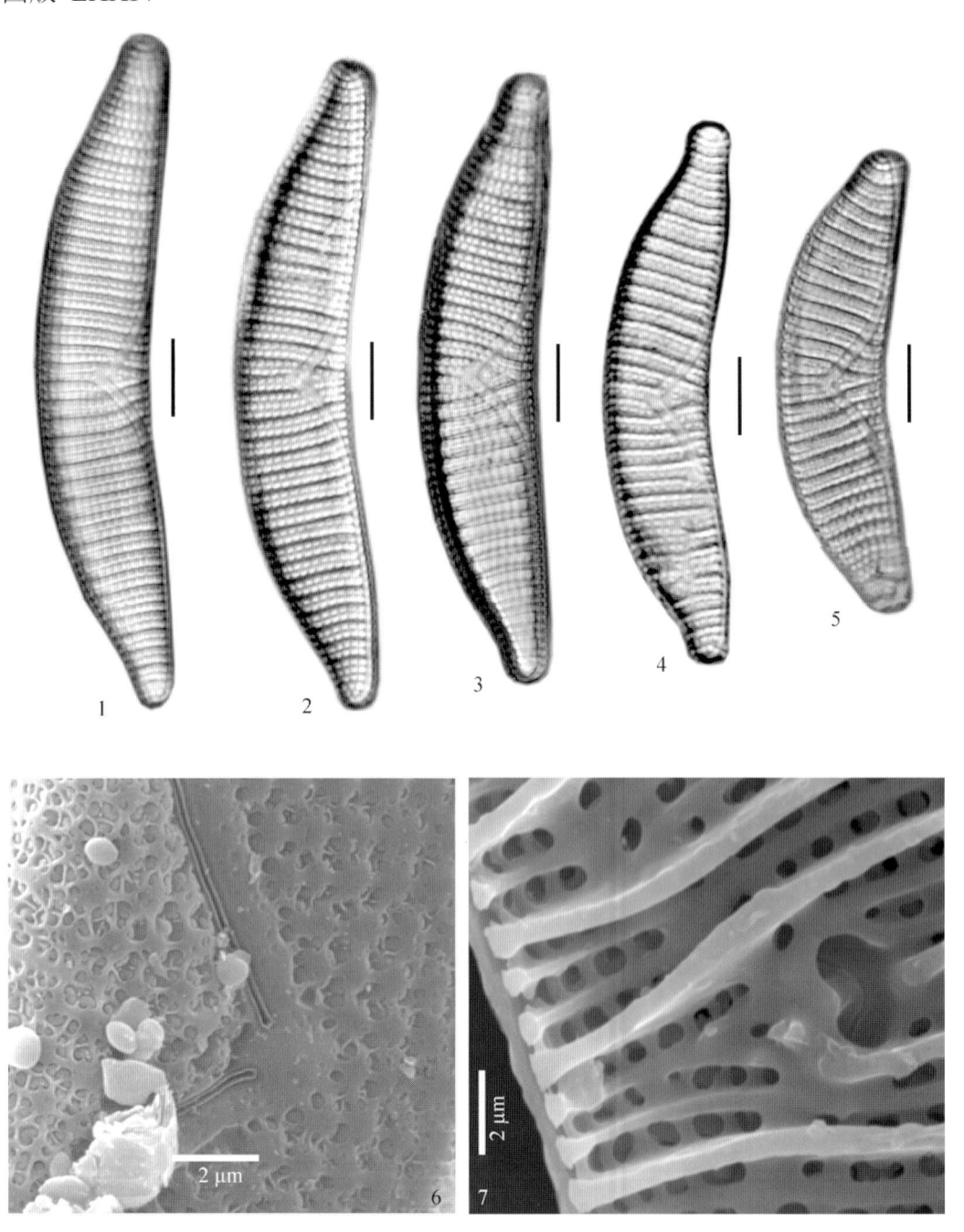

1—7. 膨大窗纹藻（原变种）*Epithemia turgida* var. *turgida* (Ehrenberg) Kützing，6. 示外壳面中部(SEM)，孔纹和近缝端形态，7. 示内壳面中部(SEM)，肋纹和近缝端形态。光镜(LM)照片，比例尺(Scale bars)=10 μm，电镜(SEM)照片比例尺见图。

图版 LXXV

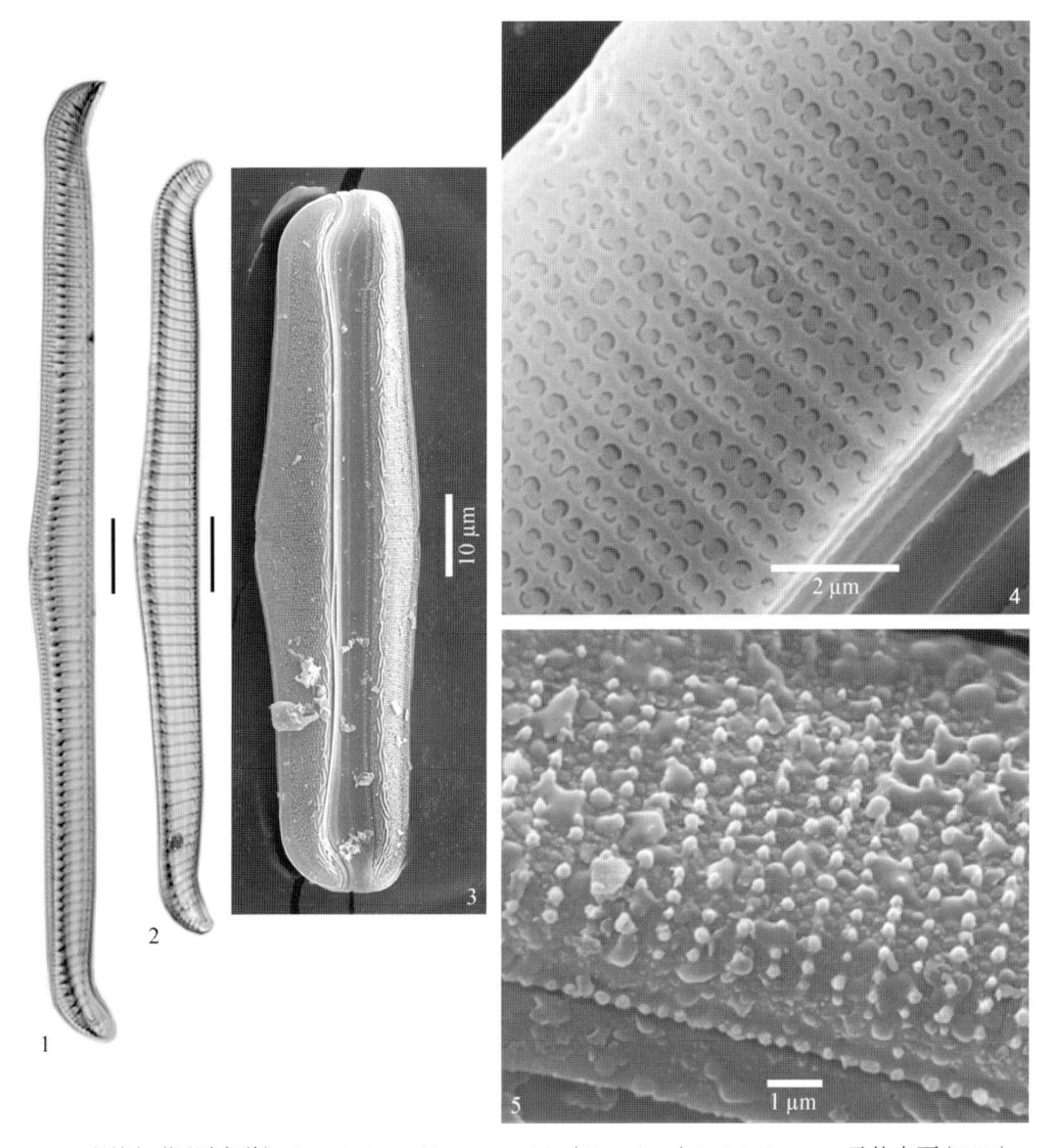

1—5. 弯棒杆藻（原变种）*Rhopalodia gibba* var. *gibba* (Ehrenberg) O. Müller，3. 示外壳面（SEM），4—5. 示外壳面中部（SEM），孔纹形态。光镜（LM）照片，比例尺（Scale bars）=10 μm，电镜（SEM）照片比例尺见图。

图版 LXXVI

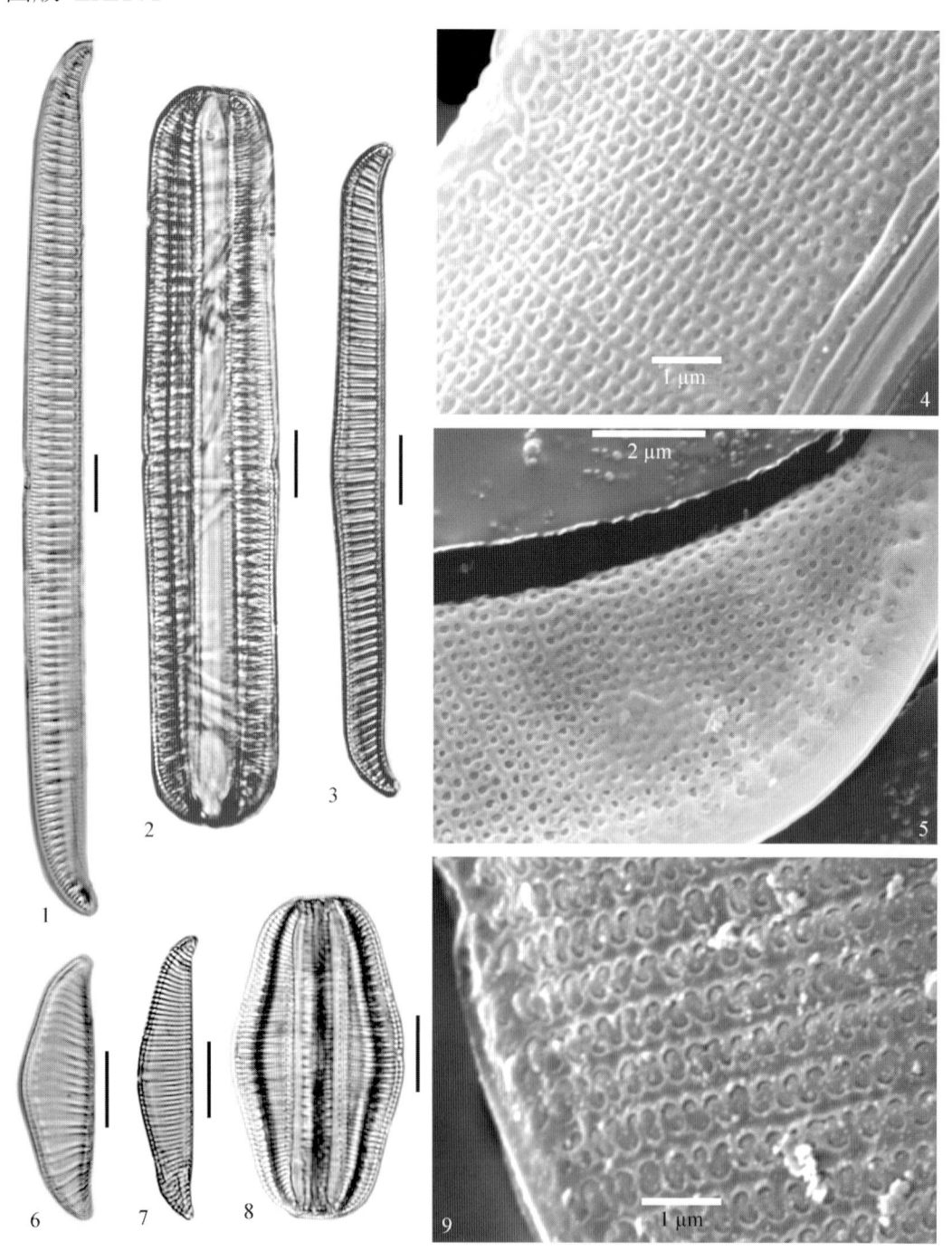

1—5. 平行棒杆藻（原变种）*Rhopalodia parallela* var. *parallela* (Grunow) O. Müller，4. 示外壳面中部 (SEM)，孔纹形态，5. 示外壳面顶端 (SEM)，孔纹形态；6—9. 弯棒杆藻偏肿变种 *Rhopalodia gibba* var. *ventricosa* (Kützing) H. & M. Paragallo, 9. 示外壳面中部 (SEM)，孔纹形态。光镜 (LM) 照片，比例尺 (Scale bars)=10 μm，电镜 (SEM) 照片比例尺见图。

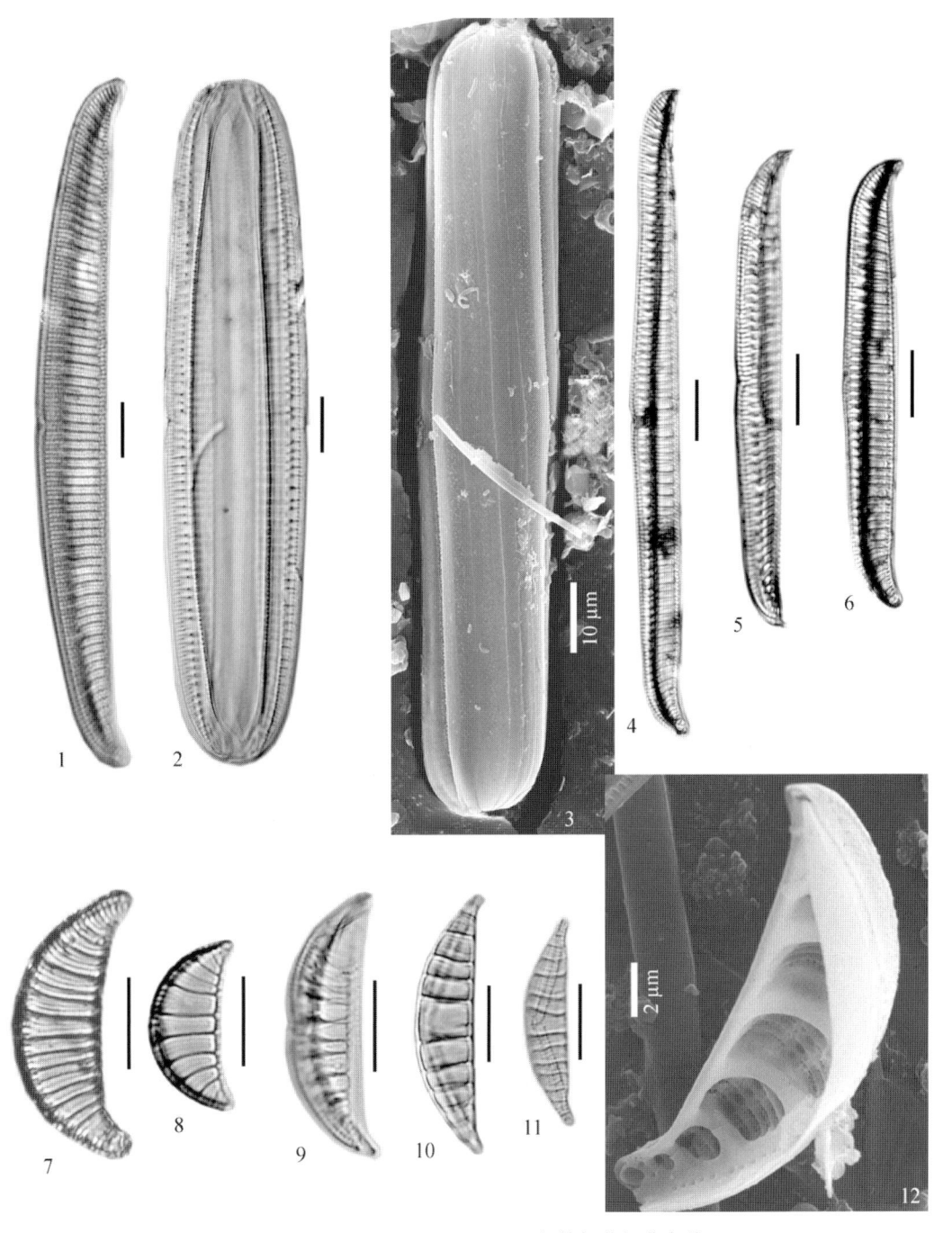

图版 LXXVII

1—2. 纤细棒杆藻 *Rhopalodia gracilis* O. Müller；3—6. 平行棒杆藻扭曲变种 *Rhopalodia parallela* var. *distorta* Fricke，3. 示壳体(SEM)；7—8. 肌状棒杆藻 *Rhopalodia musculus*(Kützing) O. Müller；9. 缢缩棒杆藻 *Rhopalodia constricta*(W. Smith) Krammer；10—12. 毕列松棒杆藻 *Rhopalodia brebissonii* Krammer，12. 示内壳面(SEM)，隔片形态。光镜(LM)照片，比例尺(Scale bars)=10 μm，电镜(SEM)照片比例尺见图。

图版 LXXVIII

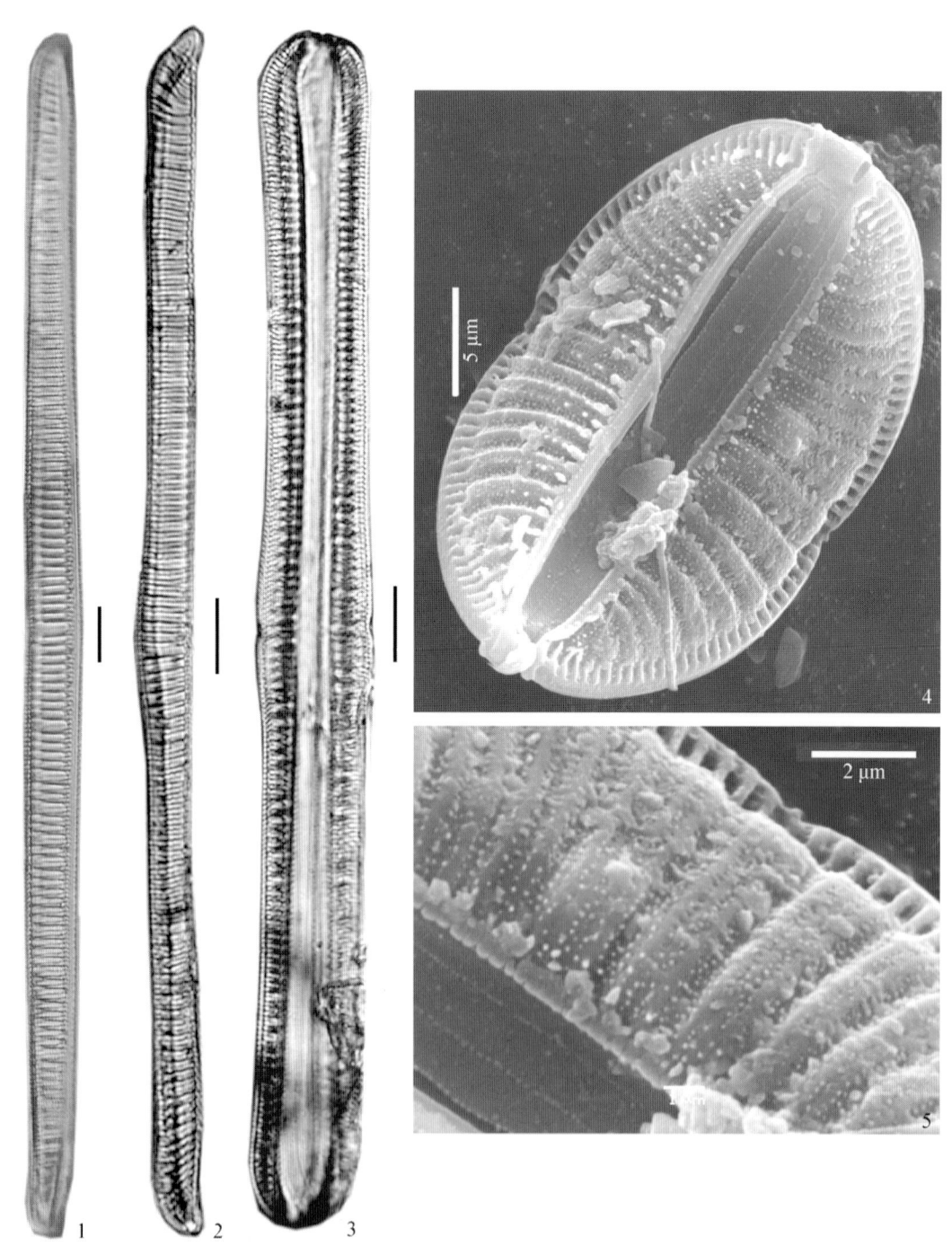

1. 平行棒杆藻巨大变种 *Rhopalodia parallela* var. *ingens* Fricke；2—3. 平行棒杆藻（原变种）*Rhopalodia parallela* var. *parallela* (Grunow) O. Müller；4—5. 缢缩棒杆藻 *Rhopalodia constricta* (W. Smith) Krammer，4. 示壳体 (SEM)，5. 示外壳面中部 (SEM)。光镜 (LM) 照片，比例尺 (Scale bars) =10 μm，电镜 (SEM) 照片比例尺见图。

图版 LXXIX

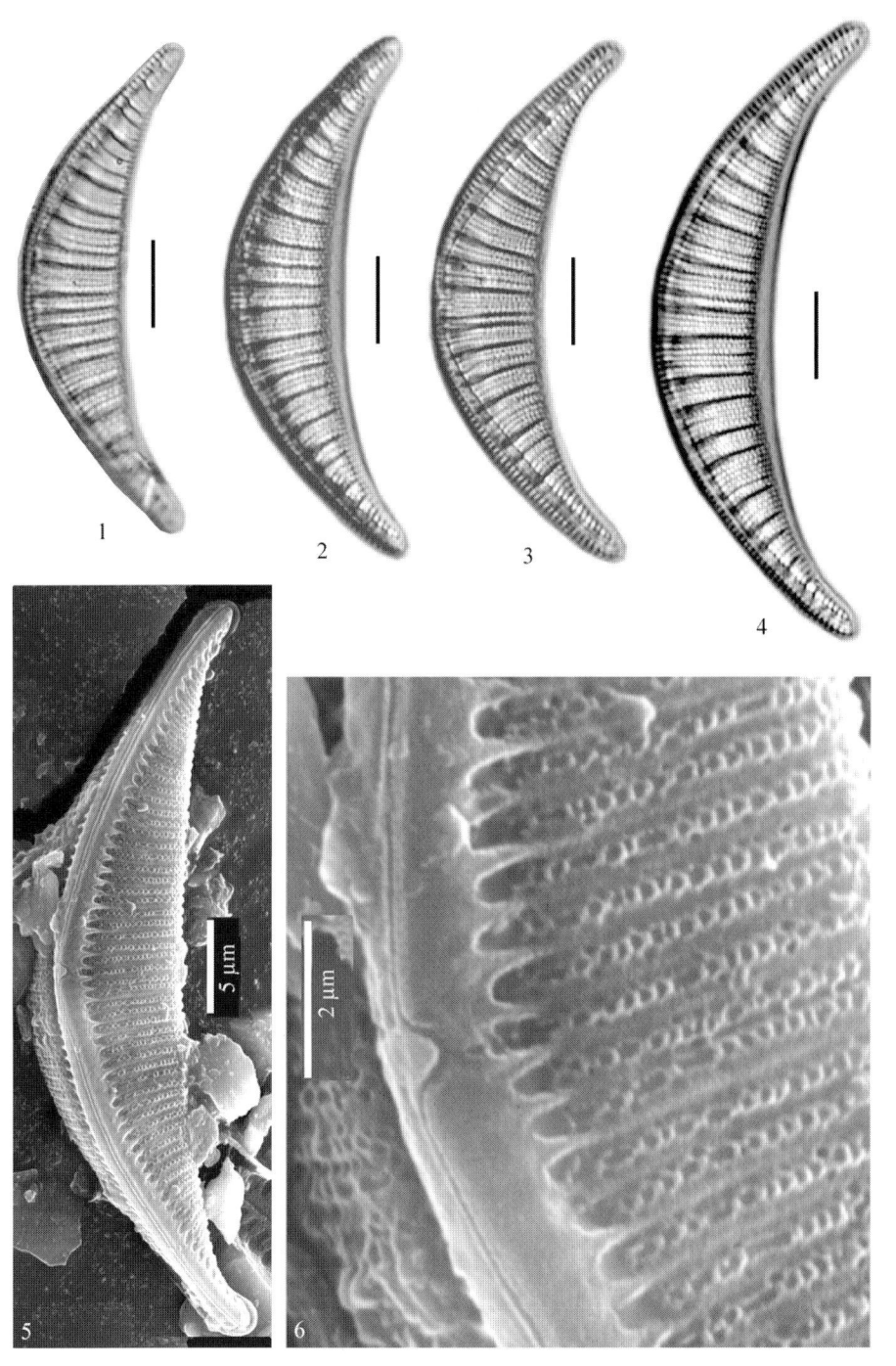

1—6. 驼峰棒杆藻 *Rhopalodia gibberula* (Ehrenberg) O. Müller，5. 示外壳面(SEM)，6. 示外壳面中部(SEM)，孔纹形态。光镜(LM)照片，比例尺(Scale bars)=10 μm，电镜(SEM)照片比例尺见图。

图版 LXXX

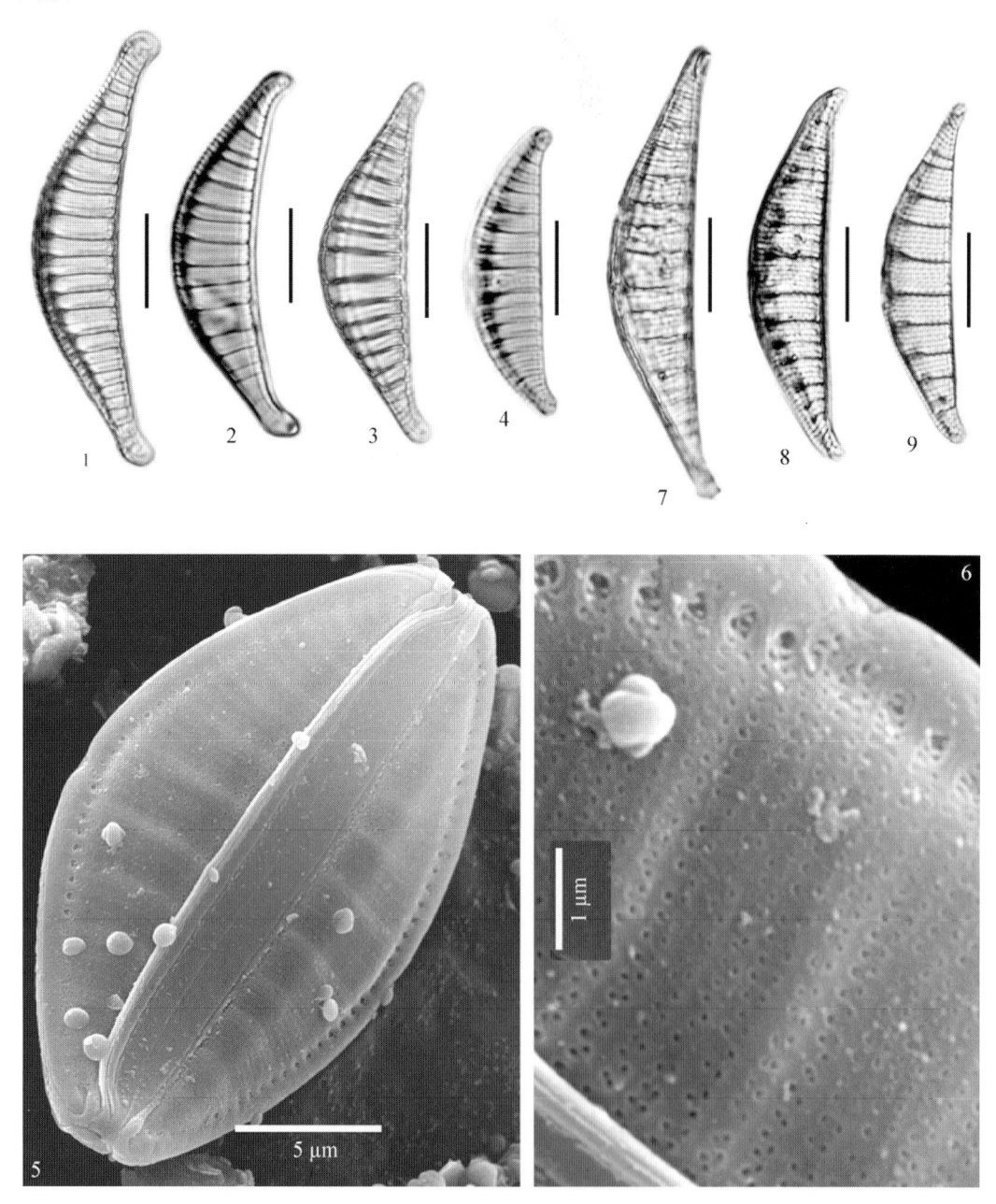

1—6. 具盖棒杆藻 *Rhopalodia operculata* (Agardh) Håkansson，5. 示外壳面(SEM)，6. 示外壳面中部(SEM)，孔纹形态；7—9. 石生棒杆藻 *Rhopalodia rupestris* (W. Smith) Krammer。光镜(LM)照片，比例尺(Scale bars)=10 μm，电镜(SEM)照片比例尺见图。

图版 LXXXI

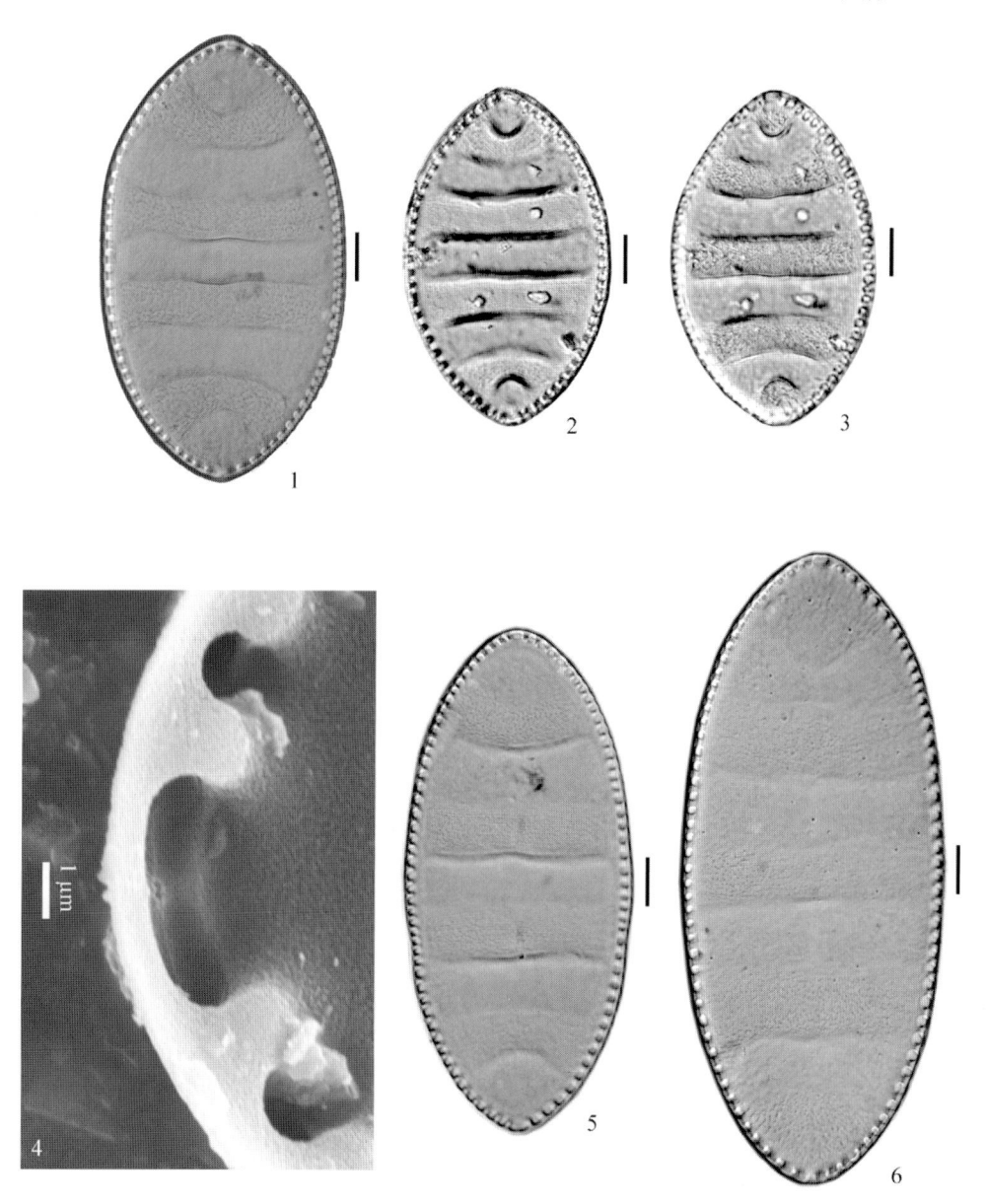

1—6. 椭圆波缘藻(原变种)*Cymatopleura elliptica* var. *elliptica* (Brébisson) W. Smith，4. 示内壳面顶端(SEM)。光镜(LM)照片，比例尺(Scale bars)=10 μm，电镜(SEM)照片比例尺见图。

图版 LXXXII

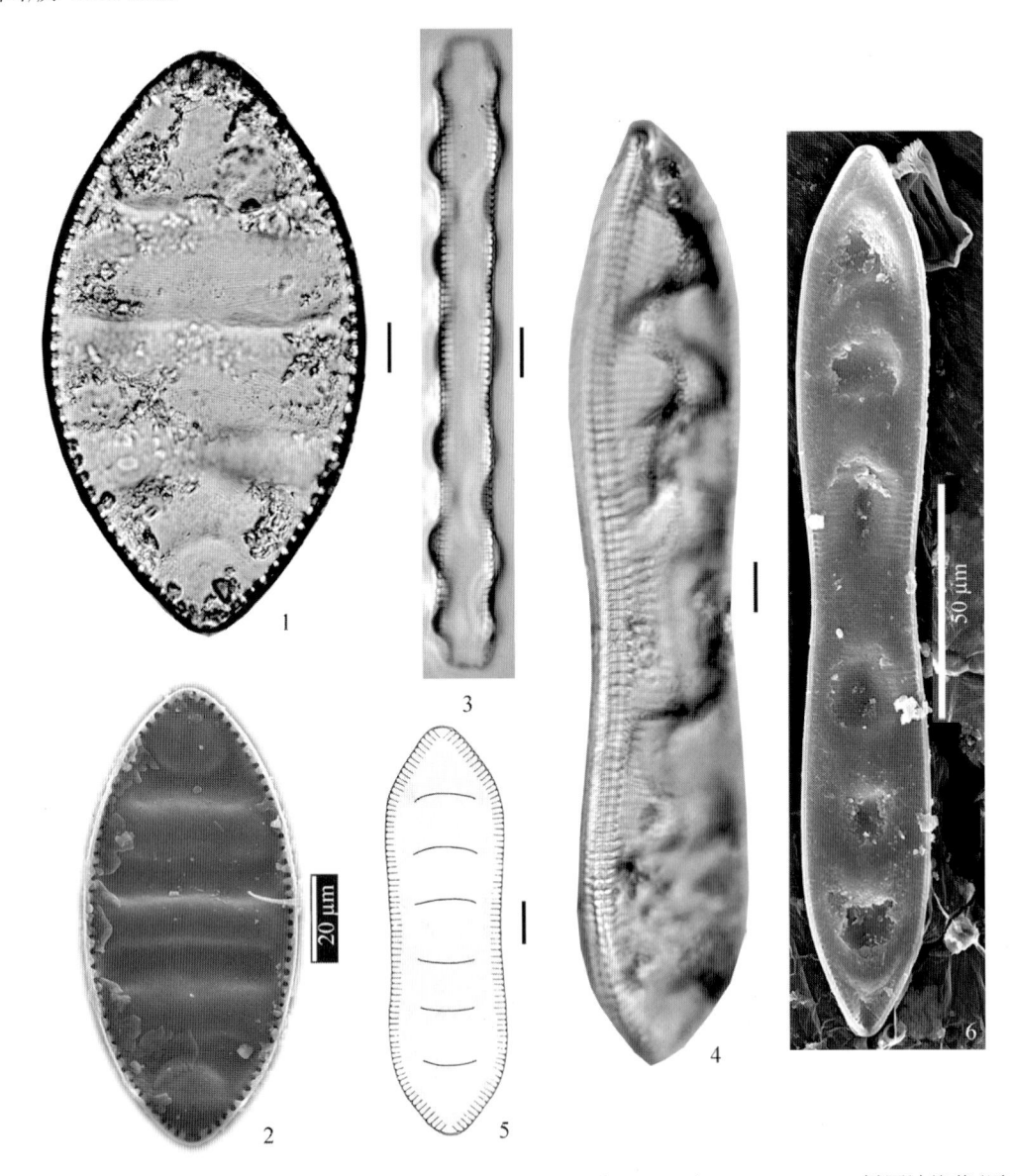

1. 椭圆波缘藻冬生变种 *Cymatopleura elliptica* var. *hibernica* (W. Smith) Van Heurck；2. 椭圆波缘藻（原变种）*Cymatopleura elliptica* var. *elliptica* (Brébisson) W. Smith，示内壳面 (SEM)；3—4. 草鞋形波缘藻（原变种）*Cymatopleura solea* var. *solea* (Brébisson) W. Smith，3. 示带面观；5. 椭圆波缘藻缢缩变种 *Cymatopleura elliptica* var. *constricta* Grunow；6. 草鞋形波缘藻细长变种 *Cymatopleura solea* var. *gracilis* Grunow，示外壳面 (SEM)。其中，5 为手绘图，引自 Zhu 和 Chen (2000)，其他除注明外，均为光镜 (LM) 照片，比例尺 (Scale bars)=10 μm，其中电镜 (SEM) 照片比例尺见图。

图版 LXXXIII

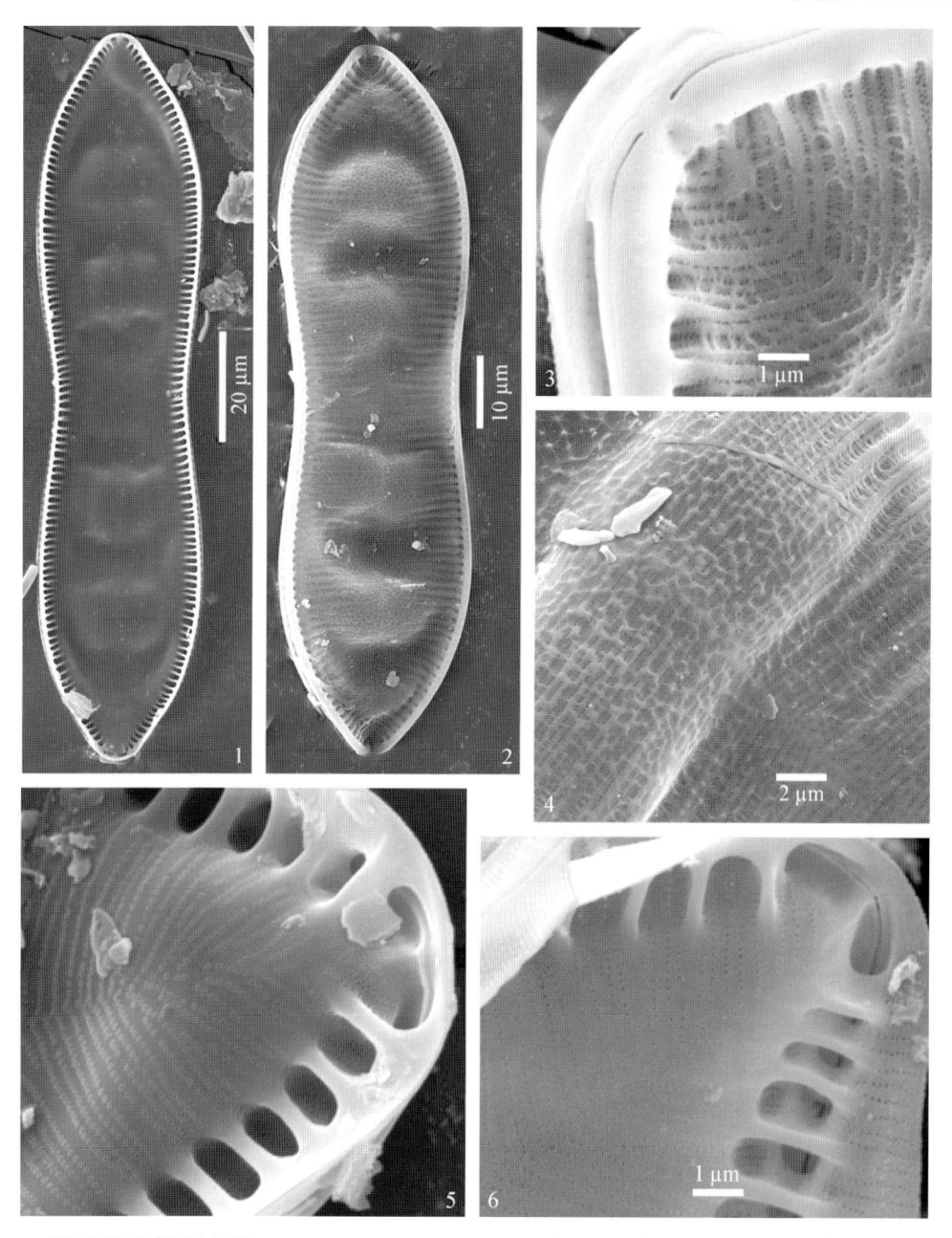

1—6. 草鞋形波缘藻（原变种）*Cymatopleura solea* var. *solea* (Brébisson) W. Smith，1. 示内壳面（SEM），2. 示外壳面（SEM），3. 示外壳面顶端（SEM），孔纹和壳缝形态，4. 示外壳面中部（SEM），5—6. 示内壳面顶端（SEM），孔纹、龙骨突和壳缝形态。所有图为电镜（SEM）照片，比例尺见图。

图版 LXXXIV

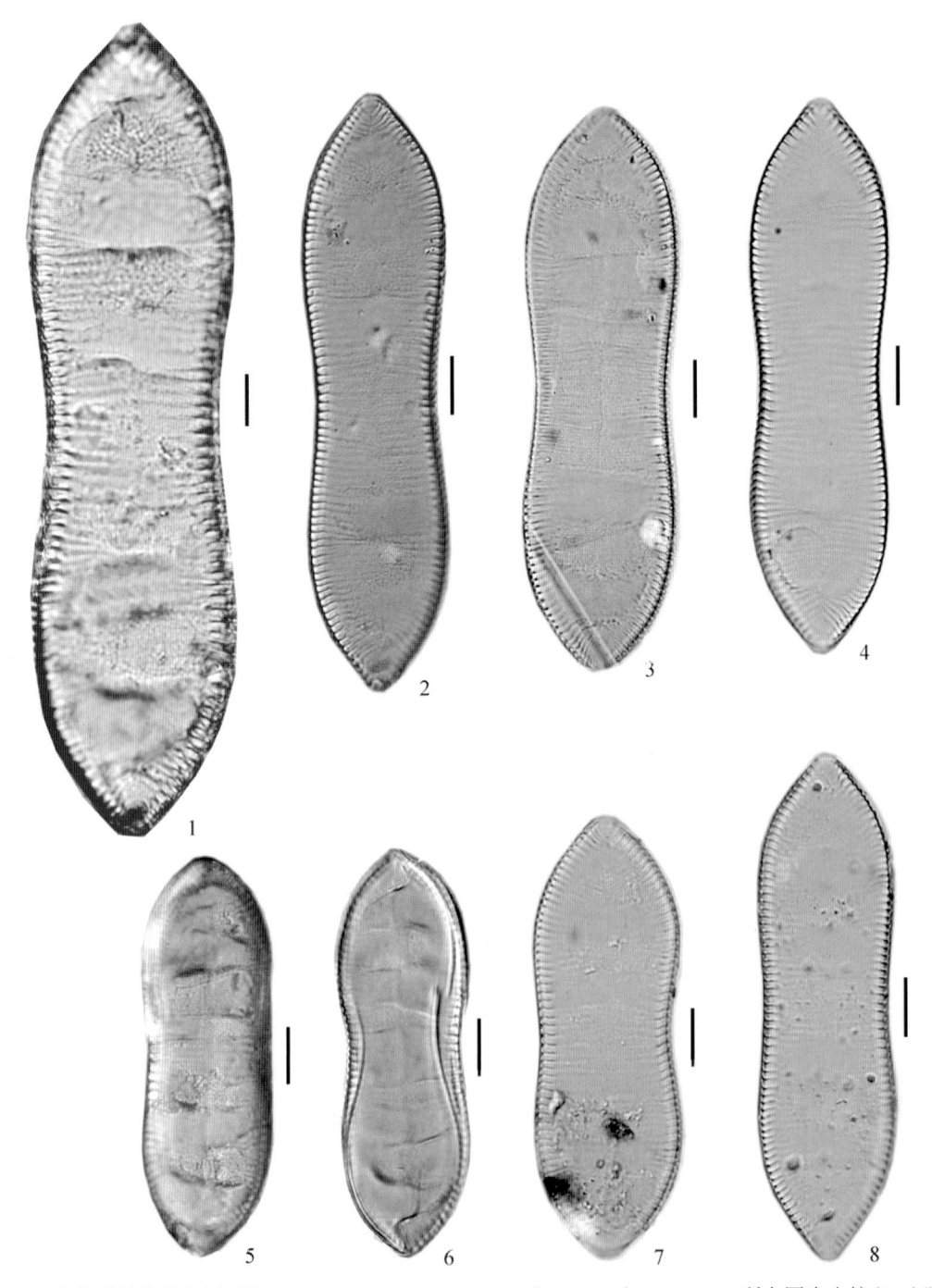

1—8. 草鞋形波缘藻（原变种）*Cymatopleura solea* var. *solea* (Brébisson) W. Smith。所有图为光镜(LM)照片，比例尺(Scale bars)=10 μm。

图版 LXXXV

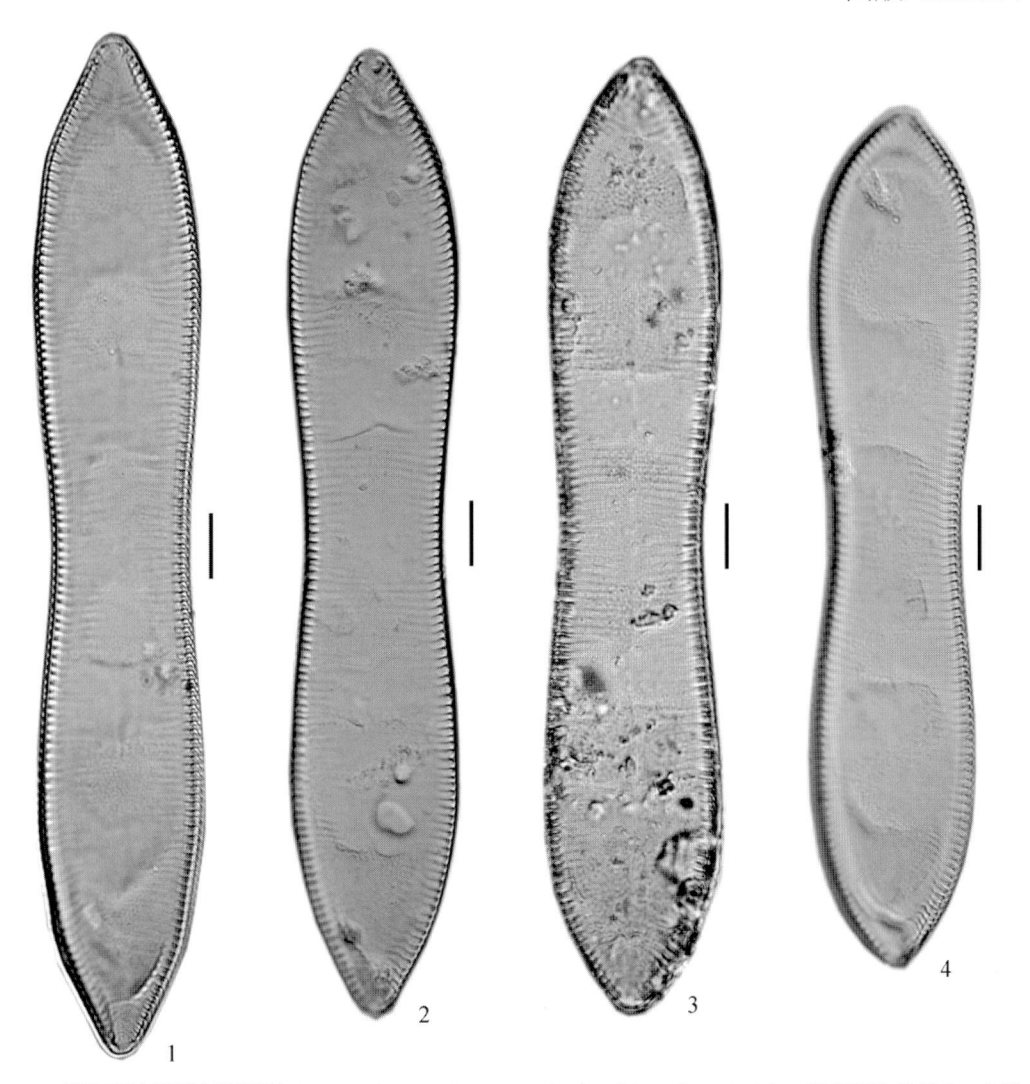

1—4. 草鞋形波缘藻（原变种）*Cymatopleura solea* var. *solea* (Brébisson) W. Smith。所有图为光镜(LM)照片，比例尺(Scale bars)=10 μm。

图版 LXXXVI

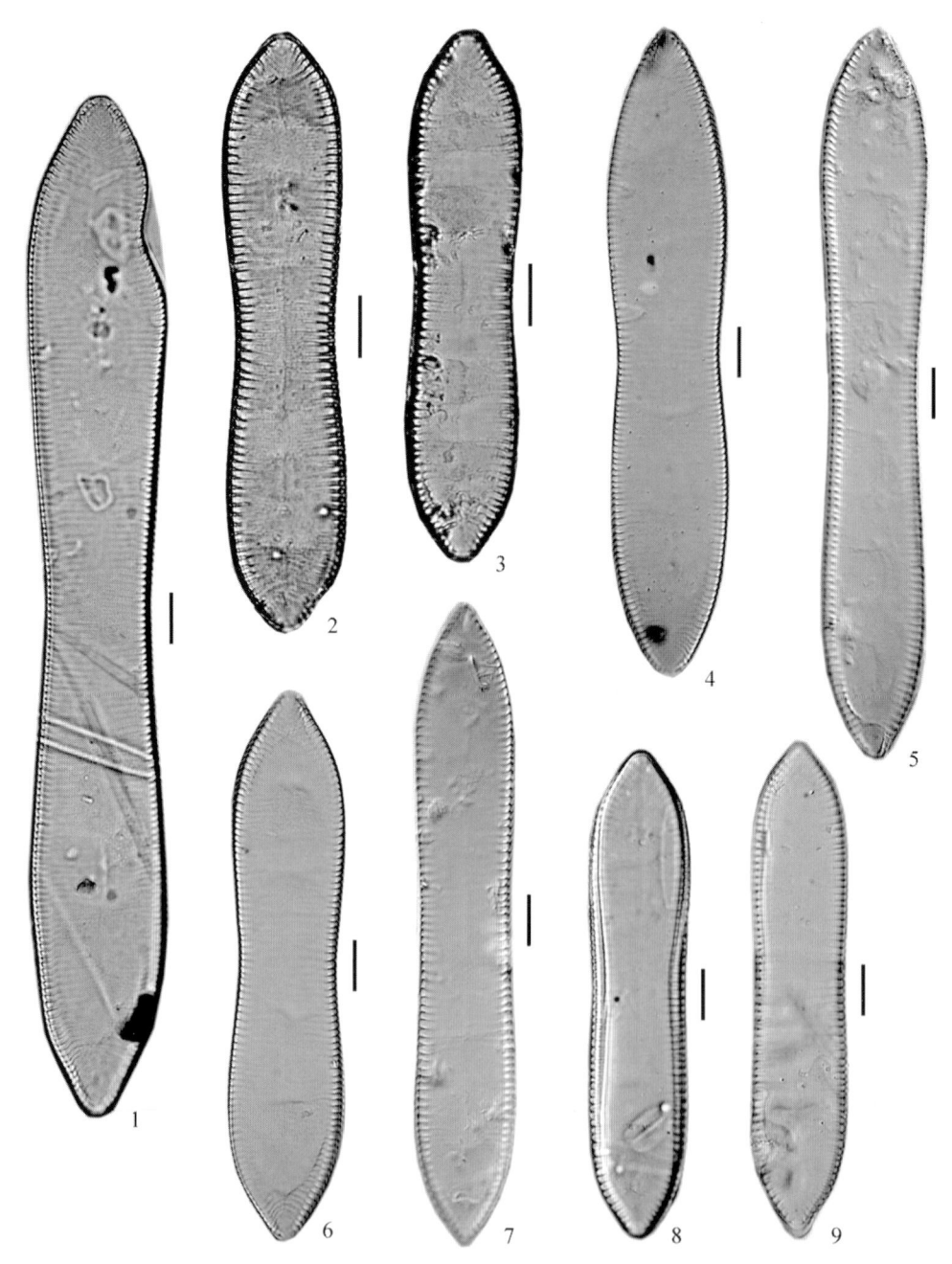

1—6. 草鞋形波缘藻（原变种）*Cymatopleura solea* var. *solea* (Brébisson) W. Smith；7—9. 草鞋形波缘藻整齐变种 *Cymatopleura solea* var. *regula* (Ehrenberg) Grunow。所有图为光镜（LM）照片，比例尺（Scale bars）=10 μm。

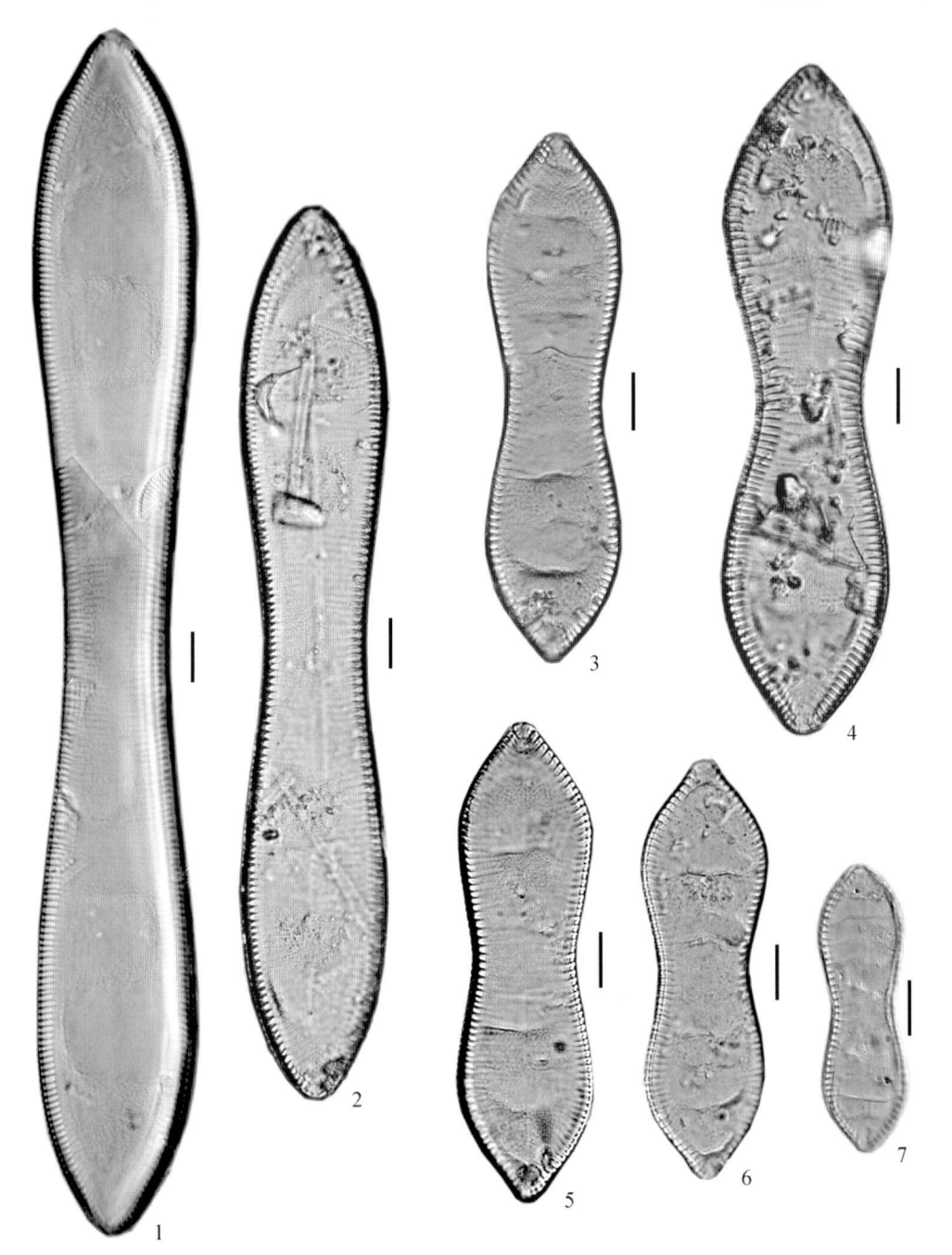

图版 LXXXVII

1—2. 草鞋形波缘藻细长变种 *Cymatopleura solea* var. *gracilis* Grunow；3—7. 草鞋形波缘藻细尖变种 *Cymatopleura solea* var. *apiculata* (W. Smith) Ralfs。所有图为光镜(LM)照片，比例尺(Scale bars)=10 μm。

图版 LXXXVIII

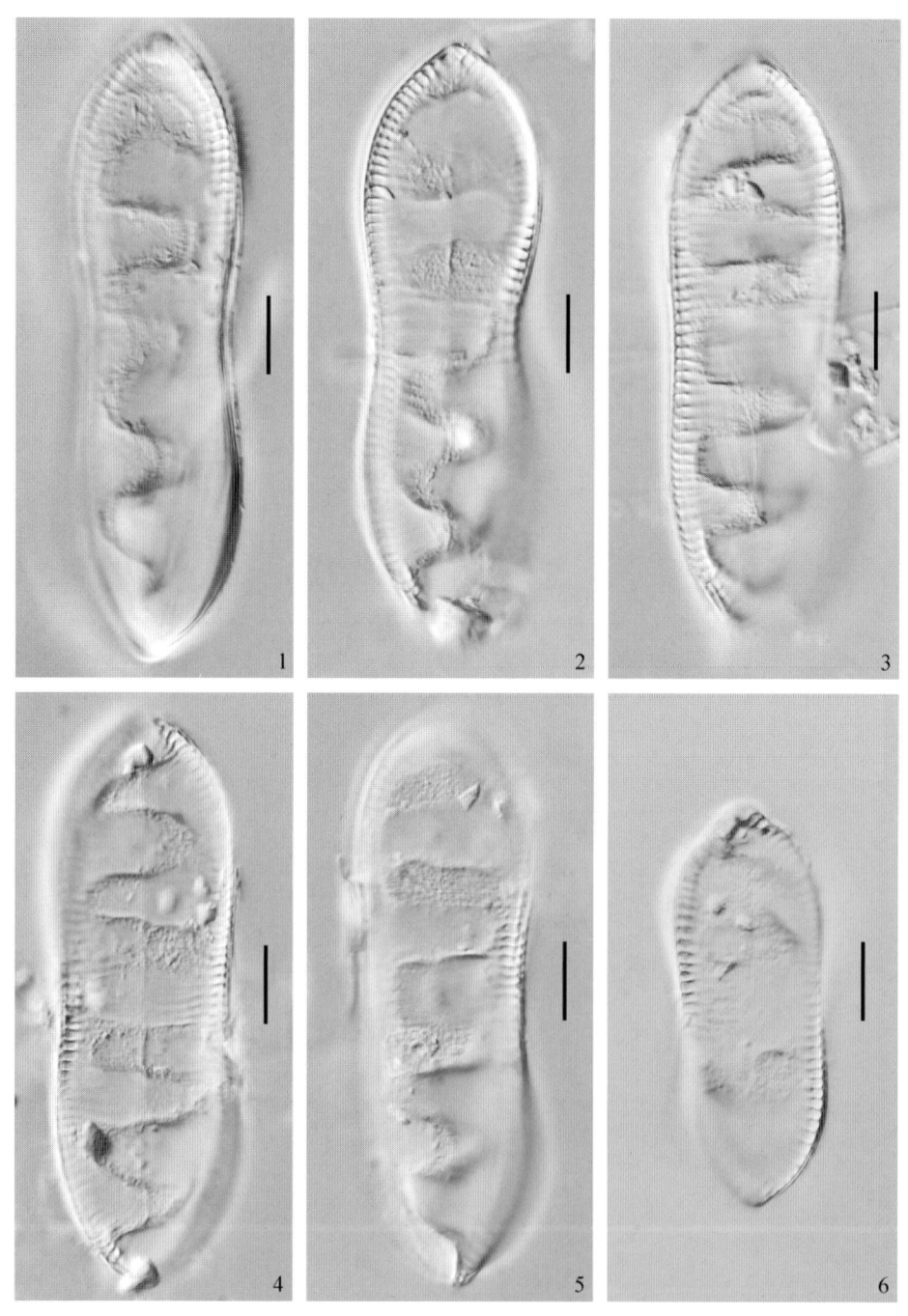

1—6. 扭曲波缘藻 *Cymatopleura aquastudia* Q-M You & J.P. Kociolek。所有图为光镜(LM)照片，比例尺(Scale bars)=10 μm。

图版 LXXXIX

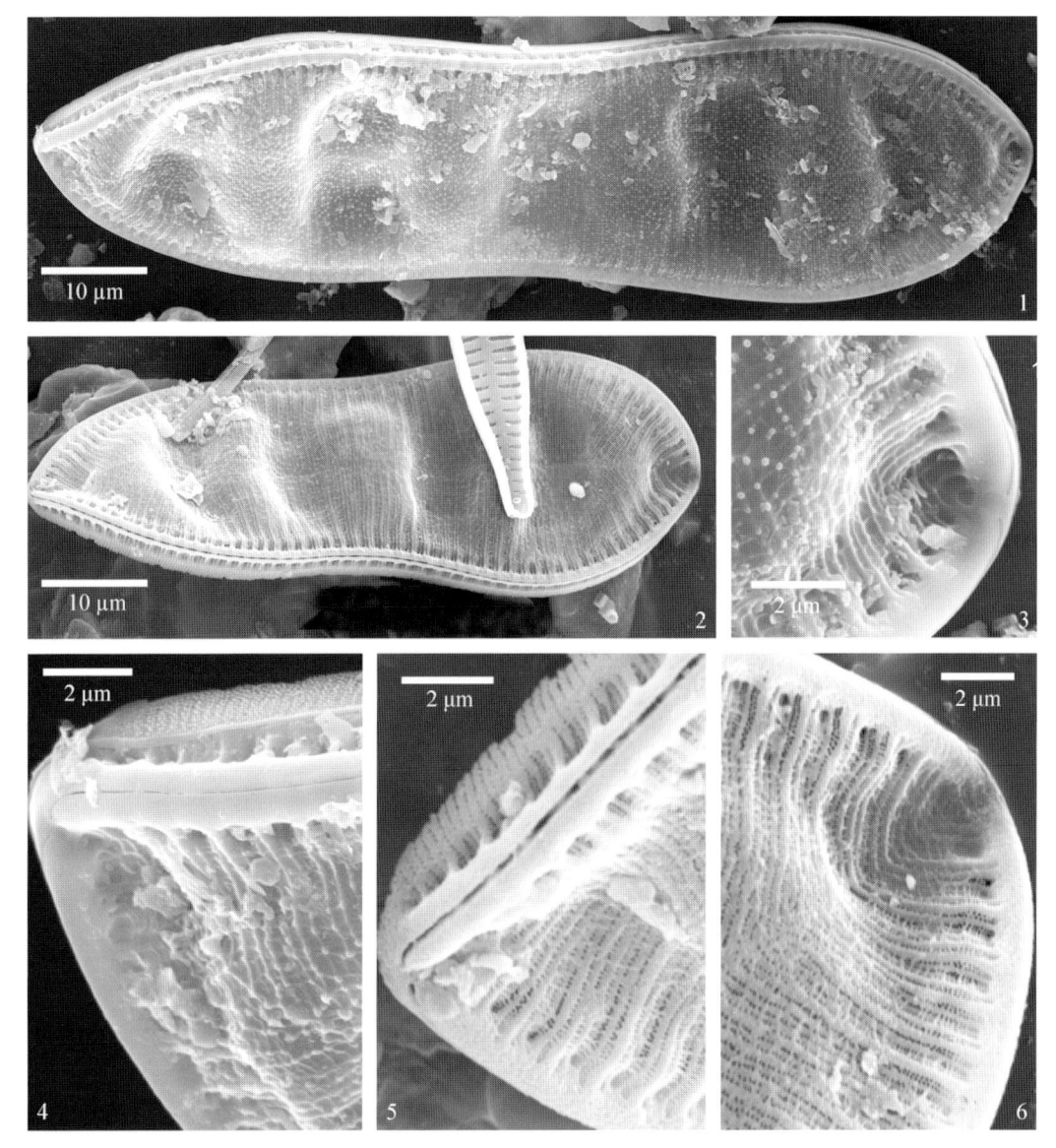

1—6. 扭曲波缘藻 Cymatopleura aquastudia Q-M You & J.P. Kociolek，1—2. 示外壳面(SEM)，3. 示外壳面顶端(SEM)，4—5. 示外壳面末端(SEM)，6. 示外壳面顶端(SEM)。所有图为电镜(SEM)照片，比例尺见图。

图版 XC

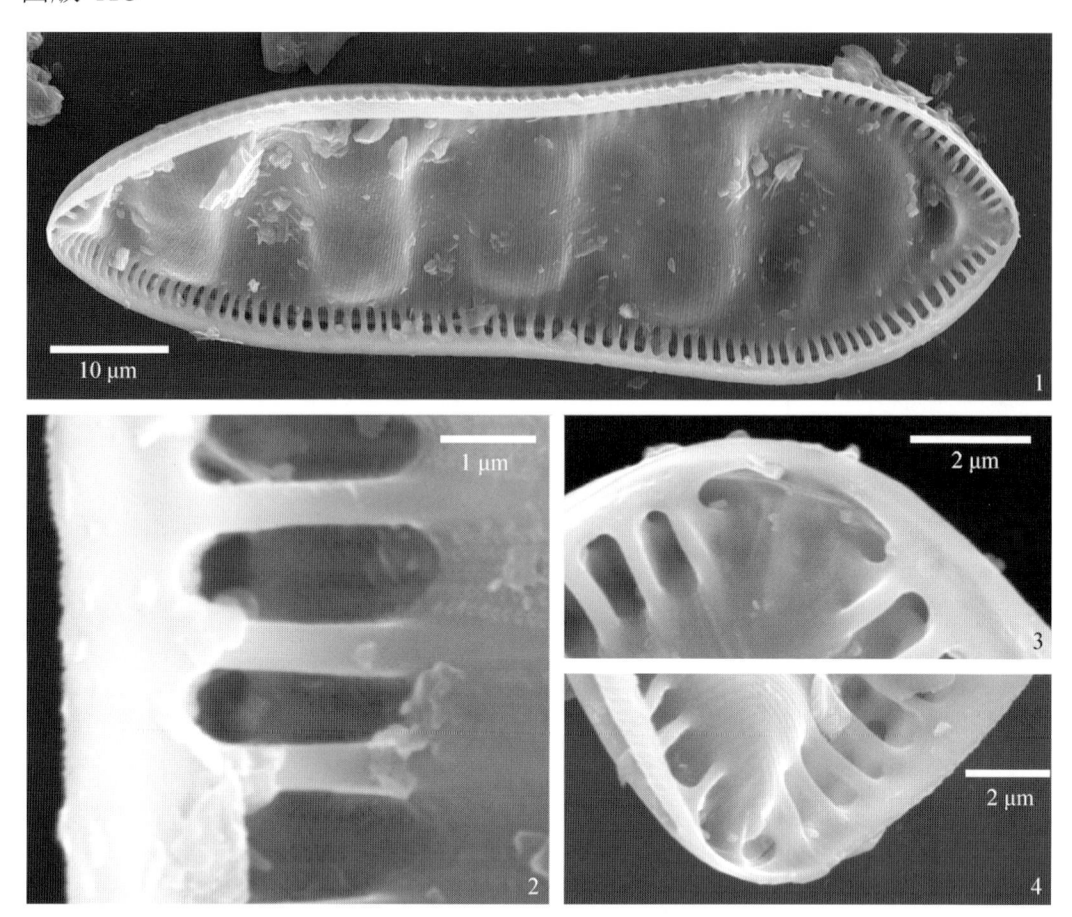

1—4. 扭曲波缘藻 *Cymatopleura aquastudia* Q-M You & J.P. Kociolek。1. 示内壳面(SEM)，2. 示内壳面边缘(SEM)，3. 示内壳面顶端(SEM)，4. 示内壳面末端(SEM)。所有图为电镜(SEM)照片，比例尺见图。

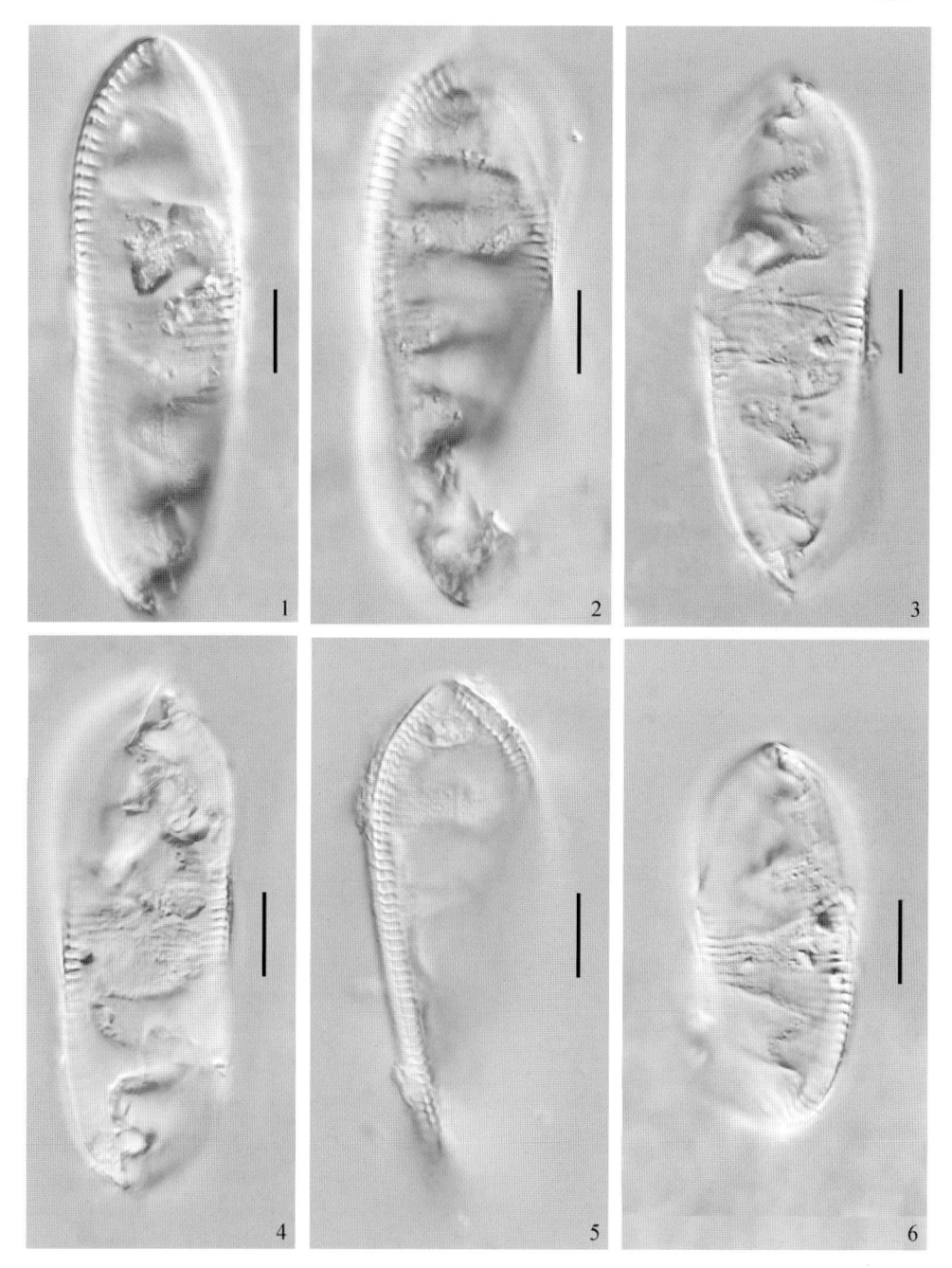

图版 XCI

1—6. 新疆波缘藻 *Cymatopleura xinjiangiana* Q-M You & J.P. Kociolek。所有图为光镜(LM)照片，比例尺(Scale bars)=10 μm。

图版 XCII

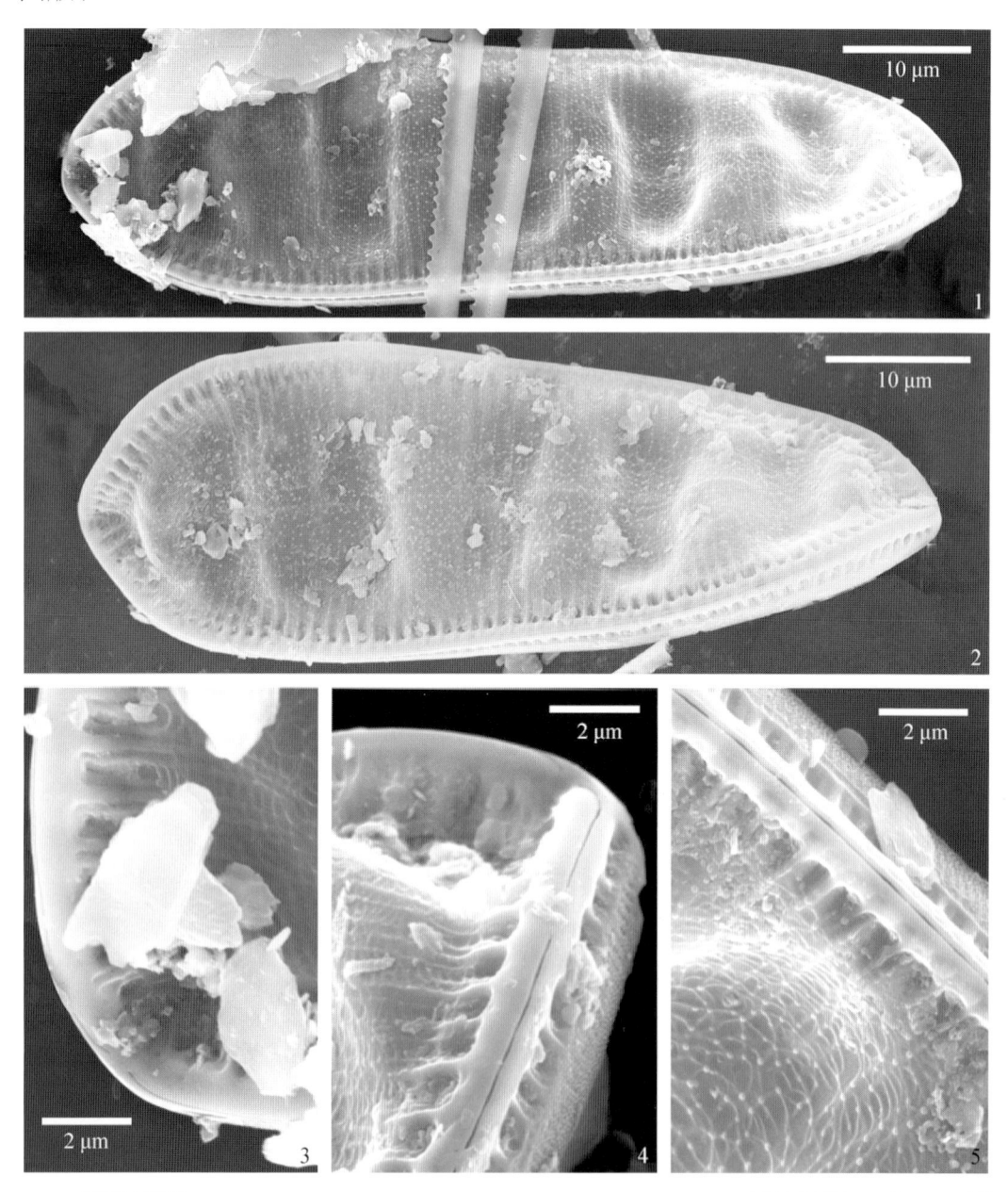

1—5. 新疆波缘藻 Cymatopleura xinjiangiana Q-M You & J.P. Kociolek，1—2. 示外壳面(SEM)，3. 示外壳面顶端(SEM)，4. 示外壳面末端(SEM)，5. 示外壳面边缘(SEM)。所有图为电镜(SEM)照片，比例尺见图。

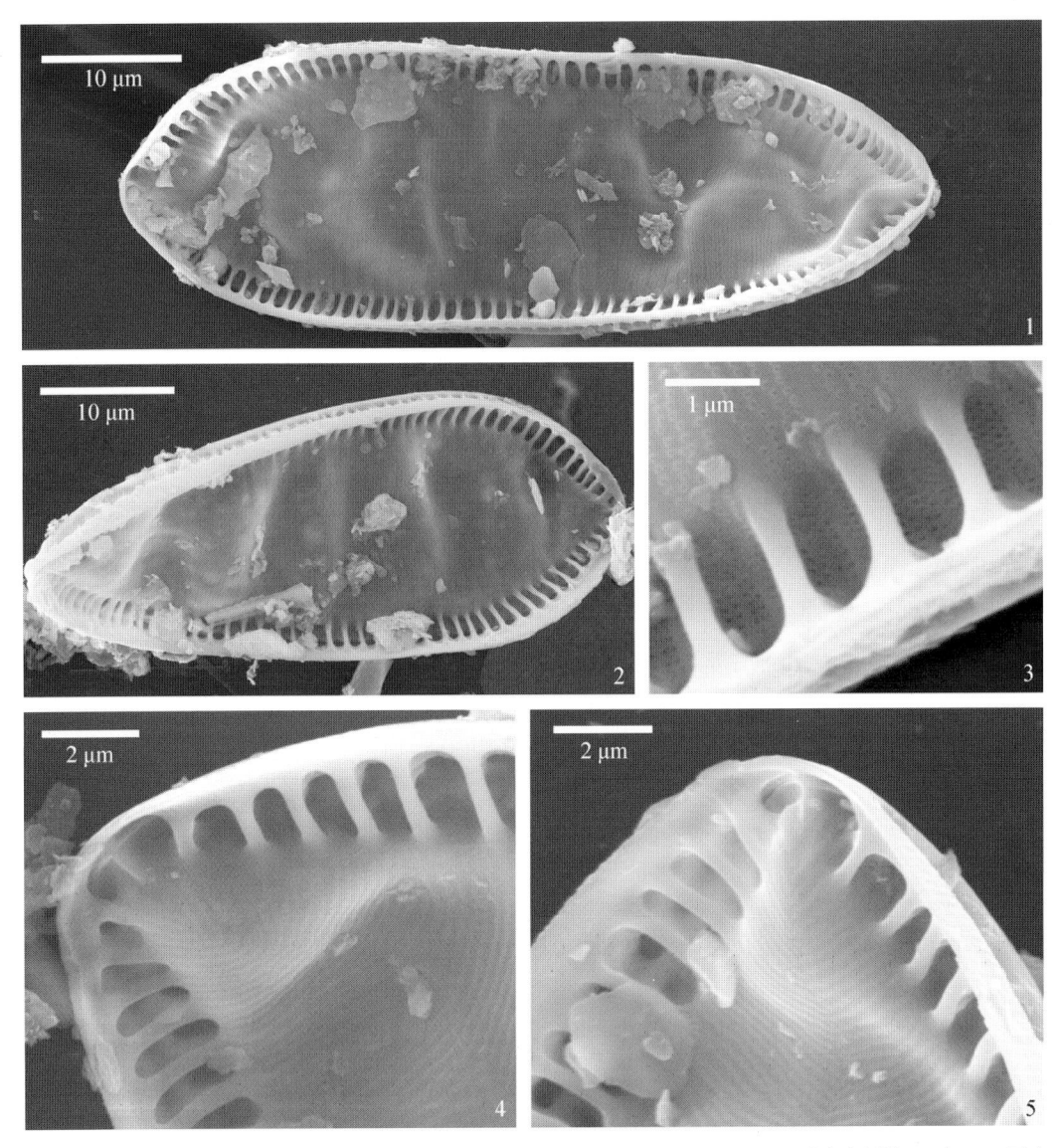

图版 XCIII

1—5. 新疆波缘藻 *Cymatopleura xinjiangiana* Q-M You & J.P. Kociolek，1—2. 示内壳面(SEM)，3. 示内壳面边缘(SEM)，4. 示内壳面顶端(SEM)，5. 示内壳面末端(SEM)。所有图为电镜(SEM)照片，比例尺见图。

图版 XCIV

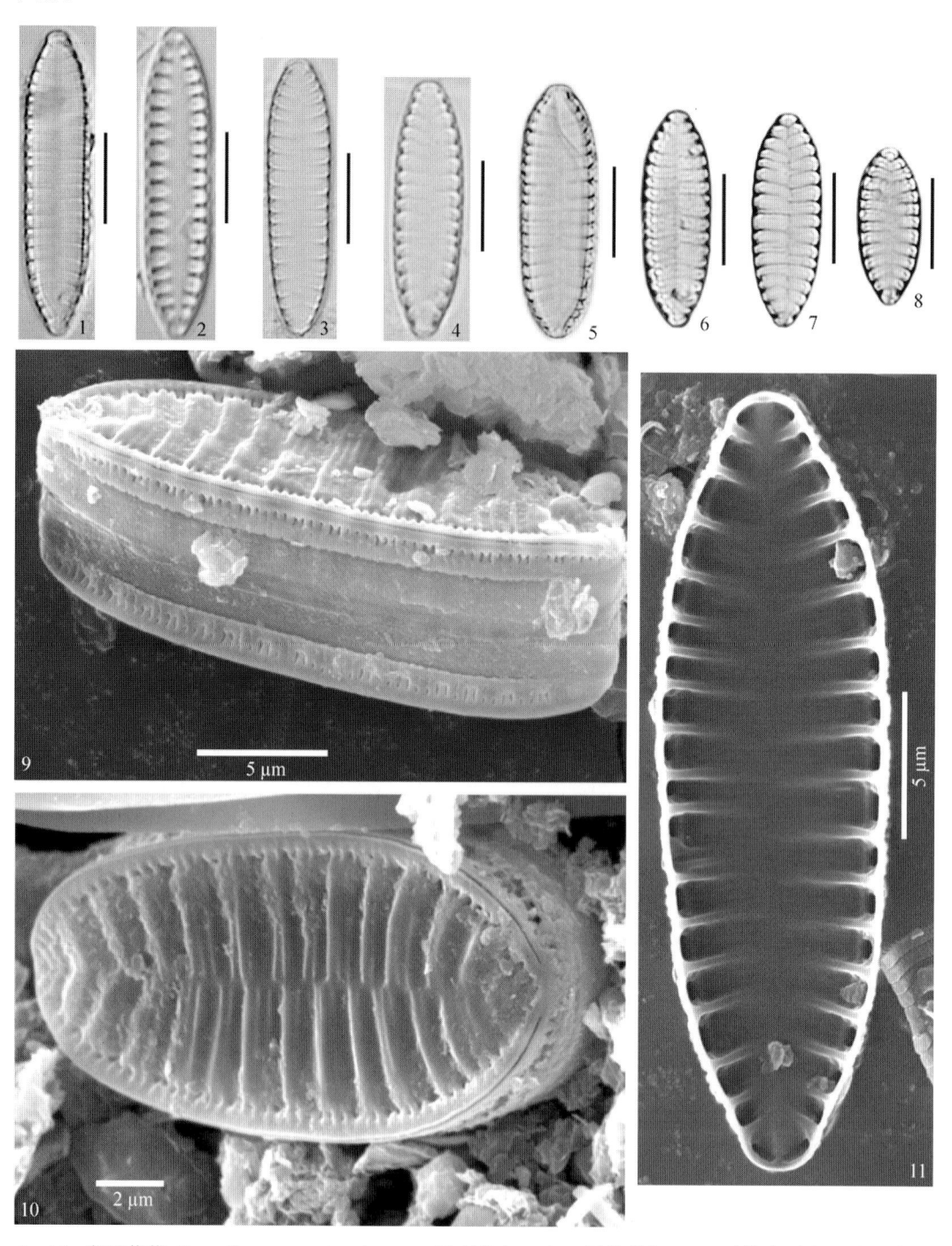

1—11. 窄双菱藻 *Surirella angusta* Kützing，9. 示壳体(SEM)，环带形态，10. 示外壳面(SEM)，肋纹和壳缝形态，11. 示内壳面(SEM)，肋纹形态。光镜(LM)照片，比例尺(Scale bars)=10 μm，电镜(SEM)照片比例尺见图。

图版 XCV

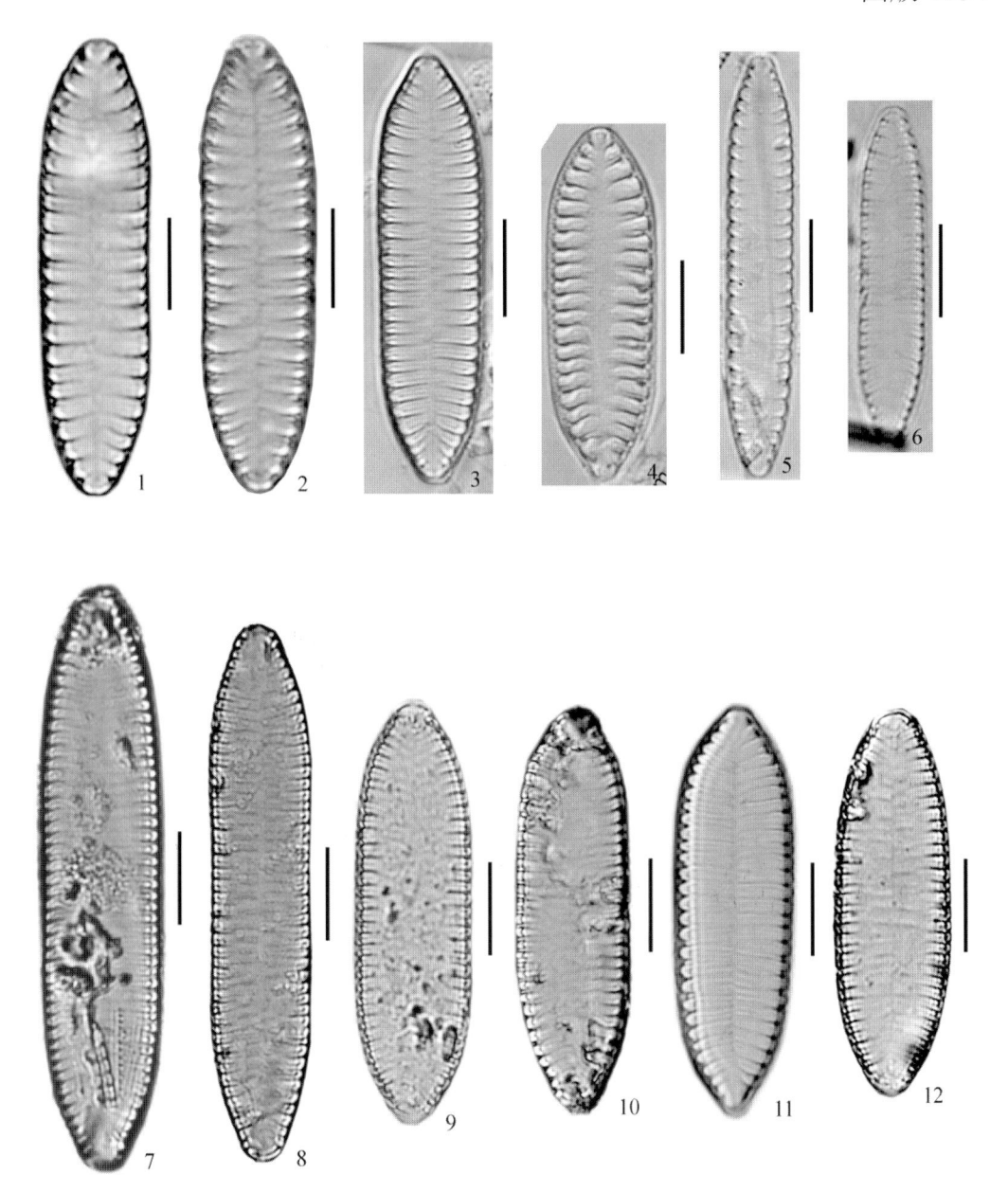

1—4. 窄双菱藻 *Surirella angusta* Kützing；5—6. ?窄双菱藻 *Surirella angusta* Kützing；7—12. 细长双菱藻 *Surirella gracilis* (W. Smith) Grunow。所有图为光镜(LM)照片，比例尺(Scale bars)=10 μm。

图版 XCVI

1. 窄双菱藻 *Surirella angusta* Kützing，示外壳面（SEM），肋纹形态；2—3. 细长双菱藻 *Surirella gracilis* (W. Smith) Grunow，2. 示外壳面（SEM），肋纹形态。光镜（LM）照片，比例尺（Scale bars）=10 μm，电镜（SEM）照片比例尺见图。

图版 XCVII

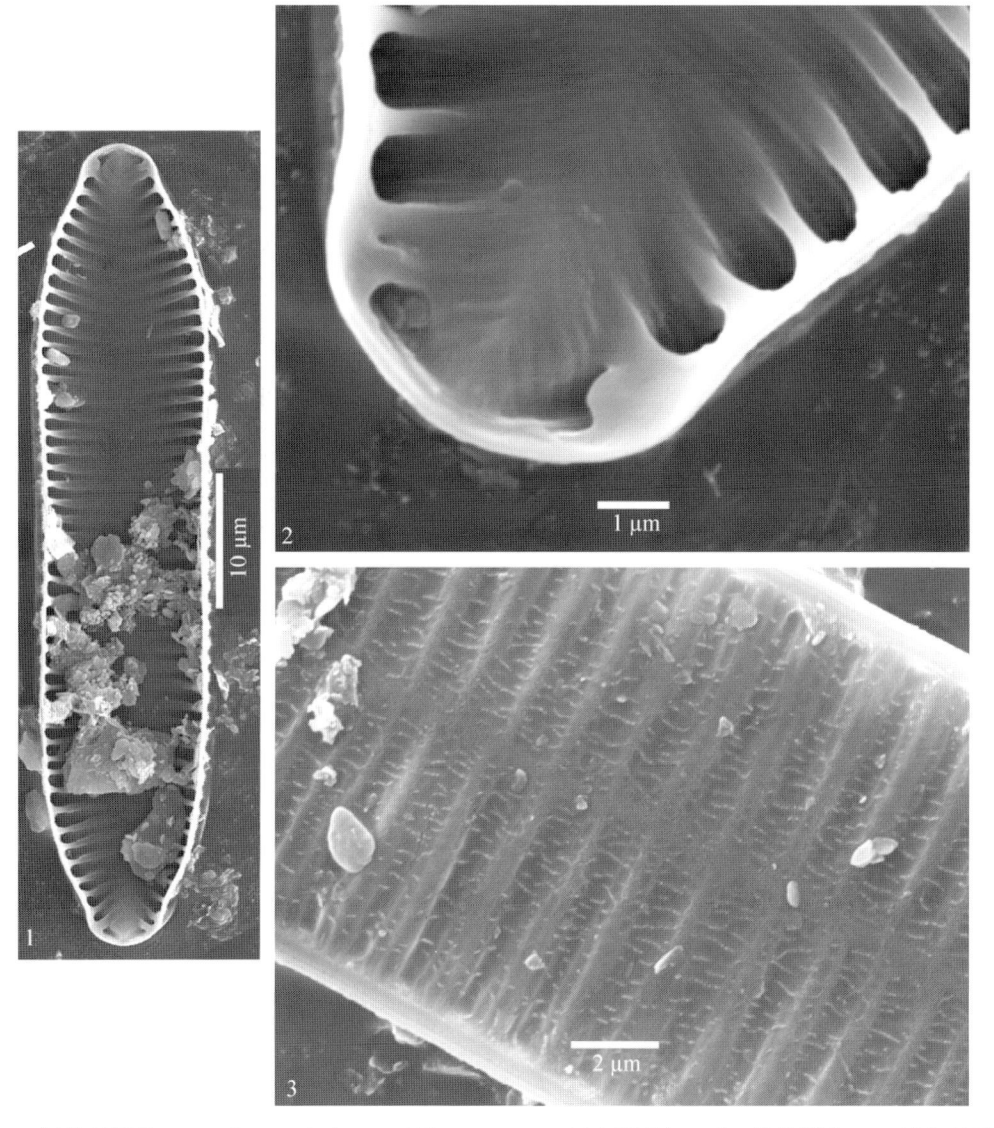

1—3. 细长双菱藻 *Surirella gracilis* (W. Smith) Grunow，1. 示内壳面(SEM)，肋纹形态，2. 示内壳面顶端(SEM)，肋纹和远缝端形态，3. 示外壳面中部(SEM)，肋纹形态。所有图为电镜(SEM)照片，比例尺见图。

图版 XCVIII

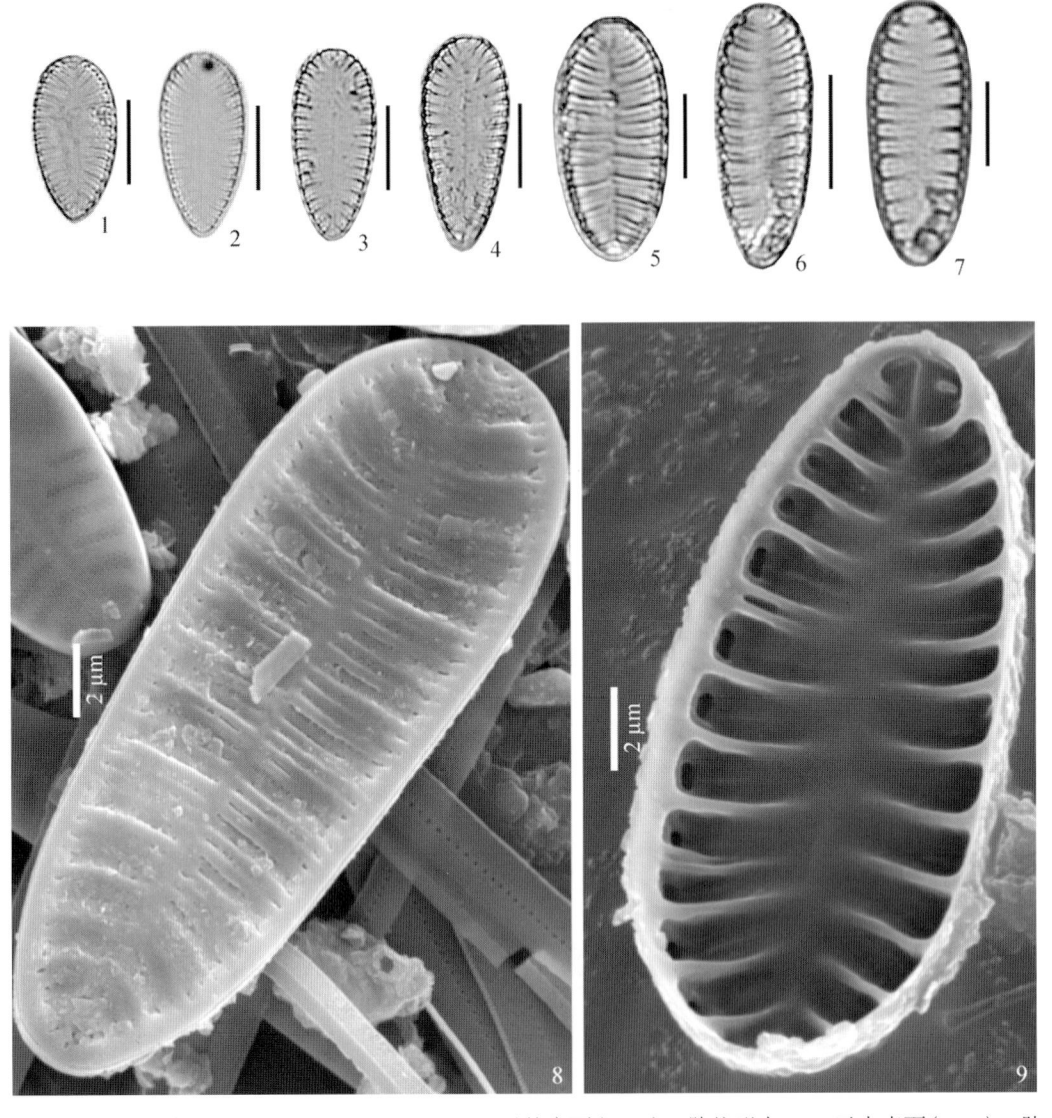

1—9. 微小双菱藻 *Surirella minuta* Brébisson，8. 示外壳面(SEM)，肋纹形态，9. 示内壳面(SEM)，肋纹形态。光镜(LM)照片，比例尺(Scale bars)=10 μm，电镜(SEM)照片比例尺见图。

图版 XCIX

1—2. 微小双菱藻 *Surirella minuta* Brébisson，示外壳面中部(SEM)，壳缝和肋纹形态；3—4. 卵圆双菱藻 *Surirella ovalis* Brébisson，3. 示外壳面中部(SEM)，肋纹形态，4. 示内壳面中部(SEM)，肋纹和孔纹形态。所有图为电镜(SEM)照片，比例尺见图。

图版 C

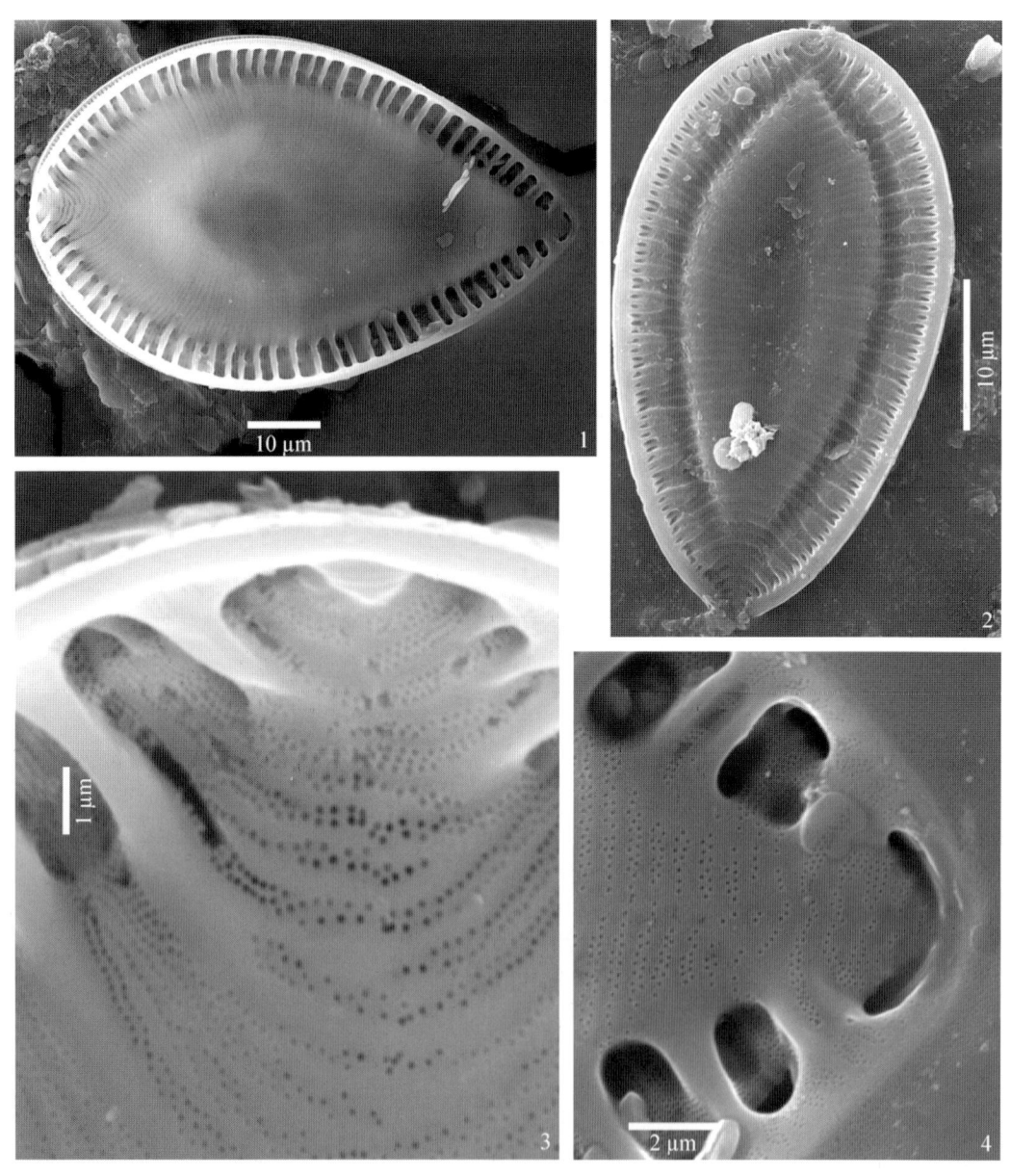

1—4. 卵圆双菱藻 *Surirella ovalis* Brébisson，1. 示内壳面(SEM)，肋纹形态，2. 示外壳面(SEM)，肋纹形态，3. 示内壳面顶端(SEM)，孔纹和壳缝形态，4. 示内壳面末端(SEM)，孔纹和壳缝形态。所有图为电镜(SEM)照片，比例尺见图。

图版 CI

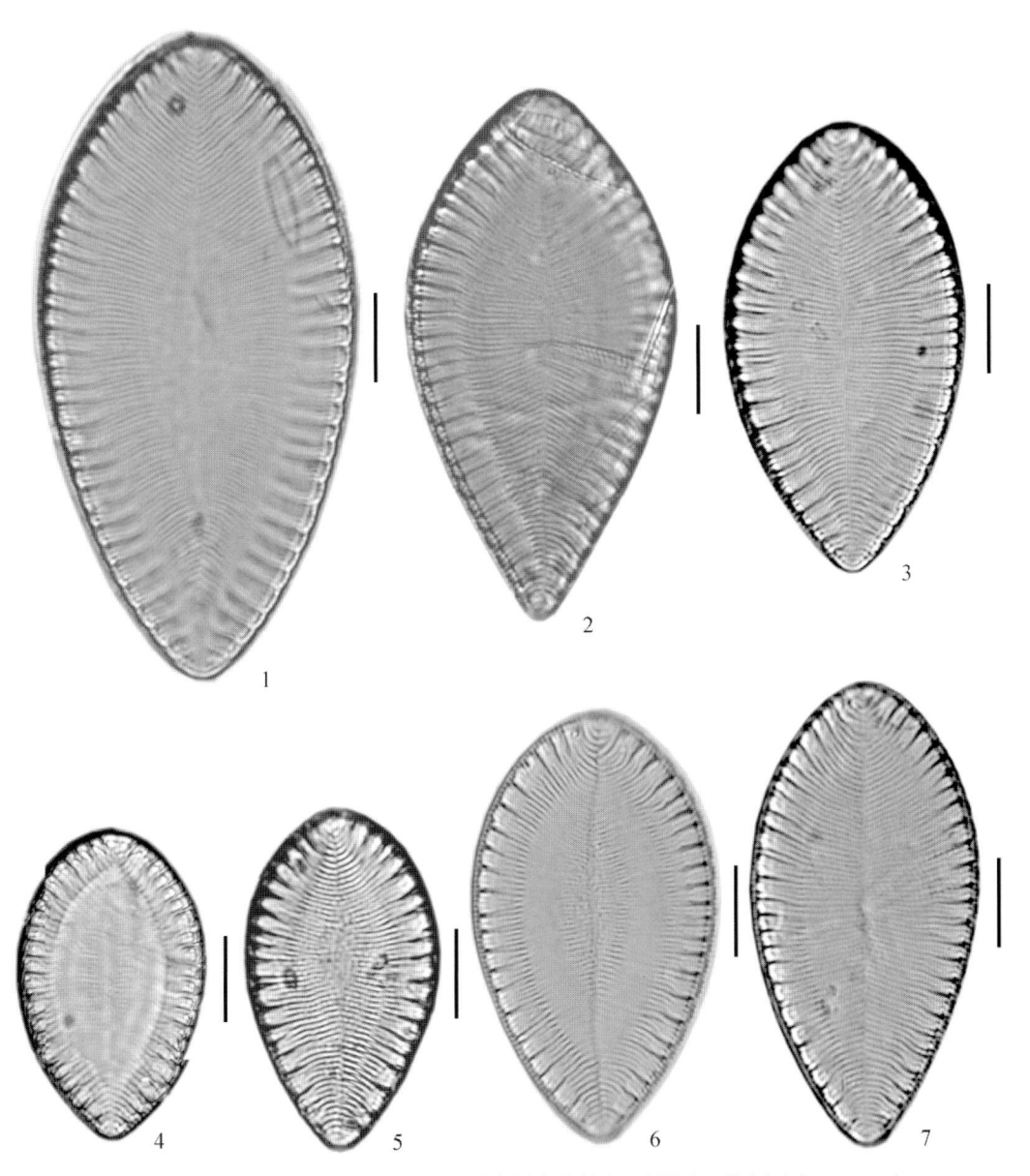

1—7. 卵圆双菱藻 *Surirella ovalis* Brébisson。所有图为光镜(LM)照片，比例尺(Scale bars)=10 μm。

图版 CII

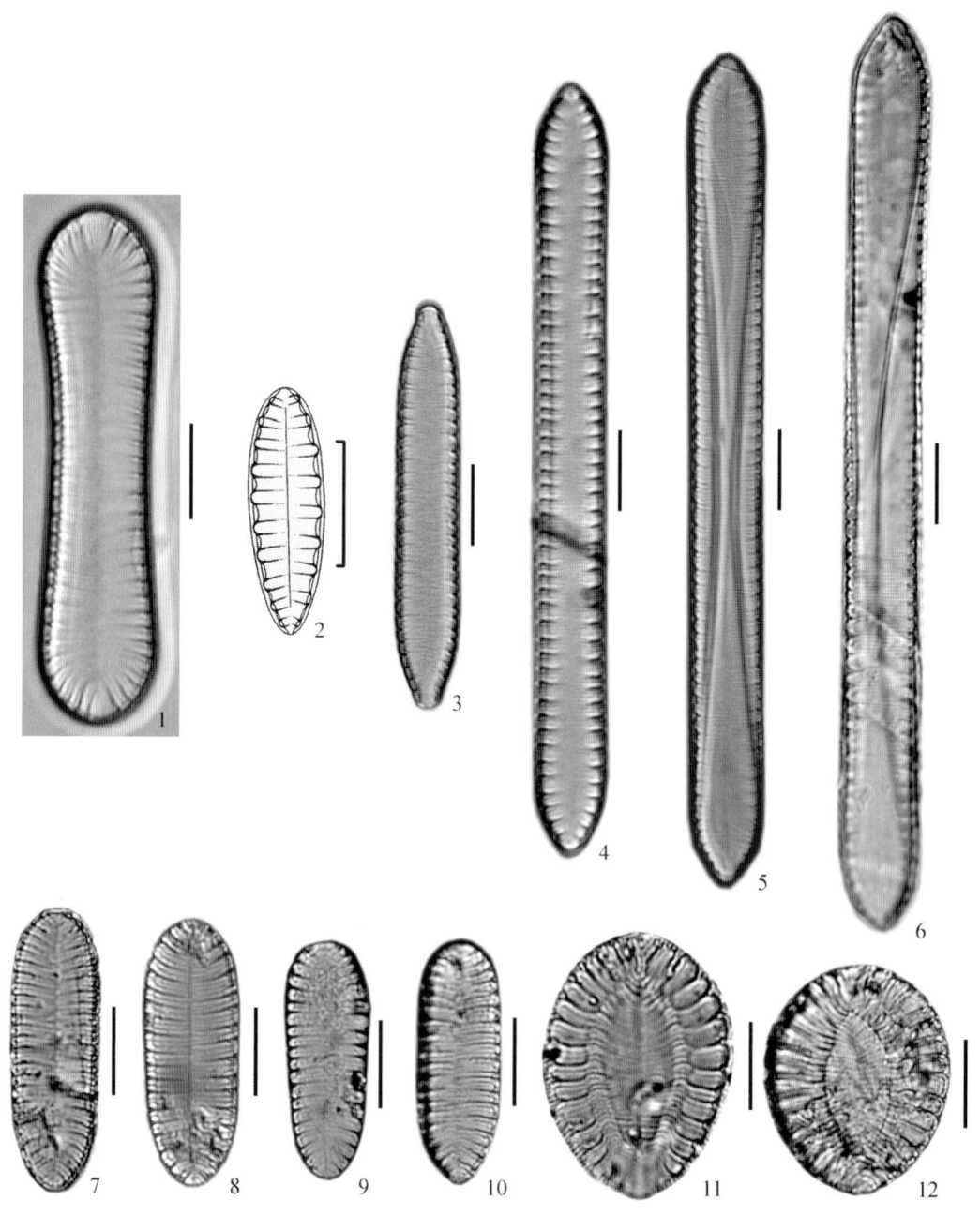

1. 泰特尼斯双菱藻 *Surirella tientsinensis* Skvortzow；2. 近盐生双菱藻 *Surirella subsalsa* W. Smith，手绘图，引自 Zhu 和 Chen(2000)；3. ? 拉普兰双菱藻 *Surirella lapponica* A. Cleve；4—6. 拉普兰双菱藻 *Surirella lapponica* A. Cleve；7—10. 维苏双菱藻 *Surirella visurgis* Hustedt；11—12. 布赖韦尔双菱藻 *Surirella brightwellii* W. Smith。所有图为光镜(LM)照片，比例尺(Scale bars)=10 μm。

1—8. 布列双菱藻 *Surirella brebissonii* Krammer & Lange-Bertalot，1—2. 示外壳面(SEM)，肋纹形态。光镜(LM)照片，比例尺(Scale bars)=10 μm，电镜(SEM)照片比例尺见图。

图版 CIV

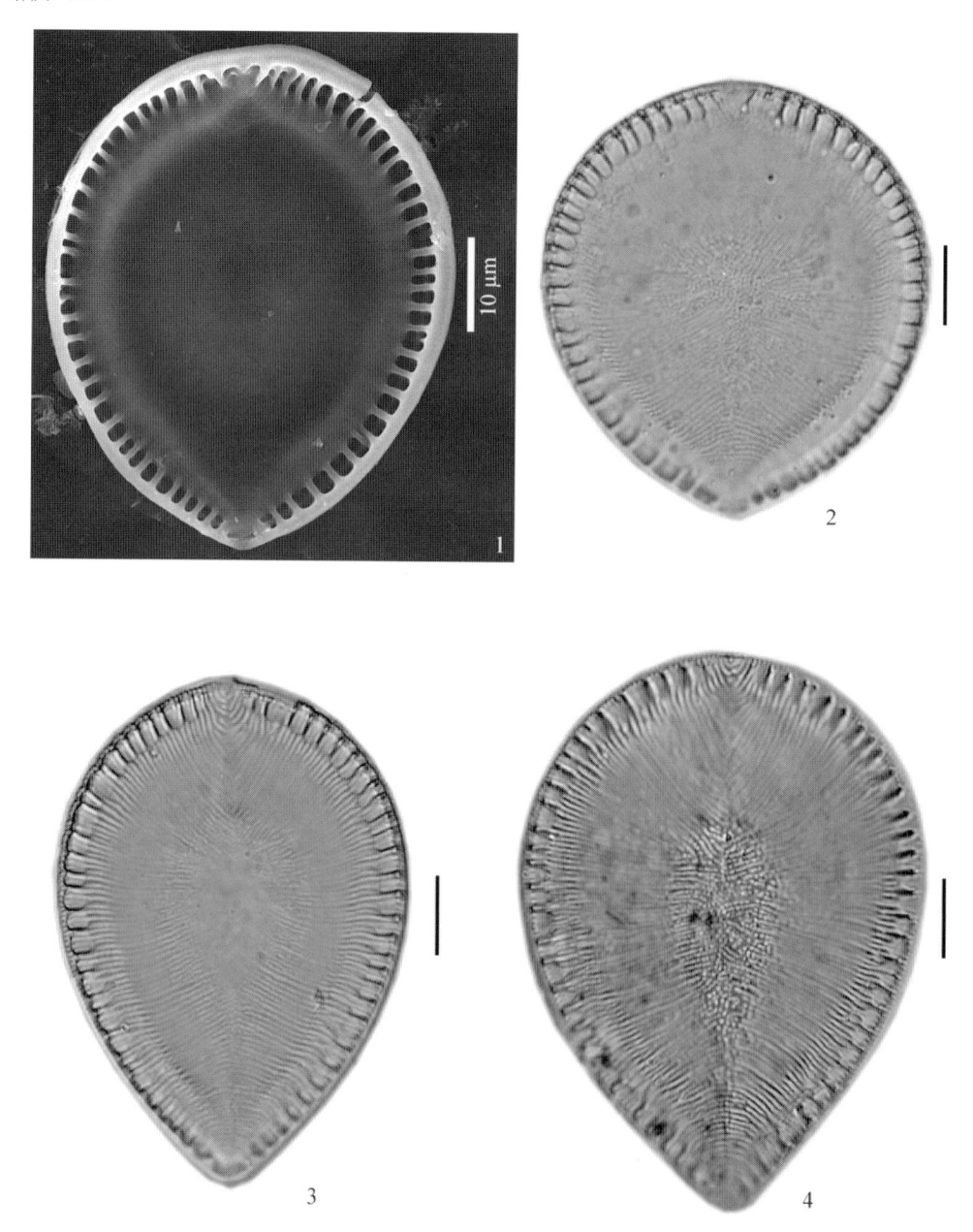

1—4. 派松双菱藻 *Surirella peisonis* Pantocsek，1. 示内壳面(SEM)，肋纹形态。光镜(LM)照片，比例尺(Scale bars)=10 μm，电镜(SEM)照片比例尺见图。

图版 CV

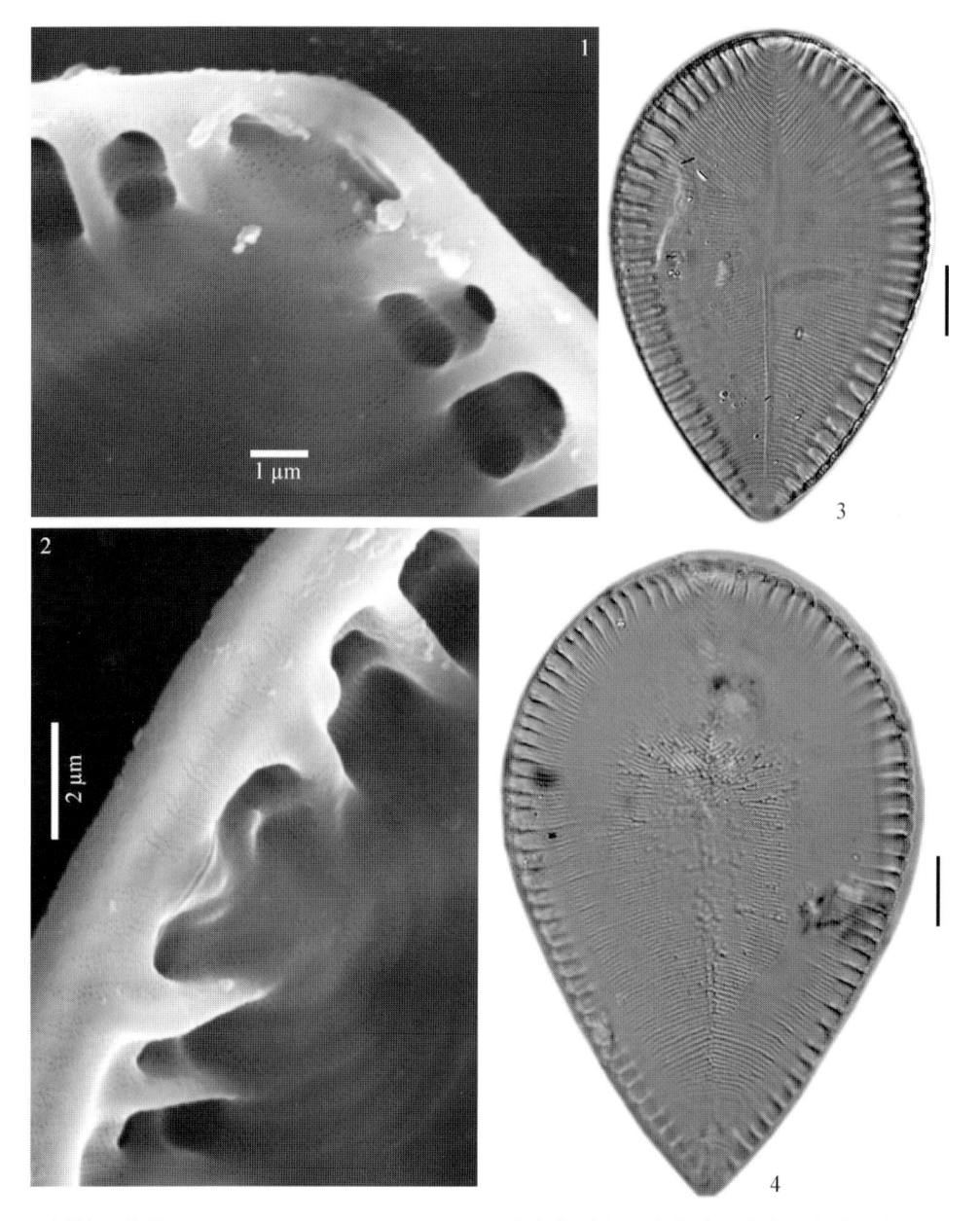

1—4. 派松双菱藻 *Surirella peisonis* Pantocsek，1. 示内壳面(SEM)末端，肋纹和壳缝形态，2. 示内壳面(SEM)顶端，肋纹和壳缝形态。光镜(LM)照片，比例尺(Scale bars)=10 μm，电镜(SEM)照片比例尺见图。

图版 CVI

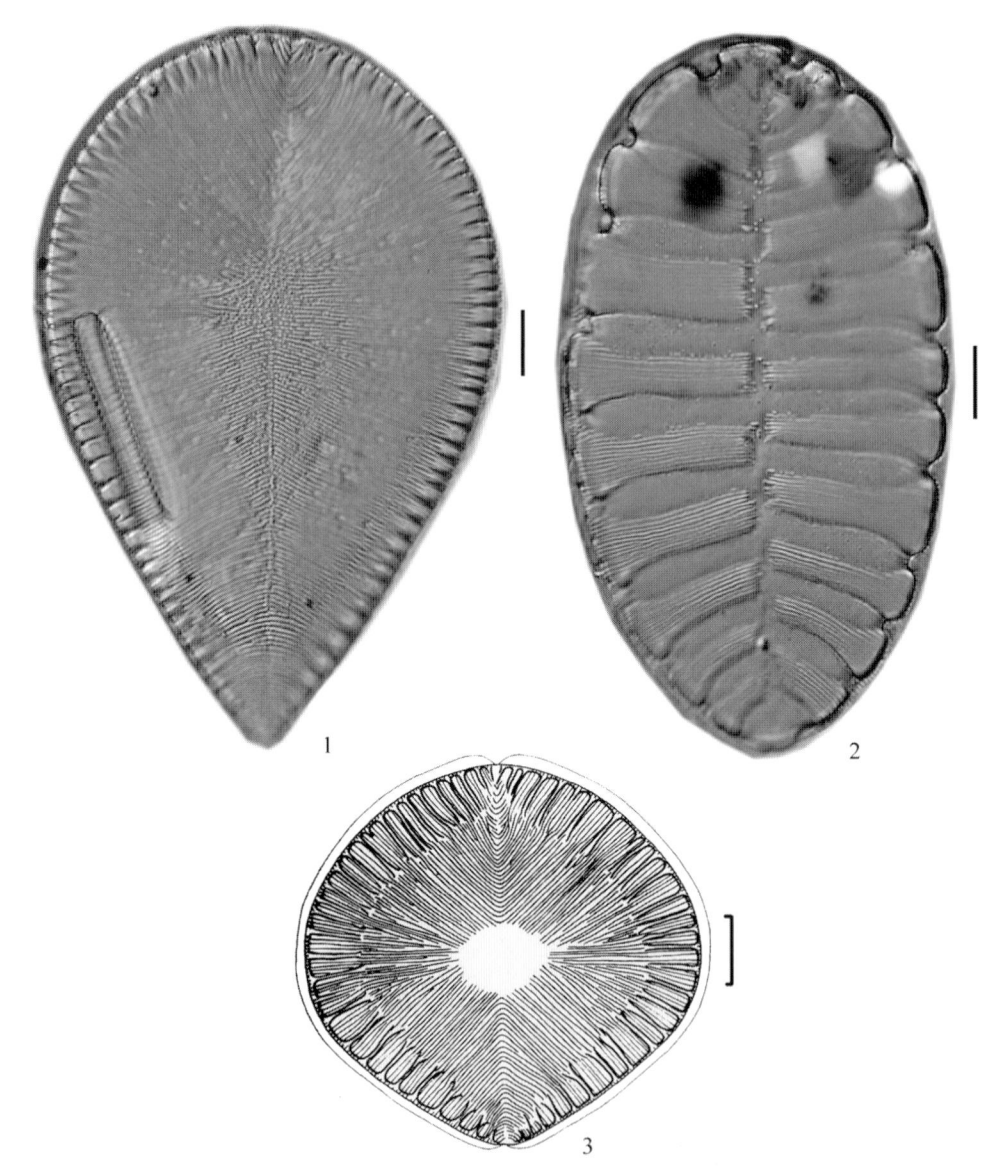

1. 派松双菱藻 *Surirella peisonis* Pantocsek；2. 具条纹双菱藻 *Surirella striatula* Turpin；3. 囊形双菱藻 *Surirella crumena* Brébisson ex Kützing，手绘图，引自 Zhu 和 Chen(2000)。所有图为光镜(LM)照片，比例尺(Scale bars)=10 μm。

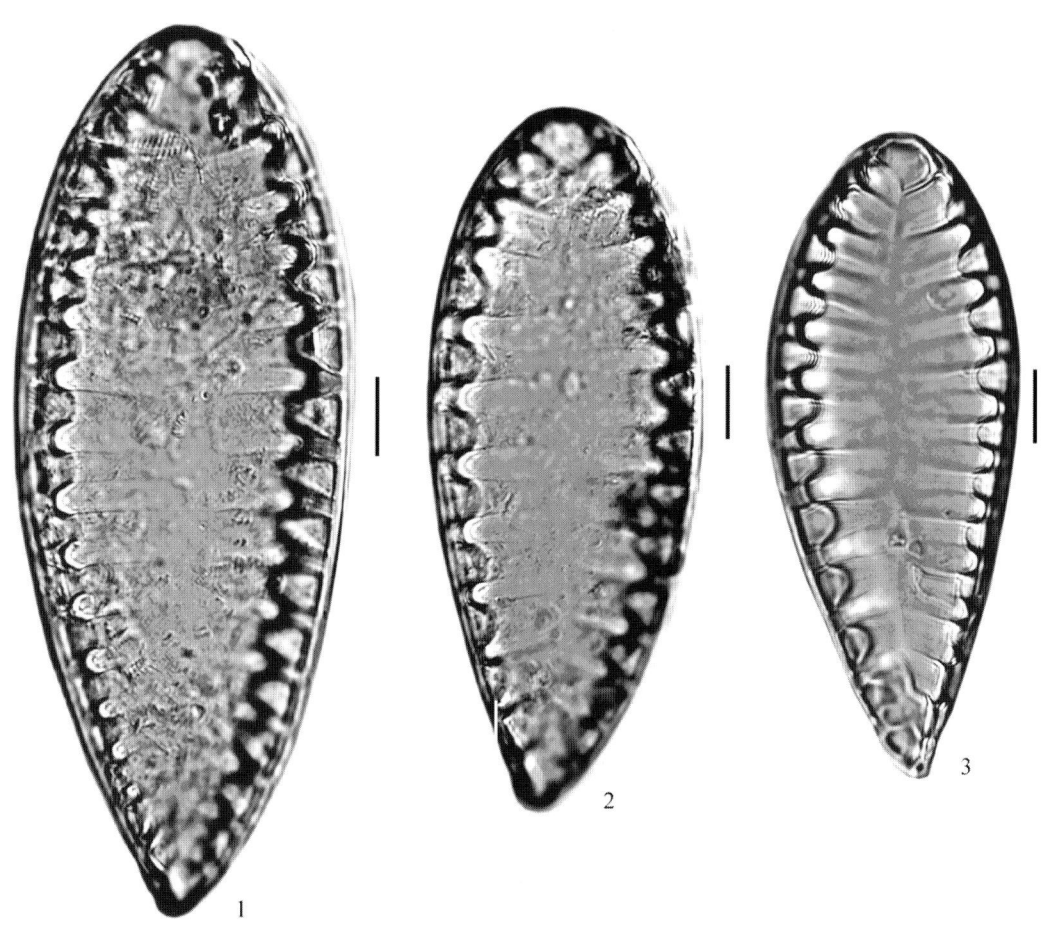

图版 CVII

1—3. 阿斯特里双菱藻 *Surirella astridae* Hustedt。所有图为光镜(LM)照片，比例尺(Scale bars)=10 μm。

图版 CVIII

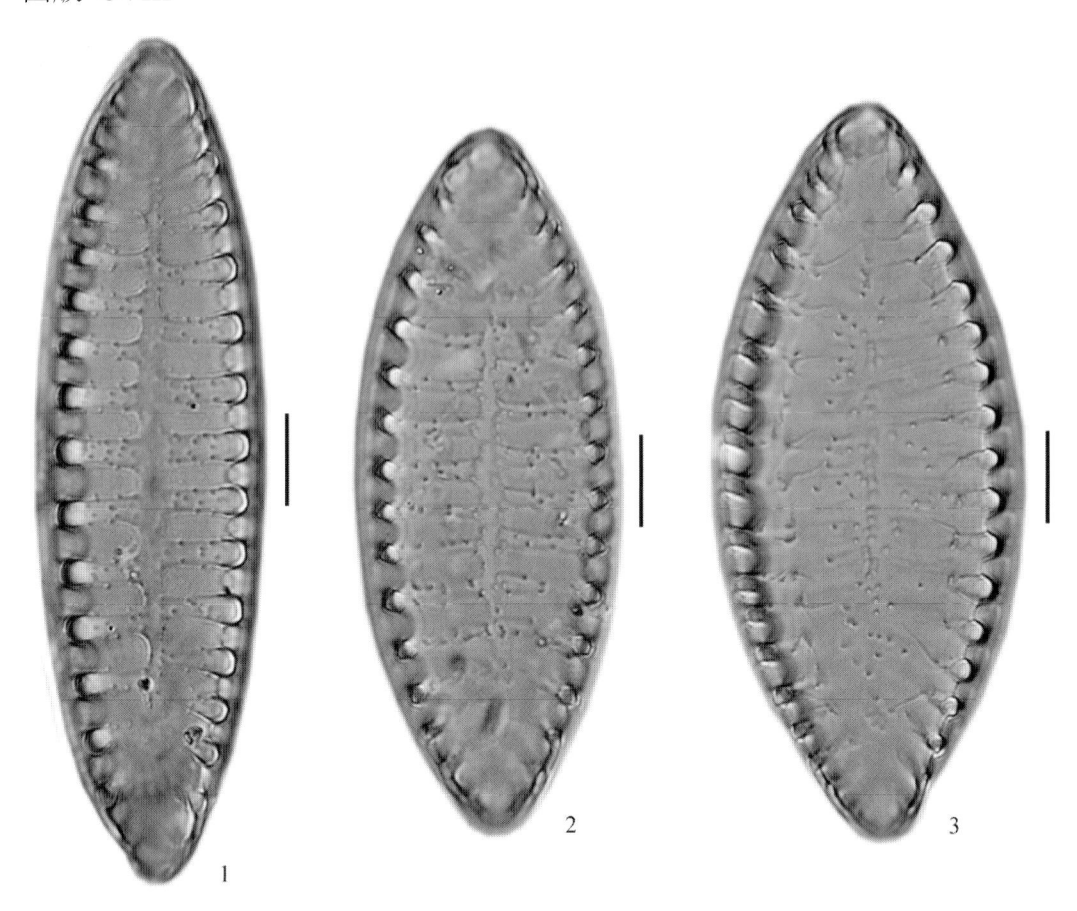

1—3. 二额双菱藻 *Surirella bifrons* Ehrenberg。所有图为光镜(LM)照片，比例尺(Scale bars)=10 μm。

图版 CIX

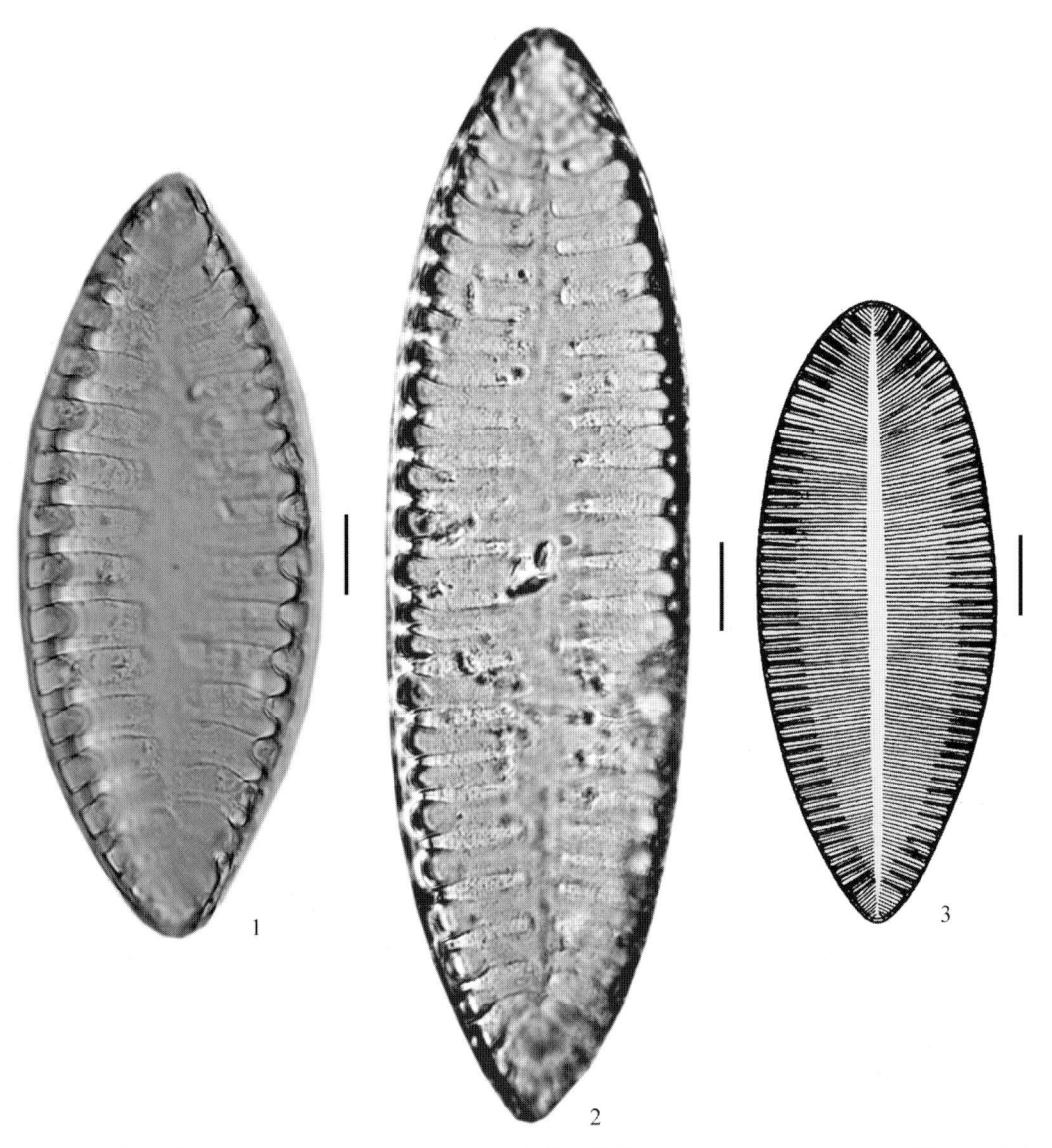

1—2. 二额双菱藻 *Surirella bifrons* Ehrenberg；3. 盘状双菱藻 *Surirella patella* Kützing，手绘图，引自 Zhu 和 Chen(2000)。所有图为光镜(LM)照片，比例尺(Scale bars)=10 μm。

图版 CX

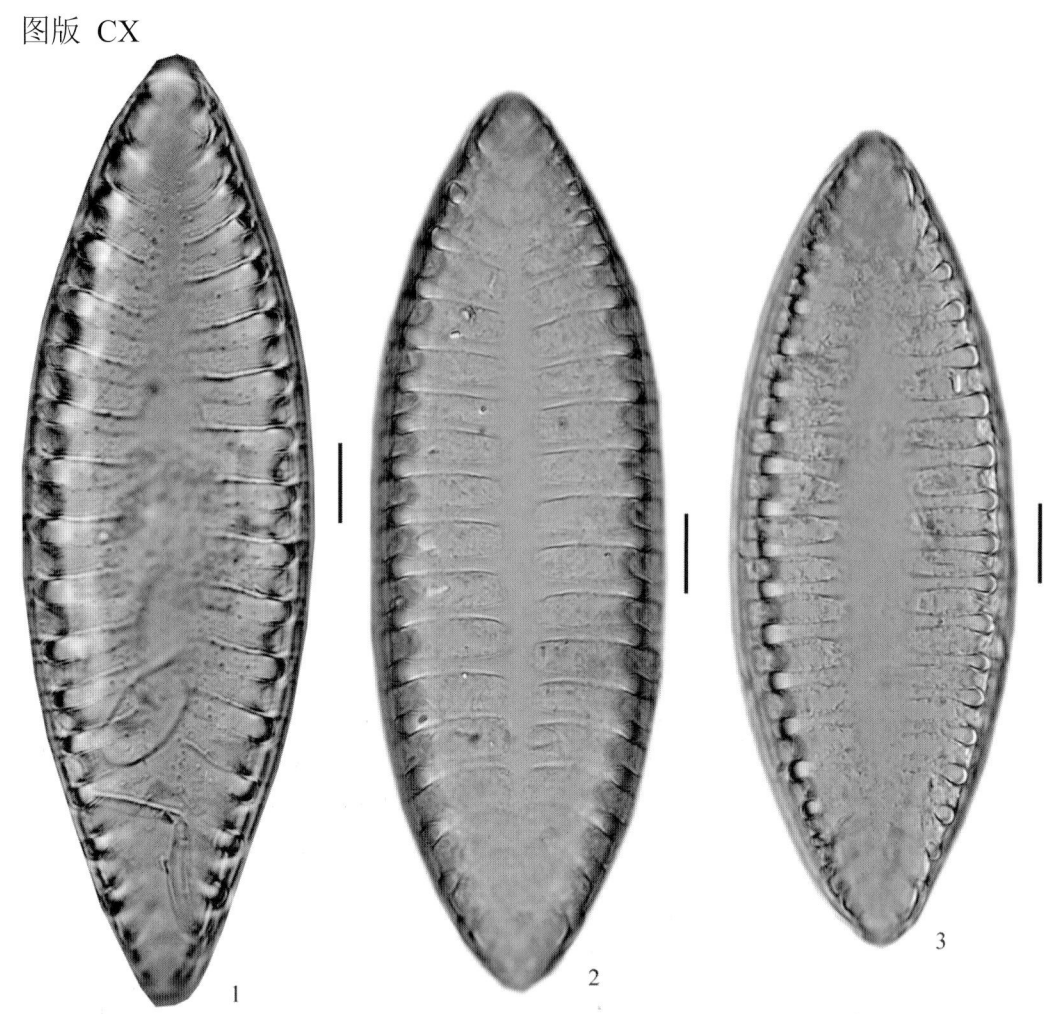

1—3. 二额双菱藻 *Surirella bifrons* Ehrenberg。所有图为光镜（LM）照片，比例尺（Scale bars）=10 μm。

图版 CXI

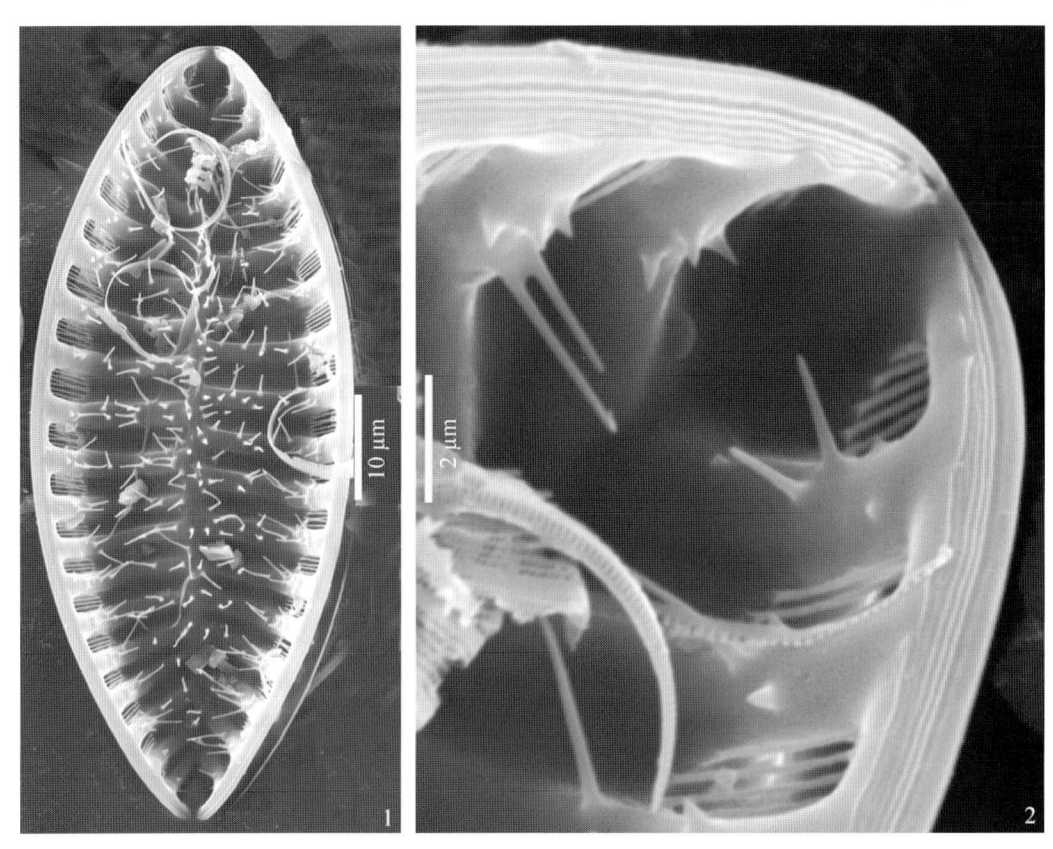

1—2. 二额双菱藻 *Surirella bifrons* Ehrenberg，1. 示外壳面(SEM)，肋纹形态。2. 示外壳面顶端，窗栏开孔(栏状孔)结构。所有图为电镜(SEM)照片，比例尺见图。

图版 CXII

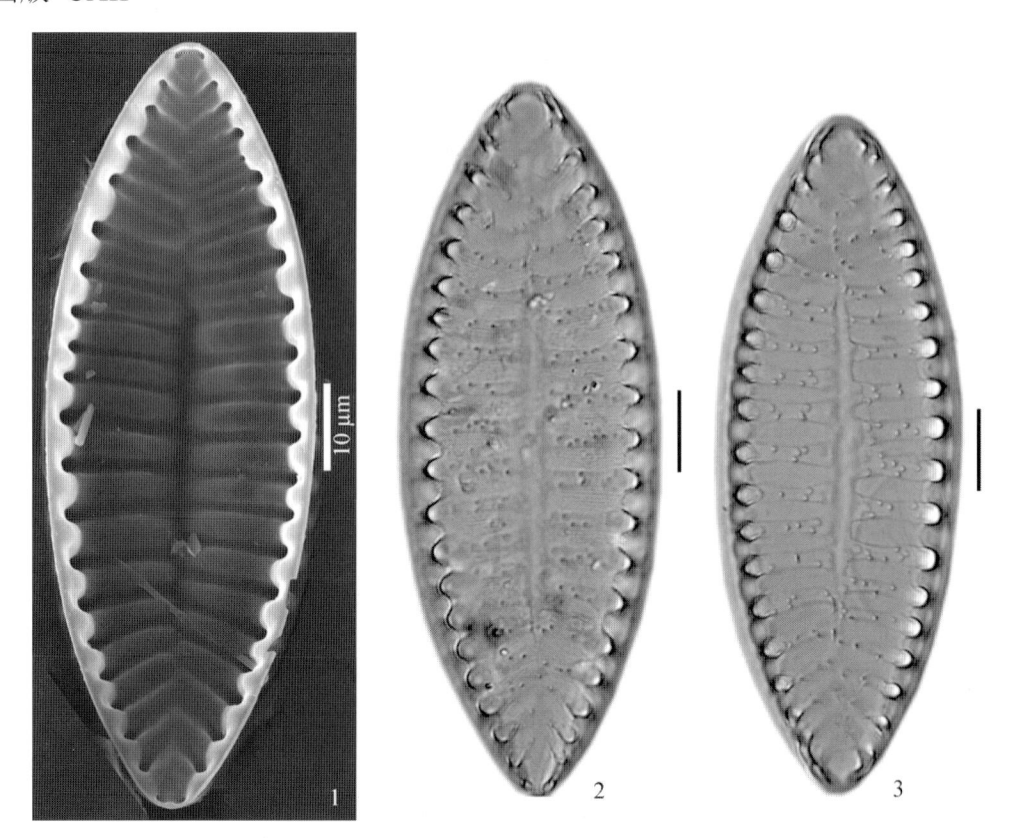

1—3. 二额双菱藻 *Surirella bifrons* Ehrenberg，1. 示内壳面（SEM），肋纹形态。光镜（LM）照片，比例尺（Scale bars）=10 μm，电镜（SEM）照片比例尺见图。

图版 CXIII

1—3. 二列双菱藻（原变种）*Surirella biseriata* var. *biseriata* Brébisson。所有图为光镜（LM）照片，比例尺（Scale bars）=10 μm。

图版 CXIV

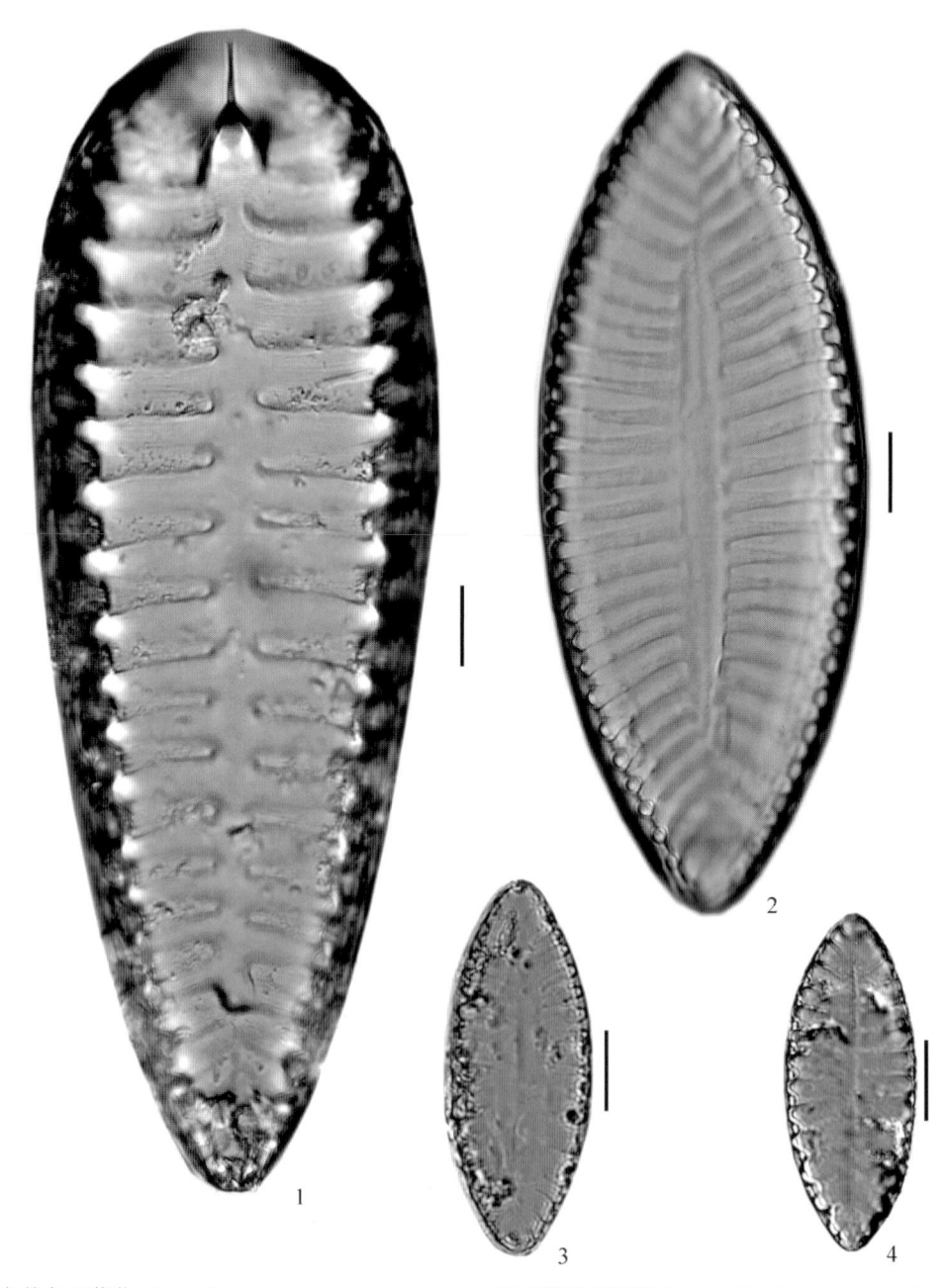

1. 卡普龙双菱藻 *Surirella capronii* Brébisson；2. 二列双菱藻（原变种）*Surirella biseriata* var. *biseriata* Brébisson；3—4. 二列双菱藻缩小变种 *Surirella biseriata* var. *diminuta* Cleve-Euler。所有图为光镜（LM）照片，比例尺（Scale bars）=10 μm。

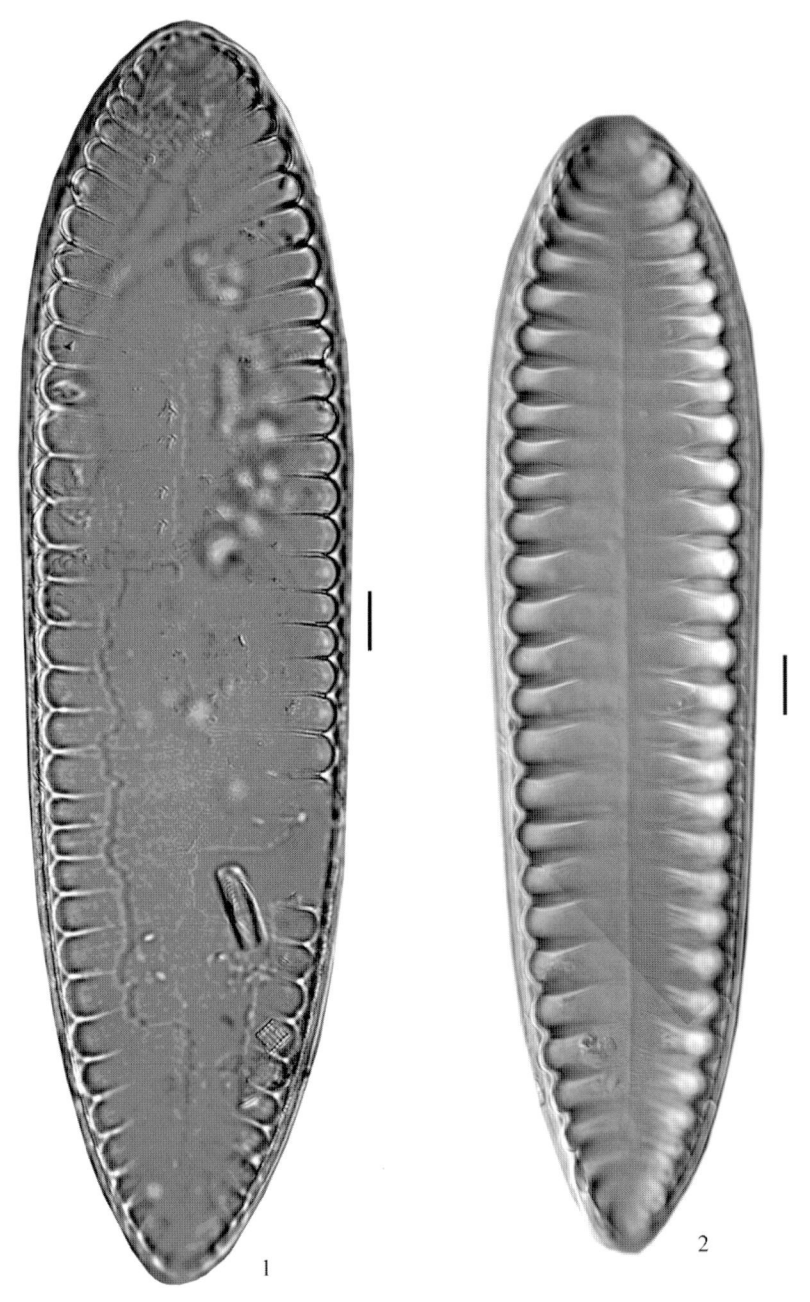

图版 CXV

1—2. 美丽双菱藻 *Surirella elegans* Ehrenburg。所有图为光镜(LM)照片，比例尺(Scale bars)=10 μm。

图版 CXVI

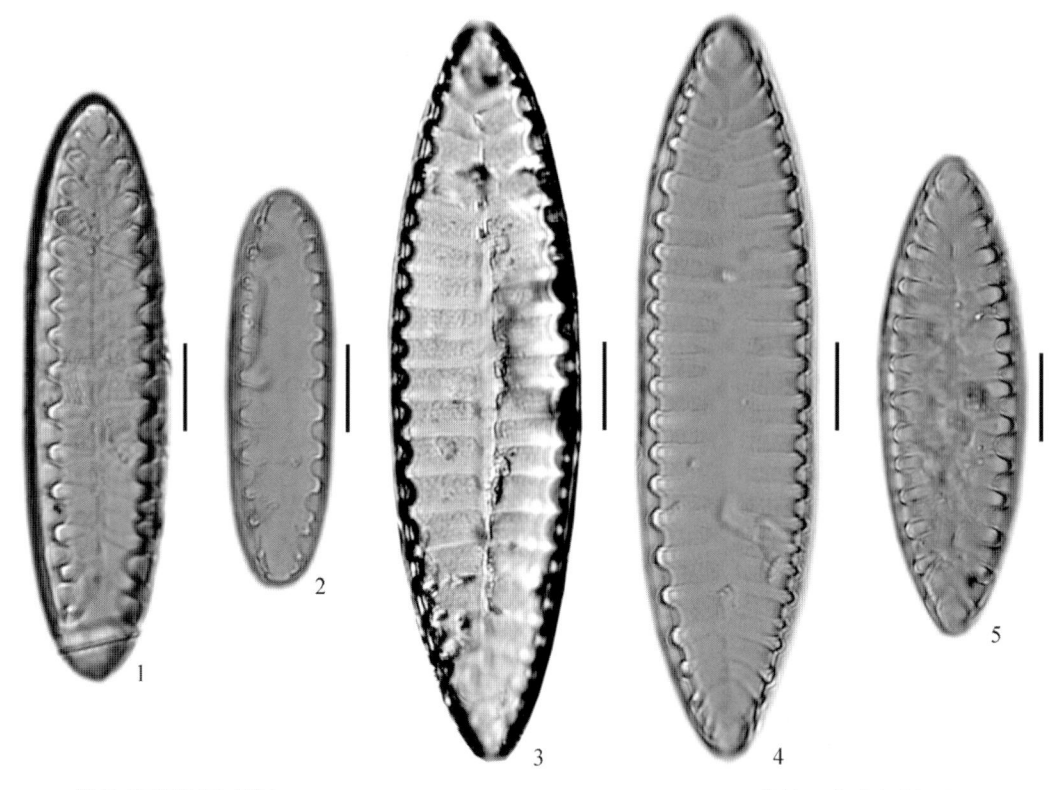

1—2. 线性双菱藻（原变种）*Surirella linearis* var. *linearis* W. Smith；3—5. ? 线性双菱藻（原变种）*Surirella linearis* var. *linearis* W. Smith。所有图为光镜（LM）照片，比例尺（Scale bars）=10 μm。

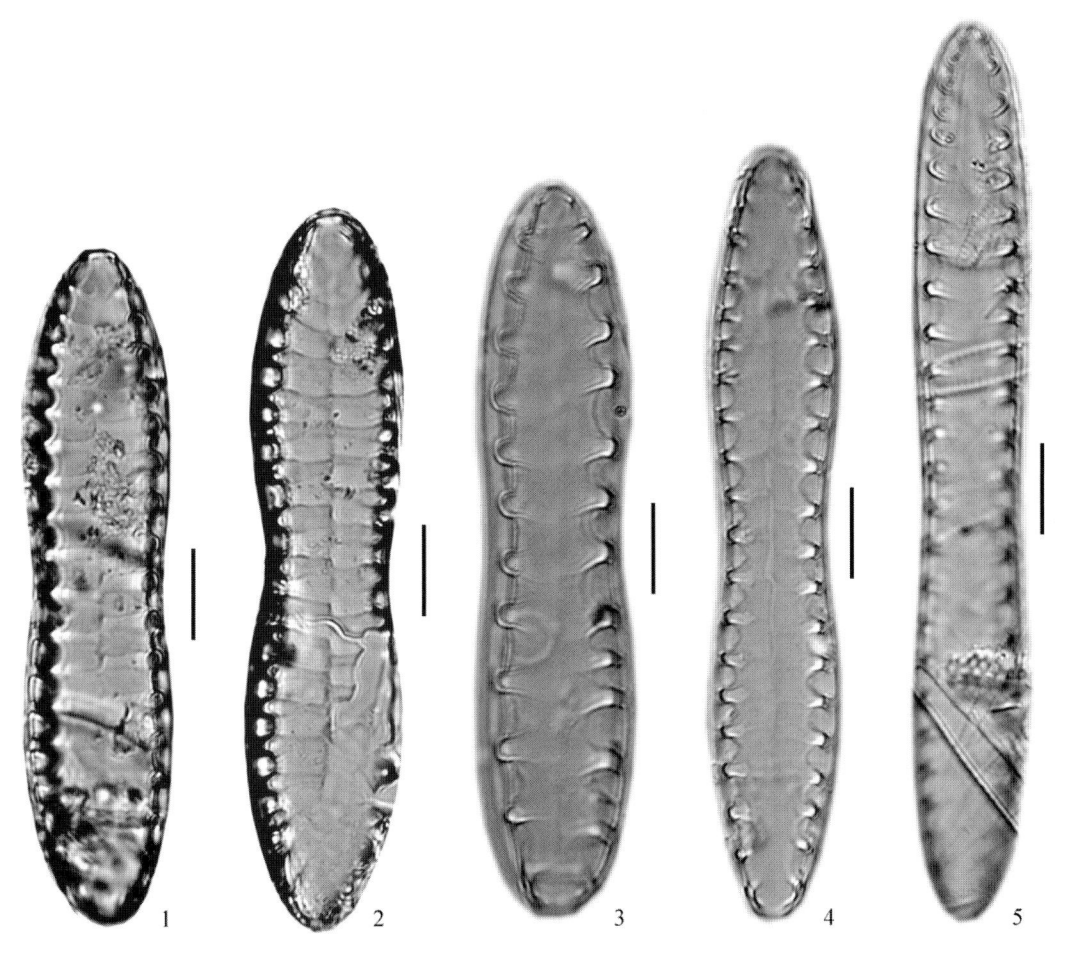

图版 CXVII

1—5. 线性双菱藻缢缩变种 *Surirella linearis* var. *constricta* Grunow。所有图为光镜（LM）照片，比例尺（Scale bars）=10 μm。

图版 CXVIII

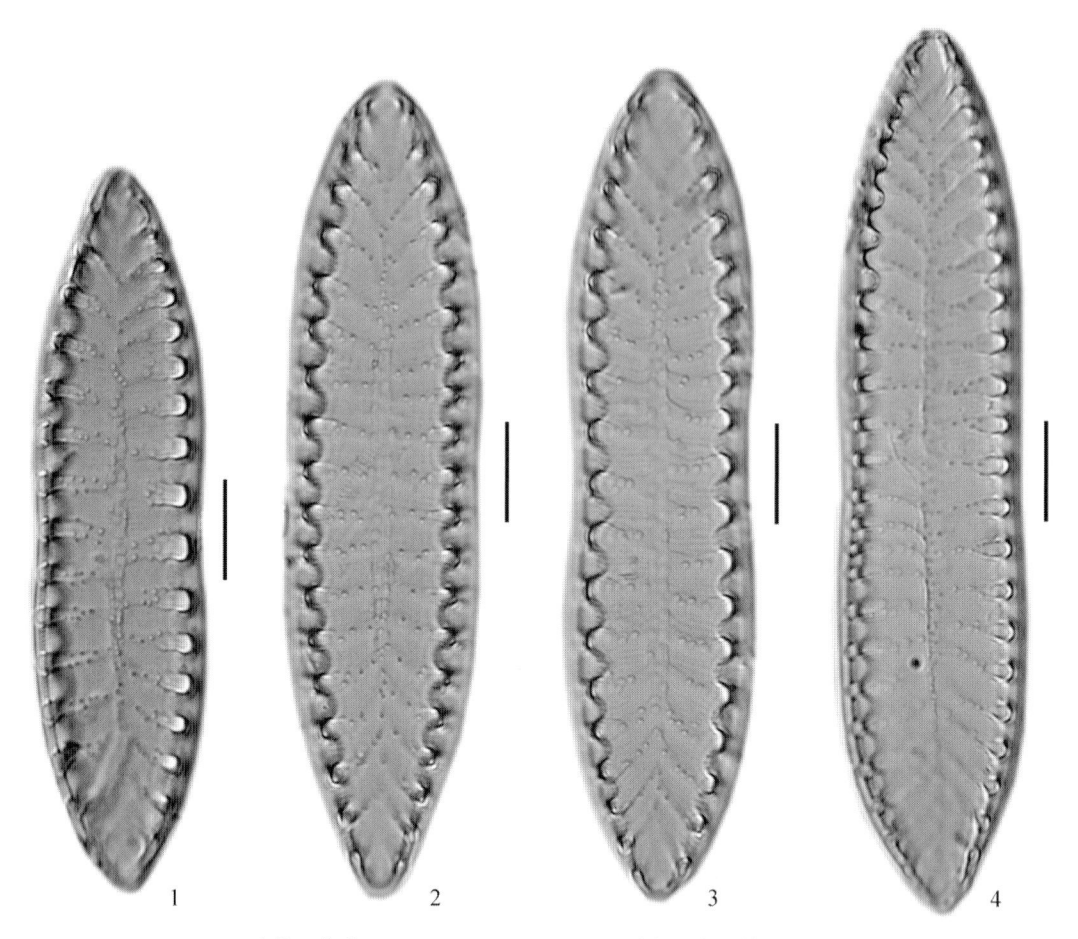

1—4. 淡黄双菱藻 *Surirella helvetica* Brun。所有图为光镜（LM）照片，比例尺（Scale bars）=10 μm。

图版 CXIX

1—2. 淡黄双菱藻 *Surirella helvetica* Brun, 1. 示外壳面(SEM), 肋纹形态, 2. 示外壳面中部(SEM), 孔纹和肋纹形态。所有图为电镜(SEM)照片, 比例尺见图。

图版 CXX

1—2. 粗壮双菱藻 *Surirella robusta* Ehrenberg；3. 粗壮双菱藻宽大变型 *Surirella robusta* f. *lata* Hustedt。
所有图为光镜(LM)照片，比例尺(Scale bars)=10 μm。

图版 CXXI

1—2. 粗壮双菱藻宽大变型 *Surirella robusta* f. *lata* Hustedt。所有图为光镜(LM)照片，比例尺(Scale bars)=10 μm。

图版 CXXII

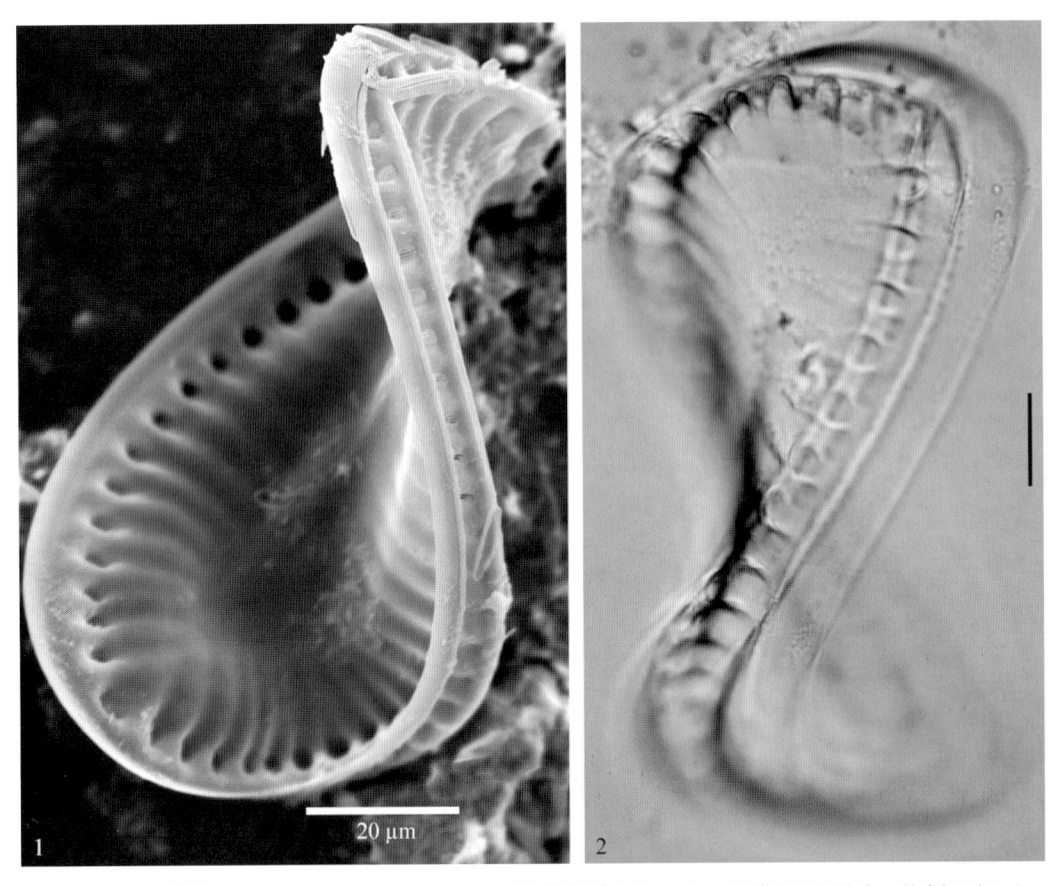

1—2. 螺旋双菱藻 *Surirella spiralis* Kützing，1. 示壳体形态(SEM)。光镜(LM)照片，比例尺(Scale bars)=10 μm，电镜(SEM)照片比例尺见图。

图版 CXXIII

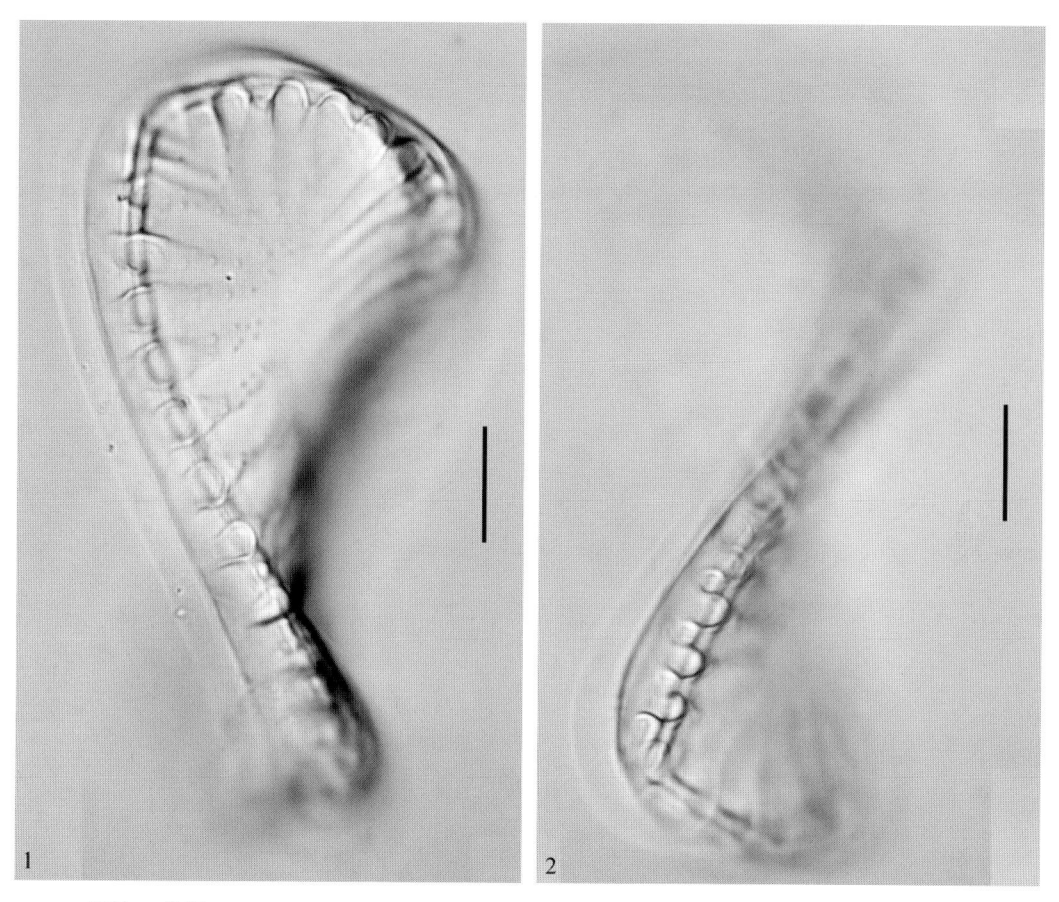

1—2. 螺旋双菱藻 *Surirella spiralis* Kützing。所有图为光镜(LM)照片，比例尺(Scale bars)=10 μm。

图版 CXXIV

1—3. 华彩双菱藻 *Surirella splendida* (Ehrenberg) Kützing。所有图为光镜(LM)照片，比例尺(Scale bars)=10 μm。

图版 CXXV

1—3. 柔软双菱藻（原变种）*Surirella tenera* var. *tenera* Greyory。所有图为光镜（LM）照片，比例尺（Scale bars）=10 μm。

图版 CXXVI

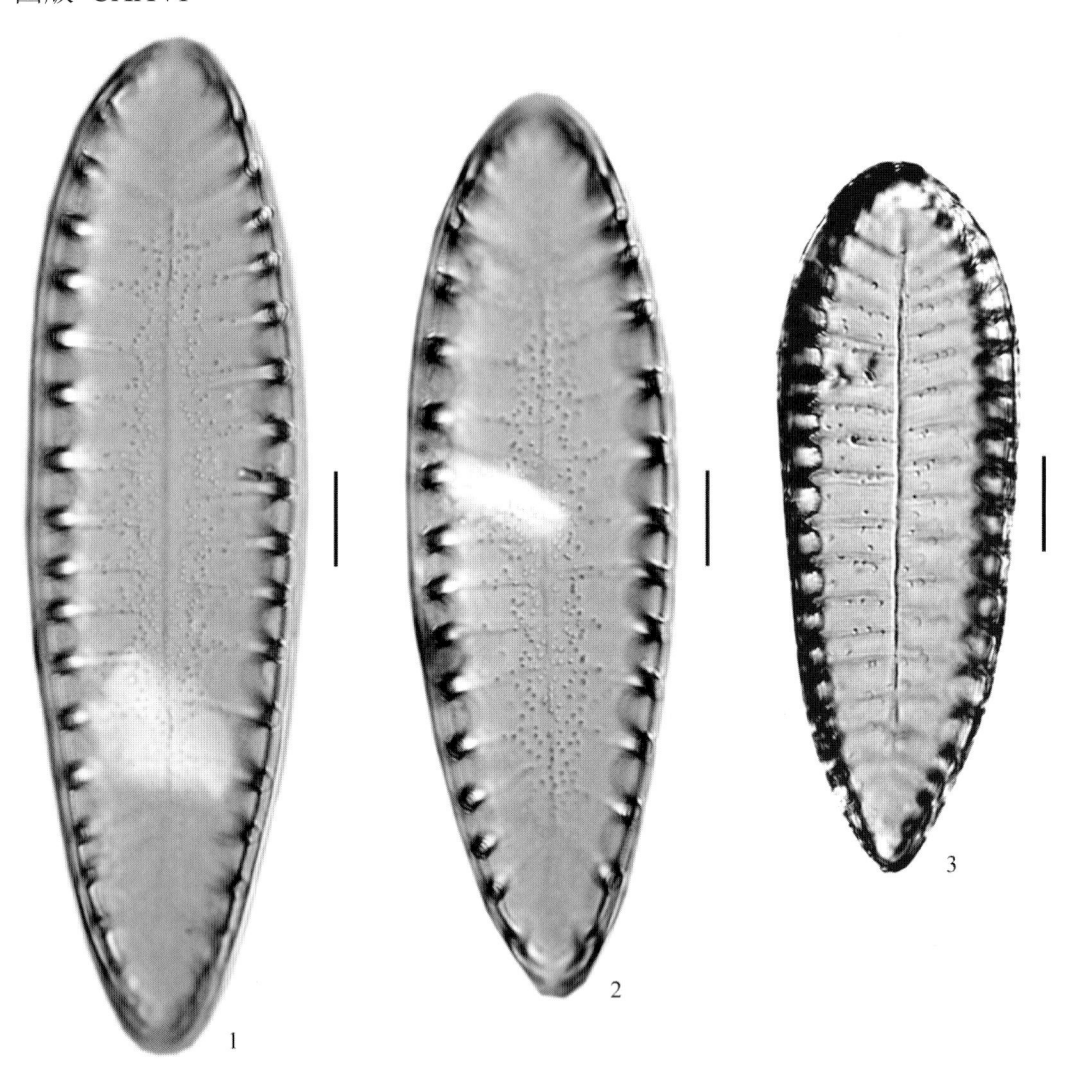

1—3. 柔软双菱藻（原变种）*Surirella tenera* var. *tenera* Greyory。所有图为光镜(LM)照片，比例尺(Scale bars)=10 μm。

图版 CXXVII

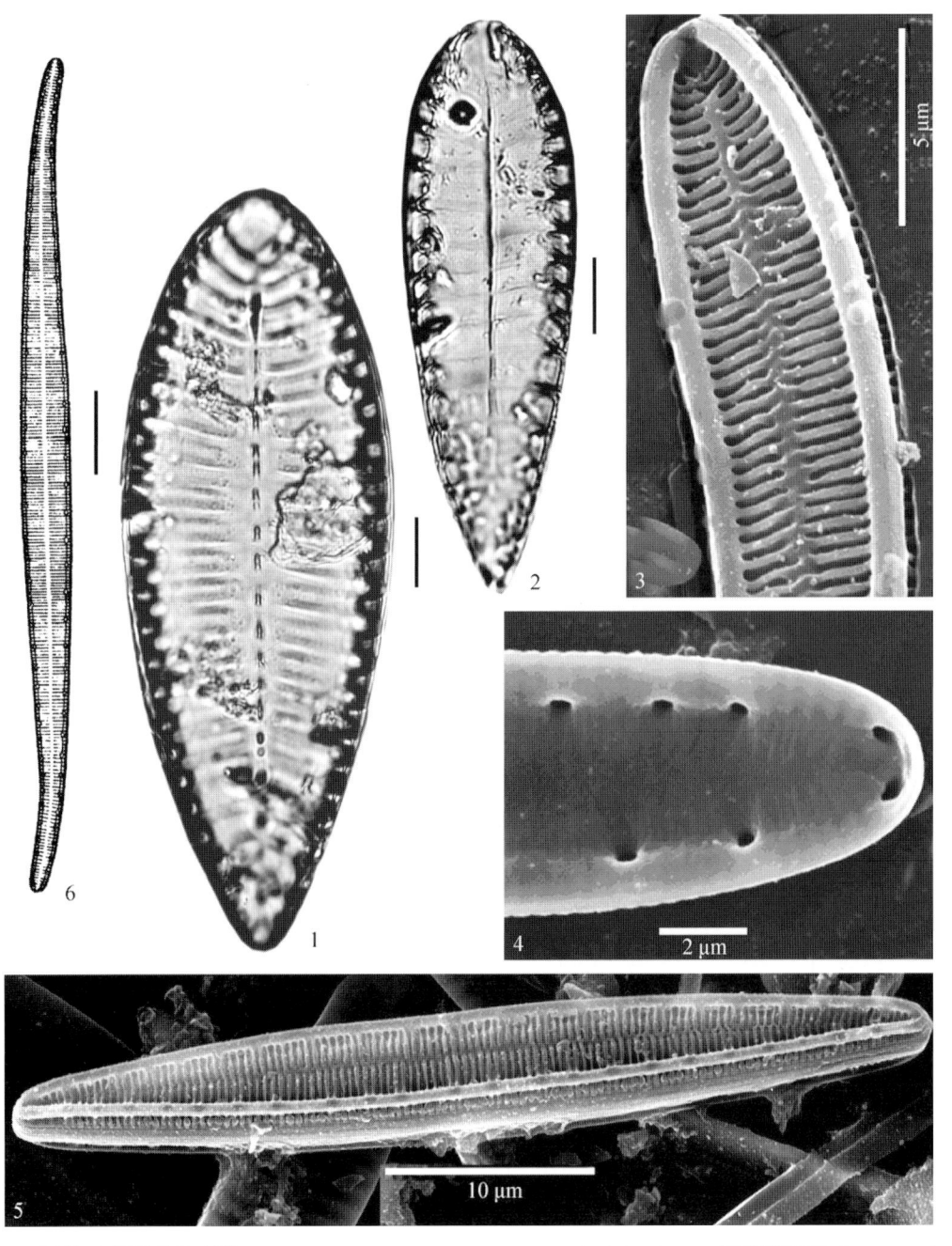

1—2. 柔软双菱藻具脉变种 *Surirella tenera* var. *nervosa* A. Schmidt；3—4. 剑形长羽藻 *Stenopterobia anceps* (Lewis) Brébisson ex Van Heurck，3. 示外壳面顶端(SEM)，肋纹形态，4. 示内壳面顶端(SEM)，肋纹形态；5. 优美长羽藻 *Stenopterobia delicatissima* (Lewis) Brébisson ex Van Heurck，示外壳面(SEM)，肋纹形态；6. 中型长羽藻 *Stenopterobia intermedia* (Lewis) Brébisson ex Van Heurck，手绘图，引自 Zhu 和 Chen (2000)。光镜(LM)照片，比例尺(Scale bars)=10 μm，电镜(SEM)照片比例尺见图。

图版 CXXVIII

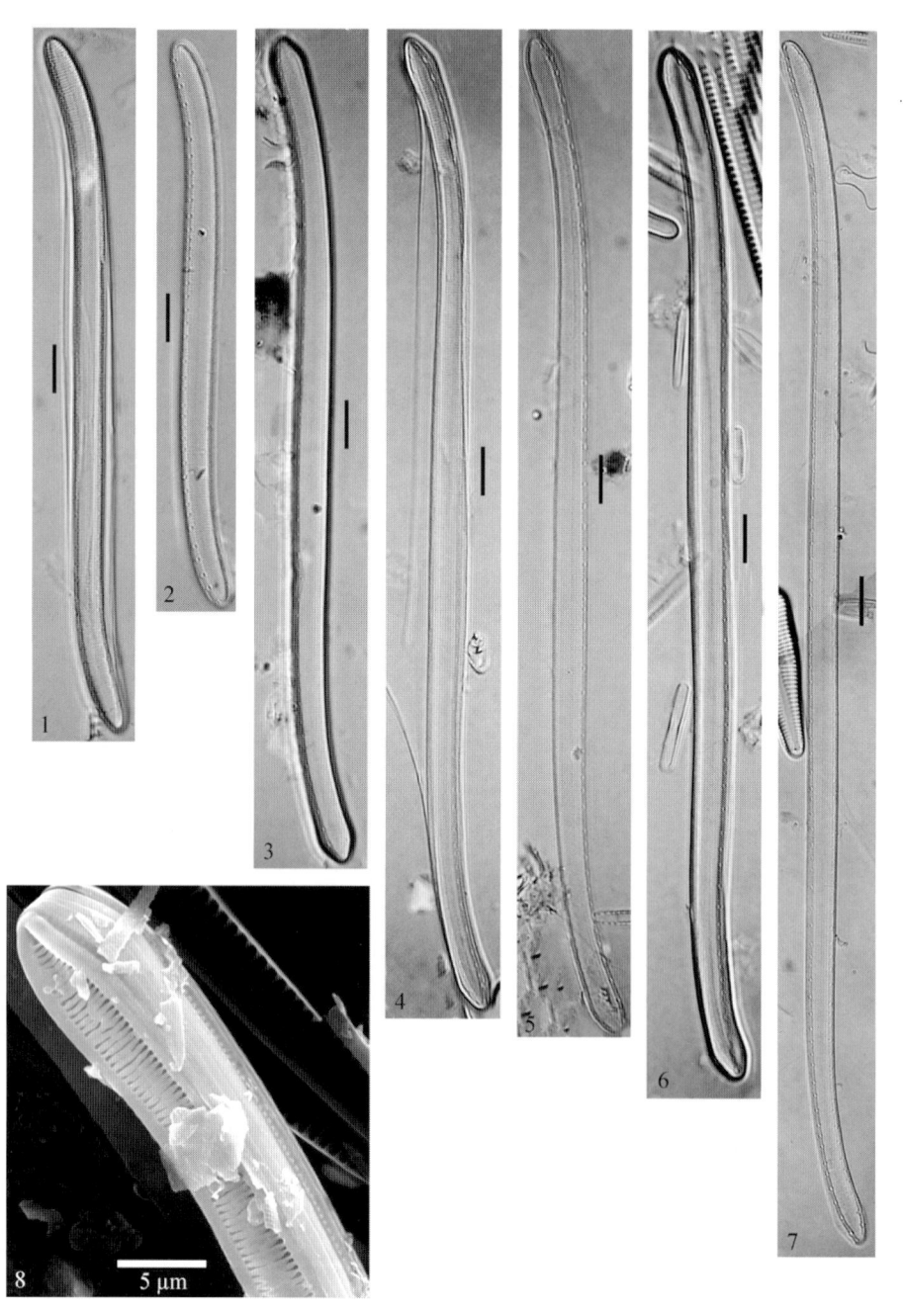

1—2. 剑形长羽藻 *Stenopterobia anceps* (Lewis) Brébisson ex Van Heurck；3—8. 中型长羽藻头端变种 *Stenopterobia intermedia* var. *capitata* Fontell，8. 示外壳面顶端（SEM），龙骨及横肋纹形态。光镜（LM）照片，比例尺（Scale bars）=10 μm，电镜（SEM）照片比例尺见图。

图版 CXXIX

1—4. 二肋纹马鞍藻 *Campylodiscus bicostatus* W. Smith，1. 示内壳面(SEM)，肋纹形态，2. 示壳体(SEM)，带面观，3. 示内壳面(SEM)，肋纹形态，4. 示外壳面(SEM)，肋纹形态。所有图为电镜(SEM)照片，比例尺见图。

图版 CXXX

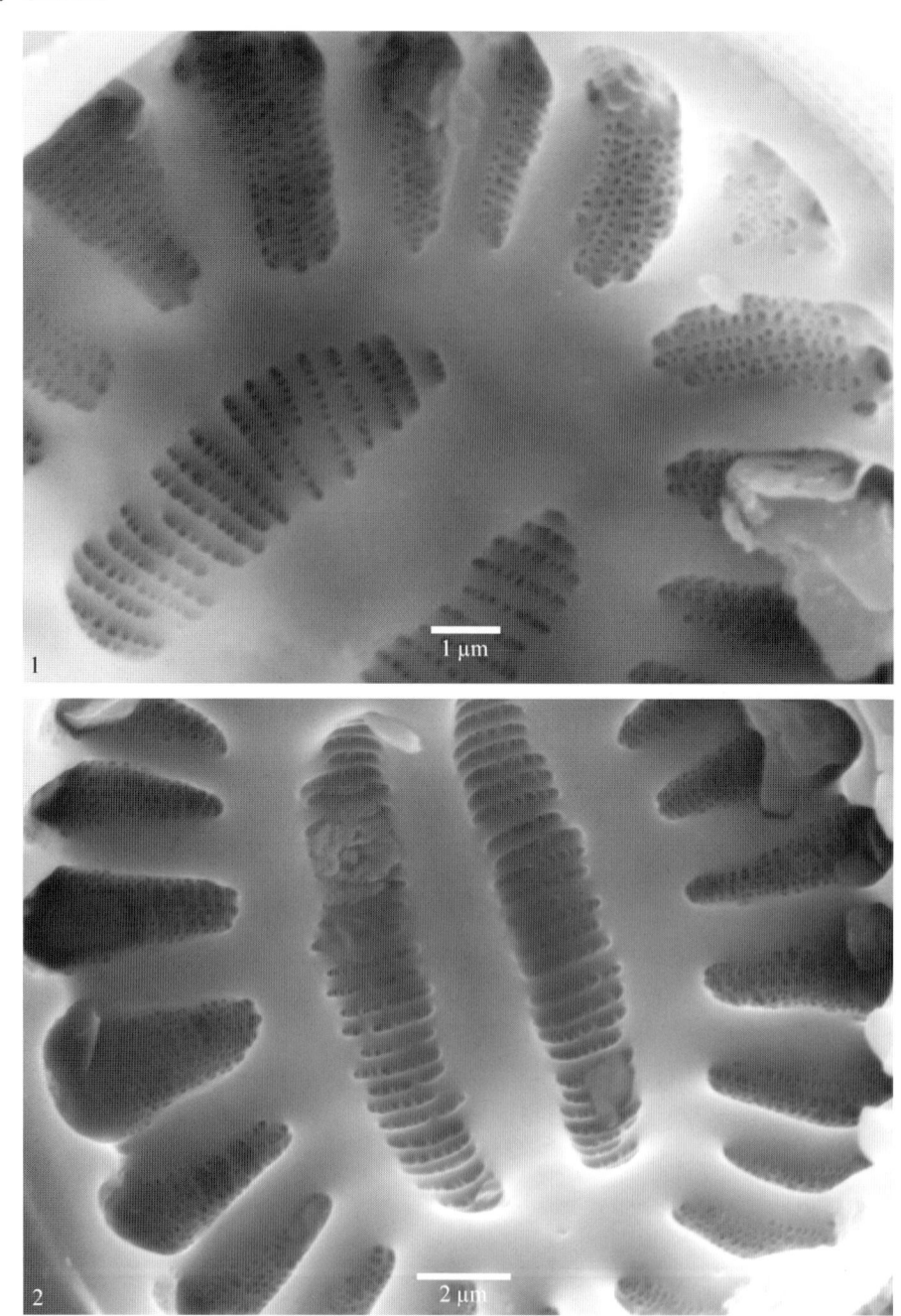

1—2. 二肋纹马鞍藻 *Campylodiscus bicostatus* W. Smith，均示内壳面(SEM)，孔纹和肋纹形态。所有图为电镜(SEM)照片，比例尺见图。

图版 CXXXI

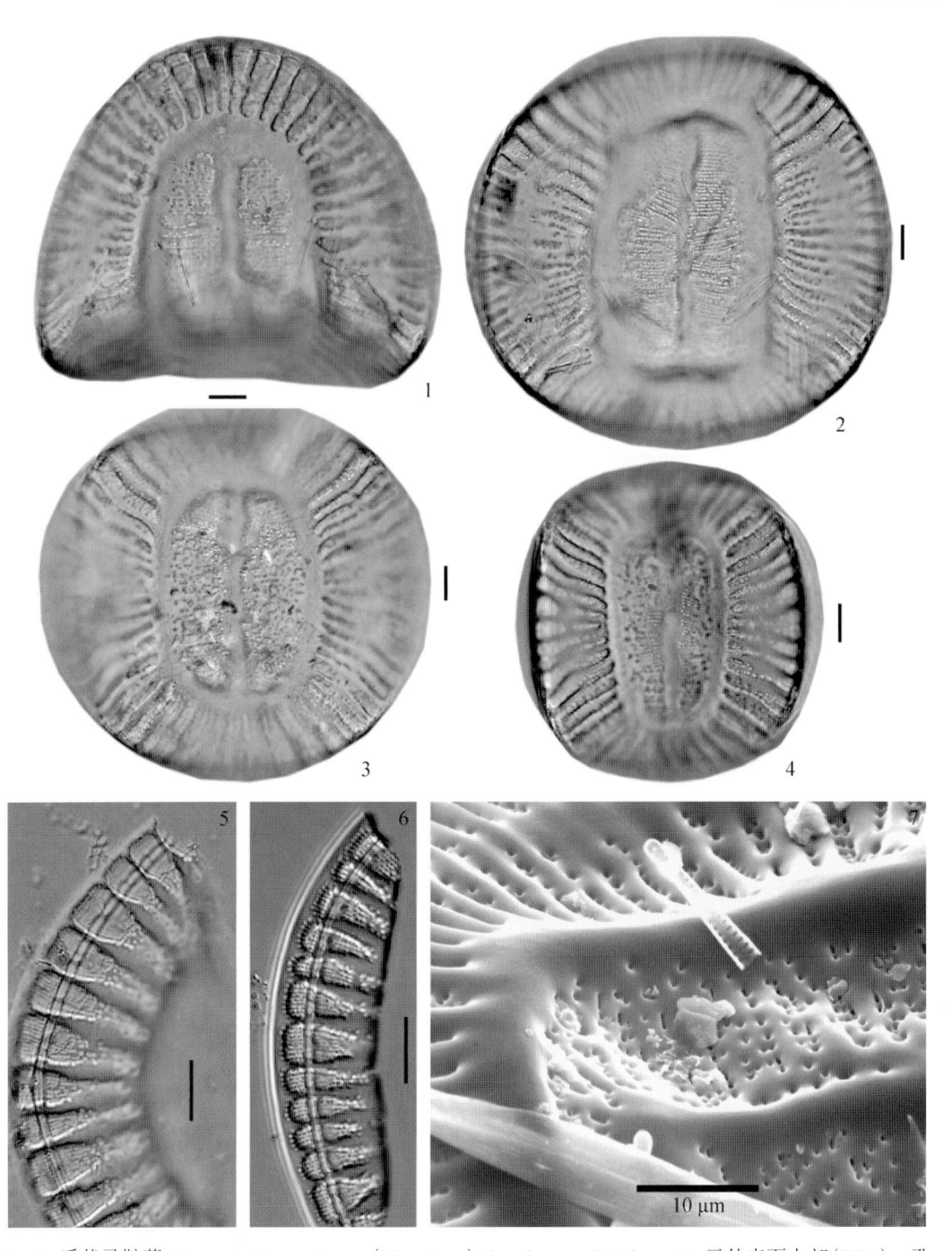

1—7. 盾状马鞍藻 *Campylodiscus clypeus* (Ehrenberg) Ehrenberg ex kützing，7. 示外壳面中部（SEM），孔纹和肋纹形态。光镜（LM）照片，比例尺（Scale bars）=10 μm，电镜（SEM）照片比例尺见图。

图版 CXXXII

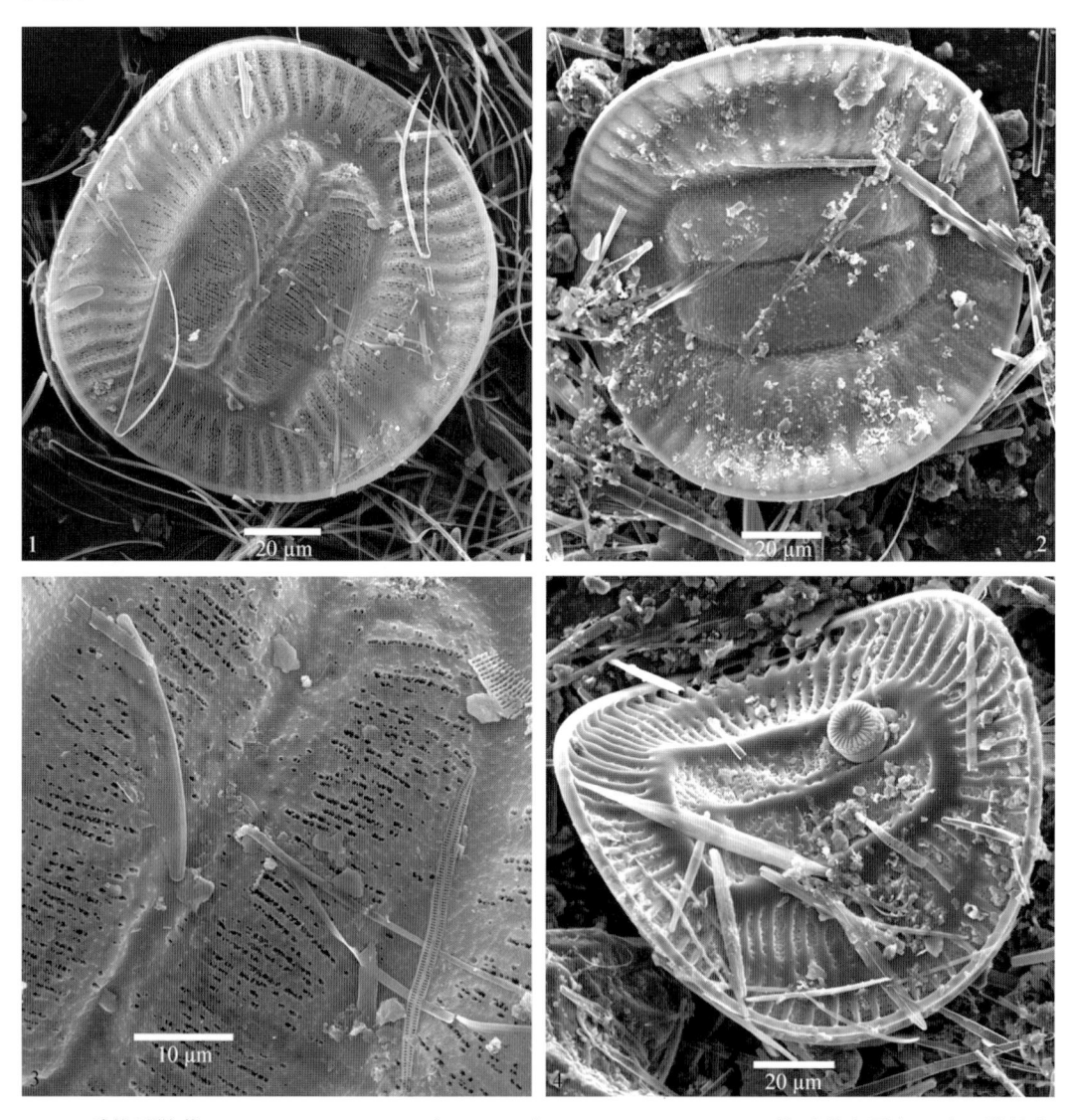

1—4. 盾状马鞍藻 *Campylodiscus clypeus* (Ehrenberg) Ehrenberg ex Kützing,均示外壳面(SEM),肋纹和孔纹形态。所有图为电镜(SEM)照片,比例尺见图。

图版 CXXXIII

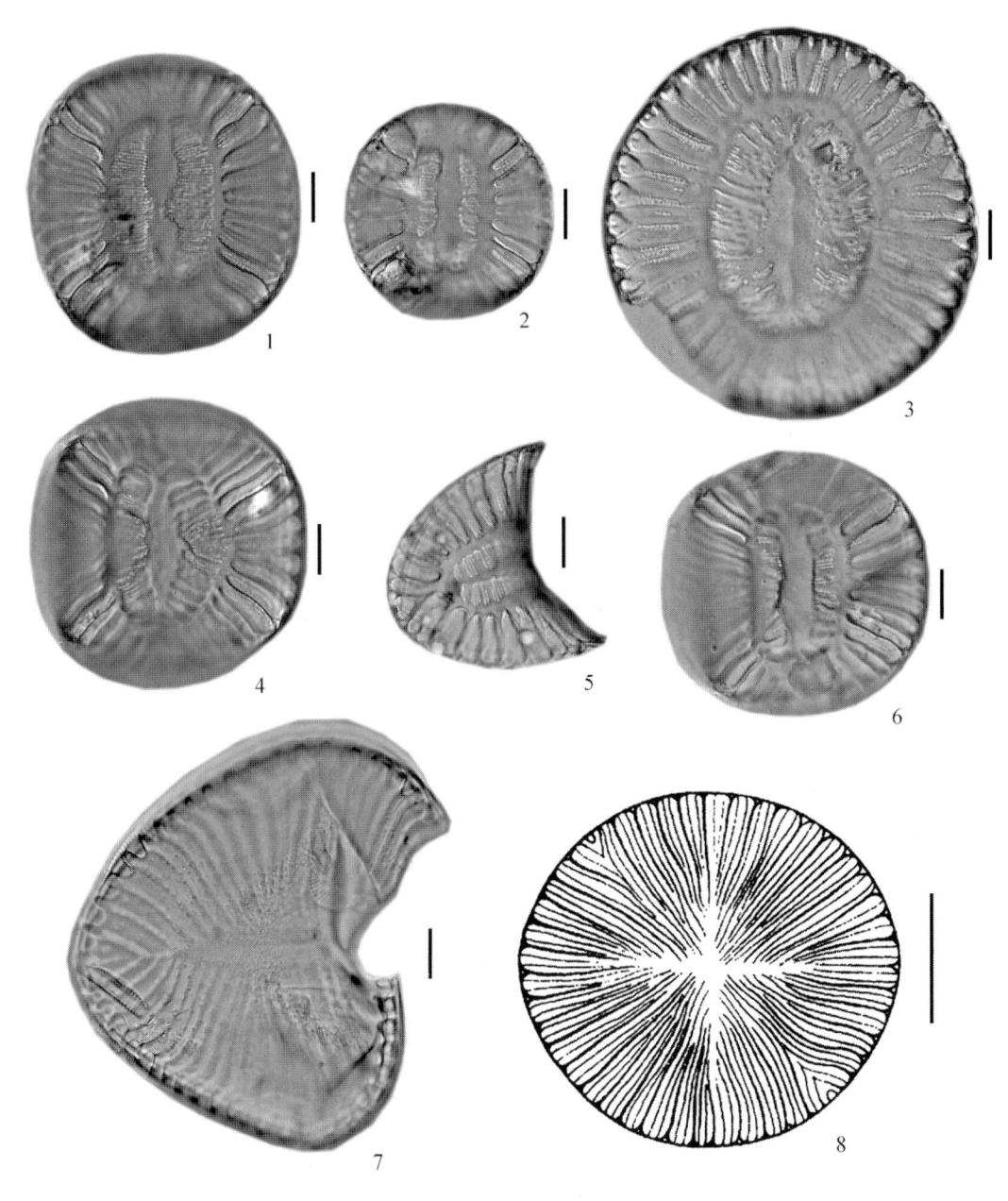

1—6. 二肋纹马鞍藻 *Campylodiscus bicostatus* W. Smith；7. 莱温德马鞍藻 *Campylodiscus levanderi* Hustedt；8. 诺克里马鞍藻 *Campylodiscus noricus* Ehrenberg，手绘图，引自 Zhu 和 Chen（2000）。所有图均为光镜（LM）照片，比例尺（Scale bars）=10 μm。

图版 CXXXIV

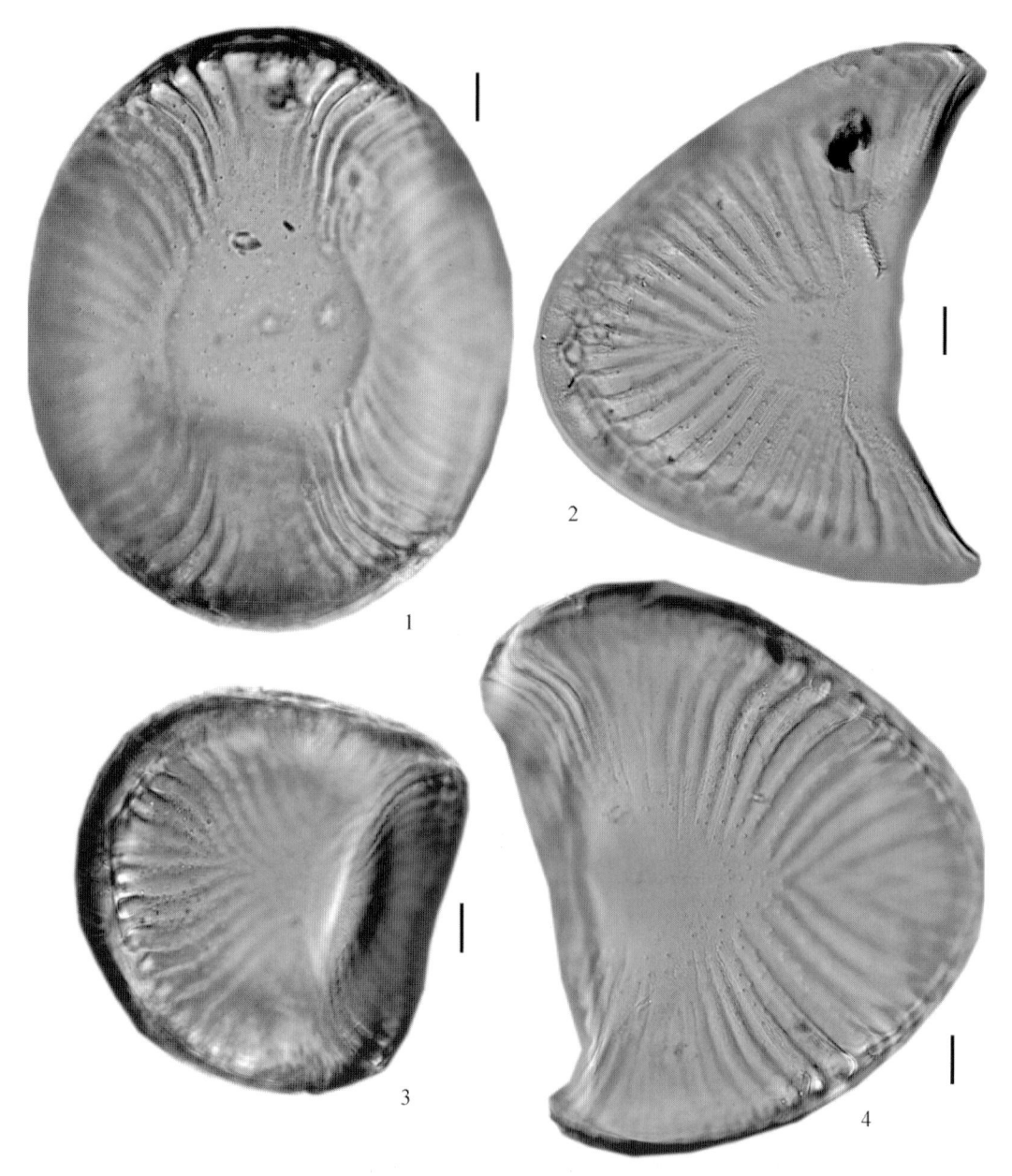

1—4. 冬生马鞍藻 *Campylodiscus hibernicus* Ehrenberg。所有图为光镜（LM）照片，比例尺（Scale bars）=10 μm。

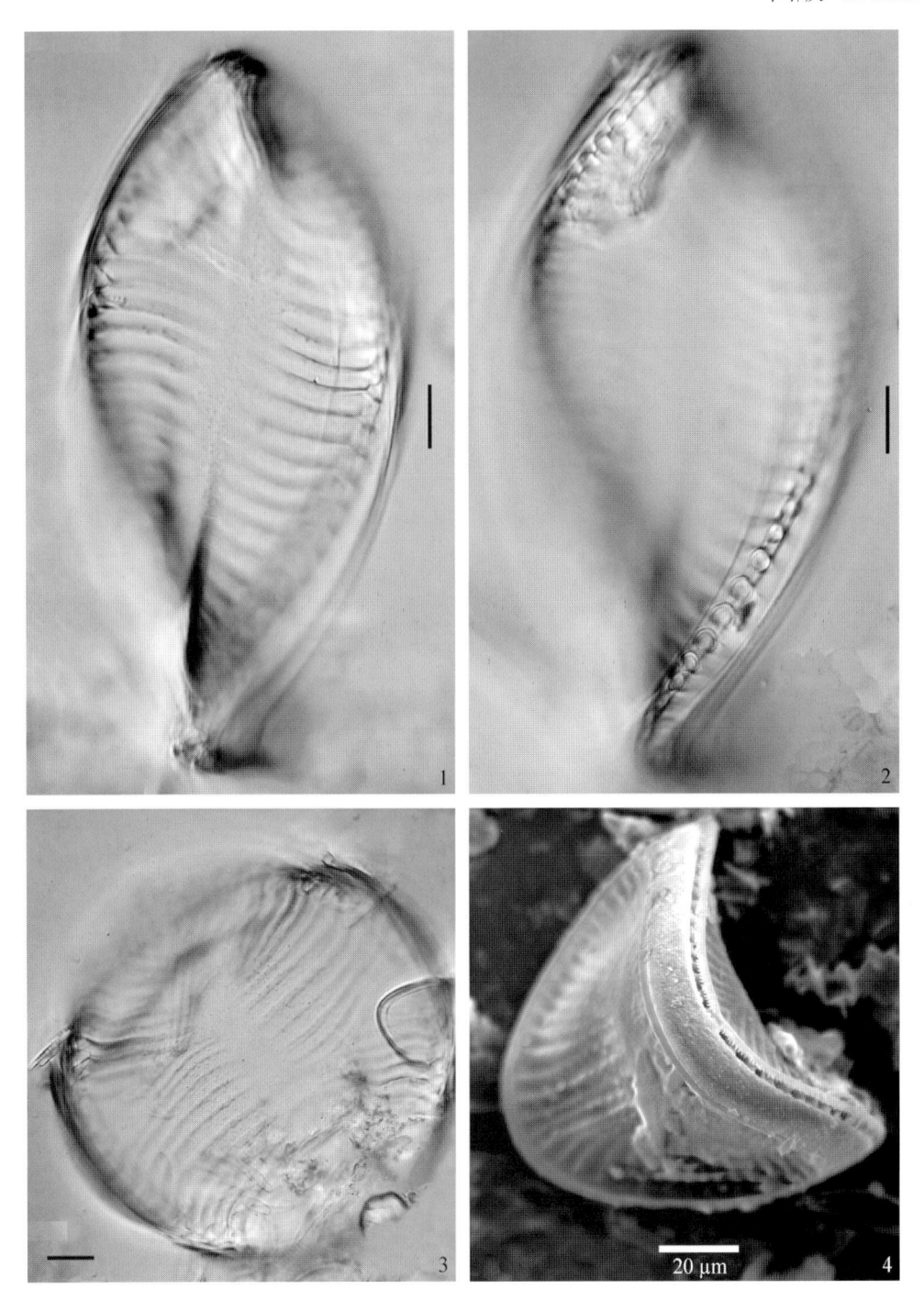

图版 CXXXV

1—2. (?) 莱温德马鞍藻 *Campylodiscus levanderi* Hustedt；3—4. 莱温德马鞍藻 *Campylodiscus levanderi* Hustedt，4. 示壳体(SEM)。光镜(LM)照片，比例尺(Scale bars)=10 μm，电镜(SEM)照片比例尺见图。

Q-4193.01

ISBN 978-7-03-057144-1

定价：198.00